**Organic Pollutants
in the Water Cycle**

*Edited by
Thorsten Reemtsma and
Martin Jekel*

Related Titles

Chiou, CT

Partition and Adsorption of Organic Contminants in Environmental Systems

2002
ISBN 0-471-23325-0

Frimmel, F. H., Abbt-Braun, G., Heumann, K. G., Hock, B., Lüdemann, H.-D., Spiteller, M. (eds.)

Refractory Organic Substances in the Environment

2002
ISBN 3-527-30173-9

Organic Pollutants in the Water Cycle

Properties, Occurrence, Analysis and
Environmental Relevance of Polar Compounds

Edited by
Thorsten Reemtsma and Martin Jekel

WILEY-VCH Verlag GmbH & Co. KGaA

The Editors

Dr. Thorsten Reemtsma
Sekr. KF 4
Technical University of Berlin
Strasse des 17. Juni 135
10623 Berlin
Germany

Prof. Dr. Martin Jekel
Sekr. KF 4
Technical University of Berlin
Strasse des 17. Juni 135
10623 Berlin
Germany

Cover
G. Schulz, Fußgönheim

■ All books published by Wiley-VCH are carefully produced. Nevertheless, authors, editors, and publisher do not warrant the information contained in these books, including this book, to be free of errors. Readers are advised to keep in mind that statements, data, illustrations, procedural details or other items may inadvertently be inaccurate.

Library of Congress Card No.
applied for

British Library Cataloguing-in-Publication Data
A catalogue record for this book is available from the British Library.

Bibliographic information published by Die Deutsche Bibliothek
Die Deutsche Bibliothek lists this publication in the Deutsche Nationalbibliografie; detailed bibliographic data is available in the Internet at http://dnb.ddb.de.

© 2006 WILEY-VCH Verlag GmbH & Co. KGaA, Weinheim

All rights reserved (including those of translation into other languages). No part of this book may be reproduced in any form – by photoprinting, microfilm, or any other means – nor transmitted or translated into a machine language without written permission from the publishers. Registered names, trademarks, etc. used in this book, even when not specifically marked as such, are not to be considered unprotected by law.

Printed in Singapore
Printed on acid-free paper

Composition Detzner Fotosatz, Speyer
Printing and Bookbinding Markono Print Media Pte Ltd, Singapore

ISBN-13: 978-3-527-31297-9
ISBN-10: 3-527-31297-8

Contents

Preface *XIII*

1 **Analytical Methods for Polar Pollutants** *1*
Thorsten Reemtsma and José Benito Quintana
1.1 Introduction *1*
1.2 The Analytical Process *1*
1.3 Sample Pretreatment and Analyte Extraction *2*
1.3.1 Sample Pretreatment *2*
1.3.2 Solid Samples *3*
1.3.3 Aqueous Samples *6*
1.3.3.1 Solid Phase Extraction *6*
1.3.3.2 Microextractions *10*
1.4 Gas Chromatographic Methods *14*
1.4.1 Derivatization *14*
1.4.1.1 Alkylation and Esterification *15*
1.4.1.2 Acylation *16*
1.4.1.3 Silylation *16*
1.4.2 Separation and Detection *17*
1.4.2.1 Separation *17*
1.4.2.2 Detection *17*
1.5 Liquid Chromatography-Mass Spectrometry *19*
1.5.1 Liquid Chromatography *20*
1.5.1.1 Ionic Analytes *20*
1.5.1.2 Non-Ionic Analytes *23*
1.5.1.3 Amphoteric Compounds *25*
1.5.1.4 Multiresidue Methods *26*
1.5.1.5 Chiral Separation *26*
1.5.2 Mass Spectrometry *27*
1.5.2.1 Ionization *27*
1.5.2.2 Mass Spectrometers and Modes of Operation *29*
1.5.2.3 Quantitation Strategies and Matrix Effects *32*
1.6 Conclusions *33*
References *34*

Organic Pollutants in the Water Cycle. T. Reemtsma and M. Jekel (Eds.).
Copyright © 2006 WILEY-VCH Verlag GmbH & Co. KGaA, Weinheim
ISBN: 3-527-31297-8

2	**Residues of Pharmaceuticals from Human Use** 41
	Thomas Heberer and Thomas Ternes
2.1	Introduction 41
2.2	Routes into the Environment 42
2.3	Wastewater 43
2.3.1	Occurrence 43
2.3.2	Removal in Municipal STPs 46
2.4	Surface Water 50
2.4.1	Occurrence 50
2.4.2	Degradation in Surface Waters 52
2.4.3	Sediments 52
2.5	Groundwater and Underground Passage 53
2.6	Drinking Water Treatment 56
2.6.1	Sorption and Flocculation 56
2.6.1.1	Flocculation 56
2.6.2	Oxidation 56
2.6.2.1	Ozonation 56
2.6.2.2	Ozonation Products 57
2.6.3	Membrane Filtration 58
2.6.4	Evaluation of the Treatment Processes 59
	References 59

3	**Antibiotics for Human Use** 65
	Radka Alexy and Klaus Kümmerer
3.1	Introduction 65
3.2	Use of Antibiotics 65
3.3	Emissions into the Environment 69
3.4	Occurrence and Fate of Antibiotics 70
3.4.1	Wastewater and Wastewater Treatment 70
3.4.1.1	Hospital Wastewater 70
3.4.1.2	Municipal Wastewater 71
3.4.2	Surface Water 74
3.4.3	Groundwater 75
3.5	Elimination and Degradation in the Aquatic Environment 76
3.5.1	Elimination by Sorption 76
3.5.2	Non-biotic Degradation 78
3.5.2.1	Photolysis 78
3.5.2.2	Hydrolysis 79
3.5.3	Biodegradation 79
3.6	Effects on Aquatic Organisms 80
3.6.1	Effects on Aquatic Bacteria and Resistance 80
3.6.2	Effects on Higher Aquatic Organisms 82
3.7	Conclusion 82
	Acknowledgments 83
	References 83

4	**Iodinated X-ray Contrast Media** 87
	Anke Putschew and Martin Jekel
4.1	Introduction 87
4.2	Source 88
4.3	Wastewater Treatment 89
4.4	Receiving Water 91
4.5	Groundwater/Exposed Groundwater 92
4.6	Treatment 94
4.7	Summary 96
	References 97

5	**Veterinary Pharmaceuticals** 99
	Gerd Hamscher
5.1	Introduction 99
5.2	Substance Classes 100
5.2.1	Aminoglycosides 101
5.2.2	β-Lactam Antibiotics 102
5.2.3	Macrolides 103
5.2.4	Quinolones and Fluoroquinolones 104
5.2.5	Sulfonamides 105
5.2.6	Tetracyclines 107
5.2.7	Various Antibiotics 108
5.3	Pathways to the Environment 110
5.3.1	Liquid Manure 111
5.3.2	Soil Fertilization 112
5.3.3	Aquaculture 113
5.4	Occurrence in Wastewater Treatment Plants 114
5.5	Surface Waters 115
5.6	Groundwater 116
5.7	Water Treatment 117
5.8	Summary 117
	Acknowledgments 118
	References 118

6	**Polar Herbicides and Metabolites** 121
	Rita Fobbe, Birgit Kuhlmann, Jürgen Nolte, Gudrun Preuß, Christian Skark, and Ninette Zullei-Seibert
6.1	General 121
6.1.1	History 122
6.1.2	Classification and Application 122
6.1.2.1	Classification 122
6.1.2.2	Application 123
6.1.3	Herbicide Classes Considered 125
6.1.3.1	Phenoxycarboxylic Acids 125
6.1.3.2	Triazines 126

6.1.3.3	Aromatic Acid Herbicides 127
6.1.3.4	Benzonitriles 128
6.1.3.5	Dinitroalkylphenol Herbicides 129
6.1.3.6	Phosphorus Compounds 130
6.1.3.7	Ureas 131
6.1.3.8	Nitrogenous Compounds 133
6.2	Entry into the Environment 134
6.3	Polar Herbicides in Wastewater 135
6.4	Polar Herbicides in Surface Water 136
6.4.1	Pathways to Surface Waters 136
6.4.2	Occurrence in Surface Water 137
6.5	Polar Herbicides in Groundwater 140
6.5.1	Pathways to Groundwater 140
6.5.2	Occurrence in Groundwater 141
6.5.3	Metabolites 144
6.6	Water Treatment 145
6.6.1	Activated Carbon 145
6.6.2	Physico-chemical Processes (Flocculation, Rapid Filtration, Sedimentation) 146
6.6.3	Oxidation and Disinfection 146
6.6.4	Natural Filtration Processes (Bank Filtration, Slow Sand Filtration, Underground Passage) 147
6.6.5	Membrane Filtration 148
6.6.6	Conclusions – Water Treatment 148
6.7	Summary 149
	References 149
7	**Aminopolycarboxylate Complexing Agents** 155
	Carsten K. Schmidt and Heinz-Jürgen Brauch
7.1	Introduction 155
7.2	Applications and Consumption 157
7.3	Occurrence and Fate in Wastewaters 162
7.3.1	Biodegradation 162
7.3.2	Other Treatment Options 166
7.3.3	Occurrence in Wastewater 167
7.4	Occurrence and Fate in Surface Waters 169
7.4.1	Occurrence and Speciation 169
7.4.2	Fate and Effects 170
7.5	Occurrence and Fate in Groundwater 172
7.6	Behavior during Drinking Water Processing 174
7.7	Conclusions 176
	References 176

8 Amines *181*
Hilmar Börnick and Torsten C. Schmidt

8.1 Introduction *181*
8.1.1 Characterization of Amines and Choice of Compounds *181*
8.1.2 Properties *182*
8.2 Production *184*
8.3 Emission *186*
8.4 Wastewater and Wastewater Treatment *189*
8.4.1 Occurrence in Wastewater *189*
8.4.2 Biological Treatment *189*
8.4.3 Oxidative Technologies *191*
8.4.4 Other Treatment Technologies *192*
8.5 Surface Water *192*
8.5.1 Occurrence *192*
8.5.2 Transformation and Degradation Processes *193*
8.5.2.1 Abiotic Processes *195*
8.5.2.2 Biological Degradation *196*
8.6 Groundwater *200*
8.6.1 Occurrence *200*
8.6.2 Sorption and Transformation Processes *201*
8.7 Drinking Water and Drinking Water Treatment *203*
8.7.1 Occurrence *203*
8.7.2 Treatment *204*
8.8 Conclusions *205*
References *206*

9 Surfactant Metabolites *211*
Thomas P. Knepper and Peter Eichhorn

9.1 Introduction *211*
9.2 Aerobic Biodegradation of Surfactants *213*
9.2.1 Introduction *213*
9.2.2 Anionic Surfactants *216*
9.2.3 Non-ionic Surfactants *221*
9.2.4 Amphoteric Surfactants *226*
9.3 Surfactants and Metabolites in Wastewater Treatment Plants *228*
9.3.1 Linear Alkylbenzene Sulfonates and their Degradation Products *229*
9.3.2 Alkylphenol Ethoxylates and their Degradation Products *230*
9.4 Surfactants and Metabolites in Surface Waters *232*
9.4.1 Linear Alkylbenzene Sulfonates and their Degradation Products *232*
9.4.2 Alkylphenol Ethoxylates and their Degradation Products *233*
9.5 Surfactants and Metabolites in Subsoil/Underground Passage *233*
9.5.1 Alkylbenzene Sulfonates and their Degradation Products *233*
9.5.2 Alkylphenol Ethoxylates and their Degradation Products *237*
9.5.3 Perfluorinated Surfactants *240*

9.6	Surfactants and Metabolites in Drinking Waters 241
9.6.1	Alkylphenol Ethoxylates (APEO) and their Degradation Products 241
9.6.2	Behavior of Sulfophenyl Carboxylates during Drinking Water Production 242
9.7	Risk Assessment 243
9.7.1	Linear Alkylbenzene Sulfonates and their Metabolites 243
9.7.2	Alkylphenol Ethoxylates and their Metabolites 244
9.8	Conclusions 244
	References 245

10 Trihalomethanes (THMs), Haloacetic Acids (HAAs), and Emerging Disinfection By-products in Drinking Water 251

Christian Zwiener

10.1	Introduction 251
10.1.1	Disinfection – Fields of Application 251
10.1.2	Disinfection in Drinking Water Treatment 253
10.2	Regulations 255
10.2.1	European Union 255
10.2.2	Germany 257
10.2.3	United States 258
10.3	Reactants Leading to DBP Formation 259
10.3.1	Disinfectants 259
10.3.1.1	Chlorine 259
10.3.1.2	Chlorine Dioxide 260
10.3.1.3	Ozone 261
10.3.1.4	Chloramine 261
10.3.2	Organic DBP Precursors 262
10.3.2.1	Natural Organic Matter (NOM) 262
10.3.2.2	Micropollutants 263
10.4	Occurrence of DBPs 264
10.4.1	Trihalomethanes (THMs) and Halogenated Acetic Acids (HAAs) 265
10.4.1.1	German Drinking Water 265
10.4.1.2	European Drinking Water 266
10.4.1.3	Canadian Drinking Water 267
10.4.1.4	United States Drinking Waters 267
10.4.2	Emerging Organic DBPs 268
10.4.2.1	Halonitriles 269
10.4.2.2	Carbonyls 270
10.4.2.3	Halogenated Hydroxyfuranones 271
10.4.2.4	Halonitromethanes 272
10.4.2.5	*N*-Nitrosamines 273
10.4.2.6	Iodinated THMs and Acids 273
10.4.2.7	Missing DBPs 274
10.4.2.8	Transformation Products of Micropollutants 275

10.4.3	Inorganic Disinfection By-products	277
10.4.3.1	Chlorite and Chlorate	277
10.4.3.2	Bromate	277
10.5	Measures to Control DBPs	278
10.5.1	Source Control	279
10.5.2	Disinfection Control	279
10.6	Conclusions	281
	Acknowledgments	281
	References	281

11 Toxicology and Risk Assessment of Pharmaceuticals *287*
Daniel R. Dietrich, Bettina C. Hitzfeld, and Evelyn O'Brien

11.1	Introduction	287
11.2	A Comparison of International Risk Assessment Procedures	290
11.2.1	The European Union Technical Guidance Document (TGD)	290
11.2.2	US-EPA	295
11.2.3	Japan	296
11.2.4	Canada	297
11.3	The Toxicological Data Set for Environmental Risk Assessment	301
11.3.1	Extrapolation from Acute to Chronic Toxicity	301
11.3.2	QSARs	305
11.3.3	"Omics"	306
11.3.4	Toxicity of Mixtures	307
11.4	Conclusions	307
	References	308

12 Assessment and Management of Chemicals – How Should Persistent Polar Pollutants be Regulated? *311*
Klaus Günter Steinhäuser and Steffi Richter

12.1	Chemicals Assessment and Management Today	311
12.1.1	Basic Legislation and Current Guidelines for Risk Assessment and Risk Management of Chemicals in Europe and Germany	311
12.1.1.1	Notification of New Substances	311
12.1.1.2	Existing Substances Legislation and Management	313
12.1.1.3	Technical Guidelines for Risk Assessment of Chemicals	314
12.1.2	Problems Impeding Effective Chemicals Management	315
12.1.3	Chemicals Management in the United States of America	318
12.1.4	Management of Specific Chemicals	321
12.1.4.1	Pesticides	321
12.1.4.2	Biocides	323
12.1.4.3	Pharmaceuticals	324
12.1.4.4	Detergents and Cleansing Agents	327
12.1.5	Reflections on Current Chemicals Management of Persistent Polar Pollutants	328

12.2	Future Chemicals Management in Europa with REACH	*329*
12.3	Persistent Organic Pollutants (POPs) and Persistent Polar Pollutants (PPPs) – A Comparison	*332*
12.3.1	Persistent Organic Pollutants (POPs)	*332*
12.3.2	Persistent Polar Pollutants (PPPs) in the Water Cycle	*334*
	Acknowledgments	*335*
	References	*336*

Subject Index   *341*

Preface

The perspective on contamination of aqueous environment by anthropogenic trace pollutants has experienced a remarkable change in the past ten to fifteen years.

Traditionally hydrophobic persistent organic pollutants (POP) that may accumulate in sediments and enrich along food chains were studied extensively. Meanwhile the awareness developed that also polar contaminants may pose a significant problem to water quality, especially if they are not well degradable.

This growing awareness of polar pollutants has several reasons, of which only a few may be mentioned here.

- studying the occurrence of polar pollutants requires that these contaminants are analytically accessible. It was only in the second half of the 1990s that the effective coupling of liquid chromatography to mass spectrometry by electrospray ionization offered a highly sensitive approach to determine polar pollutants from water (see Chapter 1). This progress in analytical chemistry was a prerequisite to direct more attention towards such polar pollutants and to study them in more detail.

- Also in the 1990s it was shown, that trace organic pollutants present in municipal wastewater effluents may have severe sub-lethal effects to aquatic biota. It was shown that xeno-estrogens may interfere with the hormon cycle of wildlife at trace level.

- Globally an increasing water demand calls for an increasing portion of indirect potable reuse of treated municipal wastewater. However, such a partial closure of water cycles at local and regional scale urges to consider new criteria for contaminant evaluation. Especially polar and persistent pollutants can be problematic as they may travel along a water cycle from wastewater to raw waters used for drinking water production. The past ten years have seen increasing evidence that such compunds are present.

The Partially Closed Water Cycle

One example of a partially closed water cycle is displayed in Figure 1. In such a cycle a polar and persistent component that is neither removed by sorption nor by

Organic Pollutants in the Water Cycle. T. Reemtsma and M. Jekel (Eds.).
Copyright © 2006 WILEY-VCH Verlag GmbH & Co. KGaA, Weinheim
ISBN: 3-527-31297-8

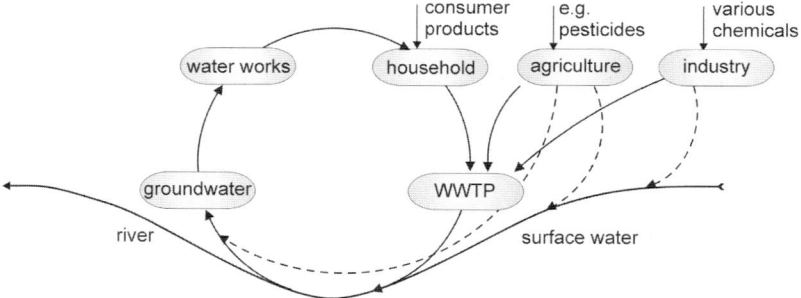

Figure 1

biodegradation could pass all barriers such as wastewater treatment or underground passage and would, then, appear in raw waters used for drinking water production. Polar pollutants may originate from consumer products used in household, pesticides applied in agriculture or chemicals used in industry. Surface runoff may also contribute. The occurrence of trace pollutants in raw waters requires an ever increasing technical effort in drinking water production.

Of the various components of such a water cycle (Fig. 1) the municipal wastewater treatment plants are, certainly, best investigated. Meanwhile, an impressive body of literature is available concerning the occurrence and removal of polar trace pollutants from municipal wastewater by biological treatment. Other processes such as the transport of pesticides applied in agriculture to groundwater are also comparatively well studied. The occurrence of polar pollutants in other compartments of this cycle, however, and their removal in or passage through other barriers than wastewater treatment or soil are less thoroughly investigated. Even less so has the occurrence and behaviour of polar pollutants in all components of a partially closed water cycle been studied systematically.

Therefore this book aims at bringing together results obtained in various studies concerning all compartments and barriers of a (hypothetic) partially closed water cycle.

The Polar Pollutants

As this book focuses on the water cycle the selection of contaminant classes that are covered is, among others, based on *polarity*. The authors agreed to select an upper limit of the octanol/water partition coefficient ($\log K_{ow}$) of 3 for inclusion into this book. Therefore, the reader may miss certain contaminant classes that he became familiar with in the past years, like endocrine disruptors. Certainly, these compounds would be an issue in a more general book on 'contaminants in water' but, due to the comparatively high $\log K_{ow}$ values of many of these compounds, they are not relevant as 'contaminants in the water cycle'.

The *production volume* is another relevant criterion as a high production volume chemical, even with an almost complete removal in wastewater treatment, could still lead to significant amounts being discharged into surface water. For this reason a number of high production volume chemicals are included in this book. The first to mention are surfactants which are used almost everywhere (Chapter 9). The occurrence of poorly degradable surfactants in surface waters and groundwaters made this class of compounds the first, for which a minimum extent of biodegradability was required by regulations in Western Europe and the United States in the early 1960s. Other important groups of polar high production volume chemicals that are covered in this book are herbicides (Chapter 6), complexing agents (Chapter 7) and amines (Chapter 8).

Also compounds used and released in substantially less amout can be problematic, if their use profile requires a certain level of *stability*. Pharmaceuticals are an example for this and the occurrence of such compounds in wastewater discharges and surface waters has received significant attention within the last years. Several chapters of this book deal with this 'dark side' of the so benefical development in pharmaceuticals (Chapters 2–5).

Finally, polar pollutants may even be generated in wastewater treatment or drinking water production as it is the case for disinfection byproducts (Chapter 10).

With improved analytical capabilities (Chapter 1) that allow to detect nanogram per litre concentrations of trace pollutants positive findings in virtually all aquatic compartments are almost inevitable. Thus, the need for proper knowledge how the occurrence of low concentrations of polar pollutants has to be evaluated is becoming more urgent. Chapter 11 deals with such aspects of ecotoxicology. A combined evaluation of physico-chemical and ecotoxicological properties of high production volume chemicals is necessary to avoid contamination and to reduce the risk related with the use of chemicals. This is the basis of the chemicals management (REACH) in the European Union (Chapter 12).

Also in the European Union the Water Framework Directive (WFD) has bundled many different regulations concerning the protection of freshwater resources. While the WFD has strengthened biological quality criteria for waters and water bodies, there is growing concern with respect to chemical quality criteria. Inter alia chemicals that may not be harmful to human health or the quality of aquatic ecosystems are not considered pollutants. Thus, the WFD may hamper rather than foster the protection of the water cycle from anthropogenic compounds that are polar and persistent and that may spread in aquatic environment.

We would be pleased if this book contributes to increasing the knowledge on and the awareness of the relevancy of polar pollutants for the quality of waters, not at least those being used as drinking waters.

We are grateful to all the authors that shared this view on polar pollutants and contributed with their expertise, time and effort in preparing the different chapters of this book.

Thorsten Reemtsma
Martin Jekel

Berlin
February 2006

List of Authors

Radka Alexy
Institute of Environmental Medicine
and Hospital Epidemiology
Hugstetter Str. 55
79106 Freiburg
Germany

Hilmar Börnick
Institute of Water Chemistry
Dresden University of Technology
Zellescher Weg 40
01217 Dresden
Germany

Heinz-Jürgen Brauch
DVGW-Water Technology Center
(TZW)
Chemical Analysis Department
Karlsruher Straße 84
76139 Karlsruhe
Germany

Daniel R. Dietrich
Environmental Toxicology
University of Konstanz
Fach X 918
78457 Konstanz
Germany

Peter Eichhorn
The State University of New York at
Buffalo
Chemistry Department
608 Natural Science Complex
Buffalo NY, 14226
United States of America

Rita Fobbe
ISAS-Institute for Analytical Sciences
Bunsen-Kirchhoff-Str. 11
44139 Dortmund
Germany

Gerd Hamscher
Centre for Food Science
Institute for Food Toxicology
University of Veterinary Medicine
Hannover Foundation
Bischofsholer Damm 15
30173 Hannover
Germany

Thomas Heberer
Federal Institute for Risk Assessment
Thielallee 88–92
14195 Berlin
Germany

Organic Pollutants in the Water Cycle. T. Reemtsma and M. Jekel (Eds.).
Copyright © 2006 WILEY-VCH Verlag GmbH & Co. KGaA, Weinheim
ISBN: 3-527-31297-8

List of Authors

Bettina C. Hitzfeld
Swiss Agency for the Environment
Forests and Landscape
Papiermühle Str. 172
3003 Bern
Switzerland

Martin Jekel
Technical University Berlin
Institute for Environmental Technology
Department of Water Quality Control,
Sekr. KF 4
Straße des 17. Juni 135
10623 Berlin
Germany

Thomas P. Knepper
Europa University of Applied Sciences
Fresenius
Limburger Straße 2
65510 Idstein
Germany

Birgit Kuhlmann
IfW Institut für Wasserforschung
GmbH
Zum Kellerbach 46
58239 Schwerte
Germany

Klaus Kümmerer
Institute of Environmental Medicine
and Hospital Epidemiology
Hugstetter Str. 55
79106 Freiburg
Germany

Jürgen Nolte
ISAS-Institute for Analytical Sciences
Bunsen-Kirchhoff-Str. 11
44139 Dortmund
Germany

Evelyn O'Brien
Environmental Toxicology
University of Konstanz
Postfach X 918
78457 Konstanz
Germany

Gudrun Preuß
IfW Institut für Wasserforschung
GmbH
Zum Kellerbach 46
58239 Schwerte
Germany

Anke Putschew
Technical University Berlin
Institute for Environmental
Technology
Department of Water Quality Control,
Sekr. KF 4
Straße des 17. Juni 135
10623 Berlin
Germany

José Benito Quintana
UIMA-University Institute of
Environment
University of A Coruña
Pazo da Lóngora, Liáns
15179 Oleiros (A Coruña)
Spain

Thorsten Reemtsma
Technical University of Berlin
Department of Water Quality Control,
Sekr KF 4
Straße des 17 Juni 135
10623 Berlin
Germany

Steffi Richter
Federal Environment Agency
Wörlitzer Platz 1
06844 Dessau
Germany

Carsten K. Schmidt
DVGW-Water Technology Center (TZW)
Chemical Analysis Department
Karlsruher Straße 84
76139 Karlsruhe
Germany

Torsten C. Schmidt
University Duisburg-Essen
Department of Chemistry
Lotharstrasse 1
47048 Duisburg
Germany

Christian Skark
IfW Institut für Wasserforschung GmbH
Zum Kellerbach 46
58239 Schwerte
Germany

Klaus Günter Steinhäuser
Federal Environment Agency
Wörlitzer Platz 1
06844 Dessau
Germany

Thomas Ternes
Federal Institute of Hydrology (BfG)
Am Mainzer Tor 1
56068 Koblenz
Germany

Ninette Zullei-Seibert
IfW Institut für Wasserforschung GmbH
Zum Kellerbach 46
58239 Schwerte
Germany

Christian Zwiener
Engler-Bunte-Institut
Bereich Wasserchemie
University of Karlsruhe (TH)
Engler-Bunte-Ring 1
76131 Karlsruhe
Germany

1
Analytical Methods for Polar Pollutants

Thorsten Reemtsma and José Benito Quintana

1.1
Introduction

The last few decades have shown that analytical chemistry and environmental chemistry are "conjoined twins". Neither can move significantly forward without the contribution and support of the other discipline. But in their conjoined development both disciplines have contributed much to our knowledge of environmental pollution, to the understanding of environmental processes, and to the development of measures and strategies to reduce contamination.

Complementary to the following book chapters that focus on the occurrence and behavior of different classes of polar pollutants in the water cycle, this chapter provides an overview of the recent status of the other half of the "conjoined twins", the analytical methods to determine these polar pollutants. This subject could easily be the topic of an independent book. Condensing it to one chapter requires considerable selectivity. Therefore, analytical strategies and approaches to the trace analysis of polar pollutants from environmental samples are outlined rather than described in detail.

1.2
The Analytical Process

A scheme of the analytical process for the determination of polar pollutants in water and particulate samples (sludge, sediment and soil) is presented in Fig. 1.1. This scheme excludes sampling, which is outside of the scope of this overview. Obviously not all the steps presented in this scheme are always necessary, and in many cases, for example, clean-up or derivatization prior to GC determination and sometimes even the enrichment step can be skipped.

The most common and important steps will be considered in separate sections, paying special attention to the most relevant techniques and to expected future developments, according to current trends in analytical chemistry: e.g., miniaturization, automation, reduction in solvent consumption, and sample manipulation.

Organic Pollutants in the Water Cycle. T. Reemtsma and M. Jekel (Eds.).
Copyright © 2006 WILEY-VCH Verlag GmbH & Co. KGaA, Weinheim
ISBN: 3-527-31297-8

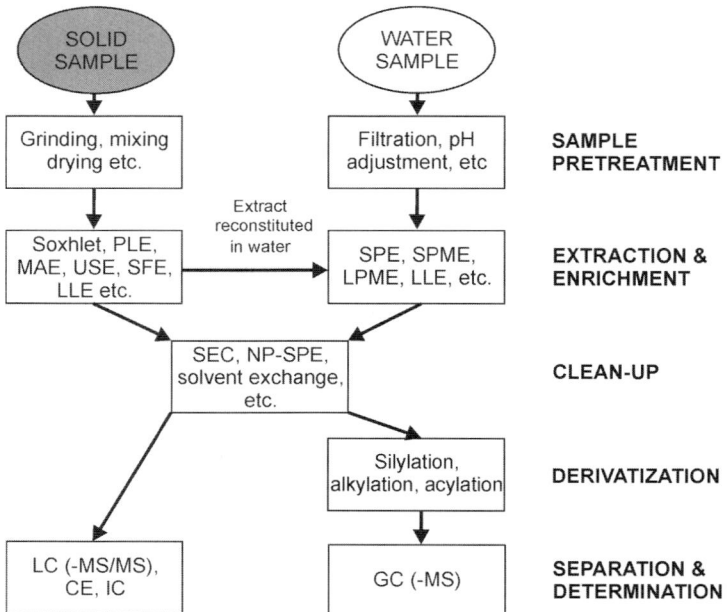

Fig. 1.1 Overview of the analytical process (excluding sampling). CE: Capillary Electrophoresis, IC: Ion Chromatography, GC: Gas Chromatography, LC: Liquid Chromatography, LLE: Liquid Liquid Extraction, LPME: Liquid Phase Microextraction, MAE: Microwave Assisted Extraction, MS: Mass Spectrometry, NP: Normal Phase, PLE: Pressurized Liquid Extraction, SEC: Size Exclusion Chromatography, SFE: Supercritical Fluid Extraction, SPE: Solid Phase Extraction, SPME: Solid Phase Microextraction, USE: Ultrasound Assisted Extraction.

Examples will be presented from different compound classes that are covered in this book and that exhibit different physicochemical properties. The analytical methods for some of these compound classes, namely surfactants [1, 2], herbicides and other pesticides [3–9], pharmaceuticals [10–13], disinfection byproducts [14], and complexing agents [15, 16], have been the subject of specific reviews, which can provide the reader with more detailed information.

1.3
Sample Pretreatment and Analyte Extraction

1.3.1
Sample Pretreatment

Several steps may be required after sampling and before analyte extraction and final determination. These steps include sample preservation, filtration, pH adjustment of aqueous samples, drying and homogenization of solid samples, etc. They are very straightforward, but if they are not performed properly the original sample composition may be seriously altered by these steps.

Even sample storage and shipping can be a critical step in sample preparation. For instance, significant losses of salicylic acid, acetaminophen, and fenofibrate were observed from a mixture of 12 acidic pharmaceuticals spiked to a treated wastewater that was stored in the dark at 4 °C [17]. It may be advisable to analyze samples as soon as possible or to store samples in the dark at −20 °C if they cannot be analyzed immediately. Sample storage may also influence the relative importance of adducts like sulfates and glucuronides as compared to the parent compound [18].

Another common step for sample preservation is acidification, but analytes may not be stable at certain pH values. For example, some fibrate drugs hydrolyze to clofibric acid and fenofibric acid at pH 2 [19], and tetracyclines may undergo epimerization [11]. Furthermore, pH adjustment should be carried out after filtration in order to avoid possible losses during filtration due to the increase in the analyte hydrophobicity.

1.3.2
Solid Samples

To date, the analysis of polar organic contaminants in the water cycle has focused on the aqueous phase, whereas particulate material has not been much considered. Therefore, analytical methods for the determination of polar compounds from solid samples, mainly sediment and sludge, are less developed [20]. Although sorption may not be considered a relevant process for many polar organic compounds, its importance gradually increases with decreasing polarity and increasing solids concentration. Moreover, complexation of ionic and ionizable polar pollutants may occur through inorganic constituents of the matrix, especially in the case of sediments and soils [8, 12]. Thus, in the development of extraction methods for sorbed compounds, one needs to consider their properties, hydrophobicity, and acid-base properties, as well as those of the particulate phase. To develop appropriate extraction conditions that are able to overcome the matrix-analyte interactions, one needs to know whether these interactions are primarily hydrophobic or electrostatic.

In contrast, methods for the determination of pesticides from soil samples are comparatively well developed [8]. Additionally, reviews have appeared recently on the determination of pharmaceuticals [11, 12], surfactants, and their metabolites [1, 2] in environmental solid samples.

The classical extraction method, both for polar and non-polar analytes, was Soxhlet extraction, which consumes large amounts of solvent as well as of the sample, and which is relatively time consuming. Therefore, current methods tend to minimize the consumption of solvents, sample amount, and extraction time by providing additional energy and/or pressure to the mixture of sample and solvent. This supports desorption and diffusion of the analytes from the sample to the solution and enhances their solubility in the extraction media. The different methods are distinguished by the way the energy is supplied to the system and the kind of extracting fluid employed, namely: microwave assisted extraction (MAE), supercritical fluid extraction (SFE), pressurized liquid extraction (PLE), and ultrasound assisted extraction (USE) [21, 22]. Most of them have been fully automated, which is another major advantage over Soxhlet extraction. For example antibiotics were extracted from agricultural soils by PLE [23] at room temperature and 1000 kPa to avoid tetracycline degradation at high temperatures, but allowing the process to be automated.

A first class of compounds that may be considered is phosphoric acid triesters. They are non-ionic and do not have ionizable groups, so they somewhat resemble classical hydrophobic organic pollutants. However, these compounds cover a broad polarity group, from relatively polar short-chain alkyl phosphates (e.g., triethylphosphate and trichloroethylphosphate; log K_{ow} 0.09 and 1.43 respectively) to quite hydrophobic long-chain alkyl phosphates and aryl phosphates (e.g., triphenylphosphate; log K_{ow} 4.76) [24]. Thus, a typical method designed for extracting PAHs or PCBs based on Soxhlet extraction with hexane or toluene works very well for the non-polar analytes but not for the most polar ones, while choosing an intermediate polarity solvent or a solvent mixture provides acceptable recoveries for the whole group of analytes. In one of the pioneering works on MAE, this was compared to Soxhlet extraction and the shake-flask system using a mixture of ethyl acetate and dichloromethane [25]. Microwave extraction yielded better recoveries except for the very polar trimethylphosphate.

More hydrophobic ionizable compounds can also be extracted by an appropriate solvent. The biocides triclosan and triclocarban have been extracted with dichloromethane [26] or acetone/methanol mixtures [27, 28]. Several pharmaceuticals can be extracted also by acetone and methanol from sediment [29], sludge [30], and suspended particulate material [31]. The clean-up by reextraction from water was achieved by using different SPE sorbents and pH values for the different pharmaceutical classes [29, 30].

In the case of more polar and ionizable analytes, however, pure organic solvents are not adequate extractants. Fluoroquinolones are amphoteric species, and their charge state depends on the pH value. Even when their net charge is zero, they are present in the zwitterionic form. For that reason, best recoveries for PLE of fluoroquinolone antibiotics from sludge and soil were obtained by a mixture of acetonitrile and water (1/1) [32]. Moreover, acidification (pH 2) further improved the extraction efficiency, and this was attributed not only to the enhanced solubility of fluoroquinolones at acidic pH but also to the protonation of the acidic sites of the matrix constituents. Finally, the clean-up of the extracts was accomplished by SPE, employing a mixed-phase cation-exchange disk cartridge, like the method for the extraction of fluoroquinolones from water samples [33]. In a similar way, Crescenzi et al. [34] extracted triazine herbicides from soil by hot water containing a phosphate buffer at pH 7.5 in a fully automated process.

The use of hot (subcritical) water extraction is an innovative way of extracting analytes of different polarity from solid matrices. Though water is a rather polar solvent at 20 °C, its dielectric constant decreases markedly as the temperature is raised to 200 °C, and it is then able to efficiently extract hydrophobic compounds, e.g., PAHs [35, 36]. Hence, the polarity of water may be matched to the analyte polarity by selecting an optimized extraction temperature. A good example of this is the extraction of surfactants from sludge. Surfactants comprise a broad group of compounds with different chemical properties, including basic, acidic, and neutral compounds. As a result, most analytical methods are dedicated to one or two compound classes [1]. However the use of subcritical water at pH 9.4 allows the efficient extraction of more than 10 different acidic and neutral chemical groups of surfactants, providing better recoveries than Soxhlet extraction for the nonylphenol ethoxy carboxylates [37]. A clear advantage of using water as a solvent is the ecological aspect and its straightforward application to reverse-phase SPE or SPME clean-up without need for solvent evaporation.

As mentioned previously, the kind of interaction (hydrophilic or hydrophobic) between the analyte and the matrix constituents is another critical point in the extraction. In the case of ionic interactions, the pH of the extraction solution may be shifted or chemicals may be added that can compete with the analytes for the matrix constituents. This technique is used in the case of phenoxyacid herbicide extraction from soils and sediments, where the addition of EDTA to the extracting solvent has been proven to improve recoveries [38–40]. The proposed mechanisms of the simultaneous extraction and derivatization of 2,4-D from soil by PLE are represented in Fig. 1.2 [39]. The same is true for the extraction of tetracycline antibiotics, where a buffer containing EDTA or an acid with chelating properties (e.g., citric acid) is employed to overcome the complexation of these analytes with sample cations [12, 23].

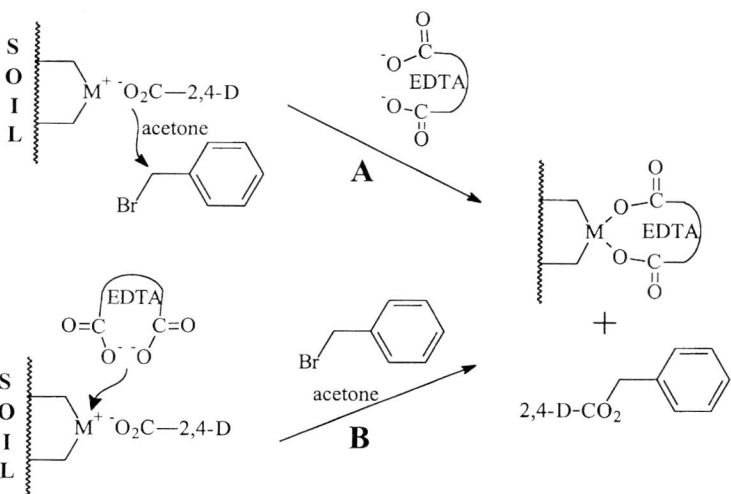

Fig. 1.2 Suggested mechanisms of the PLE-PFBBr derivatization of the herbicide 2,4-D from soil (F atoms not represented): (a) 2,4-D is released while being derivatized with PFBBr, then EDTA occupies its position at the soil surface (b) EDTA replaces 2,4-D from the active surface site, then the freely dissolved EDTA is derivatized by PFBBr. Reprinted from [39], with permission from Elsevier.

After extraction, the resulting extracts from solid samples, particularly in the case of sludge, normally need to be purified before analysis. This has been done in most cases by SPE of the extracts, either employing normal-phase materials (silica, florisil, etc.) if the analytes are relatively non-polar [26, 31, 41, 42] or by reverse and ion exchange phase sorbents if the analytes are relatively polar or possess ionic groups [23, 27, 29, 30, 32]. In many cases a method developed for the SPE of water samples was employed for this purpose after reconstituting or diluting the extract with water.

1.3.3
Aqueous Samples

The determination of polar contaminants in water samples is normally preceded by an extraction step in order to enrich the analytes of interest. This extraction should be as selective as possible in order to minimize the coextraction of matrix that may interfere with analyte detection.

Several extraction techniques for aqueous samples are available, with SPE being the standard procedure. LLE has remained important for only a few applications, e.g., the determination of haloacetic acids [14]. In fact, the US-EPA has two methods available for their determination: one based on SPE [43] and the other based on LLE [44], where, however, the volume of extracting solvent has been minimized to 4 mL of MTBE. The alternatives to SPE are microextraction techniques, namely SPME and, more recently, LPME, as they consume less organic solvent or sample volume (or virtually none in the case of SPME) [45]. Both SPE and microextractions are discussed in more detail.

Other techniques used for the analysis of volatile compounds, like headspace (HS) and purge and trap (PT), are applicable to very few of the polar target analytes considered here (e.g., some haloacetic acids [46]) because of the often ionic character and high water solubility of many polar compounds.

1.3.3.1 Solid Phase Extraction

As already mentioned, SPE is nowadays the most widely used extraction technique for polar organic analytes in water samples. SPE is very convenient; it can be automated and adapted to various analytes by a proper selection from the wide range of sorbent materials available. In the case of polar analytes, the breaking point has been the development of new polystyrene-divinylbenzene (PS/DVB) polymeric sorbent materials [47]. A scheme of the SPE sorbents and retention mechanism as a function of the analyte's properties is presented in Fig. 1.3. Obviously, some analytes can be extracted using different approaches, and selection of the most suitable extraction must take into account many factors, like experience with the specific SPE technique, simplicity of the procedure, possibilities of extending the method toward other analyte classes, and, of course, cost.

Regarding this last aspect, cost, classical silica-bonded reverse phase (RP) materials (C-18, C-8, etc.) are clearly advantageous. Nevertheless, its application towards

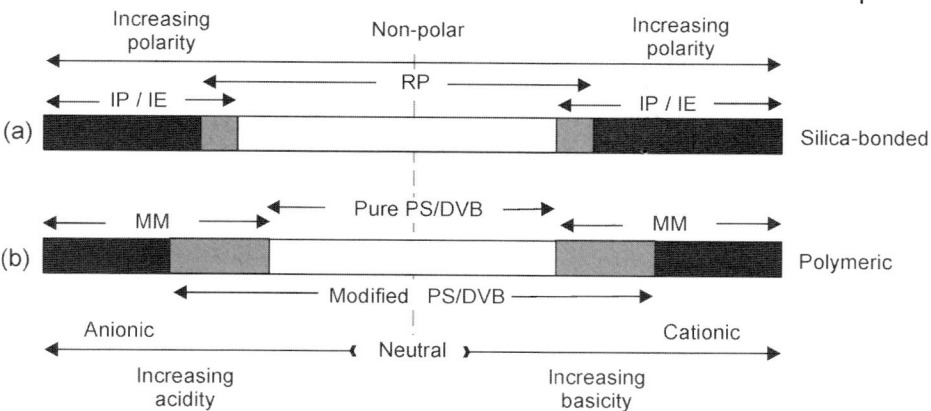

Fig. 1.3 Scheme of SPE sorbents and retention mechanisms for extracting polar pollutants from water as a function of analyte properties. IP: ion pair, IE: ion exchange, RP: reverse phase, PS/DVB: polystyrene/divinylbenzene, MM: mixed mode.

polar species is restricted to weakly acidic or basic compounds, which can be brought into the neutral species by adjusting the sample pH. Thus, C-18 cartridges and disks have been successfully employed for the extraction of acidic drugs [10, 13, 48] and pesticides [3] by adjusting the sample pH to 2–3. However, recoveries of the most polar drugs and their metabolites (e.g., salicylic acid, hydroxy-ibuprofen…) are often incomplete [10, 13, 48]. Furthermore, pH adjustment of samples is limited by the stability of the silica. Therefore, this strategy cannot be applied to strongly acidic or basic analytes or to permanently charged species (e.g., amphoteric or quaternary ammonium compounds). Other problems encountered with silica-based RP materials are the residual silanol groups, which can interact with these analytes even when end-capped cartridges are used, and traces of metals in the silica if compounds with complexing properties (e.g., tetracyclines) are to be determined. In the case of tetracyclines, the problem is solved by adding EDTA to the sample [49].

The first approach to extracting strongly acidic or basic compounds by SPE was the use of ion exchange (IE) SPE [50]. Thus, as mentioned, one of the US-EPA methods for the determination of haloacetic acids is based on anion exchange SPE [43], where the sample pH is adjusted to 5. The extraction of these compounds by an RP SPE employing silica-based materials would not be possible as the sample would need to be acidified to pH 0.5 [44], where the silica bonds would be hydrolyzed. Another official method that relies on IE-SPE is the determination of complexing agents in water samples [51]. Here the International Standards Organization offers two possibilities: either evaporation of the water sample to dryness or IP-SPE before their derivatization and GC determination. Obviously, water evaporation requires a high temperature and is a time-consuming process, while IE-SPE can provide not just preconcentration but also a clean-up that evaporation cannot.

Other applications include quaternary ammonium and acidic herbicides [3, 4]. Yet, IE-SPE has some drawbacks, the major one being that recoveries are strongly affected by the ionic strength of the sample [4]. Therefore, strong matrix effects may occur in IE-SPE of environmental samples, where this parameter may change from sample to sample. For example, it was observed that the recoveries of the complexing agents NTA and EDTA decreased by 20 and 45%, respectively, when 60 mg L^{-1} of sulfate was added to the sample [52].

The other way to extract very polar analytes on silica-bonded phases is to use ion-pair (IP) reagents with RP materials (e.g., C-18), avoiding in this way the use of IE materials, facilitating the adaptation of conventional RP methods and allowing the combined extraction of a wide range of polarities. The retention of analytes can be tuned by selecting the chain length of the IP reagent, as a wide range of these are available (Table 1.1) for both basic/cationic and acidic/anionic compounds. Some of these ion-pairing agents are volatile enough to be compatible with LC-MS. Furthermore, the ion-pairing agent can suppress interactions with silanol groups of the sorbent [3]. Applications of IP-SPE in water analyses include the determination of acidic and quaternary ammonium herbicides [3, 4], acidic phosphoric acid mono- and diesters [53], and acidic pharmaceuticals and their microbial metabolites [54]. IP-SPE is the US-EPA official method for the determination of diquat and paraquat in drinking water [55] by using a C-8 disk and sodium 1-hexanesulfonate as IP reagent for retention; the analytes are then eluted by an HCl acidified solution, which breaks the IP. The technique of IP-SPE was reviewed by Carson in 2000 [56], who nevertheless recognized that this approach was seldom used for the SPE of polar compounds, in spite of the common use of IP formation for improvement of HPLC retention of polar compounds. A reason for this may be the fact that polymeric materials have had more success for SPE than for LC and a much wider chemistry is also available for SPE. In any case, IP-SPE can also be combined with polymeric materials, and it proved useful for the reduction of phenol breakthrough

Tab. 1.1 Common ion pair reagents. Partially adapted from [56].

For basic/cationic analytes	For acidic/anionic analytes
Trifluoroacetic acid[a]	Ammonia[a]
Pentafluoropropionic acid[a]	Triethylamine[a]
Heptafluorobutyric acid[a]	Dimethylbutylamine[a]
Propanesulfonic acid salts	Tributylamine [a]
Butanesulfonic acid salts	Tetramethylammonium salts
1-Pentanesulfonic acid salts	Tetraethylammonium salts
1-Hexanesulfonic acid salts	Tetrapropylammonium salts
1-Heptanesulfonic acid sats	Tetrabutylammonium salts
1-Octanesulfonic acid salts	Tetrapentylammonium salts
1-Nonanesulfonic acid salts	Tetrahexylammonium salts
1-Decanesulfonic acid salts	Tetraheptylammonium salts
1-Dodecanesulfonic acid salts	Tetraoctylammonium salts
Dodecylsulfate, sodium salt	Hexadecyltrimethylammoinum salt
Dioctylsulfosuccinate, sodium salt	Decamethylenbis(trimethylammonum bromide)
Bis-2-ethylhexylphosphate	

a LC-MS compatible

on PS/DVB materials, thus allowing an enrichment factor increase by augmenting the sample volume [57].

A second attempt to improve the retention of polar analytes was the use of graphitized carbon black (GCB) and porous graphitic carbon (PGC) materials. These materials are able to retain compounds with a wide range of polarity, from very hydrophobic (e.g., PCBs) to very polar (e.g., some pesticides), and selectivity is gained by selection of the appropriate eluting solvent [58]. However, irreversible adsorption appears to be a frequent phenomenon, and selection of the appropriate eluting solvent is difficult as the retention mechanisms are still not very clear [58]. Carbon materials have been used for the extraction of several compound classes, alone [3, 4] or in combination with other sorbents [59]. Anyhow, the elution problems, together with the fact that sometimes low breakthrough volumes were reported [60], have hindered their popularization.

Major progress in the SPE of polar analytes came with the development of PS/DVB materials. These do not possess a silica core, and therefore all the associated problems (pH limitation, residual silanol groups and metals and the need for SPE conditioning) were overcome. They also offer π-π interactions, which increase the retention of some polar compounds [47]. Furthermore, this PS/DVB copolymer can be chemically modified by introducing more polar groups. Some researchers have investigated the introduction of vinylpyridine or vinylimidazole functional groups [61, 62], and many materials are now commercially available, like the IST Isolute ENV+ (hydroxy-modified PS/DVB), the Phenomenex Strata-X (modified PS/DVB patented composition) and the Waters Oasis HLB (polyvinylpyrrolidone /DVB, Fig. 1.4). These polar groups increase the wettability of the sorbent and offer dipole-dipole and H-bonding interaction sites, improving the retention of polar analytes [63]. Of these materials, the Oasis HLB product appears to be the most extensively used sorbent for the extraction of polar compounds and especially acidic compounds, like herbicides [64], disinfection by-products [65], and acidic drugs [66, 67]. It has also been employed for the SPE of neutral compounds like phosphoric acid triesters [68], weakly acidic analytes (e.g., triclosan [26, 27]),

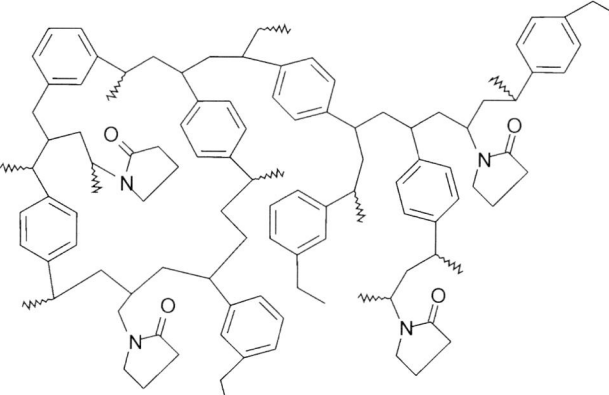

Fig. 1.4 Structure of the polymeric Oasis HLB sorbent. Redrawn from [50]; Copyright © 1998 John Wiley & Sons. With permission of John Wiley & Sons, Inc.

and for the simultaneous extraction of acidic and basic compounds like benzothiazoles and benzotriazoles [69, 70]. In a recent study, several SPE sorbents were tested for the simultaneous extraction of acidic, neutral, and basic drugs, and Oasis HLB proved to provide the best recoveries even at pH 7.8, thus avoiding the acidification step often necessary for the determination of acidic compounds with C-18 cartridges [71]. Furthermore, these sorbents can provide very high enrichment factors after pH adjustment, as no significant breakthrough was detected after the SPE of 2 L of pH 2 water for acidic herbicides [64] and pharmaceuticals [67]. However, this material is not suited for the extraction of permanently charged or very acidic or basic compounds. Here, IP-SPE or mixed-mode PS/DVB cartridges, where an ion exchange group has been introduced into the PS/DVB, or already modified-PS/DVB copolymer are required. These mixed-mode materials offer ion exchange groups to interact with the ionic parts of the analyte and the PS/DVB core for the more hydrophobic parts. This also allows the retention of a wider range of compounds. As an example, a mixed-mode anion exchange disk material has been proposed for the extraction of fluoroquinolone antibiotics in water samples [33]. On the other hand, the SPE of diquat and paraquat is achieved by using a mixed-mode cation exchange cartridge [72].

SPE can be coupled on-line with the LC(-MS) system, allowing the transfer of the whole extract into the chromatographic system, thus either reducing the sample volume sample that has to be extracted or increasing the method sensitivity. Such on-line SPE-LC-MS systems are technically more complex and require a careful adaptation of SPE elution and liquid chromatography. But given their high sensitivity and high reproducibility, with almost complete elimination of the risk of contamination, such systems are highly attractive when large numbers of samples have to be routinely analyzed for a fixed set of pollutants. Thus, most published applications are related to the determination of pesticides [7, 73, 74]. Enrichment and chromatographic separation of the analytes may also be achievable on one short column [75].

Finally, the future development of molecularly imprinted polymers (MIP) is expected to lead to an interesting alternative when a very specific SPE step is desired. These materials are synthesized in the presence of a molecule analogous to the target analyte, so that the synthesized MIP will have specific cavities and interacting sites suitable for interaction with a specific molecule or class of compounds [76]. Up to now, they have mostly been applied as a clean-up step, because the specific interactions take place under apolar organic solvent media. However, the direct application of water samples has already been demonstrated for the selective extraction of phenols [77, 78], where first the analytes are retained in an unspecific way on the MIP and then a non-polar solvent is applied to it, washing the interferences off while the analytes are retained. Despite some promising results, MIP still has to prove its broader applicability for analyte enrichment.

1.3.3.2 Microextractions

SPE has already made remarkable progress compared to LLE in terms of solvent consumption and automation. A step further was achieved by solid-phase microextraction (SPME) and liquid-phase microextraction (LPME), where either no organic solvent is employed (SPME) or only a few microliters (LPME), and also the amount of sample is reduced to ca. 10–50 mL [45].

SPME was developed in 1989 [79] and became commercially available in 1993. In this technique, the analytes are first concentrated into a sorbent coated on a fused-silica fiber that is exposed directly to the sample (direct sampling) or to its headspace (headspace sampling). After partitioning into this sorbent, the analytes can be desorbed either thermally by immersing the fiber into a GC injector or by an organic solvent if they are to be analyzed by LC or CE [80, 81]. Moreover, SPME can easily be automated for GC analysis. In the case of LC, automation can be achieved by so-called *in-tube* SPME, where the sample analytes are preconcentrated into a coated capillary that is connected on-line to the LC [82]. However, SPME hyphenation to LC is hampered by the slow desorption process, leading to band-broadening problems [45].

Nowadays, there are several sorbents available which cover a wide polarity range, and the main difficulty of the analysis of polar compounds relies on the need for derivatization of many of these compounds prior to their determination by GC. Thus, though the extraction of some polar compounds was possible using SPME, it yielded poor detection limits for many compounds due to poor chromatographic properties [83]. However, several options are available for combining SPME and derivatization, depending on the characteristics of the analytes and the derivatizating agent (Fig. 1.5) [45, 80, 84].

The most frequently employed method [45, 84] consists of adding the derivatizing agent directly to the sample. Thus, the desired derivatives are formed first and are then extracted into the SPME fiber by direct or headspace sampling for analysis. An example of this is the determination of basic psychiatric pharmaceuticals in wastewater by acetylation-SPME followed by GC-MS determination, where detection limits ranging between 15 and 75 ng L^{-1} for 10 mL sample volumes were obtained [85]. Another example is the determination of 18 aromatic primary amines,

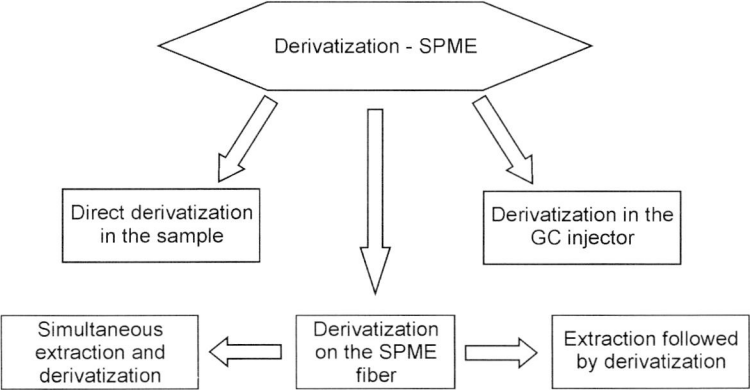

Fig. 1.5 Possible combinations of SPME and derivatization. Redrawn from [80], with permission from Elsevier.

which were converted into their iodo-derivatives in a two-step process, further extracted by SPME, and then analyzed by GC-MS [86]. The substitution of the amine by an iodine group not only reduced the polarity of the amines, but also increased the mass of the analytes by 111 amu (atomic mass units). This was advantageous for the mass spectrometric determination. However, the major limitation of the approach using direct derivatization in the sample is that it is not applicable to moisture-sensitive reactions.

The most popular method of avoiding this inconvenience is to first extract the analytes by SPME and then to derivatize them by exposing the fiber to the vapors of a derivatizing agent. In this way, SPME can be combined with silylation reactions, for example, to generate *tert*-butyldimethylsilyl-derivatives of triclosan and their related degradation products [87], for acidic drugs [88] or acidic herbicides [89]. This combination led to detection limits between 2 and 40 ng L^{-1} from sample volumes of about 20 mL. Figure 1.6 shows the GC-MS chromatogram of a wastewater sample containing 45 ng L^{-1} of mecoprop obtained by SPME extraction and on-fiber derivatization [89].

The other two combinations of derivatization and SPME shown in Fig. 1.5 (namely, "simultaneous extraction and on-fiber derivatization" and "derivatization

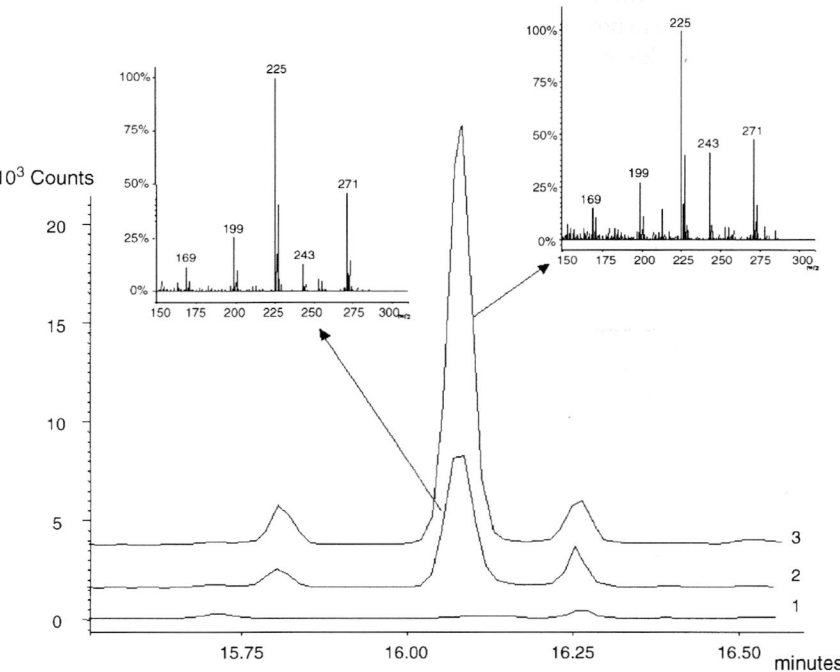

Fig. 1.6 SPME-GC-MS overlay of chromatographic signals (traces at m/z 225 + 227) and MS spectra of mecoprop in a blank sample (1), a non-spiked sewage water (46 ng/L) (2) and a sewage water spiked with the compound at 180 ng L^{-1} (3). Reprinted from [89], with permission from Elsevier.

1.3 Sample Pretreatment and Analyte Extraction | 13

in the GC injector") are only suitable for the analysis of polar organic analytes in very few concrete cases [45].

LPME is more recent than SPME and is based on partition of the analytes between the sample and a small volume (a few µL) of an acceptor solution [45]. This can be carried out by direct contact between the phases [90] or by using a porous membrane as an interface [91–93], which appears to be more robust. It is also possible to employ either flow [93] or rod configuration [91, 92] and to perform the extraction as a two-phase or three-phase system (Fig. 1.7). In the two-phase system (Fig. 1.7i), the organic solvent fills the hollow fiber and the pores of the membrane, acting both as interface and extraction medium. In the three-phase system (Fig. 1.7ii) the organic solvent is used to impregnate the membrane pores and to act as the interface, while the acceptor solution is aqueous, like the sample. Enrichment is achieved by adjusting the sample and acceptor pH in such a way that the analyte solubility is reduced in the sample and enhanced in the acceptor. For example, for the analysis of acidic analytes, the sample is acidified to obtain the neutral species, which are then extracted to the organic interface and finally back-extracted to the acceptor solution, adjusted to a basic pH, so that analytes are trapped in their anionic form. With this three-phase LPME, a very selective extraction is achieved, which may eliminate the commonly encountered matrix effects for the LC-MS/MS analysis of environmental samples [94], provided that analytes have some acidic or basic function. Up to now, home-made devices had to be used for LPME, resulting in low reproducibility and difficult handling, but a commercial device is now available (Gerstel, Mühlheim an der Ruhr, Germany), which, however, uses larger amounts of solvent (ca. 0.5–1 mL).

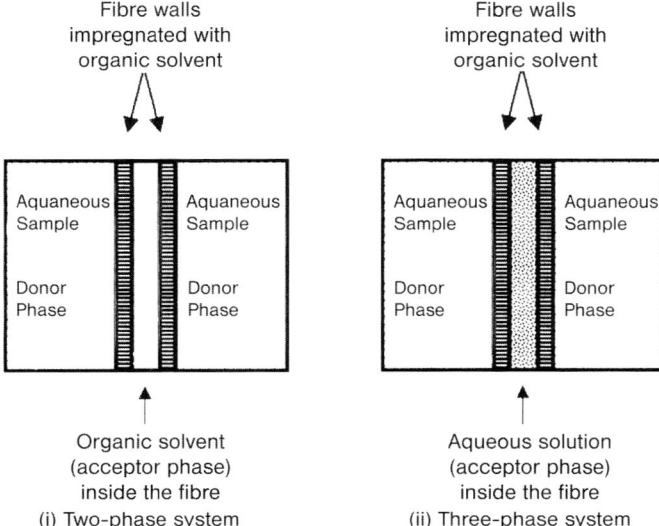

Fig. 1.7 Cross section of the hollow fiber inside the aqueous sample during (i) two-phase and (ii) three-phase LPME. Reprinted from [92], with permission from Elsevier.

1.4
Gas Chromatographic Methods

The GC determination of polar analytes covered in this book in most instances requires a prior derivatization step, with some exceptions, i.e. neutral compounds such as trialkyl phosphates, neutral drugs and disinfection byproducts. Yet even this is not possible for some analytes, and these are usually determined by LC (normally LC-MS/MS, see Section 1.5) or CE. Despite this and the high sensitivity obtainable by LC-MS/MS systems in the last decade, GC still offers some clear advantages over LC, primarily higher separation efficiency and lower costs without the problems associated with matrix effects of LC-MS/MS.

1.4.1
Derivatization

A really wide variety of derivatization reactions are available for many analyte classes. We only discuss the most important of these here, pointing out their advantages and disadvantages and giving some examples, although some have already been mentioned in Section 1.3. Table 1.2 presents an overview of the most common. More details can be found in several books and reviews [9, 95–100].

Tab. 1.2 Overview of most common derivatization reagents.

Reaction type	Reagent	Aqueous media compatible[a]	Target functional group
Alkylation/ Esterification	Diazomethane	NO	–COOH, acidic –OH and –NH group
	(Perfluoro)alkyl halides	NO	–COOH, acidic –OH and –NH groups
	Quaternary ammonium/ sulfonium salts	NO	–COOH, acidic –OH and –NH groups
	Alcohols	NO	–COOH
	(Perfluoro)acyl chloroformate	YES	–COOH
Acylation	(Perfluoro)acyl halides	NO	–OH and basic –NH groups
	(Perfluoro)acyl anhydrides	YES	–OH and basic –NH groups
	(Perfluoro)acyl chloroformate	YES	–OH and basic –NH groups
Sylilation	BSTFA, MSTFA	NO	–COOH, –OH, acidic and basic–NH groups
	MTBSTFA	NO	–COOH, acidic –OH and NH groups

[a] "NO" means that the derivatization must be carried out in an organic solvent

1.4.1.1 Alkylation and Esterification

This reaction consists of the replacement of an acidic or slightly acidic hydrogen by an alkyl group, yielding the corresponding ether or ester. Among the alkylating reagents, diazomethane is probably the most common because of its high reactivity toward most acidic groups, including carboxylic acids and several hydroxyl groups. It has been widely employed, for example, in the determination of acidic pharmaceuticals and herbicides [10, 66, 101, 102] or phenolic analytes, like triclosan [26]. However, diazomethane is toxic, carcinogenic and potentially explosive if not handled with care and must be stored at –20 °C. Hence, many other alkylation reagents have been investigated as potential replacements.

For example alkylation may be performed with alkyl halides in the presence of a base catalyst. The most frequent reagent of this type is pentafluorobenzyl bromide (PFBBr), which produces an electron-rich derivative that can be sensitively analyzed by GC-ECD or GC-NCI-MS. A good example of this is the determination of several phenolic endocrine disruptors (estrogens, surfactants, etc.) by SPE, PFBBr derivatization and GC-NCI-MS, resulting in detection limits at the 20–200 pg L^{-1} level, depending on the sample matrix and the volume employed for extraction (ranging from 1 to 5 L) [103]. Although this procedure cannot be carried out in aqueous media, it tolerates small amounts of water and can be directly coupled to the procedure for the extraction of dried solid samples, as in the PLE of acidic herbicides from soil [39], which has already been mentioned.

On-column or *in-port* alkylation can be accomplished by tetramethylammonium hydroxide (TMAH) or trimethylsulfonium hydroxide (TMSH) and other alkyl analogs, which undergo pyrolytic decomposition at the GC injection temperature, donating alkyl groups (methyl groups in these cases) and yielding trimethylamine or dimethylsulfide, respectively, as decomposition by-products [104]. While TMAH may damage the analytical column, TMSH is less reactive. They can also be used as ion-pairing reagents in order to enhance analyte extractability, for example, in combination with SPME for the determination of linear alkylbenzenesulfonates [105] and perfluorocarboxylic acids [106]. They have also been used for the derivatization of acidic pharmaceuticals and their metabolites after SPE [107–109]. Because of the limited stability of the products, analysis on the same day as the TMSH addition was recommended [108].

Carboxylic acids can also be esterified with an alcohol or by the use of alkylchloroformates. The latter compounds in fact act as alkylating agents with carboxylic acids and as acylating agents with amines and alcohols. They have the advantage that the reaction takes places even under aqueous conditions, so that derivatization can be performed on the sample before extraction [96, 110]. This derivatization scheme is also used in LC determinations to introduce a readily detectable fluorescent group or to increase the retention of analytes in the LC column (see Section 1.5.1.3). However, these reactions are not always easy to perform because they suffer from matrix effects affecting the derivatization yield [111].

1.4.1.2 Acylation

Acylation is normally the derivatization of choice for amines, and it is also very common for phenolic compounds, yielding either amides or esters. The most popular reagents are acyl halides, acyl anhydrides, and their corresponding perfluorinated analogs, when ECD or NCI-MS detection is desired. These reagents are compatible with moisture, so that the derivatization can be carried out prior to extraction in aqueous solution.

The example of the acetylation of basic drugs combined with SPME has been already discussed [85]. Acetylation has also been employed in the determination of phenolic compounds in water samples by SPE and SPME [112, 113].

1.4.1.3 Silylation

Silylation is the most widely used derivatization technique for GC analysis. This technique allows the derivatization of aromatic and also aliphatic alcohols, carboxylic acids, amines, and amides. Because of their reactivity, however, silylating agents are very sensitive to traces of water and other protic solvents.

There are two main classes of silylating agents: those producing trimethylsilyl (TMS) derivatives and those producing *tert*-butyldimethylsilyl (TBDMS) derivatives. TMS derivatives are more common and can be produced by a wide variety of reagents, like *N,O*-bis-trimethylsilyl-trifluoroacetamide (BSTFA) and *N*-methyl-*N*-trimethylsilyl-trifluoroacetamide (MSTFA). Trimethylsilylation has been applied to many polar contaminant classes, including phenols [114], acidic herbicides [114] and acidic pharmaceuticals [17, 115]. The derivatization reaction can also be performed directly on SPE disks [114], producing better yields than derivatization with alkyl halides or methanol/BF_3.

TBDMS derivatives are prepared by reaction with *N-tert*-butyldimethylsilyl-trifluoroacetamide (MTBSTFA). Compared to TMS-reagents they are less sensitive to moisture and tolerate about 1% of water [64]. Moreover EI-MS spectra of TBDMS-derivatives are normally characterized by a base peak corresponding to the loss of the *tert*-butyl (57 amu) group from the molecular ion. This helps in the identification of unknown compounds and in the confirmation of analytes by MS spectra (Fig. 1.6). Derivatization with MTBSTFA has been applied to acidic herbicides [64, 89], acidic pharmaceuticals [67, 88], phenolic compounds [87, 116], and amide type drugs [117]. This versatility is, indeed, one of the advantages of silylation reagents, as they can derivatize a wide range of compounds, including amide groups. In a comparison by Reddersen and Heberer [118] between MTBSTFA and PFBBr for the derivatization of several polar compounds and determination by GC-EI-MS, it was concluded that PFBBr provides slightly better detection limits, but MTBSTFA is able to derivatize a wider range of analytes, including carbamazepine, a compound that otherwise partially decomposes in the GC injector.

However, the drawback of TBDMS reagents as compared to TMS reagents is their lower reactivity and larger size. Thus, it was found that MTBSTFA could derivatize only the aromatic –OH groups of estrogenic compounds, while BSTFA and MSTFA react both with aromatic and aliphatic –OH groups, but only MSTFA provides a complete derivatization of α-ethynylestradiol and mestranol, because of the smaller size of MSTFA [119].

1.4.2
Separation and Detection

1.4.2.1 Separation

Gas chromatographic separation of non-polar analytes or derivatized polar analytes is normally carried out on non-polar, DB-1 or DB-5 type, columns without problems.

A special application is the separation of enantiomers, which may show different biodegradability in the environment. In earlier work, separation was achieved using laboratory-made columns, e.g., the separation of isomers of ibuprofen (after methylation), which was carried out by a column containing the chiral selector 2,6-O-dimethyl-3-O-n-pentyl-β-cyclodextrin (DMPEn) [102]. For some acidic pesticides (after derivatization) a column containing heptakis-[2, 3-dimethyl-6-*tert*-butyldimethylsilyl]-β-cyclodextrin (TBDMβ-CD) was used [101, 120]. Meanwhile, chiral GC columns became more accessible, and, for example, the separation of the pesticide metalaxyl has been accomplished on a PS086-BSCD column (BSCD: *tert*-butyldimethylsilyl-α-cyclodextrin; BGBAnalytik, Adliswil, Switzerland). As shown in Fig. 1.8, an acceptable resolution was obtained, that allowed to follow the enantiomer-specific biodegradation of this compound in soils [121].

Two very recent trends in GC are (a) the development of fast separations by the use of narrow-bore capillary columns and (b) the development of comprehensive two-dimensional gas chromatography (GCxGC). (A) Fast GC has already been applied to several hydrophobic compounds, but the application to polar analytes is still quite uncommon [122]. A method that combined SPE of 10 mL of water and fast GC-MS using a PTV injector allowed the determination of several pesticides at the ng L^{-1} level from a low sample volume and with a GC analysis time of less than 8 min [123]. Analysis times can be further reduced by so-called ultra-fast GC, which uses narrower columns and a specific heating system, in order to obtain reproducible high temperature ramps. TOF detectors can be coupled if MS detection is needed. An example of the separation of some pesticides by ultra-fast GC is shown in Fig. 1.9, where 14 analytes were separated in less than 2 min [124]. (B) GCxGC provides unsurpassed 2-dimensional chromatographic resolution and allows the separation of very complex mixtures. It has as yet been primarily applied to non-polar compounds (geochemistry, oil chemistry, PCBs, etc.) [125, 126]. Recently GCxGC has been used to separate up to 102 nonylphenol isomers in river water [127].

1.4.2.2 Detection

Electron capture detectors (ECD) and nitrogen phosphorus detectors (NPD) are still in use. However, as these detectors cannot provide additional confirmatory information, a mass spectrometer has become the detector of choice. GC-MS systems are nowadays relatively inexpensive, and especially ion trap detectors (ITD) allow the GC-MS/MS analysis of complex samples (e.g., sludge) with low limits of

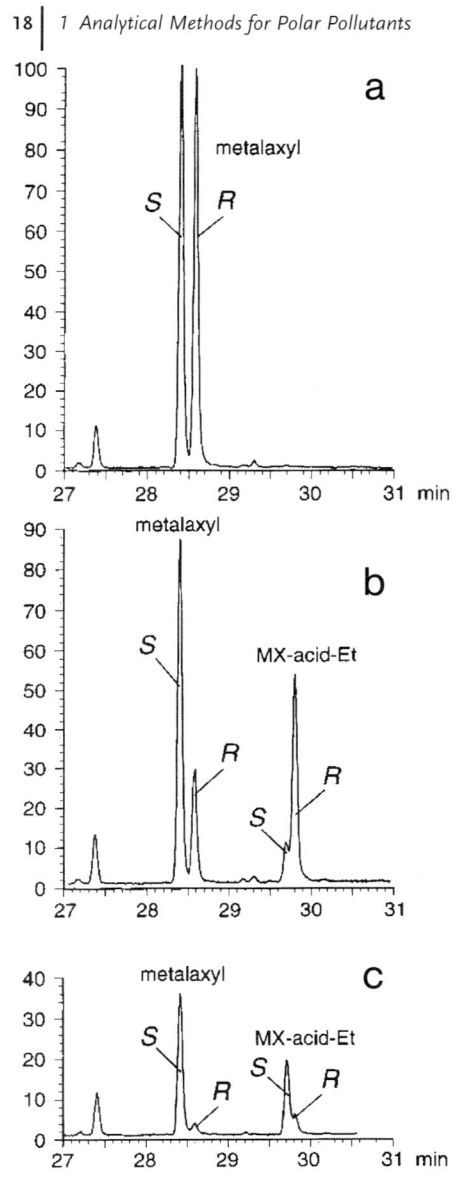

Fig. 1.8 GC EI SIM chromatograms (m/z 220) showing elution of metalaxyl and MX-acid enantiomers (analyzed as the ethyl ester) from the incubation of racemic-metalaxyl in soil after (a) 0, (b) 28, and (c) 74 d, using the chiral PS086-BSCD column. Reprinted with permission from [121]. Copyright (2002) American Chemical Society.

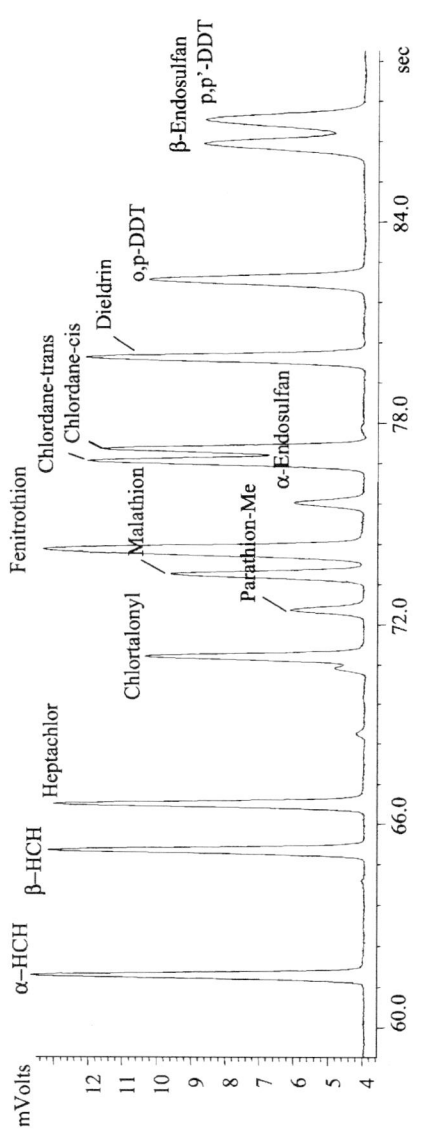

Fig. 1.9 Ultra-fast GC-FID pattern of a pesticide standard mixture. Note that the time axis is given in seconds. Reprinted from [124], with permission from Elsevier.

detection. Thus GC-ITD-MS/MS has been applied to the determination of several pharmaceuticals [101, 128] and some nonylphenol polyethoxylates [129] in river and sea water, where higher sensitivity and selectivity are needed, providing additional structural information for the identification of their degradation products [108, 129]. Another way of improving sensitivity and selectivity is by the use of negative chemical ionization (NCI) if the compounds have electrophilic groups. Sensitivity toward NCI can also be increased by forming perfluoro derivatives, e.g., for the determination of acidic pesticides in soil [39].

The use of programmed temperature vaporizers (PTV) or large volume injectors (LVI) is a final alternative for improving the detection limits. Actually, PTV / LVI are now commercially available and implemented in many GC systems and they can provide injection volumes higher than 100 µL, with the corresponding 100-fold increase in sensitivity [7] and it is expected that their use will be routine soon. However, it must be kept in mind that chemical noise increases in the same way and improved cleanup may be required.

1.5
Liquid Chromatography-Mass Spectrometry

In view of the chemical properties of polar pollutants, many of which are ionic and thermally labile, high-performance liquid chromatography (HPLC) would appear to be a more adequate chromatographic technique than gas chromatography. However, robust HPLC equipment was developed only as recently as the late 1980s. UV detectors, followed by diode-array detectors and fluorescence detectors, were most commonly employed for trace analysis by LC. Until the late 1990s, trace analysis by HPLC was hampered by the limited sensitivity and, more severely, by the limited selectivity of these detectors. Thus, much effort was spent to extend the polarity range of GC-based methods toward polar and ionic analytes, often by derivatization (see Section 1.4).

Since then, LC-based methods have rapidly gained ground in the determination of polar pollutants from aqueous samples as atmospheric pressure ionization (API) mass spectrometers came onto the market and were combined with liquid chromatography, resulting in robust and versatile LC-MS instrumentation.

LC, as well as being tailor-made for the separation of polar water-soluble analytes, offers two increasingly important advantages over GC-based methods. (A) An on-line coupling with solid-phase extraction is comparatively easy to achieve and has been widely used for many classes of compounds, especially pesticides [73, 74, 130, 131]. (B) LC allows aqueous samples to be directly injected into the chromatographic system. With the steadily increasing sensitivity of API-MS instruments, this is becoming more and more attractive.

The next section focuses solely on LC-MS-based methods and does not consider the UV, diode array, and fluorescence detectors used previously. More detailed reviews on the use of LC-MS for the trace analysis of polar pollutants have appeared in recent years [132–135].

1.5.1
Liquid Chromatography

There is a large body of literature on the LC separation of the various classes of polar contaminant. It must be noted, however, that the coupling of LC to MS poses some restrictions on the LC method, as completely volatile eluents are mandatory. Moreover, the composition of the LC eluent, its inorganic and organic modifier and its pH, may interfere strongly with the ionization of the analytes of interest and may thus have a significant effect on the sensitivity of detection. For these reasons, traditional LC methods may not be appropriate for LC-MS analyses. Unfortunately, the influence of the eluent composition on the sensitivity of MS detection turned out to be difficult to predict, and any theory on this subject cannot be a substitute for trials (and errors).

Generally, the criteria used to select the appropriate chromatographic mode are the same as those used for selecting the mode of solid-phase extraction (see Fig. 1.3a in Section 1.3.3.1). At the extremes, ion chromatography (IC, with some limitations) or ion-pair reversed-phase liquid chromatography (IP-RPLC) may be used for strongly basic or acidic analytes, whereas reversed-phase LC (RPLC) is used for most other analytes, with pH adjustment, when weakly acidic or basic compounds are analyzed. Normal-phase chromatography has never played a significant role in trace analysis of polar pollutants, but hydrophilic interaction chromatography (HILIC) may fill this gap in the future.

1.5.1.1 Ionic Analytes

Strongly Acidic Compounds
Many environmentally relevant strongly acidic compound classes are sulfonated aromatic compounds. This functional moiety occurs in technically used naphthalene mono- and di-sulfonates (NSA and NDSA), in sulfonated naphthalene formaldehyde condensates (SNFC), in anionic tensides like linear alkylbenzene sulfonates (LAS; see Chapter 9), and in many azo dyes and their degradation products. Owing to the acidity of the sulfonate group, protonation to the non-ionic form by lowering the eluent's pH value is not possible. Extraction and chromatography of such polar aromatic sulfonates are usually based on ion-pair formation, with volatile cationic counter ions (Fig. 1.10). With ammonium acetate [136] or triethylamine [137], monosulfonates are sufficiently retained, while a strong retention even of trisulfonated naphthalenes has been obtained with tributylamine (TrBA) [138]. IP-RPLC is also suited for the LC-MS analysis of degradation products of azo dyes [139, 140] or sulfonated phthalocyanine dyes [141].

Although the sulfonate group of LAS is quite acidic, the long hydrophobic alkyl chain provides sufficient retention in RPLC, so that IP-RPLC need not be used. In RPLC, LAS mixtures are separated according to the alkyl chain length [142, 143]. However, retention of the more polar carboxylated degradation products of LAS, the sulfophenyl carboxylates (SPC), requires cationic counterions such as triethyla-

$$[R'SO_3]^- + [R_3NH]^+ \xrightleftharpoons{K_{IPF}} [R'SO_3^- \cdot R_3NH^+]$$

$K_{P,A}$ ↕ mobile phase $K_{P,IP}$ ↕
stationary phase

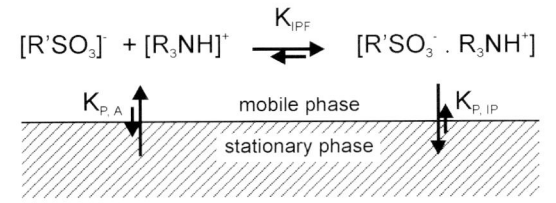

Fig. 1.10 Principle of ion-pair chromatography (IP-RPLC). The partition coefficient for the free acid anion (RSO$_3^-$) from solution (mobile phase) into the stationary phase ($K_{P,A}$) is much smaller than the partition coefficient for the ion pair ($K_{P,IP}$). Eluent conditions should ensure that the equilibrium of ion pair formation (K_{IPF}) is on the right hand side.

mine (TrEA) [143–145]. IP-RPLC with TrEA is also applicable for the analysis of branched alkylbenzenesulfonates [146] and of SPC together with their LAS precursors [144, 145]. In this case, SPCs elute before the LAS.

Only recently it was shown that IP-RPLC with TrBA is also suitable for the determination of 13 strongly acidic phosphoric acid mono- and diesters, some of which are used as plasticizers and occur in municipal wastewater [53].

Haloacetic acids (HAA) are also very acidic compounds, with pK_a values of 0.7 to 2.9. Here, dibutylamine [147] or TEA [148] have been used as ion-pairing agents. Aminopolycarboxylic acids such as EDTA used as complexing agents are important polar pollutants found in many compartments of the water cycle (see Chapter 7). LC-MS methods have been developed using ion-exchange chromatography [149–151]; however, because of the chemical differences between the eluent and the aqueous sample, the speciation of EDTA may be substantially altered and weak complexes may be destroyed [151]. The use of a suppressor (cation exchanger) post-column strongly improves the detection limits [151] and avoids formation of sodium adducts in the interface [149].

Generally, the strength of the chromatographic retention of acidic compounds in IP-RPLC can be adjusted by selecting trialkylamine with a certain number of aliphatic carbon atoms. As the amine must be volatile, a total number of 9 to 10 carbons appears to be the upper limit. The protonated trialkylamines do not tend to form adducts in modern API interfaces but act as H-donors. Therefore these amines diminish the risk of sodium adduct formation with strongly acidic analytes, and they suppress multiple charging of compounds with more than one acidic group. However, these H-donors may also influence the ionization process in ESI [152], decreasing the sensitivity of detection [138]. Thus, the concentration of alkylamines should be kept as low as is acceptable for the chromatographic retention.

Less Acidic Compounds

Many polar pollutants are weakly acidic compounds, namely aromatic and aliphatic carboxylic acids. Among these are many pesticides (see Chapter 6) and pharmaceuticals (see Chapters 2 – 5). Moreover, carboxylic acids are intermediates in the microbial or chemical oxidation of many classes of environmental contaminants. For example, SPCs are formed from LAS (see Chapter 9), alkylphenol ethoxycarboxylates (NPEC or OPEC) and dicarboxylates (CAPEC) from non-ionic alkylphenol ethoxylates (APEO: NPEO or OPEO). Oxanilic acid derivatives can be generated microbiologically from chloracetanilide herbicides [153, 154]. Carboxylic acids are also generated as breakdown products of aromatic hydrocarbons and many other aromatic compounds. There are three strategies to attain good peak shape and to increase retention of weakly acidic compounds in RPLC.

1. Mostly, the pH value of the eluent is decreased until the analytes occur in their non-dissociated state. Formic acid or acetic acid may be used, depending on the pH required. This approach was used for the analysis of X-ray contrast agents [59], succinates [155], and naphthalene carboxylates [156].

2. The addition of acid and the transformation of an analyte anion into its non-dissociated state could decrease the signal intensity for the molecular anion using ESI in negative mode. Alternatively, ammonium acetate can be added. The ion pair formed from ammonium and the analyte anion exhibits good peak shape and dissociates easily in the interface to release the molecular anion. Acidic pharmaceuticals like non-steroidal antiinflammatory drugs have long been analyzed by GC-MS after derivatization. LC-MS, using ESI in the negative ion mode with ammonium acetate as inorganic modifier, is also well suited for this purpose [157] and has also been used for acidic X-ray contrast media [158] as well as for oxanilic acid derivatives of herbicides [153, 154].

3. If the carboxylic acid analytes are small and polar, the non-ionic molecule or ion-pairs formed with ammonium experience only weak retention in RPLC. Then, organic amines in their protonated form can be added to utilize IP-RPLC. This strategy has been used for the LC-MS analysis of bile acid conjugates using dibutylamine [159], for X-ray agents with a (non-volatile) tetraalkylammonium cation [160], and for acidic pharmaceuticals (antiphlogistics and lipid regulators, Fig. 1.11) and their more polar metabolites with volatile TrBA [54, 161]. Highly polar low-molecular-weight carboxylates may also be separated by CE-MS [162] or IC-MS [163].

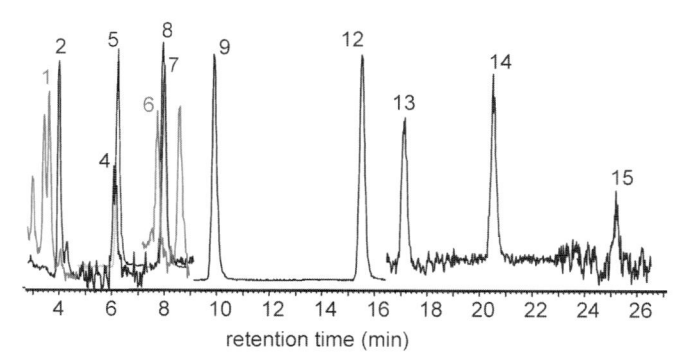

Fig. 1.11 LC-ESI-MS chromatogram of acidic pharmaceuticals in a municipal wastewater. IP-RPLC was used and analytes detected by MRM in negative ion mode. 1: salicylic acid; 2: 2,4-dichlorobenzoic acid (internal standard); 4: piroxicam; 5: clofibric acid; 6: ketoprofen; 7: naproxen; 8: bezafibrate; 9: fenoprop (internal standard); 12: diclofenac; 13: indomethacin; 14: meclofenamic acid (internal standard) 15: triclosan. Redrawn from a chromatogram shown in [161].

Strongly Basic Compounds
It is fortunate that strongly basic compounds seldom occur in the aquatic environment, because the respective LC-MS approach of using anionic counterions to retain such compounds has not resulted in methods of satisfactory sensitivity. This is due to the fact that the polyfluorinated aliphatic carboxylic acids used as counteranions, namely hexafluorobutyric acid [164, 165] or pentafluoropropionic acid [166], strongly suppress the analyte signal in positive ion mode [167]. It was hypothesized that the analyte cations and the fluorinated carboxylate anions form an ion-pair that is too strong to dissociate in the interface. Maybe HILIC will be more successful than IP-RPLC for this application.

Weakly Basic Compounds
Quaternary ammonium compounds like dialkyl dimethyl ammonium salts or benzyl alkyl dimethyl ammonium salts are used as cationic surfactants and as anti-bacterial agents. Despite the ionic character of these compounds, their retention in RPLC is no problem because of the benzyl and long-chain alkyl substituents. However, a good peak shape is more difficult to obtain. Acidification reduces the tailing to some extent, and thus most eluents are acidified with formic acid [168, 169]. However, increasing acid concentrations may severely reduce the sensitivity of detection for basic nitrogenous compounds by ESI in the positive ion mode [170]. This may be because of the anionic counterions added with increasing acid concentration. These anions may unintentionally form ion pairs with the nitrogenous cations. The ion pairs are nonionic in solution and are thus less effectively transferred into the mass spectrometer [170]. If quaternary ammonium compounds become more polar, and retention in RPLC is difficult to obtain, IP-RPLC can be used (see above).

Aromatic nitrogenous compounds like benzothiazoles [69, 171] and benzotriazoles [70] can be analyzed with ammonium acetate using ESI in positive mode. The same approach is used for basic pharmaceuticals such as β-2-sympathomimetics and beta-blockers [10, 48, 158]. A method for the determination of carbamazepine and six metabolites with ammonium acetate at pH 4 was developed [172].

Sulfonylurea, imidazolinone, and sulfonamide herbicides are quite effective and need to be determined with high sensitivity. As for many other basic analytes RPLC with a formic acid ammonium formate buffer ESI in positive ion mode has been used to determine 16 of these analytes in surface and groundwater [173].

1.5.1.2 Non-Ionic Analytes
Alkylphenol ethoxylates (APEO) are the most important class of non-ionic surfactants (see Chapter 9), and many methods for their analysis by LC-MS have been developed. Since the hydrophobic part of APEO is identical for all components, all ethoxy homologs tend to coelute in RPLC separation [174–176], and the distinction between them is only made by means of the MS detection. This coelution has often been claimed to be advantageous, since it would increase the sensitivity of detection. As a matter of fact, quantitation is severely compromised by the coelution,

because the response factors of homologs vary substantially, with a notoriously poor sensitivity for the monoethoxylate (e.g., NP1EO). Moreover, fragmentation of one ethoxy-homolog results in the increase of the response for the next lower homolog. This is not recognized and impairs quantification if these homologs are not chromatographically separated. So-called mixed-mode chromatography relying on size-exclusion and reversed-phase mechanisms provides full resolution of the homologs (Fig. 1.12) [177]. The chromatographic separation of the homologs avoids the interference of the MS signals of different homologs due to fragmentation and allows optimization the MS parameters for each of the homologs.

The chromatographic separation of alcohol ethoxylates (AEOs) by RPLC strongly depends upon the organic modifier used: with a water/methanol gradient, AEOs are separated according to the length of the hydrophobic alkyl chain [178–180], whereas a water/acetonitrile gradient also provides some separation according to the number of ethoxy groups [155, 180]. Alkylpolyglycosides (surfactants based on sugars and fatty alcohols) have a hydrophobic alkyl chain of 8 to 16 carbons that mediates retardation on RPLC [181, 182].

Neutral pharmaceuticals are not a chemically homogenous class of compounds. They are rather a mixture of environmentally relevant compounds (e.g., carbamazepine, diazepam, cyclophosphamide) that can be analyzed together at neutral pH [183]. Similar conditions can be used for macrolide antibiotics (clarithromycin, erythromycin), which are, however, more hydrophobic and require increased organic modifier concentrations [48, 184].

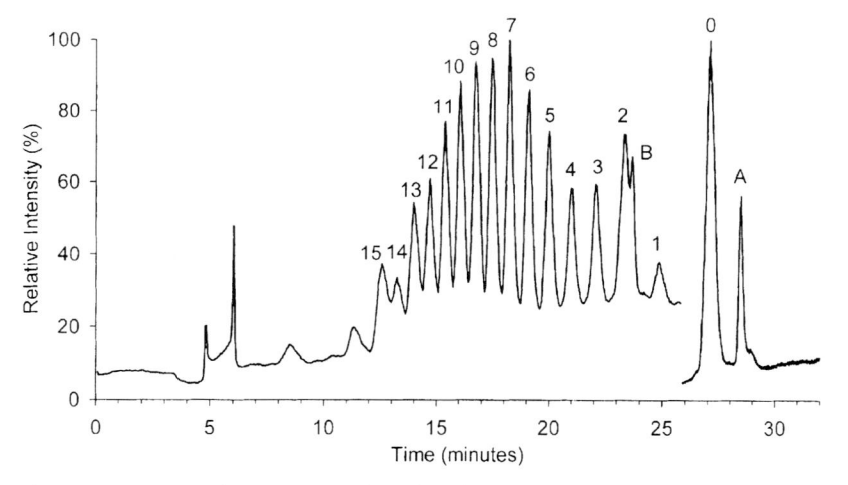

Fig. 1.12 LC-ESI-MS chromatogram of a technical mixture of NPEO. A mixed-mode separation based on SEC and RPLC was used for chromatography. Numbers denote the number of ethoxylate groups in the respective NPEO homolog, 0 denotes the NP. A and B are internal standards. Reprinted from [177], with permission from Elsevier.

1.5.1.3 Amphoteric Compounds

Many biologically active polar compounds contain nitrogenous and oxygenous functional groups and exhibit amphoteric properties. Their chromatographic separation with satisfactory peak shape can be difficult, as it may be impossible to find pH conditions at which these compounds appear as non-charged molecules. Instead they may occur as cations, zwitterions, or anions. In terms of chromatography the zwitterionic form is probably the worst.

In most cases, acidic eluents were chosen, and ESI in positive ion mode was employed for detection. This scheme was used for fluoroquinolones (norfloxacin, ciprofloxazin, and others) [33, 184] and for amphoteric cocamidopropylbetaine surfactants [155]. For the latter, IP-RPLC is also reported [185]. In order to increase the signal response of amphoteric beta-lactam antibiotics (penicillins), methanolysis to form a secondary amine was used. These exhibited a much higher signal response in the positive ion mode than that of their parent penicillines [186].

The analysis of amphoteric tetracycline antibiotics is hampered by their tendency to form stable metal complexes and to interact strongly with free silanol groups of glassware and silica materials. This can be prevented by adding EDTA before extraction (see Sections 3.2 and 3.3.1). Tetracyclines have pK_a values between 3.3 and 9.5 [187] and should be chromatographically separated with the acidic moiety in the non-dissociated form using an acidic eluent [184, 188]. Oxalic acid may be added to the eluent in order to suppress interaction with the silica core of the stationary phase [184]. Tetracycline epimers can occur that are difficult to separate by LC [189]. Detection of tetracyclines may also be performed by APCI [189], but ESI is more frequently used.

Several methods exist that combine the analysis of tetracyclines with that of sulfonamides [187, 189]. This seems consistent, as sulfonamide chromatography requires an acidic pH [190], as for tetracyclines, and ESI in positive mode is applied for both classes of analytes.

The herbicide glyphosate and its biodegradation product aminomethylphosphonic acid (AMPA) are among the most challenging analytes in terms of polarity. While glyphosate contains only three carbons per molecule, there is only one in AMPA. In the extraction of these compounds, chromatographic retention and detection have been equally difficult to achieve. Derivatization with FMOC is one strategy that circumvents these problems by introducing a fluorescent group that also increases the molecular size and hydrophobicity, and allows the use of classical RPLC coupled to MS using ESI in negative mode [111, 191]. The FMOC derivative also shows a good peak shape, which is more difficult to obtain with the original analytes. On the other hand, this derivatization reaction is not exempt of problems (see Section 1.4.1.1). Alternatively, IC-MS of anions is applicable with a cation suppressor coupled between the column and the ESI interface [149]. As for other very polar and ionic compounds, CE-MS is applicable but its sensitivity is limited because of the very small injection volumes of a few nanoliters [192].

1.5.1.4 Multiresidue Methods

Because of the substantial improvements in mass spectrometers, multi-residue methods across a range of chemically different compound classes are gaining in importance. The increasing sensitivity of mass spectrometers allows sensitivity to be sacrificed by working at non-optimal conditions, thus finding a compromise between the optimal conditions of each of the different compound classes that are combined. This has not been possible before, when optimal conditions in terms of eluent composition and instrumental settings were required for each analyte to attain sufficient sensitivity. A second instrumental improvement with respect to multiresidue methods is the possibility of rapid switching between positive and negative ion mode within one analytical run.

Presently, multiresidue methods are best developed in those areas of application where large numbers of chemically diverse analytes are found and a strong need for regular monitoring exists, such as in residue analysis of pesticides. Here, a method for the determination of 39 acidic and basic pesticides with ESI in positive and negative mode was developed comparatively early [193]. Similar approaches, now using triple quadrupole mass spectrometry in multiple reaction monitoring (MRM) instead of SIM (see Section 1.5.2.2), have been published [73]. In quite recent work, more than 100 pesticides of 20 chemical classes were analyzed within 20 min run time using ESI in positive (> 90 compounds) and negative ion mode (10 compounds) [194]. In terms of chromatography, mixtures of MeOH and AcCN or even ternary gradients are frequently used for these purposes.

Also for the trace analysis of pharmaceuticals the number of multi-methods is increasing. A method for the determination of 27 compounds, endocrine disruptors, pharmaceuticals, and personal care products using ESI in positive and negative mode has been published [195], as well as a procedure to determine 13 antibiotics belonging to 5 classes, including sulfonamides, tetracyclines, fluoroquinolones, and macrolide antibiotics using ESI in positive-ion mode [196].

1.5.1.5 Chiral Separation

Many polar pollutants found in the aqueous environment contain one or more chiral centers, especially when biological effects are intended, as in the case of pesticides and pharmaceuticals. Biotransformation may differ for the enantiomers, and a change in the enantiomeric ratio along a water path can provide an indication of the importance of biotransformation processes. Therefore, chiral separation is a valuable analytical process to study the fate of such biologically active compounds in the environment, as shown for ibuprofen [102] and pesticides [121] (see Section 1.4.2.1).

Chiral centers occur in several carboxylates, such as profen non-steroidal-antiinflammatory drugs and the chemically closely related chlorophenoxypropionic acids, tetracyclines, beta-blockers, and many other classes of compounds. Enantiomeric separation is usually based on chromatography using a stationary phase that contains one enantiomer of a so-called chiral selector. The chiral center of the polar analyte is often close to one polar or ionic moiety. Most LC enantiomeric separ-

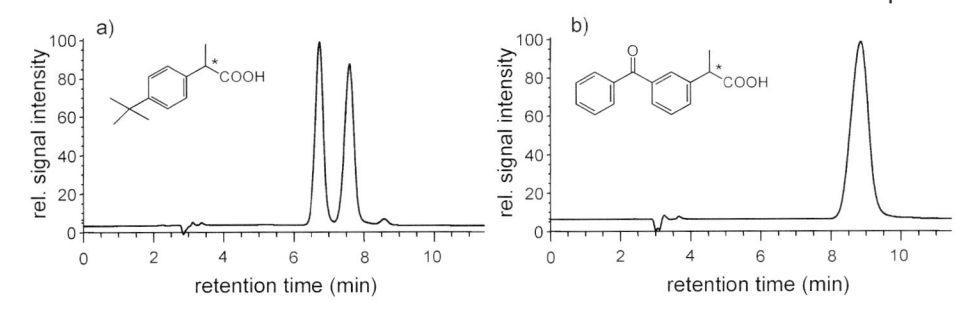

Fig. 1.13 LC-ESI-MS chromatogram of a separation of profen drug enantiomers on a chiral stationary phase (Chiralpak AD-RH using IP-RPLC). a) Ibuprofen; b) Ketoprofen. Quintana and Reemtsma, unpublished.

ations use normal-phase chromatography with only a few percent of water, because then the chiral selector can closely interact with the chiral center via the polar functional group. Several procedures of NPLC for the enantio-separation of profen drugs and beta-blockers are available [197, 198]. But, as mentioned before, NPLC is difficult to use for environmental samples: the sample solubility in pure organic solvent is often low, the samples often contain many polar matrix constituents that affect the lifetime of NPLC columns, and water may have to be added post-column to improve ionization in the MS interface. A few applications of RPLC for enantio-separation have been published, for example, for the separation of acidic profen drugs [199]. When RPLC is employed, however, hydrophobic interaction with the stationary phase governs the separation. If the hydrophobic part of a molecule becomes too large, then the influence of the polar chiral center may be too weak to obtain enantio-resolution. One such example is shown in Fig. 1.13: for ibuprofen an enantio-separation was achieved, whereas the enantiomers of the more hydrophobic ketoprofen could not be separated in IP-RPLC. To date, chiral separation of polar analytes using LC-MS is not a routine task in environmental analysis, and no applications to environmental samples are known.

Instead of liquid chromatography, capillary electrophoresis with cyclodextrins has frequently been used for enantiomeric separations, for example, for the phenoxyacid herbicides fenoprop, mecoprop, and dichlorprop [200].

1.5.2
Mass Spectrometry

1.5.2.1 Ionization

Two ionization interfaces, electrospray ionization (ESI) and atmospheric pressure chemical ionization (APCI), are available with most LC-MS instruments. There is also an ongoing effort to develop and to commercialize new interfaces, with the intention to extend the range of LC-MS toward less polar compounds that are more difficult to ionize. But, for the analysis of polar pollutants, this is quite irrelevant, and it appears that ESI and APCI will be the interfaces that accompany us for a long while.

The overwhelming majority of literature on the determination of polar pollutants from water utilizes electrospray ionization (ESI), while APCI is used only rarely. This does not fully follow the general rule that compounds already appearing as ions in aqueous solution are best analyzed by ESI, whereas atmospheric pressure chemical ionization may be preferred for moderately polar compounds that are non-ionic in solution. This rule was generally confirmed in a detailed study by Thurman and coworkers comparing APCI and ESI for 75 pesticides (Fig. 1.14) [201]. They also showed that the composition of the eluent influences the signal response in ESI more strongly than in APCI and that adduct formation may be stronger in ESI. These latter two findings are in agreement with observations from real samples that the analyte response in APCI has been found to be less affected by coeluting matrix components than in ESI [202–205]. These are some arguments in favor of APCI.

However, when using ESI, matrix effects can be reduced by decreasing the flow directed into the ESI interface, which decreases the droplet size as well as the number of molecules to be ionized at a given time. Best effects appear to be obtained by using a nano-splitting device [206], but even a flow reduction to 50 µL min^{-1}, which is possible with standard equipment, can diminish matrix effects [207].

As for the selection of ESI or APCI, the decision whether positive or negative mode works better is obvious at the chemically extreme ends. Bases are best detected as cations ([M+H]$^+$) in positive mode and acids as anions ([M–H]$^-$) in negative mode. Between these two extremes, however, there remains much scope for method optimization based on trial and error. Amphoteric compounds as well as many environmental contaminants that contain nitrogen and oxygen can be ionized in either mode, as has been found for sulfonamides and tetracyclines [189].

The final decision on whether ESI or APCI in positive or negative mode provides the best results depends not only upon the chemical properties of the analytes but also on the chromatographic system that is used, the eluent composition, and the "background" matrix of the samples that are to be analyzed. It is also important to note that the fragmentation pathway along which an ion breaks down is severely influenced by the charge it bears, as different functional elements of a molecule stabilize a positive or a negative charge. For an amino-hydroxy-naphthalene sulfonate, the transition m/z 238→80 was most intensive in the negative ion mode, as compared to

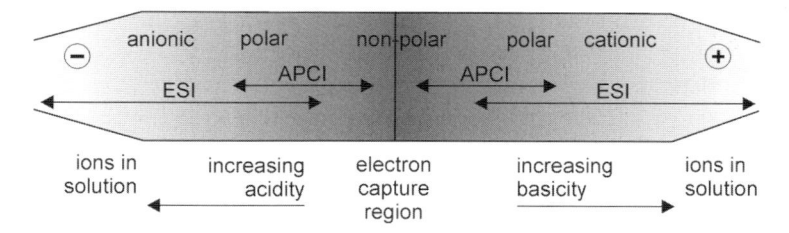

Fig. 1.14 Scheme of ionization modes (APCI: atmospheric pressure chemical ionization; ESI: electrospray ionization) selected as a function of analytes properties. Reprinted with permission from [201]. Copyright (2001) American Chemical Society.

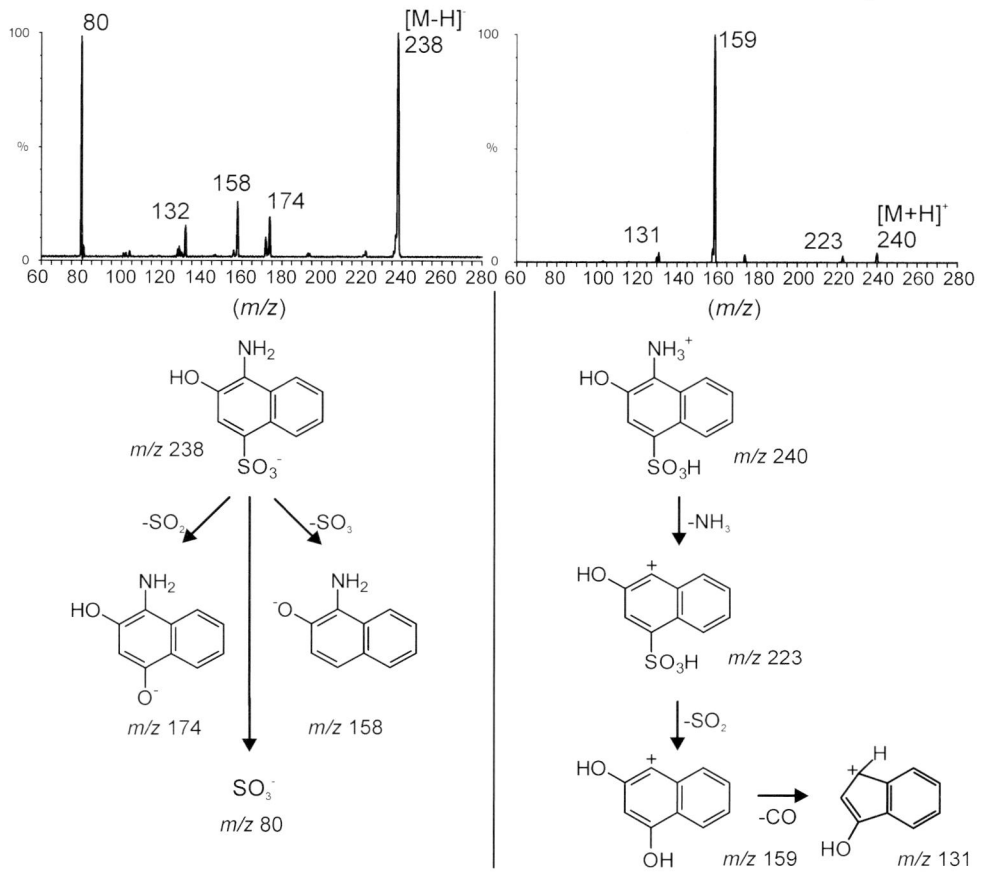

Fig. 1.15 ESI-MS spectra and suggested fragmentation schemes for the molecular ions of 1-amino-2-hydroxy-naphthalene-4-sulfonic acid in negative (left side) and positive (right side) ion mode. In negative ion mode a sulfonate anion is generated first and further fragmentations occur at this group. In positive ion mode an ammonium cation is generated, leading to fragments with the positive charge located at the aromatic system.

m/z 240→159 in the positive ion mode (Fig. 1.15). Thus, not only the intensity of the molecular ion signal, but also the intensity and the selectivity of transitions selected for multiple reaction monitoring are important criteria to be considered in method development and may influence the decision on the mode of ionization.

1.5.2.2 Mass Spectrometers and Modes of Operation

Triple-Quadrupole and Ion-Trap Mass Spectrometers
The development of mass spectrometer instrumentation has continued steadily. Single-stage quadrupole MS is no longer considered adequate for the detection of

trace pollutants from environmental samples. For routine determination of trace compounds, these instruments are replaced by triple-stage quadrupole mass spectrometers (QqQ-MS) and ion trap mass spectrometers. For quantitative analysis of target analytes, triple quad MS operated in multiple-reaction monitoring mode provides unsurpassed sensitivity and a large dynamic range of several orders of magnitude.

Meanwhile, new mass spectrometers are commercially available that are using a so-called linear ion trap coupled behind a quadrupole. These instruments can be used for high-sensitivity MRM determination of target analytes when the second mass spectrometer is operated as a quadrupole and also for high-sensitivity product ion spectra recording when the second mass spectrometer is operated as an ion trap.

Recent years have seen an increasing awareness of quality control in environmental analysis also. This development was fostered by a directive concerning quality control in residue analysis released by the Commission of the European Union [208]. This directive proposes a system of identification points and requires a minimum of three such points to confirm a positive finding. When two precursor-product ion pairs are recorded, their relative intensity is compared with that of a reference standard, and if the chromatographic retention times coincide within certain limits these criteria are fulfilled. Difficulties arise, however, when an analyte exhibits only one intense product ion and no second transition is available for confirmation.

The limited mass resolution of quadrupole- and ion-trap mass spectrometers limits their use for the identification of unknown compounds, as it does not allow the determination of elemental compositions under routine conditions. Triple-quadrupole mass spectrometers can be used to screen for compounds with a common functional group or a common structural element by using either neutral loss or precursor ion scans. By these approaches, metabolites of a known parent compound can be determined, provided that the transformation products retain the substructure or functional group which the detection was based on [209]. However, the sensitivity of detection drastically decreases in the parent ion scan, in which the first mass analyzer is scanning and, even more so, in neutral loss modes when both mass analyzers are operated in scanning mode [132].

Time-of-Flight MS
Time-of-flight mass spectrometers (TOF-MS) and quadrupole-time-of-flight mass spectrometers (Q-TOF-MS) offer a mass resolution of 5000 to >15 000 (fwhm). For small molecules, this is often sufficient to determine the molecular formulas of unknowns (TOF-MS) and of their fragment ions (Q-TOF-MS).

Despite its inherent high sensitivity for whole-spectrum recording, TOF-MS cannot yet compete with triple-quadrupole-MS operated in MRM mode in both sensitivity of detection and dynamic range (~2 orders of magnitude distance). Therefore, the main field of TOF and Q-TOF MS is qualitative analysis rather than the quantitation of known target analytes. Typical applications are metabolite identification, verification of positive findings, and screening for unknowns. Q-TOF-

MS is much more useful than TOF-MS in environmental application, because product ion spectra can be detected with high mass resolution, which is essential for the identification of unknowns [210]. In this way, not only can the elemental formula of a molecular ion be determined, but unambiguous information on some of its structural characteristics can be obtained. Conversely, exact mass information on fragment ions of an unknown molecular ion supports its molecular formula determination: (a) the exact mass difference between the product and the precursor is an additional piece of information, and (b) often only one possible combination of the proposed elemental compositions of fragment ions and the molecular ion is consistent (Table 1.3).

TOF-MS has been used to confirm the occurrence of herbicides [211] or to identify previously unknown contaminants in environmental samples [212] and industrial wastewater [213]. Q-TOF-MS is also useful to elucidate fragmentation processes of polar pollutants [214]. Because of its high resolution and good sensitivity over a full mass range, TOF-MS would seem to be advantageous for screening purposes. The use of a Q-TOF can be even more useful, as it enables product ion spectra to be recorded for further identification [210]. The high mass resolution provided by a TOF-MS can be used to reduce the chemical noise in selected ion chromatograms [212]. Though this is an illustrative application of the high resolving power of these instruments, the practical use of such chromatograms is limited.

A second advantage, besides whole-spectra recording with high sensitivity, is the recording speed of TOF-MS, which is in the range of nanoseconds for one whole spectrum. This high speed is due to the fact that no scanning is required as in

Tab. 1.3 Exact masses of the molecular anion and one fragment ion of ethylhexyl phosphate (MEHP) identified by LC-Q-TOF-MS in a municipal wastewater. Based on the determined exact masses, possible elemental compositions were calculated for the two anions and for their mass difference. The only consistent combination of elemental formulas for the three masses is highlighted in bold and corresponds to MEHP. Data from [53].

Anion	Measured (m/z)	Proposed formulas[a]	Calculated (m/z)	Error (mDa)
Molecular	209.0938	**$C_8H_{18}O_4P$**	**209.0943**	**−0.5**
		$C_6H_{16}N_3O_3P$	209.0929	0.9
		$C_9H_{17}N_5O_2$	209.0959	−2.1
		$C_{11}H_{16}NOP$	209.0970	−3.2
		$C_7H_{19}N_2O_1P_2$	209.0973	−3.5
Product	78.9623	H_2NPS	78.9646	−2.3
		PO_3	**78.9585**	**3.8**
Difference	130.1315	$C_{16}H_{16}N_3$	130.1344	−2.9
		$C_8H_{18}O$	**130.1358**	**−4.3**

[a] formulas proposed using the following conditions: mass tolerance of ±5 mDa.
Element composition: C = 0–20, H = 0–40, N = 0–5, O = 0–6, S = 0–1, and P = 0–2.

quadrupole or ion-trap mass spectrometers. This characteristic makes TOF detectors ideal for very fast chromatographic separations that result in narrow chromatographic peaks, as in ultraperformance liquid chromatography (UPLC) [215], where peak capacities of several hundreds are achievable [216].

1.5.2.3 Quantitation Strategies and Matrix Effects

The high selectivity and low chemical noise usually experienced when using LC-API-MS in MRM-mode provides unsurpassed detection limits in the analysis of polar pollutants. However, this selectivity weakens our awareness that the target analyte is only an extreme minority of the whole amount of sample injected onto the analytical column. This so-called matrix may affect the ionization of the analytes of interest and may result in an erroneous quantitation by LC-MS. A typical problem is coextracted humic material in the trace analysis of acidic pesticides from ground- and surface waters [147, 173, 217–219]. But in general, any coeluting organic compound can interfere with the ionization of a target analyte [220–222]. In quantitative analysis from sampling campaigns in the field, matrix effects may vary considerably from sample to sample and from sampling to sampling [147, 217]. As the sample matrix is also subjected to chromatographic separation, it may vary for each analyte in a multi-component analysis during its time of entry into the interface, as do the matrix effects [223]. Therefore, matrix effects are, indeed, often the major hurdle in the routine analysis of environmental samples by LC-MS.

To improve this situation, the highest priority should be given to any approach that reduces the matrix effects, either by more selective extraction or improved chromatography. More selective extraction may be achievable by two-step extraction at different pH-values [219]. For very polar analytes, a precleaning step by C_{18}-SPE can be used to remove less polar matrix constituents prior to analyte extraction with a polymeric sorbent [10, 190]. A triple-phase liquid-phase microextraction has recently proven useful for this purpose [94]. Even a clean-up of extracts before the LC-MS analysis may be used (SEC, [205]), although this introduces new sources of error. However, environmental analysis often aims at determining a larger number of analytes covering a broad range of physico-chemical properties. This substantially limits the perspectives of clean-up options, which rely on differences in the physico-chemical properties of the analytes and the sample matrix.

There may be two other options to reduce matrix effects. (A) It was recently pointed out, that signal suppression by coeluting matrix can be considerably reduced by directing lower flow volumes into the ESI source [206, 207]. Moreover, the increasing sensitivity of mass spectrometers provides increasing opportunities to omit analyte enrichment and to directly inject the aqueous sample into the LC-MS system. It may be astonishing that both these strategies can reduce matrix effects even though the concentration ratio of the target analyte to the matrix components is not changed. However, in both cases less material is entering the ionization interface at a given time. Thus, competition for whatever may limit the ionization efficacy, electrical charge or droplet surface area, can be avoided.

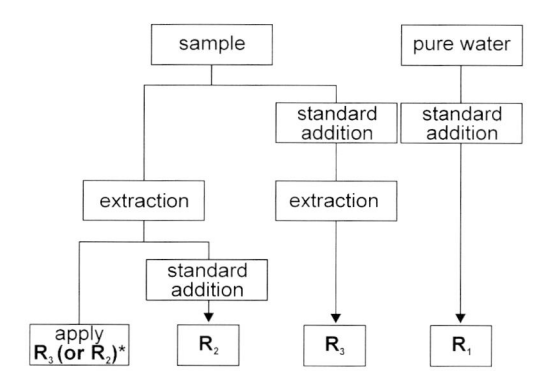

Fig. 1.16 Calibration scheme for the independent determination of matrix effects (R_2/R_1) and SPE recovery (R_3/R_2) in ESI-MS. * The response factor R_3 (or R_2, if extraction was shown to be quantitative) may finally be applied for quantitation of analytes in samples.

If matrix effects cannot be reduced, compensation is necessary to provide quantitatively accurate results. This can best be done by an isotopically labeled standard for each of the analytes included in one method. Unfortunately, such compounds are often not available for environmentally relevant analytes. Then, addition of a standard must be performed. In the frequent case that analytes are enriched prior to the LC-MS analysis, a low signal for a standard compound added to a sample prior to extraction as compared to a reference solution may result from incomplete extraction or from suppression of its signal due to matrix. Therefore, the recovery of the enrichment procedure and the matrix effects need to be distinguished. The clearest results are obtained by performing three calibrations (Fig. 1.16): one in pure aqueous solution (R_1), one by standard addition to aliquots of a sample extract (R_2), and one by standard addition to a sample before enrichment (R_3). The ratio of the response factors R_2/R_1 describes matrix effects, whereas R_3/R_2 describes the recovery of the enrichment process.

If one is confident in having a uniform matrix within a series of samples, calibration can be performed by standard addition to only one sample of a series [220]. If the matrix varies from sample to sample, standard addition to each sample would have to be performed for each sample. This, however, leads to three to four times higher sample numbers, and, although it can correct for sensitivity losses by matrix compounds, this approach cannot avoid the loss of sensitivity.

1.6
Conclusions

Recent years have seen significant progress in analytical methods for polar pollutants. Polymeric sorbents have improved the extractability of polar pollutants from aqueous samples. GC-based analyses often require derivatization of polar analytes, namely when they are ionic, but still offer the highest chromatographic resolution. With regard to mass spectrometric determination, ESI-MS appears to surmount all previous limitations with respect to analyte polarity, and ion chromatography and ion-pair chromatography have similarly tranformed chromatographic separation.

Instrumental development of mass spectrometers will continue to provide the analytical chemist with both increasing sensitivity and improved mass resolution. At the same time, the effort involved in analyzing polar pollutants in environmental samples steadily decreases while productivity increases. The development of multi-component methods across several chemical classes of compounds and the increasing use of direct injection of aqueous samples into LC-MS systems are the most recent examples of this trend.

Many of the results and much of the knowledge presented in the following chapters of this book could not have been gathered without the analytical methods for polar pollutants outlined in this chapter. Furthermore, environmental analytical chemistry provides the means to continue and extend the study of the occurrence and behavior of polar pollutants in all compartments of the water cycle.

References

1 M. Petrovic, D. Barcelo, *Trends Anal. Chem.* **2004**, 23, 762–771.
2 T. P. Knepper, D. Barceló, P. de Voogt (Ed.), *Analysis and fate of surfactants in the aquatic environment*, Elsevier, Amsterdam, **2003**.
3 M. J. M. Wells, L. Z. Yu, *J. Chromatogr. A* **2000**, 885, 237–250.
4 Y. Picó, G. Font, J. C. Moltó, J. Mañes, *J. Chromatogr. A* **2000**, 885, 251–271.
5 W. L. Budde, *Mass Spectrom. Rev.* **2004**, 23, 1–24.
6 Y. Picó, C. Blasco, G. Font, *Mass Spectrom. Rev.* **2004**, 23, 45–85.
7 R. B. Geerdink, W. M. A. Niessen, U. A. T. Brinkman, *J. Chromatogr. A* **2002**, 970, 65–93.
8 V. Andreu, Y. Picó, *Trends Anal. Chem.* **2004**, 23, 772–789.
9 M. Rompa, E. Kremer, B. Zygmunt, *Anal. Bioanal. Chem.* **2003**, 377, 590–599.
10 T. A. Ternes, *Trends Anal. Chem.* **2001**, 20, 419–434.
11 J. Beausse, *Trends Anal. Chem.* **2004**, 23, 753–761.
12 M. S. Díaz–Cruz, M. J. López de Alda, D. Barceló, *Trends Anal. Chem.* **2003**, 22, 340–351.
13 J. B. Quintana, J. Carpinteiro, I. Rodríguez in *Analysis, fate and removal of pharmaceuticals in the water cycle* (Ed. M. Petrovic, D. Barcelo), Elsevier, Amsterdam. In press.
14 S. D. Richardson, *Trends Anal. Chem.* **2003**, 22, 666–684.
15 T. P. Knepper, *Trends Anal. Chem.* **2003**, 22, 708–724.
16 M. Sillanpää, M. L. Sihvonen, *Talanta* **1997**, 44, 1487–1497.
17 R. B. Lee, K. Sarafin, T. E. Peart, M. L. Svoboda, *Water Qual. Res. J. Canada* **2003**, 38, 667–682.
18 S. Reddy, C. R. Iden, B. J. Brownawell, *Anal. Chem.* **2005**, 77, 7032–7038.
19 M. Stumpf, T. A. Ternes, K. Haberer, P. Seel, W. Baumann, *Vom Wasser* **1996**, 86, 291–303.
20 D. Barceló, *Trends Anal. Chem.* **2004**, 23, 677–679.
21 R. M. Smith, *J. Chromatogr. A* **2003**, 1000, 3–27.
22 J. R. Dean, S. L. Cresswell in *Sampling and sample preparation for field and laboratory* (Ed. J. Pawliszyn), Elsevier, Amsterdam, **2002**, pp. 559–586.
23 A. M. Jacobsen, B. Halling–Sørensen, F. Ingerslev, S. H. Hansen, *J. Chromatogr. A* **2004**, 1038, 157–170.
24 Y. Kawagoshi, I. Fukunaga, H. Itoh, *J. Mater. Cycles Waste Manag.* **1999**, 1, 53–61.
25 H. De Geus, B. N. Zegers, H. Lingeman, U. A. T. Brinkman, *Int. J. Environ. Anal. Chem.* **1994**, 56, 119–132.
26 H. Singer, S. Müller, C. Tixier, L. Pillonel, *Environ. Sci. Technol.* **2002**, 36, 4998–5004.
27 R. U. Halden, D. H. Paull, *Environ. Sci. Technol.* **2004**, 38, 4849–4855.

28 R. U. Halden, D. H. Paull, *Environ. Sci. Technol.* **2005**, 39, 1420–1426.
29 D. Löffler, T. A. Ternes, *J. Chromatogr. A* **2003**, 1021, 133–144.
30 T. A. Ternes, M. Bonerz, N. Herrmann, D. Löffler, E. Keller, B. Bagó Lacida, A. C. Alder, *J. Chromatogr. A* **2005**, 1067, 213–223.
31 M. Himmelsbach, W. Buchberger, H. Miesbauer, *Int. J. Environ. Anal. Chem.* **2003**, 83, 481–486.
32 E. M. Golet, A. Strehler, A. C. Alder, W. Giger, *Anal. Chem.* **2002**, 74, 5455–5462.
33 E. M. Golet, A. C. Alder, A. Hartmann, T. A. Ternes, W. Giger, *Anal. Chem.* **2001**, 73, 3632–3638.
34 C. Crescenzi, A. Di Corcia, M. Nazzari, R. Samperi, *Anal. Chem.* **2000**, 72, 3050–3055.
35 S. B. Hawthorne, A. Kubátová in *Sampling and sample preparation for field and laboratory* (Ed. J. Pawliszyn), Elsevier, Amsterdam, **2002**, pp. 587–608.
36 S. B. Hawthorne, Y. Yang, D. J. Miller, *Anal. Chem.* **1994**, 66, 2912–2920.
37 F. Bruno, R. Curini, A. Di Corcia, I. Fochi, M. Nazzari, R. Samperi, *Environ. Sci. Technol.* **2002**, 36, 4156–4161.
38 F. M. Guo, Q. X. Li, J. P. Alcantara-Licudine, *Anal. Chem.* **1999**, 71, 1309–1315.
39 S. Campbell, Q. X. Li, *Anal. Chim. Acta* **2001**, 434, 283–289.
40 A. Caballo-López, M. D. Luque de Castro, *Chromatographia* **2003**, 58, 257–262.
41 H. R. Buser, T. Poiger, M. Müller, *Environ. Sci. Technol.* **1998**, 32, 3449–3456.
42 K. Bester, *Water Res.* **2003**, 37, 3891–3896.
43 US-EPA, EPA Method 552.1 – *Determination of haloacetic acids and dalapon in drinking water by ion excahnge liquid-solid extraction and gas chromatography with electron capture detector*, Cincinnati, **1992**.
44 US-EPA, EPA Method 552.2 – *Determination of haloacetic acids and dalapon in drinking water by liquid-liquid extraction, derivatization and gas chromatography with electron capture detector*, Cincinnati, **1995**.
45 J. B. Quintana, I. Rodríguez, *Anal. Bioanal. Chem.* **2006**, 384, 1447–1461.
46 B. Cancho, F. Ventura, M. Galceran, A. Diaz, S. Ricart, *Water Res.* **2000**, 34, 3380–3390.
47 C. W. Huck, G. K. Bonn, *J. Chromatogr. A* **2000**, 885, 51–72.
48 F. Sacher, F. T. Lange, H. J. Brauch, I. Blankenhorn, *J. Chromatogr. A* **2001**, 938, 199–210.
49 M. Petrovic, M. D. Hernando, M. S. Diaz-Cruz, D. Barcelo, *J. Chromatogr. A* **2005**, 1067, 1–14.
50 E. M. Thurman, M. S. Mills, *Solid-Phase Extraction. Principles and Practice*, John Wiley & Sons, New York, **1998**.
51 ISO 16588: 2002. *Water quality – Determination of six complexing agents – Gas-chromatographic method*, International Standards Organization, Geneva, **2002**.
52 R. Geschke, M. Zehringer, *Fresenius J. Anal. Chem.* **1997**, 357, 773–776.
53 J. B. Quintana, R. Rodil, T. Reemtsma, *Anal. Chem.* **2006**, 78, 1644–1650.
54 J. B. Quintana, S. Weiss, T. Reemtsma, *Water Res.* **2005**, 39, 2654–2664.
55 US-EPA, EPA Method 549.1 – *Determination of diquat and paraquat in drinking water by liquid-solid extraction and high performance liquid chromatography with ultraviolet detection*, Cincinatti, **1992**.
56 M. C. Carson, *J. Chromatogr. A* **2000**, 885, 343–350.
57 E. Pocurull, M. Calull, R. M. Marcé, F. Borrull, *J. Chromatogr. A* **1996**, 719, 105–112.
58 M. C. Hennion, *J. Chromatogr. A* **2000**, 885, 73–95.
59 A. Putschew, S. Schittko, M. Jekel, *J. Chromatogr. A* **2001**, 930, 127–134.
60 O. Núñez, E. Moyano, M. T. Galcerán, *J. Chromatogr. A* **2002**, 946, 275–282.
61 N. Fontanals, M. Galia, R. M. Marce, F. Borrull, *J. Chromatogr. A* **2004**, 1030, 63–68.
62 N. Fontanals, P. Puig, M. Galia, R. M. Marce, F. Borrull, *J. Chromatogr. A* **2004**, 1035, 281–284.
63 N. Fontanals, R. M. Marcé, F. Borrull, *Trends Anal. Chem.* **2005**, 24, 394–406.
64 I. Rodríguez Pereiro, R. González Irimia, E. Rubí Cano, R. Cela Torrijos, *Anal. Chim. Acta* **2004**, 524, 249–256.

65 C. Zwiener, T. Glauner, F. H. Frimmel, *Anal. Bioanal. Chem.* **2002**, 372, 615–621.
66 S. Öllers, H. P. Singer, P. Fassler, S. R. Muller, *J. Chromatogr. A* **2001**, 911, 225–234.
67 I. Rodríguez, J. B. Quintana, J. Carpinteiro, A. M. Carro, R. A. Lorenzo, R. Cela, *J. Chromatogr. A* **2003**, 985, 265–274.
68 R. Rodil, J. B. Quintana, T. Reemtsma, *Anal. Chem.* **2005**, 77, 3083–3089.
69 A. Kloepfer, M. Jekel, T. Reemtsma, *J. Chromatogr. A* **2004**, 1058, 81–88.
70 S. Weiss, T. Reemtsma, *Anal. Chem.* **2005**, 77, 7415–7420.
71 S. Weigel, R. Kallenborn, H. Hühnerfuss, *J. Chromatogr. A* **2004**, 1023, 183–195.
72 L. Grey, B. Nguyen, P. Yang, *J. Chromatogr. A* **2002**, 958, 25–33.
73 F. Hernández, J. V. Sancho, O. Pozo, A. Lara, E. Pitarch, *J. Chromatogr. A* **2001**, 939, 1–11.
74 A. Asperger, J. Efer, T. Koal, W. Engewald, *J. Chromatogr. A* **2002**, 960, 109–119.
75 A. C. Hogenboom, P. Speksnijder, R. J. Vreeken, W. M. A. Niessen, U. A. T. Brinkman, *J. Chromatogr. A* **1997**, 777, 81–90.
76 F. Chapuis, V. Pichon, M. C. Hennion, *LC GC Eur.* **2004**, 17, 408–412, 414, 416–417.
77 E. Caro, R. M. Marcé, P. A. G. Cormack, D. C. Sherrington, F. Borrull, *J. Chromatogr. A* **2003**, 995, 233–238.
78 E. Caro, M. Masqué, R. M. Marcé, F. Borrull, P. A. G. Cormack, D. C. Sherrington, *J. Chromatogr. A* **2002**, 963, 169–178.
79 R. P. Belardi, J. B. Pawliszyn, *Wat. Pollution Res. J. Canada* **1989**, 24, 179–191.
80 J. Pawliszyn in *Sampling and sample preparation for field and laboratory* (Ed. J. Pawliszyn), Elsevier, Amsterdam, **2002**, pp. 389–477.
81 J. Pawliszyn, *Solid phase microextraction: theory and practice*, Wiley-VCH, New York, **1997**.
82 H. Kataoka, *Anal. Bioanal. Chem.* **2002**, 373, 31–45.

83 M. Moeder, S. Schrader, M. Winkler, P. Popp, *J. Chromatogr. A* **2000**, 873, 95–106.
84 E. E. Stashenko, J. R. Martínez, *Trends Anal. Chem.* **2004**, 23, 553–561.
85 J. P. Lamas, C. Salgado-Petinal, C. García-Jares, M. Llompart, R. Cela, M. Gómez, *J. Chromatogr. A* **2004**, 1046, 241–247.
86 T. Zimmermann, W. J. Ensinger, T. C. Schmidt, *Anal. Chem.* **2004**, 76, 1028–1038.
87 P. Canosa, I. Rodríguez, E. Rubi, R. Cela, *J. Chromatogr. A* **2005**, 1072, 107–115.
88 I. Rodríguez, J. Carpinteiro, J. B. Quintana, A. M. Carro, R. A. Lorenzo, R. Cela, *J. Chromatogr. A* **2004**, 1024, 1–8.
89 I. Rodríguez, E. Rubí, R. González, J. B. Quintana, R. Cela, *Anal. Chim. Acta* **2005**, 537, 259–266.
90 E. Psillakis, N. Kalogerakis, *Trends Anal. Chem.* **2002**, 21, 53–63.
91 K. E. Rasmussen, S. Pedersen-Bjergaard, *Trends Anal. Chem.* **2004**, 23, 1–10.
92 E. Psillakis, N. Kalogerakis, *Trends Anal. Chem.* **2003**, 22, 565–574.
93 J. A. Jönsson, L. Mathiasson, *LC GC Eur.* **2003**, 16, 683–690.
94 J. B. Quintana, R. Rodil, T. Reemtsma, *J. Chromatogr. A* **2004**, 1061, 19–26.
95 J. Segura, R. Ventura, C. Jurado, *J. Chromatogr. B* **1998**, 713, 61–90.
96 R. J. Wells, *J. Chromatogr. A* **1999**, 843, 1–18.
97 J. L. Little, *J. Chromatogr. A* **1999**, 844, 1–22.
98 J. M. Rosenfeld in *Sampling and sample preparation for field and laboratory* (Ed. J. Pawliszyn), Elsevier, Amsterdam, **2002**, pp. 609–668.
99 K. Blau, J. M. Halket, *Handbook of derivatives for chromatography*, John Wiley & Sons, Chichester, 1993.
100 J. Drozd, *Chemical derivatization in gas chromatography*, Elsevier, Amsterdam, **1981**.
101 H. R. Buser, M. D. Müller, N. Theobald, *Environ. Sci. Technol.* **1998**, 32, 188–192.
102 H. R. Buser, T. Poiger, M. D. Müller, *Environ. Sci. Technol.* **1999**, 33, 2529–2535.
103 H. M. Kuch, K. Ballschmiter, *Environ. Sci. Technol.* **2001**, 35, 3201–3206.

References

104. A. Zapf, H. J. Stan, *J. High Resolut. Chromatogr.* **1999**, 22, 83–88.
105. R. Alzaga, A. Peña, L. Ortiz, J. M. Bayona, *J. Chromatogr. A* **2003**, 999, 51–60.
106. R. Alzaga, J. M. Bayona, *J. Chromatogr. A* **2004**, 1042, 155–162.
107. C. Zwiener, F. H. Frimmel, *Water Res.* **2000**, 34, 1881–1885.
108. C. Zwiener, T. Glauner, F. H. Frimmel, *J. High Resolut. Chromatogr.* **2000**, 23, 474–478.
109. C. Zwiener, S. Seeger, T. Glauner, F. H. Frimmel, *Anal. Bioanal. Chem.* **2002**, 372, 569–575.
110. T. Henriksen, B. Svensmark, B. Lindhardt, R. K. Juhler, *Chemosphere* **2001**, 44, 1531–1539.
111. M. Ibáñez, O. J. Pozo, J. V. Sancho, F. J. Lopez, F. Hernández, *J. Chromatogr. A* **2005**, 1081, 145–155.
112. M. Llompart, M. Lourido, P. Landín, C. García-Jares, R. Cela, *J. Chromatogr. A* **2002**, 963, 137–148.
113. I. Rodríguez, M. I. Turnes, M. C. Mejuto, R. Cela, *J. Chromatogr. A* **1996**, 721, 297–304.
114. D. Li, J. Park, J. R. Oh, *Anal. Chem.* **2001**, 73, 3089–3095.
115. G. R. Boyd, H. Reemtsma, D. A. Grimm, S. Mitra, *Sci. Total Environ.* **2003**, 311, 135–149.
116. T. Heberer, H. J. Stan, *Anal. Chim. Acta* **1997**, 341, 21–34.
117. S. Zühlke, U. Dünnbier, T. Heberer, *J. Chromatogr. A* **2004**, 1050, 201–209.
118. K. Reddersen, T. Heberer, *J. Sep. Sci.* **2003**, 26, 1443–1450.
119. J. B. Quintana, J. Carpinteiro, I. Rodriguez, R. A. Lorenzo, A. M. Carro, R. Cela, *J. Chromatogr. A* **2004**, 1024, 177–185.
120. H. R. Buser, M. D. Müller, *Environ. Sci. Technol.* **1998**, 32, 626–633.
121. H. R. Buser, M. D. Müller, T. Poiger, M. E. Balmer, *Environ. Sci. Technol.* **2002**, 36, 221–226.
122. E. Matisová, M. Dömötörová, *J. Chromatogr. A* **2003**, 1000, 199–221.
123. M. Hada, M. Takino, T. Yamagami, S. Daishima, K. Yamaguchi, *J. Chromatogr. A* **2000**, 874, 81–90.
124. C. Bicchi, C. Brunelli, C. Cordero, P. Rubiolo, M. Galli, A. Sironi, *J. Chromatogr. A* **2005**, 1071, 3–12.
125. J. Dalluge, J. Beens, U. A. T. Brinkman, *J. Chromatogr. A* **2003**, 1000, 69–108.
126. P. Marriott, R. Shellie, *Trends Anal. Chem.* **2002**, 21, 573–583.
127. T. Ieda, Y. Horii, G. Petrick, N. Yamashita, N. Ochiai, K. Kannan, *Environ. Sci. Technol.* **2005**, 39, 7202–7207.
128. M. Stumpf, T. A. Ternes, K. Haberer, W. Baumann, *Vom Wasser* **1998**, 91, 291–303.
129. P. M. Hoai, S. Tsunoi, M. Ike, Y. Kuratani, K. Kudou, P. H. Viet, M. Fujita, M. Tanaka, *J. Chromatogr. A* **2003**, 1020, 161–171.
130. R. Bossi, K. V. Vejrup, B. B. Mogensen, W. A. H. Asman, *J. Chromatogr. A* **2002**, 957, 27–36.
131. J. Slobodnik, A. C. Hogenboom, J. J. Vreuls, J. A. Rontree, B. L. M. van Baar, W. M. A. Niessen, U. A. T. Brinkman, *J. Chromatogr. A* **1996**, 741, 59–74.
132. T. Reemtsma, *J. Chromatogr. A* **2003**, 1000, 477–501.
133. T. Reemtsma, *Trends Anal. Chem.* **2001**, 20, 533–542.
134. T. Reemtsma, *Trends Anal. Chem.* **2001**, 20, 500–517.
135. C. Zwiener, F. H. Frimmel, *Anal. Bioanal. Chem.* **2004**, 378, 851–861.
136. M. J. F. Suter, S. Riediker, W. Giger, *Anal. Chem.* **1999**, 71, 897–904.
137. M. C. Alonso, M. Castillo, D. Barcelo, *Anal. Chem.* **1999**, 71, 2586–2593.
138. T. Storm, T. Reemtsma, M. Jekel, *J. Chromatogr. A* **1999**, 854, 175–185.
139. T. Storm, C. Hartig, T. Reemtsma, M. Jekel, *Anal. Chem.* **2001**, 73, 589–595.
140. J. E. B. McCallum, S. A. Madison, S. Alkan, R. L. Depinto, R. U. R. Wahl, *Environ. Sci. Technol.* **2000**, 34, 5157–5164.
141. T. Reemtsma, *J. Chromatogr. A* **2001**, 919, 289–297.
142. A. Di Corcia, F. Casassa, C. Crescenzi, A. Marcomini, R. Samperi, *Environ. Sci. Technol.* **1999**, 33, 4112–4118.
143. E. Gonzalez Mazo, M. Honing, D. Barcelo, A. GomezParra, *Environ. Sci. Technol.* **1997**, 31, 504–510.

144 P. Eichhorn, T. P. Knepper, *Environ. Toxicol. Chem.* **2002**, 21, 1–8.
145 J. Riu, E. Gonzalez-Mazo, A. Gomez-Parra, D. Barcelo, *Chromatographia* **1999**, 50, 275–281.
146 P. Eichhorn, M. E. Flavier, M. L. Paje, T. P. Knepper, *Sci. Total Environ.* **2001**, 269, 75–85.
147 W. M. A. Niessen, *J. Chromatogr. A* **1998**, 794, 407–435.
148 R. Loos, D. Barcelo, *J. Chromatogr. A* **2001**, 938, 45–55.
149 K. H. Bauer, T. P. Knepper, A. Maes, V. Schatz, M. Voihsel, *J. Chromatogr. A* **1999**, 837, 117–128.
150 T. P. Knepper, A. Werner, G. Bogenschütz, *J. Chromatogr. A* **2005**, 1085, 240–246.
151 R. N. Collins, B. C. Onisko, M. J. McLaughlin, G. Merrington, *Environ. Sci. Technol.* **2001**, 35, 2589–2593.
152 J. A. Ballantine, D. E. Games, P. S. Slater, *Rapid Commun. Mass Spectrom.* **1995**, 9, 1403–1410.
153 R. A. Yokley, L. C. Mayer, S. B. Huang, J. D. Vargo, *Anal. Chem.* **2002**, 74, 3754–3759.
154 I. Ferrer, E. M. Thurman, D. Barcelo, *Anal. Chem.* **1997**, 69, 4547–4553.
155 L. H. Levine, J. L. Garland, J. V. Johnson, *Anal. Chem.* **2002**, 74, 2064–2071.
156 G. Ohlenbusch, C. Zwiener, R. U. Meckenstock, F. H. Frimmel, *J. Chromatogr. A* **2002**, 967, 201–207.
157 X. S. Miao, B. G. Koenig, C. D. Metcalfe, *J. Chromatogr. A* **2002**, 952, 139–147.
158 T. A. Ternes, R. Hirsch, *Environ. Sci. Technol.* **2000**, 34, 2741–2748.
159 T. Sasaki, T. Iida, T. Nambara, *J. Chromatogr. A* **2000**, 888, 93–102.
160 R. Hirsch, T. A. Ternes, A. Lindart, K. Haberer, R. D. Wilken, *Fresenius J. Anal. Chem.* **2000**, 366, 835–841.
161 J. B. Quintana, T. Reemtsma, *Rapid Commun. Mass Spectrom.* **2004**, 18, 765–774.
162 S. K. Johnson, L. L. Houk, D. C. Johnson, R. S. Houk, *Anal. Chim. Acta* **1999**, 389, 1–8.
163 W. Ahrer, W. Buchberger, *J. Chromatogr. A* **1999**, 854, 275–287.
164 R. Castro, E. Moyano, M. T. Galceran, *J. Chromatogr. A* **2001**, 914, 111–121.
165 R. Castro, E. Moyano, M. T. Galceran, *J. Chromatogr. A* **1999**, 830, 145–154.
166 M. Takino, S. Daishima, K. Yamaguchi, *J. Chromatogr. A* **1999**, 862, 191–197.
167 S. A. Gustavsson, J. Samskog, K. E. Markides, B. Langstrom, *J. Chromatogr. A* **2001**, 937, 41–47.
168 I. Ferrer, E. T. Furlong, *Environ. Sci. Technol.* **2001**, 35, 2583–2588.
169 M. J. Ford, L. W. Tetler, J. White, D. Rimmer, *J. Chromatogr. A* **2002**, 952, 165–172.
170 N. Wang, W. L. Budde, *Anal. Chem.* **2001**, 73, 997–1006.
171 T. Reemtsma, *Rapid Commun. Mass Spectrom.* **2000**, 14, 1612–1618.
172 X. S. Miao, C. D. Metcalfe, *Anal. Chem.* **2003**, 75, 3731–3738.
173 E. T. Furlong, M. R. Burkhardt, P. M. Gates, S. L. Werner, W. A. Battaglin, *Sci. Total Environ.* **2000**, 248, 135–146.
174 A. Cohen, K. Klint, S. Bowadt, P. Persson, J. A. Jonsson, *J. Chromatogr. A* **2001**, 927, 103–110.
175 P. L. Ferguson, C. R. Iden, B. J. Brownawell, *Anal. Chem.* **2000**, 72, 4322–4330.
176 M. Petrovic, A. Diaz, F. Ventura, D. Barcelo, *Anal. Chem.* **2001**, 73, 5886–5895.
177 P. L. Ferguson, C. R. Iden, B. J. Brownawell, *J. Chromatogr. A* **2001**, 938, 79–91.
178 C. Crescenzi, A. Dicorcia, R. Samperi, A. Marcomini, *Anal. Chem.* **1995**, 67, 1797–1804.
179 K. A. Krogh, K. V. Vejrup, B. B. Mogensen, B. Halling-Sorensen, *J. Chromatogr. A* **2002**, 957, 45–57.
180 G. Cretier, C. Podevin, J. L. Rocca, *Analysis* **1999**, 27, 758–764.
181 P. Eichhorn, T. P. Knepper, *J. Chromatogr. A* **1999**, 854, 221–232.
182 P. Billian, W. Hock, R. Doetzer, H. J. Stan, W. Dreher, *Anal. Chem.* **2000**, 72, 4973–4978.
183 T. Ternes, M. Bonerz, T. Schmidt, *J. Chromatogr. A* **2001**, 938, 175–185.
184 X. S. Miao, F. Bishay, M. Chen, C. D. Metcalfe, *Environ. Sci. Technol.* **2004**, 38, 3533–3541.
185 P. Eichhorn, T. P. Knepper, *J. Mass Spectrom.* **2001**, 36, 677–684.

186 F. Bruno, R. Curini, A. Di Corcia, M. Nazzari, R. Samperi, *Rapid Commun. Mass Spectrom.* **2001**, 15, 1391–1400.

187 S. W. Yang, J. Cha, K. Carlson, *Rapid Commun. Mass Spectrom.* **2004**, 18, 2131–2145.

188 G. Hamscher, S. Sczesny, H. Hoper, H. Nau, *Anal. Chem.* **2002**, 74, 1509–1518.

189 M. E. Lindsey, M. Meyer, E. M. Thurman, *Anal. Chem.* **2001**, 73, 4640–4646.

190 C. Hartig, T. Storm, M. Jekel, *J. Chromatogr. A* **1999**, 854, 163–173.

191 R. J. Vreeken, P. Speksnijder, I. Bobeldijk-Pastorova, T. H. M. Noij, *J. Chromatogr. A* **1998**, 794, 187–199.

192 L. Goodwin, J. R. Startin, B. J. Keely, D. M. Goodall, *J. Chromatogr. A* **2003**, 1004, 107–119.

193 A. Di Corcia, M. Nazzari, R. Rao, R. Samperi, E. Sebastiani, *J. Chromatogr. A* **2000**, 878, 87–98.

194 J. Klein, L. Alder, *J. AOAC Int.* **2003**, 86, 1015–1037.

195 B. J. Vanderford, R. A. Pearson, D. J. Rexing, S. A. Snyder, *Anal. Chem.* **2003**, 75, 6265–6274.

196 A. L. Batt, D. S. Aga, *Anal. Chem.* **2005**, 77, 2940–2947.

197 B. Kafkova, Z. Bosakova, E. Tesarova, P. Coufal, *J. Chromatogr. A* **2005**, 1088, 82–93.

198 G. D'Orazio, Z. Aturki, M. Cristalli, M. G. Quaglia, S. Fanali, *J. Chromatogr. A* **2005**, 1081, 105–113.

199 P. S. Bonato, M. Del Lama, R. de Carvalho, *J. Chromatogr. B* **2003**, 796, 413–420.

200 K. Otsuka, C. J. Smith, J. Grainger, J. R. Barr, D. G. Patterson, N. Tanaka, S. Terabe, *J. Chromatogr. A* **1998**, 817, 75–81.

201 E. M. Thurman, I. Ferrer, D. Barcelo, *Anal. Chem.* **2001**, 73, 5441–5449.

202 S. Zuehlke, U. Duennbier, T. Heberer, *Anal. Chem.* **2004**, 76, 6548–6554.

203 R. King, R. Bonfiglio, C. Fernandez-Metzler, C. Miller-Stein, T. Olah, *J. Am. Soc. Mass Spectrom.* **2000**, 11, 942–950.

204 B. K. Matuszewski, M. L. Constanzer, C. M. Chavez-Eng, *Anal. Chem.* **1998**, 70, 882–889.

205 M. P. Schlüsener, K. Bester, *Rapid Commun. Mass Spectrom.* **2005**, 19, 3269–3278.

206 E. T. Gangl, M. Annan, N. Spooner, P. Vouros, *Anal. Chem.* **2001**, 73, 5635–5644.

207 A. Kloepfer, J. B. Quintana, T. Reemtsma, *J. Chromatogr. A* **2005**, 1067, 153–160.

208 Commission Decision (2002/657/EC): Implementing Council directive 96/23/EC concerning the performance of analytical methods and the interpretation of results., *Official J. Eur. Communities* **2002**, L067, 8–36.

209 R. Steen, I. Bobeldijk, U. A. T. Brinkman, *J. Chromatogr. A* **2001**, 915, 129–137.

210 I. Bobeldijk, J. P. C. Vissers, G. Kearney, H. Major, J. A. van Leerdam, *J. Chromatogr. A* **2001**, 929, 63–74.

211 M. Maizels, W. L. Budde, *Anal. Chem.* **2001**, 73, 5436–5440.

212 A. C. Hogenboom, W. M. A. Niessen, D. Little, U. A. T. Brinkman, *Rapid Commun. Mass Spectrom.* **1999**, 13, 125–133.

213 T. P. Knepper, *J. Chromatogr. A* **2002**, 974, 111–121.

214 K. Klagkou, F. Pullen, M. Harrison, A. Organ, A. Firth, G. J. Langley, *Rapid Commun. Mass Spectrom.* **2003**, 17, 2373–2379.

215 R. Plumb, J. Castro-Perez, J. Granger, I. Beattie, K. Joncour, A. Wright, *Rapid Commun. Mass Spectrom.* **2004**, 18, 2331–2337.

216 S. A. C. Wren, *J. Pharm. Biomed. Anal.* **2005**, 38, 337–343.

217 R. Steen, A. C. Hogenboom, P. E. G. Leonards, R. A. L. Peerboom, W. P. Cofino, U. A. T. Brinkman, *J. Chromatogr. A* **1999**, 857, 157–166.

218 E. Dijkman, D. Mooibroek, R. Hoogerbrugge, E. Hogendoorn, J. V. Sancho, O. Pozo, F. Hernandez, *J. Chromatogr. A* **2001**, 926, 113–125.

219 R. B. Geerdink, A. Attema, W. M. A. Niessen, U. A. T. Brinkman, *LC GC Int.* **1998**, 11, 361–372.

220 J. Zrostlikova, J. Hajslova, J. Poustka, P. Begany, *J. Chromatogr. A* **2002**, 973, 13–26.

221 S. Ito, K. Tsukada, *J. Chromatogr. A* **2002**, 943, 39–46.
222 R. Pascoe, J. P. Foley, A. I. Gusev, *Anal. Chem.* **2001**, 73, 6014–6023.
223 B. K. Choi, D. M. Hercules, A. I. Gusev, *Fresenius J. Anal. Chem.* **2001**, 369, 370–377.

2
Residues of Pharmaceuticals from Human Use

Thomas Heberer and Thomas Ternes

2.1
Introduction

In recent years, the occurrence and fate of pharmaceutical residues in the environment has gained much scientific and public attention [1–3]. Residues of pharmaceuticals administered in large quantities to humans or animals have been detected in all compartments of the aquatic environment [2, 3]. Although such residues only appear at trace-level concentrations, there is some concern that even traces of these biologically active compounds may hamper or at least influence aquatic organisms. Since 1997, veterinary pharmaceuticals newly registered for animal health have to undergo an environmental risk assessment (ERA) according to the EMEA guidelines [4] in the course of their approval for registration (see Chapter 11). A similar ERA will in the near future also be applied to pharmaceuticals for human use [5]. This ERA procedure will, however, not apply to those pharmaceuticals that have already been registered. Therefore, those pharmaceuticals currently found in the aquatic environment will not be affected by this new regulation. Since pharmaceuticals are also excluded from REACH (Registration, Evaluation, Authorization of Chemicals), environmental studies for these older pharmaceuticals do not have to be provided (see Chapter 12). Although data showing environmental effects are very scarce, the mere occurrence of such residues, often deriving from municipal sewage effluents, may not be acceptable to the public. Because of the very limited data about environmental effects of pharmaceutical residues, it cannot be ruled out that, in the future, the assessment of environmental hazards will be similar to the assessment of 17α-ethinylestradiol, which has effects on fish down to lower ng L^{-1} concentrations [6]. Thus, there is a large public and scientific need to learn more about the extent of the occurrence of pharmaceutical residues, about their fate during sewage treatment and in the environment, and about natural and technological processes that are able to remove such residues from sewage or raw waters used for drinking water supply. This chapter will address all these issues, but it will exclude those compounds such as antibiotics or X-ray contrast media that are described in Chapters 3 and 4 of this book. Excluding antibiotics also

Organic Pollutants in the Water Cycle. T. Reemtsma and M. Jekel (Eds.).
Copyright © 2006 WILEY-VCH Verlag GmbH & Co. KGaA, Weinheim
ISBN: 3-527-31297-8

means excluding the majority of drugs prescribed in animal health (see Chapter 5). Thus, this chapter mainly focuses on the environmental aspects of residues from human medicine.

2.2
Routes into the Environment

The administration of medication to humans and feed additives to livestock have been identified as the main sources of pharmaceutical residues in the environment [7, 8]. The disposal of unused medication via the toilet seems to be of minor importance. However, pharmaceuticals applied to humans or animals are not completely metabolized by the body. Often they are excreted only slightly transformed, frequently conjugated to polar molecules (e.g., as glucoronides) or even unchanged. Resorption is another key issue when considering the occurrence of pharmaceutical residues in urine or feces. Heberer and Feldmann [9] emphasized the importance of the inclusion of pharmacokinetic data in modeling calculations for the estimation of concentrations or "loads" of drug residues in municipal sewage. They also pointed out the dependence of the resorption rate on the mode of application [9]. Excreted conjugates of pharmaceuticals can easily be cleaved by enzymes/bacteria that are present in manure/sewage and during sewage treatment [10]. The original drugs may then be released into the aquatic environment, e.g., by effluents from municipal sewage treatment plants (STPs). Figure 2.1 compiles possible sources and pathways for the occurrence of drug residues in the environment.

Several investigations have shown that substances of pharmaceutical origin are often not removed during wastewater treatment and also not biodegraded in the environment [11–16]. Drug residues may also leach into groundwater aquifers and

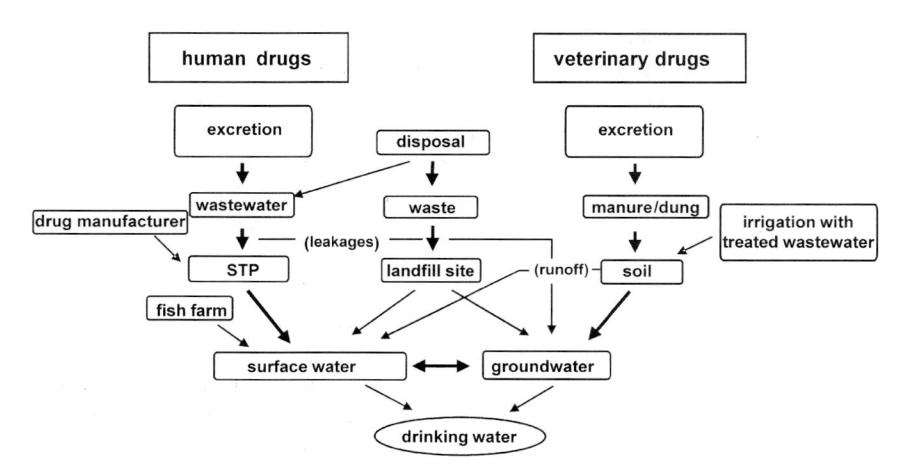

Fig. 2.1 Pathways for the contamination of the environment by pharmaceuticals (adapted from Ternes [11]).

have already been reported to occur in ground- and drinking water samples collected from waterworks using bank filtration or artificial groundwater recharge downstream from municipal STPs [14, 15, 21–24, 32]. Additional sources of drug residues in groundwater are the leaching of such residues from manure applied to agricultural areas, landfill leachates [31, 33, 34], reuse of wastewater by soil-aquifer treatment (SAT) [21, 35, 36], leaking sewer lines, or spillages of production residues [37].

To date, residues from more than 100 pharmaceuticals have been identified and quantified at concentrations up to the µg L^{-1} level in sewage or surface water [2, 3, 7]. In total, 34 compounds have also been detected in groundwater samples and residues of 16 drugs or drug metabolites have been reported to occur at low ng L^{-1} levels in various drinking water samples [8, 31]. Eight out of these sixteen compounds are antiphlogistic drugs and their metabolites, two are anti-epileptic drugs, and three are blood lipid regulators including a pharmacologically active metabolite. Thus, the following sections cover data on the occurrence, fate, and removal of most compounds relevant to public-water supply.

2.3
Wastewater

2.3.1
Occurrence

Residues of pharmaceuticals from human health applications are frequently detected in influents and effluents from municipal STPs. Today's analytical methods, such as capillary gas chromatography-mass spectrometry (GC-MS) or high-performance liquid chromatography coupled to tandem-MS (LC-MS/MS), are able to detect residues of trace organics at low ng L^{-1} concentrations even in very complex matrices such as untreated sewage. The amounts of pharmaceuticals administered to the patients and the degree of elimination from the human body are decisive for the probability of detection and the concentration levels of such residues in raw sewage.

In the mid 1970s, clofibric acid, the active metabolite of the blood-lipid regulators clofibrate, etofyllin clofibrate, and etofibrate, was found in investigations of samples collected from STPs in the United States [38, 39], while the parent compounds themselves were generally not detected. In the early 1990s, clofibric acid was "rediscovered" as an environmentally important residue occurring in the water cycle [40, 41]. In some subsequent investigations, clofibric acid was identified as a refractory contaminant almost ubiquitously present in municipal sewage influents and effluents [11, 14, 15, 42–45]. Thus, findings of clofibric acid in sewage water have been reported from Austria, Brazil, Canada, Germany, Italy, Sweden, Switzerland, and the United States. In addition to clofibric acid a variety of lipid-regulating agents, including gemfibrozil, bezafibrate, and fenofibric acid have been identified in STP effluents by investigators in both Germany and North

America [3, 11, 17, 46, 83]. In German STP effluents, maximum concentrations of bezafibrate were found of up to 4.6 µg L^{-1} and a median of 2.2 µg L^{-1}. Lipid regulators from the "statin" class (e.g., atorvastatin, lovastatin, pravastatin, simvastatin) are widely prescribed in North America and hence have been found in STP effluents in Canada with concentrations up to 117 ng L^{-1} in raw sewage and up to 59 ng L^{-1} in treated sewage [47].

The two most popular *pain killers, acetaminophen* (sold as paracetamol in Europe) and *acetylsalicylic acid* (ASA), mainly sold as OTC drugs (OTC: over the counter) at estimated rates of more than 500 t year^{-1} each in Germany alone [31], are easily biodegradable during conventional sewage treatment. In investigations of sewage effluents in Germany, Ternes [11] detected acetaminophen in less than 10% of all sewage effluent samples. In the same study, ASA was detected at a median concentration of only 0.22 µg L^{-1} [11]. Its pharmacologically active metabolite salicylic acid and two other metabolites (*ortho*-hydroxyhippuric acid and gentisic acid) are found at µg L^{-1} concentrations in sewage influents [11, 48] but salicylic acid only was also detected at ng L^{-1} concentrations in effluent samples [11, 14, 49]. Much higher concentrations, up to 13 µg L^{-1}, have been measured in sewage effluent samples collected in Greece and Spain [32, 50]. However, residues of salicylic acid do not necessarily derive from the metabolism of ASA but may be from other sources including its natural formation [3].

The *analgesic and anti-rheumatic (prescription) drug diclofenac* is sold in comparatively small amounts of approximately 75 t year^{-1} in Germany [31]. Nevertheless, it was identified as one of the most important drug residues in the water cycle [14]. In long-term monitoring investigations of municipal sewage influents and effluents carried out in Berlin, Germany, diclofenac was found with average concentrations of 3.02 and 2.51 µg L^{-1}, respectively [14]. It was also frequently detected, mostly at µg L^{-1} concentrations, in Austria, Brazil, Canada, Czech Republic, Finland, France, Germany, Greece, Italy, Spain, Sweden, Switzerland, and the United States [3, 11, 8, 44, 51]. Heberer and Feldmann [9] carried out some modeling calculations and several field trials to estimate and measure the loads for diclofenac in the sewage of a large municipality with a catchment area of about one million inhabitants. For diclofenac, an average load of 4.4 kg per week was measured in the municipal sewage effluents, which is equal to 226 kg per year or 0.23 g per inhabitant and year [9]. Interestingly, hospital effluents contributed less than 10% to the total loads of diclofenac [9].

The *analgesic drug ibuprofen* is sold in amounts equal to or even higher than those of diclofenac. It has also frequently been detected in sewage effluents in Austria, Brazil, Canada, Finland, France, Germany, Greece, Italy, Spain, Sweden, Switzerland, Taiwan, and the United States [3, 7, 51, 52]. Usually, it is detected at concentrations lower than those measured for diclofenac. Its better degradability during sewage treatment was reported by several authors, such as Zwiener and Frimmel [13] and Buser et al. [53]. In Spain, Switzerland, and Sweden, ibuprofen was, however, also measured at µg L^{-1}-concentrations [44, 50, 54, 55]. The highest concentration was reported by Farré et al. [50], who detected up to 85 µg L^{-1} of ibuprofen in sewage effluent samples from Spain. In the United States, sewage effluent samples contained up to 3.38 µg L^{-1} of ibuprofen, and its concentrations were in most cases higher than those of diclofenac [35, 36, 56]. In raw sewage, ibuprofen is found together with its metabolites, hydroxy-ibuprofen, carboxy-ibuprofen, and carboxy-hydratropic acid [53, 57]. Out of these metabolites, only hydroxy-ibuprofen may [57] but need not necessarily be found in sewage effluents.

Additionally to the above-mentioned compounds, a *large number of other antiphlogistic* drugs such as 4-aminoantiyrine, codeine, fenoprofen, flurbiprofen, hydrocodone, indometacin, ketoprofen, meclofenamic acid, naproxen, phenazone, phenylbutazone, propyphenazone, the metabolites N-acetyl-4-aminoantipyrine (AAA) and N-formyl-4-aminoantipyrin (FAA) (metabolites of metamizole), and the phenazone-type metabolites 1-acetyl-1-methyl-2-dimethyl-oxamoyl-2-phenylhydrazide (AMDOPH), 1-acetyl-1-methyl-2-phenylhydrazide (AMPH), 4-(2-methylethyl)-1,5-dimethyl-1,2-dehydro-3-pyrazolone (PDP) can also be found in municipal sewage or surface water samples [3, 7, 51, 52, 58].

Carbamazepine and primidone are two anti-epileptic drugs frequently detected in municipal sewage in Europe and North America [2, 3, 7, 8, 46, 83]. Carbamazepine has also been detected in sewage effluents in Taiwan [52]. In sewage effluents, concentrations for carbamazepine are at the low µg L^{-1} level [3, 7], whereas primidone is only found at levels below 1 µg L^{-1} [14, 15]. The loads of carbamazepine in municipal sewage were also modeled and measured in the above-mentioned study by Heberer and Feldmann [9]. An average load of 2.0 kg carbamazepine per week was measured in the municipal sewage effluents of an STP serving about one million inhabitants. This is equal to 105 kg per year or 0.11 g of carbamazepine per inhabitant and year [9]. Again, the hospital effluents contributed only a low percentage (less than 15%) to the total loads [9].

Cytostactic drugs are administered in cancer therapy. Their residues are mainly discharged via hospital effluents. Thus, residues of the anti-cancer drugs ifosfamide and cyclophosphamide have been found in hospital effluents at concentrations up to 24 and 146 ng L^{-1}, respectively [59, 60]. In the influents and the effluents of STPs, cyctostatic drugs have, with one exception (one effluent sample containing 2.9 µg L^{-1} of ifosfamide [11]), only occasionally been found with low ng L^{-1}-concentrations [11, 60]. Even though they only appear sporadically at trace-level concentrations, residues of such highly potent and toxicologically active compounds may be of concern for the aquatic environment.

Various authors have reported the occurrence of residues of *beta-blockers* such as metoprolol, acebutolol, oxprenolol, propranolol, betaxolol, bisoprolol, and nadolol in municipal sewage effluents up to the µg L^{-1} level [11, 55, 56, 61–63]. The sporadic occurrence of residues of *bronchodilator drugs* (salbutamol, terbutaline, clenbuterol ,and fenoterol), the *tranquilizer* diazepam, *the antidiabetic drug* glibenclamide, *the calcium influx inhibitor* nifedipine, the *calcium antagonist* verapamil, *pheneturide*, the *hemorheologic agent* pentoxifylline, the antidiabetic metformin, the antidepressant fluoxetine, the antihypertensive diltiazem and the antacid cimetidine in sewage influents or effluents at low ng L^{-1} concentrations were also reported by different authors [11, 64–67, 83, 84].

2.3.2
Removal in Municipal STPs

In general, municipal STPs have not been designed to remove residues of trace organics but to remove or to decrease the concentrations of pathogens and the loads of bulk organics or inorganic constituents that may otherwise pollute the receiving waters and lead to eutrophication. The ability of drug residues to pass through STPs depends on various factors, including the chemical and biological persistence of the individual compound, its sorption behavior, its evaporation via water vapor ,and the technology used for sewage purification. For most pharmaceuticals considered in this chapter, the extent of stripping by aeration is negligible because of their low volatility [20]. Sorption to suspended solids in the wastewater and subsequent removal by sedimentation as primary and secondary sludge comprise another removal process which is relevant for lipophilic compounds (log P_{ow} >3 at ambient pH) or for compounds possessing specific interactions, as has been reported for fluoroquinolones and the tetracyclines [68]. Based on the high polarity of most pharmaceuticals and the corresponding metabolites, significant sorption by non-specific interactions can mainly be ruled out. Hence, the transformation or even mineralization of pharmaceuticals by microorganisms in activated sludge treatment remains as the decisive removal process. Whether trace pollutants can be eliminated in an STP strongly depends upon the treatment technology and operation. Several compounds, such as ASA or paracetamol, are readily biodegradable during conventional sewage treatment, while others pass through STPs without reduction of their concentrations.

With respect to pharmaceuticals, it is also crucial to consider the metabolites excreted by humans when describing the mass flux of a compound during wastewater treatment. Many pharmaceuticals are conjugated with glucuronic acid or sulfate in order to enhance the polarity prior to excretion. For instance, the conjugates of the natural hormones estrone and estradiol are generally present in the same range as the free compounds in the raw wastewater [69]. However, the conjugates can be cleaved in STPs, and then the active pharmaceuticals are released and increase the concentrations of the free compounds [10]. Since, except for estrogens, no data on the conjugates have been published and all authors analyze only for the non-conjugate forms in the raw wastewater and STP effluents due to a lack of appropriate analytical methods, most elimination rates given in the literature are underestimated. In order to determine an accurate mass balance, the load of the conjugated form(s) needs be added to the load of the non-conjugated forms. Furthermore, for a total mass balance of a pharmaceutical it is crucial to consider also the excreted metabolites which are formed by the phase I reaction. These phase I metabolites cannot be converted back into the original pharmaceutical. The observed biological removal in an STP varies strongly from compound to compound, with currently no evident correlation to the compound structure [70]. The present authors have reported, for instance, the following results. Ibuprofen is often removed beyond the quantification limit (≥90%); Naproxen shows significant removal (50–80%). Partial removal is also seen for diclofenac (20–40%). Finally, no removal is found for the antiepileptic drug carbamazepine. Comparable results were found by other authors, whereas a high variation in the elimination rates has been determined for individual pharmaceuticals. Several reasons may be responsible for these differences. One major drawback could be the incomplete mass balance due to the presence of conjugates which are cleaved during passage through the STP and which are not considered in the respective studies. Another very likely reason is the high variability of operational conditions (e.g., temperature, redox potential) and treatment technologies in STPs. For example the flow conditions may be highly variable, including several recirculations which are difficult to predict. Finally, the determination of trace concentrations (sub-µg L^{-1} concentrations) of polar organic compounds in a difficult matrix such as raw or partially treated wastewater is very challenging. Matrix effects are likely to occur in the determination (see Chapter 1).

One example of the high variations in elimination rates is the *antiphlogistic drug diclofenac*, which was identified frequently as being highly persistent during conventional sewage purification. In studies of composite sewage influent and effluent samples from different municipal STPs applying conventional secondary sewage treatment, Heberer [14] observed a removal of only 17% of the diclofenac residues. This result was also more or less comparable with investigations reported by other authors [9, 12, 13, 42, 51, 71], whereas Ternes [11] reported in his study a removal rate of 69% for diclofenac. However, in the next sampling period of the same STP a removal of only 31% was obtained [72], which was within the range of 20–40% identified in the POSEIDON project [21, 70]. What might be reason? Diclofenac is not excreted as conjugates to a great extent, but it could be sorbed at a pH < 7 on suspended solids because of the protonation of the carboxylic moiety. The removal via sorption at pH 6.6 as detected in the primary sludge of the Wiesbaden STP was estimated to be about 10% [21]. Hence, higher elimination rates are likely to be explained by biological degradation. The sludge age was shown to play a major role in the biological degradation. Significant decomposition was only observed when the aerobic sludge was over 8 or even 12 days old [53, 73, 74].

Other residues of pharmaceuticals have been shown to be remarkably persistent, such as the anti-epileptic drugs carbamazepine and primidone, the antiphlogistic drugs phenazone and propiphenazone, and the drug metabolites AAA, AMPH, AMDOPH, clofibric acid, FAA, and PDP [7, 11, 58, 64]. Even STPs applying nitrification, denitrification, and biological phosphorus removal with elevated sludge ages did not achieve significant removal of these pharmaceuticals. However, it has to be mentioned that some of these compounds, such as clofibric acid, are excreted to a major extent as conjugates, and hence the elimination rate may also be underestimated. The removal efficiency of pharmaceuticals depends strongly on treatment conditions, which, together with different application patterns, may also explain the differences in the elimination rates and the concentration levels of the individual compound in the sewage effluents.

The influence of these conditions can be seen in two studies [11, 75] in which rainfall events led to a dilution of the wastewater and hence significantly increased the flow rate in the STP. As a consequence, the elimination rates of several pharmaceuticals decreased significantly. Bezafibrate, diclofenac, naproxen, and clofibric acid were not further eliminated (Fig. 2.2). About 2 days were required to reestablish the initial removal efficiency of the STP. In another study, the concentrations of ibuprofen detected in the secondary effluent of a municipal STP applying nitrification and denitrification increased from <1 ng L^{-1} to 630 ng L^{-1} after a heavy rainfall event [75]. In general, removal rates of more than 90% are achievable for ibuprofen [51, 75].

These results demonstrate very clearly that the operational conditions of an STP are crucial for the elimination of many pharmaceuticals. Obviously, the rainfall events disturb the microbial activity of the STP enormously.

Membrane Bioreactor

In general, the membrane bioreactor plant shows comparable removal rates to those of plants equipped with secondary clarification, confirming the expectation that the micro- and ultrafiltration membranes are inappropriate for the removal of micropollutants directly by sieving (the molecular size is at least 100 times smaller than the pore size of the membranes) [70, 76]. However, results of other studies indicate that the higher sludge age often reached in membrane bioreactors may significantly improve the removal of specific compounds.

Kimura et al. [77] examined the ability of submerged membrane bioreactors to remove several pharmaceutical residues including clofibric acid, diclofenac, ibuprofen, ketoprofen, mefenamic acid, and naproxen. Experiments were conducted at an existing municipal wastewater treatment facility, and the performance of the membrane bioreactors was compared with that of a conventional activated sludge

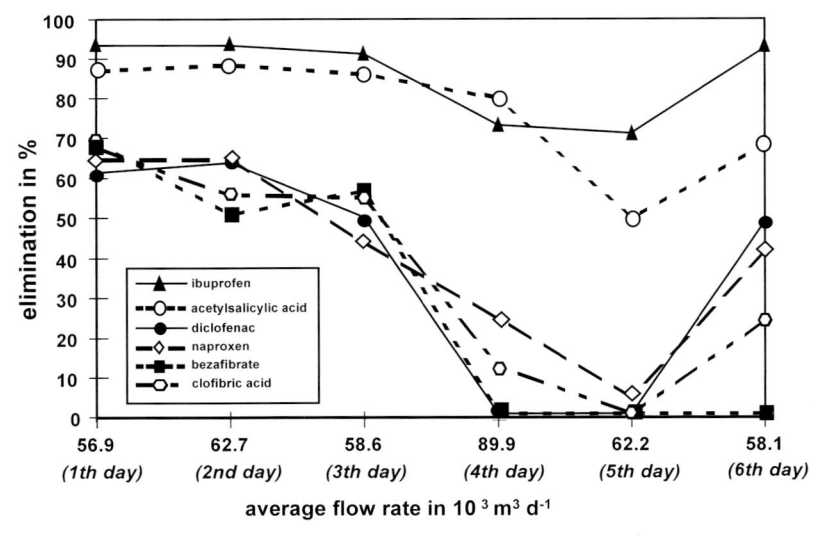

Fig. 2.2 Elimination of selected pharmaceuticals in a municipal STP with a rainfall event on the fourth day increasing the flow rate from about 60 000 m^3 d^{-1} to about 90 000 m^3 d^{-1}. Fig. reproduced from Ternes [11] with kind permission from Elsevier.

process. Membrane bioreactors exhibited much better removal of ketoprofen and naproxen than that of the conventional activated sludge process. For the other compounds, comparable removal rates by the two types of treatment were observed. Kimura et al. [77] also observed that removal efficiencies of the drug residues were dependent on their molecular structures, such as the number of aromatic rings or the presence of chlorine.

Ozonation

In a pilot study, a German municipal STP effluent was ozonated in order to test the effect of ozonation on the removal of pharmaceuticals, iodinated X-ray contrast media (ICM), and musk fragrances from municipal wastewater (Table 2.1) [63]. In

Tab. 2.1 Concentrations and percent elimination (in brackets) of pharmaceuticals (LOQ: 0.050 µg L^{-1}) and estrone (LOQ: 0.003 µg L^{-1}) present in a municipal STP effluent before (n=6) and after ozonation (n=2) at doses of 5, 10, 15 mg L^{-1} O_3 (taken from [63]) (LOQ ≡ limit of quantification).

	Untreated µg L^{-1}	5 mg L^{-1} O_3 µg L^{-1} (%)	10/15 mg L^{-1} O_3 µg L^{-1} (%)
Antibiotics			
Trimethoprim	0.34 ± 0.04	n.d. (> 85)	n.d. (>85)
Sulfamethoxazole	0.62 ± 0.05	n.d. (>92)	n.d. (>92)
Clarithromycin	0.21 ± 0.02	n.d. (>76)	n.d. (>76)
Erythromycin	0.62 ± 0.24	n.d. (>92)	n.d. (>92)
Roxithromycin	0.54 ± 0.04	n.d. (>91)	n.d. (>91)
Beta-blockers			
Atenolol	0.36 ± 0.01	0.14 (61)	n.d. (>86)
Sotalol	1.32 ± 0.14	n.d. (>96)	n.d. (>96)
Celiprolol	0.28 ± 0.01	n.d. (>82)	n.d. (>82)
Propranolol	0.18 ± 0.01	n.d. (>72)	n.d. (>72)
Metoprolol	1.68 ± 0.04	0.37 (78)	n.d. (>93)
Antiepileptic			
Carbamazepine	2.08 ± 0.04	n.d. (>98)	n.d. (>98)
Antiphlogistics			
Ibuprofen	0.13 ± 0.03	0.067 (48)	n.d. (>62)
Naproxen	0.10 ± 0.01	n.d. (>50)	n.d. (>50)
Indomethacin	0.10 ± 0.04	n.d. (>50)	n.d. (>50)
Diclofenac	1.3 ± 0.1	n.d. (>96)	n.d. (>96)
Lipid regulators			
Clofibric acid	0.12 ± 0.02	0.060 (50)	n.d. (>59)
Fenofibric acid	0.13 ± 0.04	0.060 (>62)	n.d. (>62)
Natural estrogen			
Estrone	0.015 ± 0.002	n.d. (>80)	n.d. (>80)
Caffeine	0.22 ± 0.03	0.11 (50)	n.d. (>87)

n.d.: not detectable, below LOQ

the original STP effluent, 5 antibiotics, 5 beta-blockers, 4 antiphlogistics, 2 lipid regulator metabolites, the antiepileptic drug carbamazepine, 4 ICMs, the natural estrogen estrone, and 2 musk fragrances were determined. ICMs, from radiological examinations, were present at the highest concentrations (diatrizoate: 5.7 µg L^{-1}, iopromide: 5.2 µg L^{-1}). After applying 10–15 mg L^{-1} ozone (contact time: 9 min), the pharmaceuticals investigated, the musk fragrances (HHCB, AHTN), and the estrone were no longer detected. Only, ICMs (diatrizoate, iopamidol, iopromide, and iomeprol) were recalcitrant. Advanced oxidation processes (O_3/UV-low pressure mercury arc, O_3/H_2O_2), which were non-optimized for wastewater treatment, did not lead to a significantly higher removal efficiency for the ICM than ozone alone.

2.4
Surface Water

2.4.1
Occurrence

In Europe, the first comprehensive studies of the occurrence of pharmaceuticals in rivers and streams were reported in the mid 1980s by Watts et al. [78], Waggott [79], and Richardson and Bowron [80]. These investigations in the United Kingdom revealed that pharmaceuticals were present in the aquatic environment at concentrations of up to approximately 1 µg L^{-1}, although the concentrations of some drugs could not always be accurately detected because LC-MS/MS was not available at that time (see Chapter 1). Meanwhile, numerous studies appeared describing the presence of pharmaceuticals and contrast media in treated wastewater and the aquatic environment in Europe and North America. In Germany, the first monitoring investigations of surface waters, carried out in the mid 1990s [11, 81], revealed that residues of pharmaceuticals are occurring as widespread pollutants in the aquatic environment. Especially in and downstream from urban areas, concentrations of drug residues are higher than those of pesticides and several other priority pollutants. During the period 1996–1998, in a German study [11, 72], the occurrence of 55 pharmaceuticals, six hormones, and nine metabolites were monitored in the receiving water bodies. For 24 pharmaceuticals, one hormone, and five metabolites, 90 percentiles were found above the limit of quantification. Although most concentrations were detected in the ng L^{-1} range, in a few cases concentrations even exceeding 1 µg L^{-1} (e.g., beta-blocker metoprolol, anti-epileptic agent carbamazepine) were measured in receiving water bodies. Furthermore, between 1998 and 2000, various drug residues were monitored and found to be ubiquitous in investigations of surface water samples collected from the river Elbe and its tributaries [82].

In the years 1999 and 2000, Kolpin et al. [83] conducted a study on the occurrence of pharmaceuticals, hormones, antioxidants, biocides, and other organic wastewater contaminants in surface waters across 30 states of the United States. They focused on locations downstream of intense urbanization and livestock production in 139 rivers and streams. The authors detected 82 of the 95 compounds in at least one stream sample. In total, residues of 29 pharmaceuticals were detected in these samples [83]. In a later study carried out in 2001, Kolpin et al. [84] investigated the urban contributions of pharmaceuticals and related compounds to streams during differing flow conditions by analyzing 76 water samples collected upstream and downstream of selected towns and cities in Iowa (USA) during high-, normal-, and low-flow conditions. They found many more positive detections during low-flow conditions [84]. This result is not surprising, because dilution of almost constant inputs of drug residues may decrease surface water concentrations below today's analytical detection limits.

In Germany, a nationwide monitoring program was initiated and carried out between 2000 and 2001 to obtain a comprehensive picture on the contamination of Germany's surface waters [85]. Between 2000 and 2003, an extended surface water monitoring program was conducted in the city of Berlin [15]. The aim of this program was to calculate the annual loads of selected drug residues and to monitor possible seasonal differences.

It can be concluded that the situation in North America regarding pharmaceutical pollution is similar to that in Europe. Nevertheless, these comprehensive monitoring studies, as well as a multitude of subsequent individual studies, included only a small subset (< 15%) of the predicted number of pharmaceuticals that may potentially enter the environment.

Most of the drug residues found in municipal STP effluents were also detected in the receiving waters. Therefore, the percentage of treated or even raw wastewater is the crucial parameter for the contamination of surface waters with pharmaceuticals. One example is illustrated in Fig. 2.3.

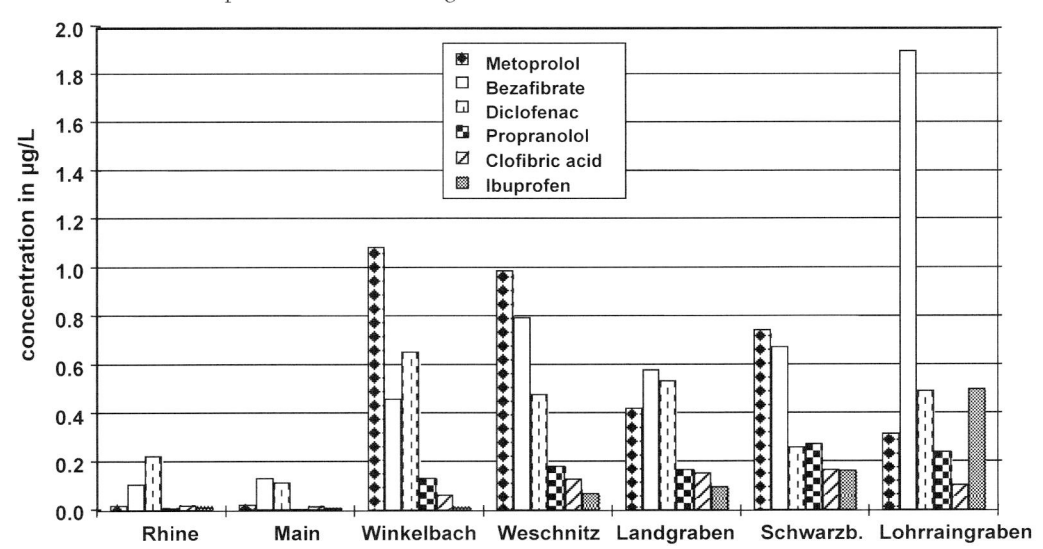

Fig. 2.3 Concentrations of selected pharmaceuticals (beta-blockers, lipid regulators, antiphlogistics) in small rivers and streams (Winkelbach, Weschnitz, Landgraben, Schwarzbach) of the Hessian Ried area in comparison with the large rivers Rhine (Mainz) and Main (Bischofsheim). Fig. reproduced from Ternes [11] with kind permission from Elsevier.

2.4.2
Degradation in Surface Waters

A significant elimination of diclofenac in the water of a lake in Switzerland was observed by Buser et al. [71]. This led to the conclusion that a photolytic degradation of diclofenac residues occurs in surface waters. This assumption was confirmed in laboratory experiments with spiked lake water, which showed a rapid and extensive photodegradation by sunlight irradiation [71]. Several photoproducts of diclofenac have been characterized but have not been detected under natural conditions [71]. In the meantime, photodegradation of diclofenac has also been observed in other studies conducted in Germany and Switzerland [15, 44] and in irradiation experiments conducted by Huber et al. [86] and Andreozzi et al. [87]. In laboratory studies, Winkler et al. [88] observed that ibuprofen and predominantly its pharmacologically inactive stereoisomer were readily degraded in a river biofilm reactor. Their metabolites hydroxy- and carboxy-ibuprofen were further degraded in the biofilm reactor. Abiotic transformation and adsorption were not relevant [88].

The abiotic transformation of carbamazepine by solar photodegradation was studied by Andreozzi et al. [89]. Carbamazepine can be transformed photochemically in distilled and in river waters, with nitrate promoting and humic acids inhibiting its degradation [89]. This assumption is in line with results from monitoring investigations by Heberer and Reddersen [90], who found a significant decrease in carbamazepine loads in Berlin surface waters in summer periods even though the discharge from WWTPs did not decrease [90].

2.4.3
Sediments

Few data are available concerning the fate of the pharmaceuticals in the water/sediment compartment. In a recent study [91], the environmental fate of ten selected pharmaceuticals and pharmaceutical metabolites was investigated in water/sediment systems using a laboratory batch set-up according to OECD 308. Biological and chemical transformations as well as sorption can occur, whereas photolytic degradation is excluded. The antiphlogistic ibuprofen, its metabolite 2-hydroxy-ibuprofen, and the antiphlogistic acetaminophen displayed a low persistence with DT_{50}-values (dissipation time 50%) ≤ 20 d. Because of the rapid formation of bound residues, acetaminophen is likely to be first biologically transformed and then bound to the sediment. Moderate persistence was found for the parasiticide ivermectin and the tranquilizer oxazepam, with DT_{50}-values of 15 d and 54 d respectively. For the tranquilizer diazepam, the antiepileptic carbamazepine, its corresponding metabolites 10,11-dihydro- and 10,11-dihydroxy-carbamazepine and the lipid regulator clofibric acid, DT_{90}-values of > 365 d were found, this extremely high persistence being exhibited in the water/sediment system. An elevated level of sorption onto the sediment was observed for ivermectin, diazepam, oxazepam, and carbamazepine. The respective K_{OC} values calculated from the experimental data ranged from 1172 L kg^{-1} for ivermectin down to 83 L kg^{-1} for carbamazepine.

These K_{OC} values were in good agreement with estimated and experimental data previously reported by other authors. Based on these results it can be concluded that biodegradation is the major process leading to the removal of pharmaceuticals, whereas sorption plays only a minor role for most of the compounds. However, the formation of bound residues can be important, as shown for acetaminophen.

2.5 Groundwater and Underground Passage

When surface water under the influence of sewage effluents is used for groundwater recharge, polar drug residues such as carbamazepine and primidone or drug metabolites such as AMDOPH or clofibric acid can leach into the groundwater [3, 7, 22]. Findings of pharmaceutical residues have, however, also been reported in groundwater contaminated by landfill leachates [31, 33, 34, 92] when wastewater is reused via SAT [35, 36] or as a result from spills of production residues [37, 93].

In the past few years, research has focused on investigating natural removal processes occurring during the passage of drug residues through soil when entering the groundwater body. Besides sorption and partition studies [94, 95], laboratory studies have been conducted with soil columns under defined saturated or non-saturated conditions, applying spiked solutions of environmentally relevant pharmaceuticals [96–98]. In laboratory soil column experiments, significant removal of ibuprofen was observed, most likely caused by microbial degradation [97]. Clofibric acid exhibited no degradation and almost no retardation (retardation factor: Rf = 1.1) [96]. In contrast, diclofenac (Rf = 2.0) and propyphenazone (Rf = 1.6) were retarded, but for both compounds no significant degradation was observed under the conditions prevailing in the soil column [96]. Drewes et al. [35, 36] and Sedlak and Pinkston [56] investigated the fate of several pharmaceutical residues during SAT of sewage effluents after secondary treatment. They observed a high efficacy for the removal of diclofenac, ibuprofen, ketoprofen, naproxen, fenoprofen, or metoprolol. Other compounds such as propyphenazone, carbamazepine, and primidone were, however, not removed by SAT [35, 36].

Further information on the transport and removal of drug residues was also derived from field studies carried out at groundwater recharge sites [15, 35, 36, 99–102]. In Berlin, Germany, different pharmaceutical residues including antibiotics and several compounds mentioned in this chapter were investigated in the years between 2002 and 2005 [99–102]. The transport and fate of drug residues was investigated across different bank filtration sites and at a facility for groundwater replenishment (GWR) using surface water under the influence of municipal wastewater for groundwater recharge. Diclofenac, propyphenazone, carbamazepine, primidone, and the drug metabolites clofibric acid and AMDOPH were found to leach from the surface water into the groundwater aquifers during bank filtration and also occurred at low ng L^{-1} concentrations in the receiving water-supply wells [99, 100]. However, other compounds such as indometacin and bezafibrate, which were also detected at concentrations up to 100 ng L^{-1} in the surface water were efficiently removed by bank filtration [99, 100]. In conclusion, a complete removal of all potential pharmaceutical residues by bank filtration can, however, not be guaranteed [22]. Similar results were also obtained at the GWR site [102]. Fig. 2.4 shows the hydrogeological cross-section of the field sites and Table 2.2 compiles the results for several pharmaceutical residues that were detected in samples collected from lake Tegel, the monitoring wells, and/or water-supply well No. 20.

As shown in Table 2.3, the compounds could be categorized according to their removal behavior. Drug residues with log P_{OW} values of >3 such as bezafibrate or indomethacin (group 3) are readily removed by bank filtration, whereas compounds with lower log P_{OW} values (groups 1 and 2) are incompletely removed. In particular, the drug metabolite AMDOPH and the two anti-epileptic drugs carbamazepine and primidone were identified as highly persistent and very mobile compounds that easily leach through the sub-soil into groundwater. This example also demonstrates that in areas with surface waters under the influence of municipal sewage effluents, such compounds have to be considered as being relevant to the quality of potable water when groundwater recharge is used in drinking water production.

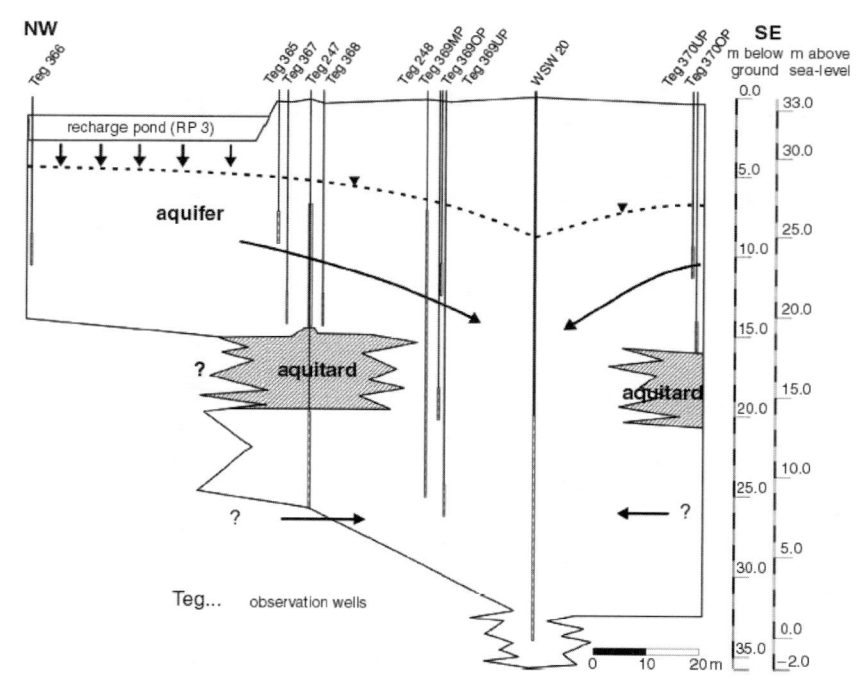

Fig. 2.4 Hydrogeological cross-section of a groundwater recharge transect (recharge pond RP3, monitoring well field and water-supply well WSW 20) located at the groundwater replenishment site in Berlin-Tegel, Germany. (Figure reproduced from Heberer and Adam [102] with kind permission from CSIRO.)

Tab. 2.2 Concentrations of pharmaceutical residues in ng L^{-1} detected in recharge pond RP3, in the monitoring wells (for well assignments please refer to Fig. 2.4) and in water-supply well WSW20. Arrows indicate groundwater flow direction toward WSW20. Samples were collected between July 2002 and June 2003, n = 6–12. Table reproduced from Heberer and Adam [102] with kind permission from CSIRO.

	Pond RP3	Teg 366	Teg 365	Teg 247	Teg 368	Teg 248	Teg 369 OP	Teg 369 UP	WSW 20	Teg 370 OP	Teg 370 UP
n =	12	12	12	8	6	8	6	6	11	6	6
Flow direction											
AMDOPH	455	440	395	425	390	315	300	330	**1570**	1085	3915
Carbamazepine	470	545	430	385	460	430	220	230	**210**	20	20
Primidone	135	140	125	115	170	95	80	90	**100**	30	70
Propyphenazone	120	20	20	30	15	20	10	10	**40**	10	55
Clofibric acid	20	5	5	10	10	5	5	5	**5**	15	n.d.
Diclofenac	135	15	45	5	15	10	<5	<5	**10**	n.d.	n.d.
Indomethacin	20	<10	<10	n.d.	n.d.	n.d.	n.d.	n.d.	**n.d.**	n.d.	n.d.
Bezafibrate	30	n.d.	n.d.	n.d.	n.d.	n.d.	n.d.	n.d.	**n.d.**	n.d.	n.d.

n.d.: not detected; limits of detection (LODs) = <1 ng L–1 for all analytes

Tab. 2.3 Classification of target compounds according to their attenuation rates during groundwater recharge at a groundwater replenishment plant in Berlin-Tegel (Fig. 2.4). Table adapted from Heberer and Adam [102] with kind permission from CSIRO.

Group	Compound	log P_{OW}	Mean removal at Teg 248	Mean attenuation at WSW20
1: low removal rates (0–50%)	AMDOPH	unknown	31%	–245% (exceptional case*)
	Carbamazepine	2.45	9%	55%
	Primidone	0.91	30%	26%
2: medium removal rates (51–95%)	Propyphenazone	2.05	83%	67% (exceptional case)[a]
	Clofibric acid	3.1	75%	75%
	Diclofenac	1.13	93%	93%
3: high removal rates (>95%)	Indomethacin	4.27	>95%	>95%
	Bezafibrate	4.2	>97%	>97%

[a] Increased concentrations caused by former contamination of the groundwater body.

2.6
Drinking Water Treatment

2.6.1
Sorption and Flocculation

2.6.1.1 Flocculation

Flocculation with Fe(III)chloride and aluminum sulfate exhibited no significant elimination of the selected pharmaceuticals from raw water [21, 103]. For instance, the relative concentration levels (c/c_0) using flocculation in the laboratory scale Jar test were 96±11% for diclofenac, 87±10% for clofibric acid, 111±15% for bezafibrate, 87±12% for carbamazepine, and 110±14% for primidone. The transference of these results from laboratory scale to waterworks conditions was shown by monitoring scaled-up flocculation processes in two waterworks yielding similar results.

GAC Filtration

Filtration over granular activated carbon (GAC) has been shown to be a very efficient removal process [21, 103]. Even relatively high concentrations of pharmaceuticals such as carbamazepine, diclofenac, and bezafibrate could be almost completely removed at specific throughputs of over 70 m^3 kg^{-1} with the exception of clofibric acid. Clofibric acid is less prone to adsorption, but could be removed completely at a specific throughput of 15 to 20 m^3 kg^{-1}. On the pilot scale, an activated carbon adsorber was operated according to the previous description. Carbamazepine showed the highest adsorption capacity of the selected pharmaceuticals and could be removed at a specific throughput of about 50 m^3 kg^{-1} in a carbon layer of 80 cm and more than 70 m^3 kg^{-1} in a layer of 160 cm. After GAC filtration in a full-scale waterworks, diclofenac and bezafibrate were not detected above LOQ and the concentrations of carbamazepine and primidone were reduced by more than 75% and that of clofibric acid by ca. 20%.

2.6.2
Oxidation

2.6.2.1 Ozonation

Huber et al. [104] determined second-order rate constants for the reactions of selected pharmaceuticals with ozone (k_{O3}) and OH radicals (k_{OH}) in bench-scale experiments. High reactivities with ozone (k_{O3}) were found for carbamazepine (~3×10^5 M^{-1}s^{-1}) and diclofenac (~1×10^6 M^{-1} s^{-1}), indicating that these compounds are very rapidly transformed during ozonation. Lower reactivities were found for bezafibrate (590±50 M^{-1} s^{-1}), diazepam (0.75±0.15 M^{-1} s^{-1}), ibuprofen (9.6±1 M^{-1} s^{-1}), and iopromide (<0.8 M^{-1} s^{-1}). Finally, the authors concluded that ozonation and advanced oxidation processes (AOPs) are promising for the efficient removal of pharmaceuticals in drinking waters. Although ozone appreciably reduced the concentration levels of bezafibrate and primidone, it exhibited limited efficiency in removing clofibric acid [103]. The efficiency of the ozonation process for the removal of the pharmaceuticals turned out to be very compound specific. At a small ozone dose of 0.5 mg L^{-1} the concentrations of diclofenac and carbamazepine were reduced by more than 97%, while clofibric acid decreased by only 10–15% using the same ozone dose. Even extremely high ozone doses up to 2.5–3.0 mg L^{-1} reduced clofibric acid by ≤40% only. Primidone and bezafibrate were reduced by 50% at ozone concentrations of about 1.0 mg L^{-1} and 1.5 mg L^{-1}, respectively.

2.6.2.2 Ozonation Products

With the ozone doses used in waterworks for disinfection, only a partial transformation of the organic compounds can be expected. The toxicity of the products is almost totally unknown. However, the oxidation processes increase the polarity of the compounds, which is frequently associated with reduced biological toxicity and enhanced microbial degradability. For diclofenac, a main oxidation product was detected with a mass spectrum showing an increase in the molecular weight of 16 amu, which is evidence for substitution of a hydrogen by a hydroxy moiety. A hydroxylation of the secondary amino group is likely, but has to be confirmed, e.g., by NMR. Because of the missing active sites that are susceptible to ozone attack, for clofibric acid, reactions of ozone with this structure are expected to be very slow. Thus, OH radical reactions should be predominant, with $k_{OH} \approx 5 \times 10^9$ $M^{-1} s^{-1}$ [104].

In a recent study, formation of the anti-epileptic drug carbamazepine (CBZ) during reactions with ozone and OH radicals was investigated [105]. Three new oxidation products were identified (Fig. 2.5): 1-(2-benzaldehyde)-4-hydro-(1H,3H)-quinazoline-2-one (BQM), 1-(2-benzaldehyde)-(1H,3H)-quinazoline-2,4-dione (BQD), and 1-(2-benzoic acid)-(1H,3H)-quinazoline-2,4-dione (BaQD). Additional kinetic studies of the ozonation products showed very slow subsequent oxidation kinetics with ozone (second-order rate constants, k_{O_3} = ~7 $M^{-1} s^{-1}$ and ~1 $M^{-1} s^{-1}$ at pH = 6 for BQM and BQD, respectively). Rate constants for reactions with OH radicals, k_{OH}, were determined as ~ 7×10^9 $M^{-1} s^{-1}$ for BQM and ~5×10^9 $M^{-1} s^{-1}$ for BQD. Thus, mainly reactions with OH radicals lead to their further oxidation. BQM and

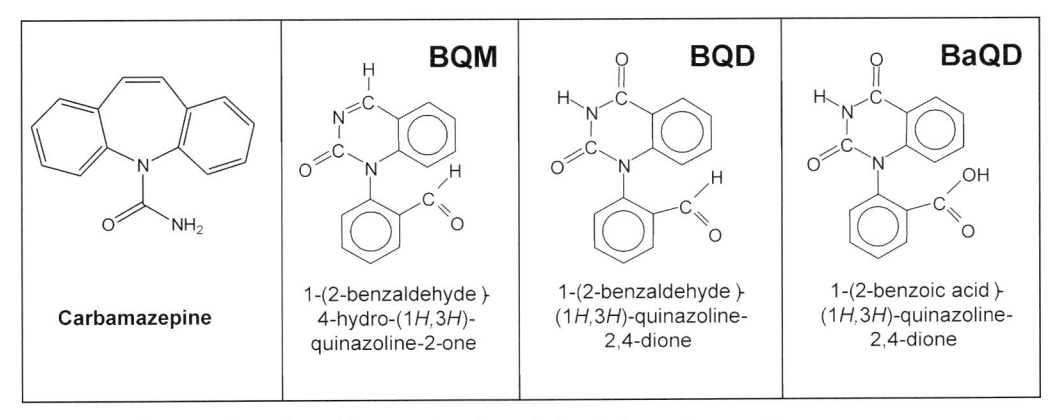

Fig. 2.5 Oxidation products formed by ozonating the antiepileptic drug carbamazepine.

BQD were also identified in ozonated water of a German waterworks containing CBZ in its raw water at 0.07–0.20 µg L^{-1}. Currently there is no data available on the biological effects of the formed oxidation products.

2.6.3
Membrane Filtration

High-pressure-driven membranes such as nanofiltration (NF) or reverse osmosis (RO) membranes might be applicable to remove pharmaceutical residues from contaminated raw waters which are or might be used for drinking water production [106].

Drewes et al. [35] investigated different wastewater treatment technologies including activated sludge, trickling filter, SAT, NF, and RO for their capability to remove drug residues at full-scale facilities in Arizona and California used for indirect potable reuse. They concluded that, in contrast to SAT, none of the investigated pharmaceutical residues was found in tertiary effluents after NF or RO treatment. The rejection of drug residues by polyamide NF/RO membranes was also investigated by Kimura et al. [107]. Their investigations were based on a protocol established for determination of rejection efficiency of NF/RO membranes [108]. Kimura et al. [107] observed that negatively charged compounds (e.g., the analgesic drug diclofenac) can be rejected to a great extent (i.e. >90%) by electrostatic repulsion, regardless of other physical/chemical properties of the test compounds. In contrast, the rejection of non-charged compounds was mainly influenced by the size of the compounds, but solute affinity for the membrane also influenced the rejection efficiency. In several cases, NF was almost unable to remove non-charged compounds (e.g., phenacetin), whereas primidone, another non-charged compound, was always rejected by more than 70%. This result also suggests that additional processes are responsible for efficient rejection of such compounds [107]. In another study, Kimura et al. [106] examined the retention behavior of pharmaceutical residues with RO membranes of two different materials (polyamide and cellulose acetate). In total, the polyamide membrane exhibited a better performance in terms of rejection but often retention was not complete (57–91%). Kimura et al. [106] state that the molecular weight of the test compounds could generally indicate the tendency of rejection for the polyamide membranes (size exclusion dominated the retention by the polyamide membrane), while polarity was better able to describe the retention trend of the tested compounds by the cellulose acetate membrane. However, salt rejection or molecular weight cut-off (MWCO), often used to characterize membrane rejection properties, does not provide quantitative information on drug residue rejection by NF/RO membranes [106].

Xu et al. [109] investigated the rejection of pharmaceutical residues by a variety of commercially available RO, NF, and ultra-low-pressure RO (ULPRO) membranes to simulate operational conditions for drinking water treatment and wastewater reclamation. In general, the presence of effluent organic matter improved the rejection of ionic organics by tight NF and RO membranes, likely as a result of a decreased negatively charged membrane surface. Rejection rates of ionic pharmaceutical residues ranged from 90 to 95%. Kimura et al. [107] reported another interesting phenomenon: the rejection percentage was lower at lower feed water concentrations of 100 ng L^{-1} as compared to 100 µg L^{-1}.

In two field trials [75, 110], the performance of membrane-based mobile drinking water purification units was investigated, applying different pre-filtration techniques such as bag, slit, or ultrafiltration, and final purification by RO. In both trials, potable water was generated from highly contaminated raw-water sources such as contaminated surface water or municipal sewage effluents simulating "worst-case" conditions which may occur in civil disaster operations or in military out-of-area missions. Even under these "worst-case" scenarios, all pharmaceutical residues were efficiently removed and their concentrations decreased from micrograms per liter to below their analytical limits of detection (<1–10 ng L^{-1}).

2.6.4
Evaluation of the Treatment Processes

For waterworks applying only flocculation and sand filtration, substantial removal of most of the pharmaceutical residues cannot be expected. According to present knowledge, contamination of raw waters by polar pharmaceuticals can only be removed using more advanced techniques such as ozonation, AOP, or activated carbon filtration. An alternative to ozonation and GAC filtration can be seen in membrane filtration by NF, RO, or ULPRO. The contamination of the raw water is influenced mainly by its percentage of treated wastewater. In Germany and many other European countries, waterworks using surface water or bank filtrates of rivers are equipped with GAC, ozonation, or even both. In these cases, contamination of the drinking water by the investigated pharmaceuticals is rather unlikely.

Although pharmaceuticals are currently not regulated in drinking water directives world-wide, precautionary measures should be employed to remove pharmaceuticals as efficiently as possible through improved or existing treatment techniques. More care has to be taken when contaminated groundwater is used as the raw water for drinking water production. Waterworks using groundwater are in general not equipped with advanced treatment techniques such as GAC, ozonation, or membrane filtration like those generally found in waterworks using surface waters.

References

1 J. Raloff, Science News **1998**, 153, 187–192. Can be found under http://www.sciencenews.org/sn_arc98/3_21_98/bob1.htm.

2 C.G. Daughton, T.A. Ternes, *Environ. Health Perspect.* **1999**, 107, 907–938.

3 Th. Heberer, Toxicol. Lett. **2002**, 131, 5–17.

4 EMEA (European Medicines Agency), Environmental risk assessment for veterinary medicinal products other than GMO-containing and immunological products. EMEA/CVMP/055/96. London, **1997**, can be found under http://www.emea.eu.int/pdfs/vet/regaffair/005596en.pdf.

5 EMEA (European Medicines Agency), Committee for Medicinal Products for Human Use (CHMP), Guideline on the Environmental Risk Assessment of Medicinal Products for Human Use. CHMP/SWP/4447/00 draft. London, **2005**, can be found under http://www.emea.eu.int/pdfs/human/swp/444700en.pdf.
6 Routledge, E.J., Sheahan, D., Desbrow, C., Brighty, G.C., Waldock, M., Sumpter, J.P. *Environ. Sci. Technol.* **1998**, 32, 1559–1565.
7 Boxall, A.B.A., Fogg, L.A., Blackwell, P.A., Kay, P., Pemberton, E.J., Croxford, A. *Rev. Environ. Contam. Toxicol.* **2004**, 180, 1–91.
8 Th. Heberer, M. Adam in *Advances in Pharmacology: Hot Spot Pollutants Vol. 52 – Pharmaceuticals in the Environment* (Eds.: D.R. Dietrich, T. Petry and S. Webb), Academic Press, **2005**, pp. 11–36.
9 Th. Heberer, D. Feldmann, *J. Hazard. Mater.* **2005**, 122, 211–218.
10 T.A. Ternes, P. Kreckel, J. Mueller, *Sci. Total Environ.* **1999**, 225, 91–99.
11 T.A. Ternes, *Water Res.* **1998**, 32, 3245–3260.
12 C. Zwiener, T. Glauner, F.H. Frimmel, *J. High Res. Chromatogr.* **2000**, 23, 474–478.
13 C. Zwiener, F.H. Frimmel, *Sci. Total Environ.* **2003**, 309, 201–211.
14 Th. Heberer, *J. Hydrol.* **2002**, 266, 175–189.
15 Th. Heberer, K. Reddersen, A. Mechlinski, *Water Sci. Technol.* **2002**, 46, 81–88.
16 S.J. Khan, J.E. Ongerth, *Water Sci. Technol.* **2002**, 46, 105–113.
17 X.S. Miao, F. Bishay, M. Chen, C.D. Metcalfe, *Environ. Sci. Technol.* **2004**, 38, 3533–3541.
18 B. Halling-Sorensen, S.N. Nielsen, P.F. Lanzky, F. Ingerslev, H.C.H. Lutzhoft, S.E. Jorgensen. *Chemosphere* **1998**, 36, 357–394.
19 C.S. McArdell, E. Molnar, M.J.F Suter, W. Giger, *Environ. Sci. Technol.* **2003**, 37, 5479–5486.
20 T.A. Ternes, A. Joss, H. Siegrist. *Environ. Sci. Technol.* **2004**, 38, 393A–399A.
21 T.A.,Ternes, M.-L. Janex-Habibi, T. Knacker, N. Kreuzinger, H.-S. Siegrist, Final report of the EU project POSEIDON, Contract No. EVK1–CT–2000–00047, http://www.poseidon.bafg.de.
22 I.M. Verstraeten, Th. Heberer, T. Scheytt, in *Riverbank Filtration: Improving Source-Water Quality* (Eds.: C. Ray, G. Melin, R.B. Linsky), Dordrecht, Kluwer Academic Publishers, **2002**, pp. 175–227.
23 Th. Heberer, A. Mechlinski, *Hydroplus* **2003**, 137 (10), 53–60.
24 P.E, Stackelberg, E.T. Furlong, M.T. Meyer, S. D. Zaugg, A. K. Henderson, D. B. Reissman, *Sci. Total Environ.* **2004**, 329, 99–113.
25 Th. Heberer, H.J. Stan, *Vom Wasser* **1996**, 86, 19–31.
26 Th. Heberer, H.J. Stan, *Int. J. Environ. Anal. Chem.* **1997**, 67, 113–124.
27 H.J. Brauch, F. Sacher, E. Denecke, T. Tacke, *gwf Wasser Abwasser* **2000**, 14, 226–234.
28 W. Kuehn, U. Mueller, J. AWWA **2000**, 12, 60–69.
29 Th. Heberer, U. Dünnbier, Ch. Reilich, H.J. Stan, *Fresenius' Environ. Bull.* **1997**, 6, 438–443.
30 F. Sacher, F. Th. Lange, H.-J. Brauch, I. Blankenhorn, *J. Chromatogr. A* **2001**, 938, 199–210.
31 T.A.Ternes, in *Pharmaceuticals and Personal Care Products in the Environment: Scientific and Regulatory Issues. Symposium Series 791* (Eds.: C.G. Daughton, T. Jones-Lepp), American Chemical Society, Washington D.C., **2001**, pp. 39–54.
32 Th. Heberer, B. Fuhrmann, K. Schmidt-Bäumler, D. Tsipi, V. Koutsouba, A. Hiskia, in *Pharmaceuticals and Personal Care Products in the Environment: Scientific and Regulatory Issues. Symposium Series 791* (Eds.: C.G. Daughton, T. Jones-Lepp), American Chemical Society, Washington D.C., **2001**, pp. 70–83.
33 W.P. Eckel, B. Ross, R.K. Isensee, *Ground Water* **1993**, 31, 801–804.
34 J.V. Holm, K. Rügge, P.L. Bjerg, T.H. Christensen, *Environ. Sci. Technol.* **1995**, 29, 1415–1420.
35 J. Drewes, Th. Heberer, K. Reddersen, *Water Sci. Technol.* **2002**, 46 (3), 73–80.

36 J. Drewes, Th. Heberer, T. Rauch, K. Reddersen, *Ground Water Monit. Remediat.* **2003**, 23, 64–72.
37 K. Reddersen, Th. Heberer, U. Dünnbier, *Chemosphere* **2002**, 49, 539–545.
38 A.W. Garrison, J.D. Pope, F.R. Allen, in *Identification and Analysis of Organic Pollutants in Water. Chapter 30* (Ed.: C.H. Keith) Ann Arbor Science Publishers, Ann Arbor, **1976**, 517–556.
39 C. Hignite, D.L. Azarnoff, *Life Sci.* **1977**, 20, 337–342.
40 H.J. Stan, M. Linkerhägner, *Vom Wasser* **1992**, 79, 75–88.
41 H.J. Stan, Th. Heberer, M. Linkerhägner, *Vom Wasser* **1994**, 83, 57–68.
42 M. Stumpf, T.A. Ternes, R.-D. Wilken, S.V. Rodrigues, W. Baumann, *Sci. Total Environ.* **1999**, 225, 135–141.
43 D.B. Patterson, W.C. Brumley, V. Kelliher, P.L. Ferguson, *Am. Lab.* **2002**, 34, 20–28.
44 C. Tixier, H.P. Singer, S. Oellers, S.R. Müller, *Environ. Sci. Technol.* **2003**, 37, 1061–1068.
45 B. Soulet, A. Tauxe, J. Tarradellas, *Int. J. Environ. Anal. Chem.* **2002**, 82, 659–667.
46 C.D. Metcalfe, B.G. Koenig, D.T. Bennie, M. Servos, T.A. Ternes, R. Hirsch, *Environ. Toxicol. Chem.* **2003**, 22, 2872–2880.
47 X.S. Miao, C.D. Metcalfe, *J. Chromatog. A* **2003**, 998, 133–141.
48 T.A. Ternes, M. Stumpf, B. Schuppert, K. Haberer, *Vom Wasser* **1998**, 90, 295–309.
49 S. Flaherty, S. Wark, G. Street et al., *Electrophoresis* **2002**, 23, 2327–2332.
50 M. Farré, I. Ferrer, A. Ginebreda, M. Figueras, L. Olivella, L. Tirapu, M. Vilanova, D. Barcelo, *J. Chromatogr. A* **2001**, 938, 187–197.
51 N. Lindqvist, T. Tuhkanen, L. Kronberg, *Water Res.* **2005**, 39, 2219–2228.
52 W.C. Lin, H.C. Chen, W.H. Ding, *J. Chromatogr. A* **2005**, 1065, 279–285.
53 H.-R. Buser, T. Poiger, M.D. Müller, *Environ. Sci. Technol.* **1999**, 33, 2529–2535.
54 I. Rodriguez, J.B. Quintana, J. Carpinteiro, A.M. Carro, R.A. Lorenzo, R. Cela, *J. Chromatogr. A* **2003**, 985, 265–274.
55 R. Andreozzi, R. Marotta, N. Paxéus, *Chemosphere* **2003**, 50, 1319–1330.
56 D.L. Sedlak, K.E. Pinkston, *Water Resources Update* **2001**, 56–64.
57 M. Stumpf, T.A. Ternes, K. Haberer, W. Baumann, *Vom Wasser* **1998**, 91, 291–303.
58 S. Zühlke, U. Dünnbier, Th. Heberer, *Anal. Chem.* **2004**, 76, 6548–54.
59 T. Steger-Hartmann, K. Kümmerer, J. Schecker, *J. Chromatogr. A* **1996**, 726, 179–184.
60 K. Kümmerer, T. Steger-Hartmann, M. Meyer, *Water Res.* **1997**, 31, 2705–2710.
61 ARGE (Arbeitsgemeinschaft für die Reinhaltung der Elbe) Arzneistoffe in Elbe und Saale; ARGE self-published, **2003**, Hamburg, Germany.
62 D.B. Huggett, I.A. Khan, C.M. Foran et al., *Environ. Pollut.* **2003**, 121, 199–205.
63 T.A. Ternes, J. Stüber, N. Herrmann et al., *Water Res.* **2003**, 37, 1976–1982.
64 T.A. Ternes, M. Bonerz, T. Schmidt, *J. Chrom. A.* **2001**, 938, 175–185.
65 O. Gans, R. Sattelberger, S. Scharf, *Vom Wasser* **2002**, 98, 165–176.
66 E. Möhle, S. Horvath, W. Merz, J.W. Metzger, *Vom Wasser* **1999**, 92, 207–223.
67 S.T. Glassmeyer, E.T. Furlong, D.W. Kolpin, J.D. Cahill, S.D. Zaugg, S.L. Werner, M.T. Meyer, D.D. Kryak, *Environ. Sci. Technol.* **2005**, 39, 5157–5169.
68 E. Golet, I. Xifra, H. Siegrist, A. Alder, W. Giger, *Environ. Sci. Technol.* **2003**, 37, 3243–3249.
69 P. Adler, T. Steger-Hartmann, W. Kalbfus, *Acta Hydrochim. Hydrobiol.* **2001**, 29, 227–241.
70 A. Joss, E. Keller, A.C. Alder, A. Göbel, Ch. McArdell, T.A. Ternes, H. Siegrist, *Water Res.* **2005**, 39, 3139–3152.
71 H.R. Buser, T. Poiger, M.D. Müller, *Environ. Sci. Technol.* **1998**, 32, 3449–3456.
72 T.A. Ternes 2000. Residues of pharmaceuticals, diagnostics and antiseptics in wastewater, rivers and ground water – a new challenge for water technology. (Rückstände an Arzneimitteln, Diagnostika und Antiseptika in Abwasser, Flüssen und Grundwasser –

eine neue Herausforderung für die Wasserwirtschaft), habilitation thesis, Johannes Gutenberg-University Mainz.
73 F. Tilton, W.H. Benson; D. Schlenk, *Aquat. Toxicol.* **2002**, 61(3–4), 211–224.
74 N. Kreuzinger, M. Clara, B. Strenn, H. Kroiss, *Water Sci. Technol.* **2004**, 50(5), 149–156.
75 Th. Heberer, D. Feldmann, in *Pharmaceuticals in the Environment* (Ed.: K. Kümmerer), 2nd Edition, Springer Verlag, Berlin, **2004**, pp. 391–410.
76 M. Clara, B. Strenn, M. Ausserleitner, N. Kreuzinger, *Water Sci. Technol.* **2004**, 50(5) 29–36.
77 Kimura K., Hara H., Watanabe Y., *Desalination* **2005**, 178, 135–140.
78 Watts C.D., M. Crathorne, M. Fielding, C.P. Steel, Identification of Non-volatile Organics in Water Using Field Desorption Mass Spectrometry and High Performance Liquid Chromatography. In: Analysis of Organic Micropollutants in Water. Ed: G. Angeletti, A. Bjorseth, pp 120–131. D.D. Reidel Publishing Co., Dodrecht **1983**.
79 Waggott A., Trace organic substances in the River Lee [Great Britain]. *Chem. Water Reuse*, Volume 2, 55–99. Editor: Cooper, William J. Ann Arbor Sci.: Ann Arbor, Mich., **1981**.
80 M.L. Richardson, J.M. Bowron, *J. Pharm. Pharmacol.* 1985, 37, 1–12.
81 Th. Heberer, K. Schmidt-Baeumler, H.-J. Stan, *Acta Hydrochim. Hydrobiol.* **1998**, 26, 272–278.
82 S. Wiegel, A. Aulinger, R. Brockmeyer, H. Harms, J. Loffler, H. Reincke, R. Schmidt, B. Stachel, W. von Tumpling, A. Wanke, *Chemosphere* **2004**, 57, 107–126.
83 D.W. Kolpin, E.T. Furlong, M.T. Meyer, E.M. Thurman, S.D. Zaugg, L.B. Barber, H.T. Buxton, *Environ. Sci. Technol.* **2002**, 36, 1202–1211.
84 D.W. Kolpin, M. Skopec, M.T. Meyer, E.T. Furlong, S.D. Zaugg, *Sci. Total Environ.* **2004**, 328, 119–130.
85 Bund/Länderausschuss für Chemikaliensicherheit (BLAC) Arzneimittel in der Umwelt Auswertung der Untersuchungsergebnisse; self-published, Hamburg, **2003**.
86 M.M. Huber, S. Canonica, G.Y. Park, U. von Gunten, *Environ. Sci. Technol.* **2003**, 37, 1016–1024.
87 R. Andreozzi, R. Marotta, N. Paxéus, *Chemosphere* **2003**, 50, 1319–1330.
88 M. Winkler, J.R. Lawrence, T.R. Neu, *Water Res.* **2001**, 35, 3197–3205.
89 R. Andreozzi, R. Marotta, G. Pinto, A. Pollio, *Water Res.* **2002**, 36, 2869–2877.
90 Th. Heberer, K. Reddersen, in preparation.
91 D. Löffler, J. Römbke, M. Meller, T.A. Ternes. *Environ. Sci. Technol.* **2005**, 39, 5209–5218.
92 M. Ahel, I. Jelicic, in *Pharmaceuticals and Personal Care Products in the Environment: Scientific and Regulatory Issues. Symposium Series 791* (Eds.: C.G. Daughton, T. Jones-Lepp), American Chemical Society, Washington D.C., **2001**, pp. 100–115.
93 S. Zühlke, U. Dünnbier, Th. Heberer, *J. Chromatogr. A* **2004**, 1050, 201–209.
94 T. Scheytt, P. Mersmann, R. Lindstädt, Th. Heberer, *Chemosphere* **2005**, 60, 245–253.
95 T. Scheytt, P. Mersmann, R. Lindstädt, Th. Heberer, *Water, Air, Soil Pollut.* **2005**, 165, 3–11.
96 G. Preuss, U. Willme, N. Zullei-Seibert, *Acta Hydrochim. Hydrobiol.* **2001**, 29, 269–277.
97 P. Mersmann, T. Scheytt, Th. Heberer, *Acta Hydrochim. Hydrobiol.* **2002**, 30, 275–284.
98 T. Scheytt, P. Mersmann, M. Leidig, A. Pekdeger, Th. Heberer, *Ground Water* **2004**, 42, 767–773.
99 Th. Heberer, A. Mechlinkski, *Hydroplus* **2003**, 137; *Hydrosciences* October **2003**, 53–60.
100 Th. Heberer, A. Mechlinkski, B. Fanck, A. Knappe, G. Massmann, A. Pekdeger, B. Fritz. *J. Ground Water Monit. Remediat.* 24, 70–77.
101 S. Zühlke, U. Dünnbier, Th. Heberer, B. Fritz. *J. Ground Water Monit. Remediat.* 24, 78–85.
102 Th. Heberer, M. Adam, *Environ. Chem.* **2004**, 1, 22–25.
103 T.A. Ternes, M. Meisenheimer, D. McDowell, F. Sacher, H.-J. Brauch, B.,

Haist-Gulde, G. Preuss, U., Wilme, N. Zulei-Seibert, *Environ. Sci. Technol.* **2002**, 36, 3855–3863.

104 M.M. Huber, S. Canonica, G.-Y. Park, U. von Gunten, *Environ. Sci. Technol.* **2003**, 37, 1016–1024.

105 D.C. McDowell, M.M. Huber, M. Wagner, U. von Gunten, T.A. Ternes, *Environ. Sci. Technol.* **2005**, 39, 8014–8022.

106 Kimura K., Toshima S., Amy G., Watanabe Y. *J. Membrane Sci.* **2004**, 245, 71–78.

107 Kimura K., Amy G., Drewes J., Heberer T., Kim T.U., Watanabe Y. *J. Membrane Sci.* **2003**, 227, 113–121.

108 Kimura K., Amy G., Drewes J., Watanabe Y. *J. Membrane Sci.* **2003**, 221, 89–101.

109 Xu P., Drewes J.E., Bellona C., Amy G., Kim T.U., Adam M., Heberer Th., *Water Environ. Res.* **2005**, 77, 40–48.

110 Th. Heberer, D. Feldmann, K. Reddersen, H.J. Altmann, T. Zimmermann. *Acta Hydrochim. Hydrobiol.* **2002**, 30, 24–33.

3
Antibiotics for Human Use

Radka Alexy and Klaus Kümmerer

3.1
Introduction

Antibiotics are among the most important compounds used in medicine. Although antibiotics have been applied in large quantities for some decades, the existence of these substances in the environment has attracted little attention until recently. It is only lately that pharmaceuticals, of which antibiotics are an important group, have been identified as environmental contaminants [1, 2]. Antibiotics are a diverse group of chemicals that can be divided into sub-groups such as β-lactams, quinolones, tetracyclines, macrolides, sulfonamides, and others. Their possible effects in and on the environment are of particular interest because of their properties as bio-active chemicals. Because of their often complex chemical structure, their behavior in the environment is not always easy to assess. There is concern that antibiotics emitted into the aquatic environment can reach drinking water, or that they may affect non-target organisms in the environment. In contrast to the properties and effects desired from their therapeutic application, these same properties are often disadvantageous for target and non-target organisms in the environment. Antibiotics are by definition active against bacteria. As a consequence, bacteria have developed mechanisms for resisting them. Therefore, bacterial resistance promoted by the input of antibiotics into the environment is also under discussion [3].

In this chapter we focus on the input, occurrence, and fate of antibiotics in the aquatic environment.

3.2
Use of Antibiotics

The volume of use is usually only calculated on a nationwide scale. Wise [4] estimated antibiotic consumption worldwide to lie between 100 000 and 200 000 t year^{-1}. In 1996, about 10 200 t of antibiotics were used in the EU, of which approx-

Organic Pollutants in the Water Cycle. T. Reemtsma and M. Jekel (Eds.).
Copyright © 2006 WILEY-VCH Verlag GmbH & Co. KGaA, Weinheim
ISBN: 3-527-31297-8

imately 50% was applied in veterinary medical therapy and as growth promoters for animals [5]. According to data supplied by the European Federation of Animal Health [6], in 1999, 13 216 t of antibiotics were used in the European Union and Switzerland, of which 65% was applied in human medicine. According to differences in legislation (e.g., permission to sell antibiotics over the counter or not) and the differing degrees of importance ascribed to the use of antibiotics, reliable data providing information on the use and the use patterns of antibiotics and their per capita consumption exist for a few countries only (see Table 3.1). In most countries, β-lactam antibiotics, including the sub-groups of penicillins, cephalosporins and, as a marginal fraction, carbapenems, make up the largest share of human-use antibiotics. They account for approximately 50–70% of total antibiotic use in human medicine. In most countries, sulfonamides, macrolides, and fluoroquinolones follow in order of decreasing use [7].

Antibiotic prescription rates vary markedly between countries (see Table 3.1 [6–8]). In a study performed on a local scale, the mass balance of 42 antibiotics and 15 antimycotics was investigated over a period of one year [10]. It was found that the local use pattern differs from the nationwide one (Fig. 3.1). Surprisingly, seasonal changes were small [10].

In the EU, consumption of antibiotics for medical purposes amounts to a total of 22 g per capita and year, while in the Unites States the available data give a figure of about 17 g. According to other published data (see Table 3.1 [9]) the consumption of antibiotics per capita and year in human medicine in various European countries ranges between 4 g and 8 g. Expressed on a per capita basis, local sales may be high or low. In 2002, per capita consumption in a small German municipality was 2.6 g, compared with the nationwide per capita consumption of 4.95 g calculated for 1998 [8, 10]. In Germany, use outside hospitals in the community is about two thirds to three quarters of the total amount used in human medicine. This agrees with estimated ratios for other countries. Community use is reported to be 70% in the UK [11] and 75% in the United States [4].

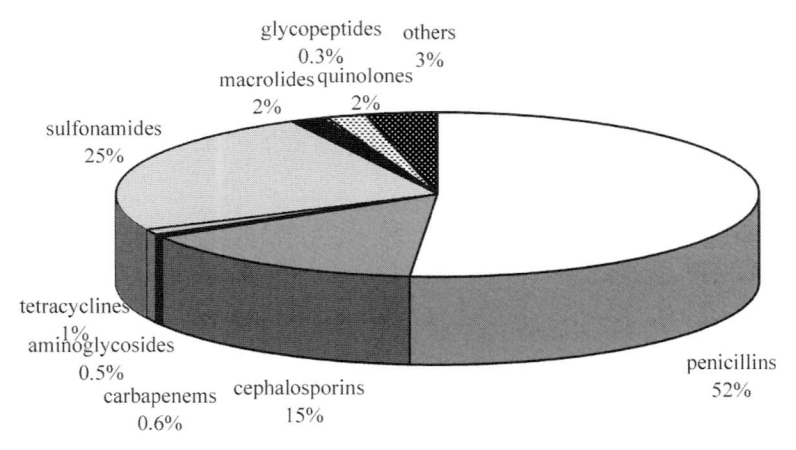

Fig. 3.1 Release of antibiotics into effluent in Germany according to groups, excretion rates included [8].

3.2 Use of Antibiotics | 67

Tab. 3.1 Use, input and concentration of antibiotics in different countries (quoted concentrations are maximum values for different compounds).

Region/country	Total volume used in human medicine [t year⁻¹]	Volume used in human medicine [g per capita]	Of which in hospitals [%]	Unused drug [%]	Measured in sewage up to [µg L⁻¹]	Measured in surface water up to [µg L⁻¹]	Reference
Worldwide	100000–200000 ?	?	?	?	?	?	4
European Union and Switzerland	8637	22.4	?	?	?	?	6
Austria	38	4.7	?	20–30 ?	?	?	19
Canada	?	?	?	?	?	0.87 treated waste water, no dilution	14
Denmark	40	7.4	?[a]	?	5 ?[b]	?	17, 18
Germany	412	4.95	25	20–40 ?	6	1.7	8, 16, 24, 26
Italy	283	4.88	?	?	0.85	0.25	24–27
Switzerland	34.2	4.75	20–40 ?	?	0.57	0.2	15
The Netherlands	40.9	3.9	20	8.3	4.4	0.11–0.85	20–23
U.S.A.	4,860 ?	17?	70?	?	1.9	0.73	12, 25

[a] data only available as daily defined dose
[b] only sulfamethizole and penicillin V which have not been measured or detected in other studies
? data not available or quality/significance not clear

The proportion of antibiotics used in hospitals differs by compound because of differences in water solubility and activity spectrum. Compounds of low solubility are administered intravenously, which is normally only done in hospitals. Data for Germany are shown in Fig. 3.2.

Besides being used for treating infections of humans, large amounts of antibiotics are used for veterinary purposes (see Chapter 5). For example 70% of the approximately 16 200 t of antibiotics produced in the United States in 2000 were used in healthy livestock [12]. This is eight times the amount used in human medicine. In the EU, in 1999, 29% of antibiotics were used for treatment of animals and 6% as growth promoters [6]. Since most growth promoters have now been banned within the EU, only four compounds remained in this group of feed additives, and these were banned in 2006. This may explain the decline in the use of antibiotics in animal husbandry compared with human medicine. The World Health Organization advises abandonment of the use of antibiotics as growth promoters, as data show that there is no need for this [13]. Therefore, the amounts used in livestock will decline, as already shown by the examples of Sweden, Denmark, and Switzerland. In intensive fish farming, infections are treated by feeding antimicrobial agents directly into the water. This results in high local concentrations in the water compartment and adjoining sediments [31, 38]. In fish farming antibiotic use is increasingly being diminished by the use of vaccines.

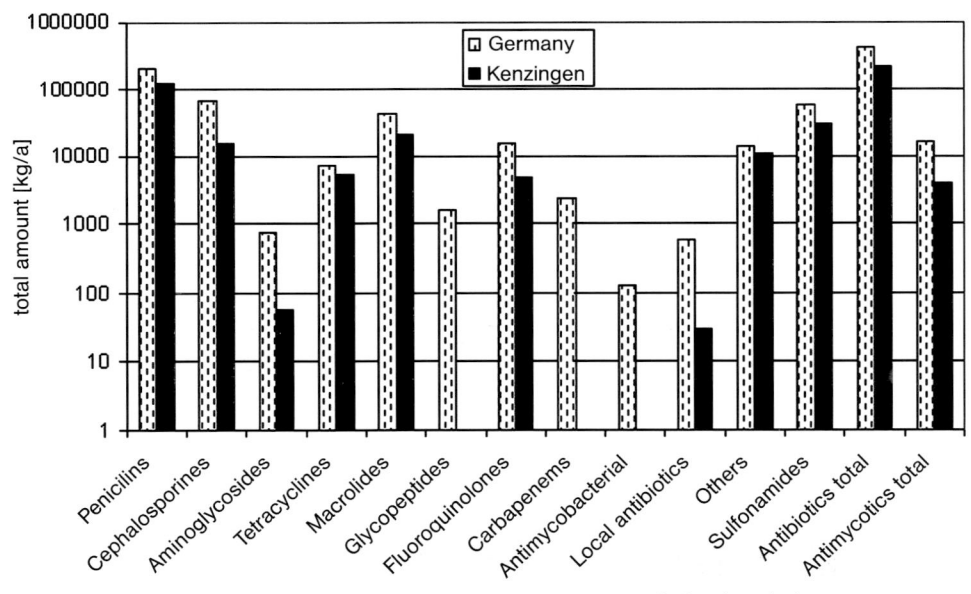

Fig. 3.2 Consumption of antibiotics (groups) in Germany calculated on the basis of data collected locally (Kenzingen): 21.1 kg year^{-1}, of which 13.5 kg year^{-1} emitted into sewage (solid bars) versus the German nation wide average [10].

3.3
Emissions into the Environment

There are two main ways in which antibiotics reach the environment. Antibiotics for human use end up in wastewater (Figs. 3.3 and 3.5). Antibiotics used for veterinary purposes or as growth promoters are excreted by the animals and end up in manure (Fig. 3.3). Emissions from production are negligible according to present knowledge.

A detailed balance in Germany has shown that three quarters of the volume used in medicine is excreted into wastewater [8]. Antibiotics are often bio-transformed within the human body. Excretion rates for the unchanged active compound cover a broad range (10–90%) [8]. Phase I metabolites result from modification of the active compound, e.g., by hydrolysis (e.g., of ester bonds) or oxidation (including hydroxylation, deamination, desulfuration, and dealkylation) [28]. Phase II metabolites are mostly phase I metabolites that have been modified by glucuronidation, acetylation, glycination, methylation, or sulfatation ("coupling reactions") to facilitate excretion. On a mass basis, the average metabolic rate of all antibiotics is 30% [8]. Little is known about the fate and activity of such metabolites. An important question to be addressed is whether the glucuronides, methylates, glycinates, acetylates, and sulfates are still active and whether they can be cleaved by bacteria during sewage treatment and in the environment. This would result in the active com-

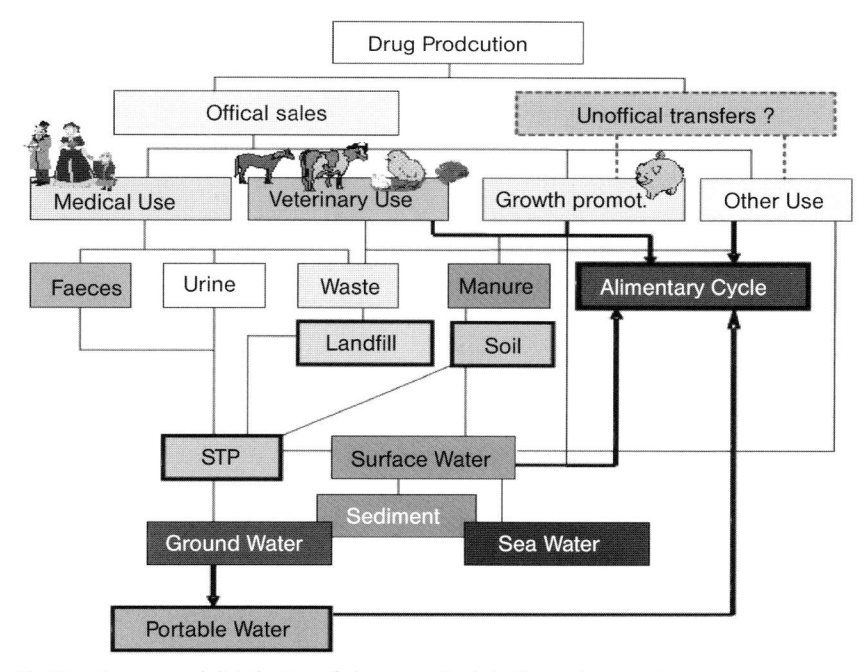

Fig. 3.3 Sources and distribution of pharmaceuticals in the environment (STP: sewage treatment plant) [2].

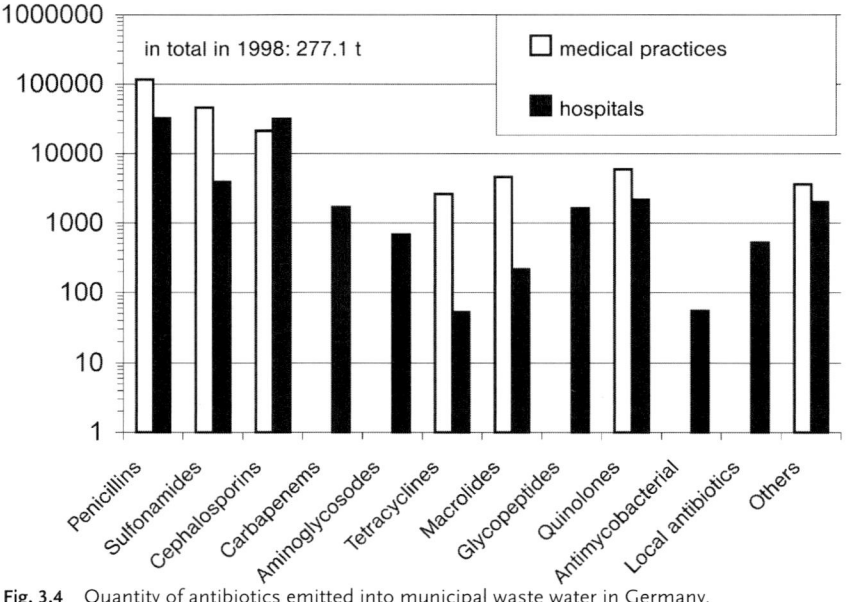

Fig. 3.4 Quantity of antibiotics emitted into municipal waste water in Germany, according to origin (excretion rates included) [8].

pound being set free. Indeed, in the veterinary field the deacetylation of sulfa drugs is a well-known process.

The emission of antibiotics from hospitals and households into municipal sewage results in an annual average expected concentration of about 71 µg L^{-1} in Germany. Of this, 38.3 µg L^{-1} is attributable to penicillins, 12.8 µg L^{-1} to sulfonamides, 13.5 µg L^{-1} to cephalosporins, and 2.1 µg L^{-1} to quinolones [8].

Outdated or leftover drugs are sometimes disposed of down household drains. The proportion which takes this route is not known. The total load released per compound depends on its use and excretion rate (Fig. 3.4).

3.4
Occurrence and Fate of Antibiotics

3.4.1
Wastewater and Wastewater Treatment

3.4.1.1 Hospital Wastewater

A few frequently used antibiotics have been analyzed in the effluent of pharmaceutical companies and hospitals in concentrations high enough and potent enough to cause adverse effects on wastewater bacteria [32, 59–61]. The antibiotic load in municipal wastewater is lower. Therefore, upon discharge into the sewer system, hospital wastewater is strongly diluted by sanitary wastewater (typically >100-fold). It is then treated, together with municipal wastewater, in municipal sewage treatment plants. Separate collection of patients' excretions is often impossible for organizational, economic, and technical reasons. In addition, the antibiotics which are emitted from hospitals to the sewage treatment plants represent only 30% of the total antibiotic influent to STPs.

Some investigations on antibiotics concentrations in hospital wastewater are available [59–61, 65]. The amoxicillin concentration measured in the wastewater of a large German hospital was between 28 and 82.7 µg L^{-1} [60]. Färber [61] demonstrated the presence of penicillins, macrolides, sulfonamides, fluoroquinolones, and tetracyclines in concentrations of up to 15 µg L^{-1} (sum of the compounds detected) in a hospital wastewater and in concentrations of up to 1 µg L^{-1} in the influent of the sewage treatment plant to which the hospital wastewater was discharged. Hartmann et al. [59] found ciprofloxacin in hospital effluent in concentrations between 3 and 89 µg L^{-1}.

3.4.1.2 Municipal Wastewater

In general, sulfonamides (sulfamethoxazole), macrolides (roxythromycin and a decomposition product of erythromycin: dehydrato-erythromycin), tetracyclines, and fluoroquinolones (ciprofloxacin) were the most frequently analysed antibiotics in municipal wastewater (Table 3.2). Studies conducted in different countries have detected a number of antibiotics in the low µg L^{-1} or ng L^{-1} range in municipal wastewaters and effluents from sewage treatment plants (STPs) [14, 15, 37, 62, 64]. β-Lactams are detected far less frequently despite the fact that these compounds are used in the greatest quantities. Alder et al. [15] studied quinolones and other antibiotics such as macrolides and sulfonamides. Individual concentrations ranged from 255 to 568 ng L^{-1} in raw sewage and from 36 to 106 ng L^{-1} in final wastewater effluents (Fig. 3.7). In another study, antibiotics were detected in the same range in the influent and in the effluent of three smaller municipal sewage treatment plants [62]. Additionally, the antibiotics clindamycin, metronidazole, and ofloxacin were found [64]. A summary of the results for effluent of STP, surface water, and groundwater is presented in Table 3.2.

Studies have shown that most antibiotics are not fully eliminated in conventional wastewater treatment and are then released into the environment. The overall removal efficiency of a Spanish STP was 60% for the antibiotic sulfamethoxazole [67]. In three German STPs of varying size and technology, Alexy and co-workers [10] found an elimination rate of 50% for the sulfonamide sulfamethoxazole and the macrolide clarithromycin, whereas it was 90% for trimethoprim and 25% for the macrolide roxithromycin.

A few studies on antibiotics used in medicine have calculated a mass balance comparing local or regional use with the antibiotic concentrations determined in municipal sewage treatment plants [10, 37]. On a regional scale and using ciprofloxacin as an example, it was found that the substance was eliminated to some extent in Swiss STPs, in the river Glatt, and in Lake Greifenegg [37].

3 Antibiotics for Human Use

Table 3.2 Concentrations of different antibiotics in ng L^{-1} measured in the aquatic environment [55].

Antibiotics	Effluent of STP	Surface water	Ground water (* = bank filtration)	Reference
Penicillins				
Penicillins	Up to 200	Up to 3		[61]
Flucloxacillin		7		[32]
Piperacillin		48		[32]
Macrolides				
Macrolides	Up to 700	Up to 20	Up to 2*	[61]
Azithromycin		Up to 3		[32]
Erythromycin-H$_2$O	Up to 287			[63,101]
			Up to 49	[100]
	Up to 6,000	Up to 1,700		[62]
		Up to 190		[32]
		Up to 1,700		[74]
		Up to 15.9		[24]
	Up to 400			[10]
Clarithromycin	Up to 328	Up to 65		
	Up to 240	Up to 260		
		Up to 37		
		Up to 20.3		[63, 101]
				[62]
				[32]
				[24]
	Up to 38			[10]
Roxithromycin	Up to 68			[10]
	Up to 72			[63, 101]
			Up to 26	[100]
	Up to 1,000	Up to 560		[62]
		Up to 14		[32]
		Up to 180		[74]
Quinolones				
Fluorquinolones	Up to 100	Up to 5		[61]
Fluorquinolones (ciprofloxacin, norfloxacin)	Up to 106	Up to 19		[63, 101]
Ciprofloxacin		9		[32]
		Up to 30		[74]
		Up to 26.2		[24]
Norfloxacin		Up to 120		[74]
Ofloxacin	Up to 82			[10]
		20		[32]

Table 3.2 Concentrations of different antibiotics in ng L^{-1} measured in the aquatic environment [55] (Continued).

Antibiotics	Effluent of STP	Surface water	Ground water (* = bank filtration)	Reference
Sulfonamides				
Sulfonamides	Up to 1,000	Up to 40	Up to 20*	[61]
Sulfamethoxazole	Up to 370			[10]
		Up to 163	Up to 410	[100]
	Up to 2,000	Up to 480	Up to 470	[62]
		Up to 52		[32]
		Up to 1,900		[74]
Sulfamethazine			Up to 160	[62]
		Up to 220		[74]
Sulfamethizole		Up to 130		[74]
Sulfadiazine			Up to 17	[100]
Sulfadimidine			Up to 23	[100]
		Up to 7		[32]
Sulfadimethoxine		Up to 60		[74]
Tetracyclines				
Tetracyclines	Up to 20	Up to 1		[61]
Tetracycline		Up to 110		[74]
Chlortetracycline		Up to 690		[74]
Oxytetracycline		Up to 340		[74]
		Up to 19.2		[24]
Others				
Trimethoprim	Up to 38			[10]
		Up to 24		[100]
	Up to 660	Up to 200		[62]
		Up to 12		[32]
		Up to 710		[74]
Ronidazol			Up to 10	[100]
Chloramphenicol	Up to 68			[10]
	Up to 560	Up to 60		[62]
Clindamycin	Up to 110			[10]
		Up to 24		[32]
Lincomycin		Up to 730		[74]
		Up to 248.9		[24]
Spiramicin		Up to 74.2		[24]
Oleandomycin		Up to 2.8		[24]
Tylosin		Up to 280		[74]
		Up to 2.8		[24]

In another study connecting use and presence in an STP, the concentrations measured in the STP were in general in the expected range from balances of local consumption [10]. However, there were clear differences in the species of antibiotics detected as well as in their concentration, within and between the three sampling periods. The pattern of compounds present in the STP influent in spring, autumn, and winter did not always correlate with the pattern of prescription. In most cases, the concentration in the effluent was lower than in the influent. Some elimination was due to sorption.

Golet et al. [37] and Giger et al. [63] have confirmed the hypothesis that fluoroquinolones become highly enriched in sewage sludge (concentrations ranging from 1.4 to 2.4 mg kg^{-1} of dry matter). First results of analysis of conventional sewage sludge using the method described by Thomsen et al. [66] show that sulfathiazole, sulfapyridin, sulfamethoxazole, azithromycin, clarithromycin, and roxithromycin can be determined in concentrations of up to 75 µg kg^{-1}. Alexy et al. [10] found concentrations between a few µg kg^{-1} for trimethoprim and macrolides, and several hundred µg kg^{-1} for fluoroquinolones.

Certain operational parameters can improve elimination of antibiotics in municipal wastewater treatment. It was shown that the elimination of roxithromycin and sulfamethoxazole increases with sludge age [68]. In this study it was also found that membrane bioreactors remove micropollutants to an extent similar to that achieved by conventional activated sludge plants run with comparable sludge retention times.

Advanced treatment processes such as membrane technology and (photo)oxidation are under discussion to improve the elimination of pharmaceuticals and other trace compounds in sewage treatment. Antibiotics in the effluent of a production facility were eliminated by photochemical treatment with and without ozone [69]. Photochemical treatment of the effluent, from wastewater treatment plants with and without ozone and microwave-assisted photo-oxidation resulted in the elimination of selected antibiotics such as sulfamethoxazole, some β-lactams, and ciprofloxacin [67, 70, 71].

3.4.2
Surface Water

After passing through the wastewater treatment plant, antibiotics are emitted into surface water. Run-off due to rain after the application of manure to agricultural land could be an additional factor in the input of antibiotics into surface waters (Fig. 3.5) [30].

In consequence, investigations that have been conducted in various countries have detected a number of antibiotics in the low µg L^{-1} or the ng L^{-1} range in different environmental compartments [15, 23–27, 37, 51, 61–64, 73–74] (Table 3.2). A variety of antibiotics have been detected in concentrations of up to 1.9 µg L^{-1} (sulfamethoxazole) in different studies in surface water. The most frequently detected antibiotics were erythromycin-H$_2$O (up to 1,7 µg L^{-1}), clarithromycin (up to 260 ng L^{-1}), roxithromycin (up to 260 ng L^{-1}, sulfamethoxazole (up to 1,9 µg L^{-1})

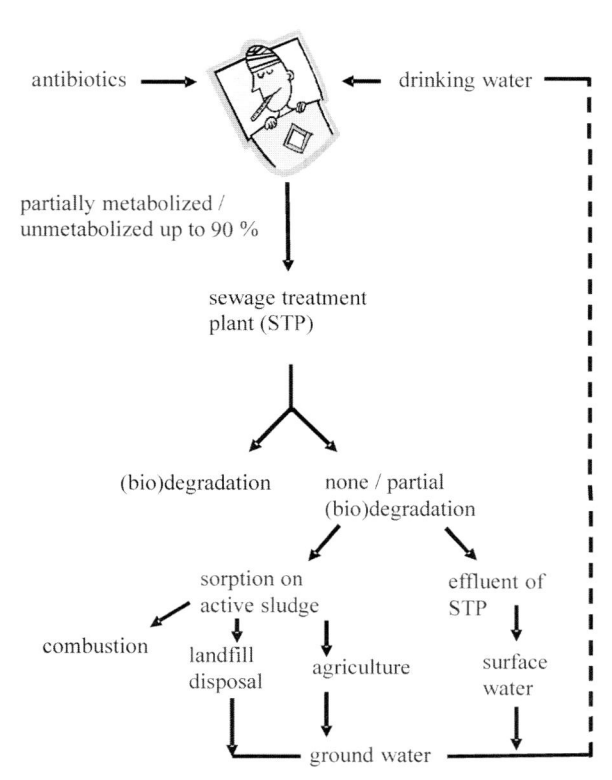

Fig. 3.5 Distribution of pharmaceuticals used in human medicine in the environment.

and trimethoprim (up to 710 ng L^{-1}) (see Table 3.2). In the Glatt River the compounds were present at concentrations below 19 ng L^{-1} [37]. The fate of ciprofloxacin was studied in its catchment.

Concentrations found in surface water are a result of the input into STPs and the elimination processes in sewage treatment as well as in surface water, such as degradation and sorption onto particulate matter.

Some investigations have demonstrated the presence of antibiotics in sediments beneath fish farms, where certain compounds are applied extensively [75–77] (see Chapter 5). Quinolones have been found to sorb to sediments [78]. The sorbed compounds had long half-lives in soil and sediment [75, 78, 79–83].

3.4.3
Groundwater

Contamination of groundwater by antibiotics may occur via spreading of manure in agriculture or from infiltration of surface water or treated wastewater (Fig. 3.5).

In a comprehensive monitoring program the occurrence of a large number of pharmaceuticals in groundwater samples was studied at 105 monitoring wells in Baden-Württemberg, southwest Germany [73]. Of the 60 pharmaceuticals analyzed eight compounds were detected in at least three groundwater samples. Two of these were antibiotics. Erythromycin-H_2O and sulfamethoxazole were detected 10 and 11 times, respectively. Other antibiotics such as sulfadiazine, sulfadimidine, sulfamethoxazole, ronidazol, dapson and roxithromycin were detected less frequently. Accompanying investigations with boron as the indicator parameter pointed to wastewater as the major source of these pharmaceuticals. Furthermore, other tracer compounds for wastewater such as nonylphenols or bisphenol A have often been found in these samples [73].

Only sulfamethoxazole has been detected in groundwater samples of canton Zürich, Switzerland [72]. In another study, sulfamethoxazole was found in concentrations up to several µg L^{-1} [86]. Derksen et al. found erythromycin and sulfamethoxazole in groundwater at concentrations below 10 ng L^{-1} [20].

Local emission of antibiotics may result in high concentrations even in groundwater, as demonstrated for a groundwater downgradient from a landfill originating from a pharmaceutical production site [85].

Manure is used as an agricultural fertilizer. Moreover, stabilized sludge from activated sludge municipal wastewater treatment may be utilized in agriculture. In Germany, 31% of the 2.4×10^6 t of dry activated sludge produced [29] were utilized in agriculture. Antibiotics included in these sludges may seep through the soil and enter groundwater in case they are not retained by or degraded in the soil (Fig. 3.3) (see Chapter 5). However, in an investigation performed by Hirsch et al. [62] the load of antibacterial agents found in groundwater in rural areas with high concentrations of livestock proved to be insignificant. Antimicrobial residues were found in only two groundwater samples out of 51 tested. According to analytical field data only a few sulfonamides may not be strongly sorbed. Other antibiotics, such as tetracyclines or quinolones are strongly sorbed. Tetracyclines have not been detected under soil treated with manure containing tetracycline [84].

3.5
Elimination and Degradation in the Aquatic Environment

In the aquatic environment, organic compounds may be eliminated by biodegradation, by non-biotic degradation processes such as hydrolysis and photolysis, or by sorption onto suspended particles, sediments and sludge.

3.5.1
Elimination by Sorption

Binding to particles or the formation of complexes with metal cations may cause a loss in detectability, as well as a loss in antibacterial activity. The loss of antibacterial activity, for example, was demonstrated for an aquaculture antimicrobial in seawater driven by the formation of complexes with the magnesium and calcium ions naturally present in marine water [31]. Tetracyclines are able to form complexes with divalent cations, such as calcium or magnesium [32].

In general, the sorption behavior of an organic compound strongly depends upon its chemical structure. Unlike highly lipophilic compounds such as PCBs, most antibiotics are complex chemical molecules which may contain acidic and basic groups within the same molecule (Fig. 3.6). Therefore, the distribution of such antibiotics between water and particulate matter (log K_D) depends on pH.

For example the acid constants of ciprofloxacin (pK_a) are 6.16 and 8.63. At a pH of 7.04, the isoelectric point of ciprofloxacin, the molecule carries both a negative and a positive charge, i.e. it is neutral as an entity despite the charges within the molecule. Therefore, solubility, hydrophobicity, and distribution between different phases (water/sludge or sediment) are pH dependent. Ionic interactions may contribute to sorption. Therefore, the sorption of antibiotics is governed by pH, stereochemistry, and chemical nature of both the sorbent and the sorbed molecule [33, 34]. Thus, the logK_{ow} of the neutral molecule is not an appropriate parameter to assess the sorption and distribution behavior of antibiotics [33]. This has been proven for ciprofloxacin: its logK_{ow} has been calculated to be about –1.74 [35] and was experimentally determined to be –0.28 [35]. This value suggests that only a small proportion would partition into sludge or sediments. However it was found that ciprofloxacin sorbs well onto active sludge and sediments [36, 37], as would be expected for compounds with a log K_{ow} above 3 or 4. As a consequence, more than 90% of cipofloxacin entering an STP was removed by sorption onto the activated sludge (Fig. 3.7). Some antibiotics such as tetracyclines and quinolones also contain planar aromatic structures which are favorable for intercalation, e.g., into the layers of some clay minerals.

Antibiotics can also diffuse into biofilms, present in sewage pipes, sludge flocks, or on stones in rivers and lakes. This may result in a biased risk estimate, as concentration in such "reservoirs" may be much higher than in the free water phase. The effects and behavior of antibiotics in such bio-solids with high bacterial density and special conditions has not yet been investigated. It is not known how

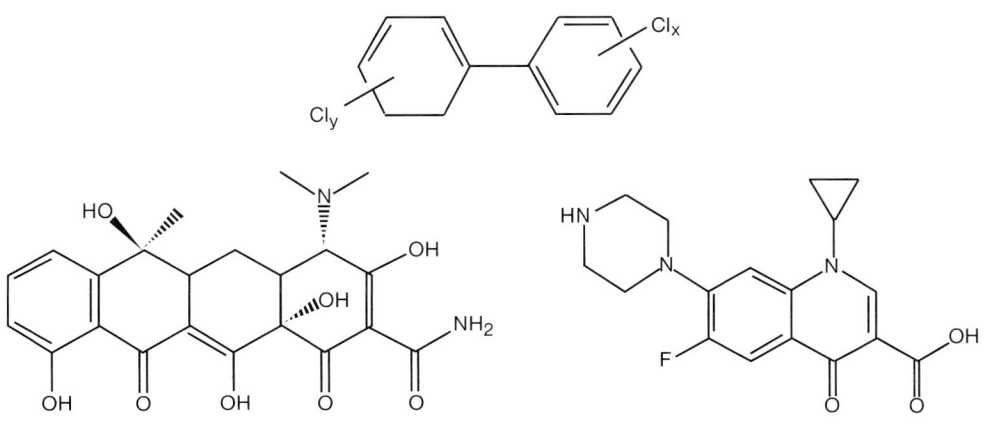

Fig. 3.6 Chemical structures of polychlorinated biphenyls (top) and the antibiotics tetracycline (bottom left) and ciprofloxacin (bottom right).

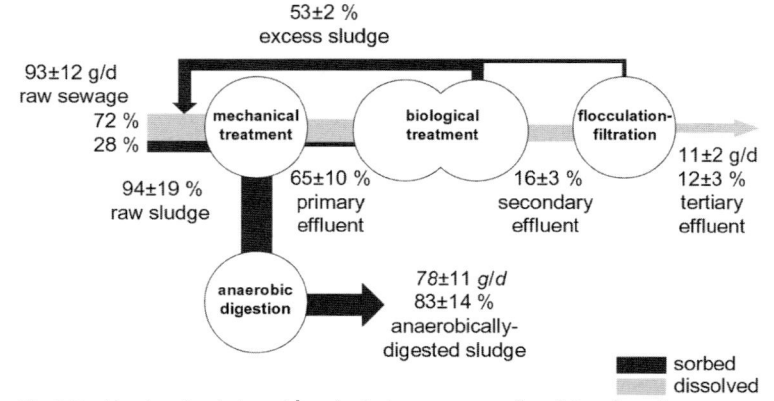

Fig. 3.7 Absolute loads in g d^{-1} and relative mass transfer of ciprofloxacin through mechanical treatment, biological treatment, flocculation-filtration, and anaerobic sludge digestion in Zurich-Werhölzli wastewater treatment plant (100% equal to total ciprofloxacin-input mass flow entering the WWTP) [15].

strongly the antibiotics are sorbed onto sludge, particulate matter, bio-solids such as sewage sludge, and sediments, and under what circumstances they are (bio)available and active after sorption. Little is known about the relevant properties of conjugates and other metabolites.

3.5.2
Non-biotic Degradation

3.5.2.1 Photolysis

If a substance is sensitive to light, photodecomposition can be of major significance in the elimination process, as reported by Lunestad et al. [38]. Data from drug registration may give guidance on compounds where photodegradation can be expected. Some antibiotics such as quinolones, tetracyclines, metronidazole, and sulfonamides are sensitive to light. Fluoroquinolones are insensitive to hydrolysis and increased temperatures, but are degraded by UV light [39–43]. This is of significance for surface water as an additional elimination pathway. Samuelsen [42], for example, investigated the persistence of oxytetracycline and its sensitivity toward light in seawater as well as on the surface of sediments. The compound was unstable in seawater but stable in sediments. As no decomposition mechanism other than photodegradation is known for this antimicrobial agent [44], it remains in the sediment for a long period of time, as proven by Lunestad and Goksøyr [31]. Pouliquen et al. [45] studied the elimination of oxytetracycline in seawater. The half-lives varied because of differences in temperature, light intensity, and flow rate from one test tank to another. Sulfanilic acid was found as a degradation product common to most of the sulfa drugs. Photodegradation of these drugs in natural water samples (e.g., Lake Superior) was attributed solely to direct photolysis [98].

3.5.2.2 Hydrolysis

Another kind of abiotic elimination of substances is hydrolysis. Instability in water has been demonstrated for some antibiotics [31, 45]. The β-lactam ring of β-lactam antibiotics, for example penicillins, can be opened by β-lactamase, an enzyme present in bacteria, or by chemical hydrolysis. The significance of this process for the elimination of β-lactams in the aquatic environment has not yet been studied extensively. The finding that many of the most frequently applied penicillins are difficult to detect in the environment may be due to such a hydrolysis process, especially in alkaline sewage [32, 45].

3.5.3
Biodegradation

The potential of chemicals to undergo biotransformation is an important aspect of assessing their environmental fate and the risk they present to the aquatic organisms. As outlined above, several different abiotic processes may lead to the removal of chemicals from sewage, surface water, groundwater, and soil. Biologically mediated processes (i.e. biotransformation) can result in the partial transformation or total mineralization of chemicals in the aquatic environment. Organic compounds may be utilized by microorganisms for energy and as building blocks for their growth. Some compounds are biodegraded without any energy gain, provided that another compound is available whose biodegradation supplies the necessary energy. Sometimes total degradation does not take place and the process is stopped before mineralization has been completed. These biodegradation intermediates, i.e. the products of biotransformation, can be even more stable than the parent compounds. They often also vary in their toxicity. Bacteria and fungi are the two major groups of organisms degrading organic compounds. Fungi are particularly important in soils, but do not play an important role in the aquatic environment. Therefore, in sewage treatment plants (STPs), surface water, groundwater, and seawater bacteria are assumed to be responsible for most biodegradation processes.

Laboratory biodegradation tests are frequently used to investigate the biodegradability of chemicals. However, the handling of the bacteria used in such a test may have a significant impact on the test results and, thus, on the classification of a chemical under investigation. Bacteria may or may not have been pre-adapted to a substrate and acclimatized to the test conditions. Pre-adapted bacteria normally provide better biodegradation results. Since antibiotics are designed to be active against bacteria, this point is of particular importance for the biodegradability testing of antibiotics.

Monitoring bacterial toxicity before conducting a biodegradation test is essential if negative results are to be avoided. If a compound is toxic against some of the bacteria, selection of certain bacteria may take place in toxicity tests as well as in biodegradability tests. The diversity of the microbial consortium employed in a test may change because of action of the test compound, and this change depends both on the test organisms and the activity profile of the test compound. In the standardized respiration inhibition test, no effects of antibiotics against STP bacteria were found in concentrations up to 100 mg L^{-1}. Prolongation of the test resulted in EC$_{50}$ values between 1 and 100 µg L^{-1} for some antibiotics [46, 47]. Halling-Soerensen reported difficulties with the nitrification inhibition test [48]: the use of a complex inoculum in the nitrification inhibition test resulted in dose-effect curves that could not be interpreted. In some cases enhancement of nitrification occurred, in some there was no effect, while in others the process was inhibited [51]. Conducting the test with a single species resulted in the usual S-shaped dose-effect curves [48]. Monitoring changes in the composition of a microbial test consortium is possible either by counting colony-forming units or by other means such as chemotaxonomy fluorescence *in situ* hybridization (FISH) or the use of a toxicity control [49]. Different parameters are used for each method by which different groups of bacteria of varying activity are monitored [50]. Therefore, one cannot speak of *the* bacterial toxicity of a compound. Rather, toxic effects have to be specified in relation to the test conditions.

It can be assumed that microbial degradation of trace pollutants will be slower in surface water than in the sewage system because of its lower bacterial density and lower bacterial diversity. In our own investigations, more than 20 compounds representing the most important groups of antibiotics such as β-lactams, tetracyclines, macrolides, lincosamides, aminoglycosides, quinolones, sulfonamides, glycopeptides, carbapenems, and nitroimidazoles were found not to be readily biodegradable in the Closed Bottle Test [36, 51–53]. Only a few β-lactams are biodegraded to a modest extent [51]. Halling-Sørensen et al. [56] and Ingerslev et al. [57, 58] also reported that antibiotics are not easily biodegradable. They do not biodegrade well under anaerobic conditions [42, 51, 75, 76]. Of 16 antibiotics from the groups mentioned above, only benzylpenicillin (penicillin G) was completely mineralized in a combination test (combination of OECD 302 B and OECD 301 B) [51]. Trials with radio-labeled compounds revealed that approximately 25% of benzylpenicillin was mineralized within 21 days, whereas ceftriaxone and trimethoprim were not mineralized at all [54].

3.6
Effects on Aquatic Organisms

3.6.1
Effects on Aquatic Bacteria and Resistance

Antibiotics are designed to be active against micro-organisms. They often exhibit different spectra of activity and mechanisms that mediate this activity. As a result, bacteria have developed a broad range of mechanisms to respond to the antibiotic threat. Examples are the use of efflux pumps and chemical modification for inactivation of the antibiotic such as cleavage of β-lactams by β-lactamases or the modification of binding sites in the bacterium.

Bacteria appear to be mostly unaffected by antibiotics if the toxicity of the antibiotics is tested according to ISO or OECD guidelines. However, it is difficult to determine whether all bacterial species remained unaffected or were affected to the same extent or whether resistant bacteria developed. It is known that antibiotics in sub-inhibitory concentrations can have an impact on cell functions and that they can change the genetic expression of virulence factors or the transfer of antibiotic resistance [88]. Therefore, as in biodegradation tests, the results of toxicity tests do not only depend upon the test compound and its mode of action, but also on the test conditions and the species used for effect testing (see above).

There has been growing concern about antimicrobial resistance for some decades [89]. A vast amount of literature is available on the emergence of resistance and the use of antimicrobials in medicine, veterinary medicine, and animal husbandry. There are numerous reports on the occurrence of resistant bacteria in the aquatic environment [3]. Bacteria are either resistant by nature or they may have become so through the use of antibiotics, or by uptake of genetic material encoding resistance (e.g., in hospital effluent) in the environment before they reach a sewage treatment plant or the soil. The transfer of resistant bacteria to humans may occur via water or food if plants are watered with surface water or sewage sludge, if manure is used as a fertilizer, or if resistant bacteria are present in meat. The selection pressure against microbial biocoenosis due to the presence of antibiotics above a certain concentration is an important factor in the selection and spread of resistant bacteria. Transfer of resistance genes as well as of bacteria already themselves resistant is particularly favored by the presence of antibiotics over a long period and in sub-therapeutic concentrations.

The development of resistance through the input of antibiotics into the environment is a new issue in this discussion. In general, the emergence of resistance is a highly complex process which is not yet fully understood in respect of the significance of the interaction of bacterial populations and antibiotics, even in a medicinal environment. On the one hand, knowledge of effects of sub-inhibitory concentrations of antimicrobials on environmental bacteria is scarce and contradictory, especially with respect to resistance. On the other hand, there is a huge volume of evidence that antibiotic resistance is already present in natural environments and that it is exchanged between bacteria [90]. The transfer, as well as the new combination of resistance genes, will preferably happen in compartments with high bacterial density. Such conditions can be found at the interface between water and particulate matter, in biofilms such as sewage sludge flocs, at the inner surface of sewage pipes, and generally at any surface in natural waters.

The link between the presence of antimicrobial agents and the favoring of resistant bacteria, as well as the transfer of resistance at concentrations as low as those found for antimicrobials in the environment has not been shown, yet. On the contrary, it has been shown that the transfer of resistance and the selection of resistant bacteria are not favored at the high antibiotic concentrations found in hospital effluents [88] and the lower levels found in the aquatic environment. Gentamicin improved the transfer rate of resistance in staphylococci at a concentration of 100 µg L^{-1}. Others such as macrolides, quinolones, or vancomycin did not have such an impact. In our own investigations we found that multi-resistant *Acinetobacter* was no longer detectable after two weeks in the presence of a mix of several important antibiotics at a concentration 100-fold above that expected in the aquatic environment [91]. Therefore, according to the present state of our knowledge, the input of bacteria that have already become resistant through the application of antibiotics seems to be responsible for the spread of resistant bacteria in the environment. The development of resistance in the environment at environmental concentrations of antibiotics is far less likely.

3.6.2
Effects on Higher Aquatic Organisms

So far, the effects of antibiotics on aquatic organisms have only been studied for a few compounds and on only a few types of organisms. Because of the use of antibiotics in aquaculture, the body of knowledge covering their effects on fish is adequate, but few data are available on what happens to other aquatic organisms, such as daphnia or algae. In most cases, it is not known whether the standardized tests usually applied are applicable to antibiotics. Furthermore, little is known about the physiology of these organisms in relation to the mode of action of environmental chemicals. Acute effect concentrations (EC_{50}) are in the mg L^{-1}-range [92]. Algae seem to be the aquatic organisms most sensitive to antibiotics [48, 92–93]. Chronic EC_{50}-values in the lower µg L^{-1}-range have been reported for algae and daphnia (for effects on bacteria see above). Therefore, a better understanding of the effects of antibiotics on non-target environmental organisms is needed for improved risk assessment. This in turn will allow for better and more targeted risk management.

3.7
Conclusion

It has been shown that antibiotics used in human and veterinary medicine as well as for other purposes enter the aquatic environment. The measured concentrations range from lower µg L^{-1} in raw sewage to higher ng L^{-1} in surface water. Therefore, one could assume a steady state between input and elimination in the different environmental compartments such as lakes and rivers. This could be called pseudo persistence or second-order persistence [94]. The concentrations found in the environment are the result of the relative stability, the amounts used, and the metabolism that is the rate at which the compound is excreted. That is, even if the concentration in a specific environmental compartment is the same, the contribution of the above-mentioned factors may be different.

Most of the antibiotics are not satisfactorily biodegraded either in STPs or in the aquatic environment. To reduce the discharge of antibiotics with municipal wastewater, the treatment of effluent using (photochemical) oxidation processes and membrane filtration technology as well as filtration and reverse osmosis have been described. It has been found that none of these techniques works well for all compounds.

Therefore, it is advisable to minimize the input of antibiotics into the environment. Several approaches can be used. The appropriate training of medical doctors and pharmacists is necessary, and the issue should be included in the curricula for medicine and pharmaceutical sciences. The public has to be better informed. The pharmaceutical industry could help to promote this. It has been estimated that a 50% reduction in the use of antibiotics is possible in human medicine [4, 95]. It is recommended that microbiological analysis should be improved, leading to improvements in prescription practice based on these results. Additionally, drug remains should not be disposed of down drains. A cradle-to-the-grave stewardship is necessary [94, 96]. The issue of "pharmaceuticals in the environment" should be incorporated into the curricula of medical doctors and pharmacists. Chronopharmacology has shown that pharmaceuticals are much more effective if they are administered at the right moment than on a three times daily basis [97]. Fewer drugs may be required to obtain the same effect if administered at the proper time. It should be established whether the concept of chronopharmacology [97] is applicable to antibiotics. It would offer advantages to both patients and the environment. In the long run, the development of new galenic forms giving improved drug targeting and on-site delivery may also help to reduce the input of antibiotics into the aquatic environment.

Acknowledgments

The authors thank Gerd Hamscher and Christian Daughton for fruitful discussions. Bent Halling-Sørensen, Mark Montforts, Chris Metcalfe, Michael Focazio, and Ettore Zuccato provided additional information on the use of antibiotics in different countries.

References

1. B. Halling-Sørensen, N. Nilesen, P. F. Lanzky, F. Ingerslev, C. Holten-Lutzhöft, S. E. Jørgensen, *Chemosphere* **1998**, 36, 357–393.
2. K. Kümmerer, *Pharmaceuticals in the Environment: Sources, Fate, Effects and Risks*. 2nd ed. (Ed.: K. Kümmerer) Springer Heidelberg Berlin **2004**, pp. 3.
3. K. Kümmerer, *J. Antimicrob. Chemother.* **2004**, 54, 311–320.
4. R. Wise, *J. Antimicrob. Chemother.* **2002**, 49, 585–586.
5. FEDESA (European Federation of Animal Health), *Antibiotics and Animals*. FEDESA/FEFANA Press Release Brussels, **1997**, 8 September.
6. FEDESA (European Federation of Animal Health), *Antibiotic Use in Farm Animals does not threaten Human Health*. FEDESA/FEFANA Press Release Brussels, **2001**, 13 July.
7. H. Goossens, M. Ferech, R. V. Stichele, Lancet **2005**, 365, 579–587.
8. K. Kümmerer, A. Henninger, Clin. Microbiol. Inf. **2003**, 9, 1203–1214.
9. S. Mölstad, C. S. Lundborg, A. K. Karlsson, O. Cars, *Scand. J. Infec. Dis.* **2002**, 34, 366–371.
10. R. Alexy, A. Sommer, F. T. Lange, K. Kümmerer, submitted.
11. House of Lords, Select Committee on Science and Technology. 7th Report. The Stationary Office London **1998**.

12 Union of Concerned Scientists, *70 percent of all antibiotics given to healthy lifestock*. Press Release. 8 January **2001**. Cambridge, MA, USA.

13 D. Ferber, Science **2003**, 301, 1027.

14 X-S. Miao, F. Bishay, M. Chen, C. D. Metcalfe, Environ. Sci Technol. **2004**, 38, 3533–3541.

15 A. C. Alder, C. S. McArdell, E. M. Golet, H-P. E. Kohler, E. Molnar, Ngoc Anh Pham Thi, H. Siegrist, M. J.-F. Suter, W. Giger, *Pharmaceuticals in the Environment: Sources, Fate, Effects and Risks*. 2nd ed. (Ed.: K. Kümmerer) Springer. Heidelberg Berlin, **2004**, 55–66.

16 P. Greiner, I. Rönnefahrt, *Management of environmental risks in the life cycle of pharmaceuticals. European Conference on Human and Veterinary Pharmaceuticals in the Environment*. Lyon, 14–16 April **2003**, http://www.envirpharma.org/ presentation/speeche/ ronnefahrt.pdf (access 29 June 2005).

17 L. Guardabassi, A. Dalsgaard, *Occurrence and fate of antibiotic resistant bacteria in sewage*. Environmental Project No. 722 002 Miljøprojekt. Danish Environmental Protection Agency.

18 F. Bager, H.-D. Emborg, DANMAP 2000. *Consumption of antimicrobial agents and occurrence of antimicrobial resistance in bacteria from food animals, foods and humans in Denmark*. Copenhagen **2001**.

19 S. Sattelberger, *Arzneimittelrückstände in der Umwelt, Bestandsaufnahme und Problemstellung*. Report des Umweltbundesamtes Österreich, Wien, **1999**.

20 J. G. M. Derksen, G. M. Eijnatten, J. Lahr, P van der Linde, A. G. M. Kroon, *Environmental Effects of Human Pharmaceuticals. The presence and Risks*. RIZA Report 2001.051, Association of River Water Works – RIWA and Institute for Inland Water Management and Wastewater Treatment – RIZA, July **2002**.

21 Ministerie van Verkeer en Waterstaat, (Ed.) *Humane en veterinaire geneesmiddelen in Nederlands oppervlaktewater en afvalwater*. RIZA rapport 023, **2003**, Lelystadt.

22 H. A. Verbrugh, A. J. de Neeling, Eds. *Consumption of antimicrobial agents and antimicrobial resistance among medically important bacteria in the Netherlands*. SWAB NETHMAP, **2003**.

23 F. Sacher, P. G. Stoks, *Pharmaceutical Residues in Waters in the Netherlands. Results of a monitoring programme for RIWA*. Nieuwegen, **2003**.

24 D. Calamari, E. Zuccato, S. Castiglioni, R. Bagnati, R. Fanelli, Environ Sci. Technol. **2003**, 37, 1241–1248.

25 E. Zuccato, D. Calamari, M. Natangelo, R. Fanelli, Lancet. **2000**, 355, 1789–1790.

26 S. Castiglioni, R. Fanelli, D. Calamari, R. Bagnati, E. Zuccato, Regul. Toxicol. Pharmacol. **2004**, 39, 25–32.

27 S. Castiglioni, E. Zuccato, D. Calamari, *Comparison of the concentrations of pharmaceuticals in STPs and rivers in Italy as a tool for investigating their environmental distribution and fate*. SETAC Europe 14th Annual Meeting, Prague, Czech Republic, 18–22 April **2004**, Proceedings p. 61.

28 H. Lüllmann, K. Mohr, *Taschenatlas der Pharmakologie*. Thieme Stuttgart, **2001**.

29 E. A. Stadlbauer. GIT Labor-Fachzeitschrift. **2004**, 1, 8–10.

30 A. B. A. Boxall, P. Kay, P. A. Blackwell, L. A. Fogg in *Pharmaceuticals in the Environment: Sources, Fate, Effects and Risks*. 2nd ed. (Ed.: K. Kümmerer) Springer, Heidelberg, Berlin, **2004**, pp. 166.

31 B. T. Lunestad, J. Goksøyr, Dis. Aquat. Org. **1990**, 9, 67–72.

32 T. Christian, R. J. Schneider, H. A. Färber, D. Skutlarek, M. T. Meyer, H. E. Goldbach, Acta Hydrochim. Hydrobiol. **2003**, 31, 36–44.

33 J. Tolls, Environ. Sci. Technol. **2001**, 35, 3397–3406.

34 V. L. Cunningham in *Pharmaceuticals in the Environment: Sources, Fate, Effects and Risks*. 2nd ed. (Ed.:K. Kümmerer), Springer Heidelberg Berlin **2004**, pp. 13.

35 Howard & Meylan, J. Pharm. Sci. **1995**, 84, 83–92.

36 J. Wiethan, A. Al-Ahmad, A. Henninger, K. Kümmerer, Vom Wasser **2000**, 95, 107–118.

37 E. M. Golet, A. C. Alder, W. Giger, Environ. Sci. Technol. **2002**, 36, 3645–3651.

38 B. T. Lunestad, O. B. Samuelsen, S. Fjelde, A. Ervik, *Aquaculture* **1995**, 134, 217–225.
39 J. Burhenne, M. Ludwig, P. Nikoloudis, M. Spiteller, *Environ. Sci. Pollut. Res.* **1997**, 4, 10–15.
40 J. Burhenne, M. Ludwig, M. Spiteller, *Environ. Sci. Pollut. Res.* **1997**, 4, 61–71.
41 A. L. Boree, W. A. Arnold, K. McNeill, *Environ. Sci. Technol.* **2004**, 38, 3933–3940.
42 O. B. Samuelsen, *Aquaculture* **1989**, 83, 7–16.
43 B. Halling–Sørensen, G. Sengelov, J. Tjørnelund, *Arch. Environ. Contam. Toxicol.* **2002**, 42, 263–271.
44 H. Oka, Y. Ikai, N. Kawamura, M. Yamada, K. Harada, S. Ito, M. Suzuki, *J. Agric. Food Chem.* **1989**, 37, 226–231.
45 H. Pouliquen, H. Le Bris, L. Pinault, *Mar. Ecol. Progr. Ser.* **1992**, 89, 93–98.
46 K. Froehner, T. Backhaus, L.H. Grimme, *Chemosphere* **2000**, 40, 821–828.
47 K. Kümmerer, R. Alexy, J. Hüttig, A. Schöll, *Water Res.* **2004**, 38, 2111–2116.
48 B. Halling-Sørensen, *Chemosphere* **2000**, 40, 731–739.
49 K. Kümmerer, A. Al-Ahmad, A. Henninger. *Acta Hydrochim. Hydrobiol.* **2002**, 30, 171–178.
50 A. Al-Ahmad, M. Wiedmann-Al-Ahmad, G. Schön, F. D. Daschner, K. Kümmerer. *Bull. Environ. Contam. Toxicol.* **2000**, 64, 764–770.
51 K. Kümmerer, R. Alexy, T. Kümpel, A. Schöll, W. Kalsch, T. Junker, F. Moltmann, T. Knacker, W. Theis, C. Weihs, E. Urich, S. Gartiser, M. Metzinger, M. Wenz, T. Lange, C. Beimfohr, (2003): *Eintrag von Antibiotika in die aquatische Umwelt. Prüfung der biologischen Abbaubarkeit ausgewählter Antibiotika, ihr Vorkommen im Abwasser und ihr möglicher Einfluss auf die Reinigungsleistung kommunaler Kläranlagen. Identifizierung von Risikofeldern. Abschlussbericht,* F&E–Vorhaben **2003**, 298 63 722.
52 A. Al-Ahmad, F. D. Daschner, K. Kümmerer, *Arch. Environ. Contam. Toxicol.* **1999**, 37, 158–63.
53 K. Kümmerer, A. Al-Ahmad, V. Mersch-Sundermann, *Chemosphere* **2000**, 40, 701–10.
54 T. Junker, R. Alexy, T. Knacker, K. Kümmerer. *Environ. Sci. Technol.* **2006**, 40, 318–326.
55 R. Alexy, K. Kümmerer, *Korresp. Wasser-Abwasser* **2005**, 52, 563–571.
56 B. Halling-Sørensen, H. Ch. Lützhoft, H. R. Andersen, F. Ingerslev, *J. Antimicrob. Chemother.* **2000**, 46, 53–58.
57 F. Ingerslev, B. Halling-Sørensen, *Environ. Tox. Chem.* **2000**, 19, 2467–2473.
58 F. Ingerslev, L. Torang, M. L. Loke, B. Halling-Sørensen, N. Nyholm, *Chemosphere* **2001**, 44, 865–872.
59 A. Hartmann, A. C. Alder, T. Koller, R. M. Widmer, *Environ. Toxicol. Chem.* **1998**, 17, 377–382.
60 K. Kümmerer, *Chemosphere* **2001**, 45, 957–69.
61 H. Färber, *Hyg. Med.* **2002**, 27, 35.
62 R. Hirsch, T. Ternes, K. Haberer, K. L. Kratz, *Sci. Total Environ.* **1999**, 225.
63 W. Giger, A. C. Alder, E. M. Golet, H.-P. E. Kohler, C. S. McArdell, E. Molnar, H. Siegrist, M. J. F. Suter, *Chimia* **2003**, 57, 9, 485–491.
64 E. M. Golet, A. Strehler, A. C. Alder, W. Giger, *Anal. Chem.* **2002**, 74, 5455–5462.
65 K. Ohlsen, T. Ternes, G. Werner, D. Löffler, W. Witte, J. Hacker, In: T. Track, G. Kreysa (Ed.) *Spurenstoffe in Gewässern. Pharmazeutische Reststoffe und endokrin wirksame Substanzen.* Wiley-VCH GmbH & Co. **2003**, 197–209.
66 A. Thomsen, A. Göbel, C. S. McArdell, A. C. Alder, W. Giger, Poseidon Symposium Braunschweig, 4–5 November **2003**, proceedings, p. 71.
67 M. Carballa, F. Omil, J. M. Lema, M. Llompart, C. García, Poseidon Symposium Braunschweig, 4–5 November **2003**, proceedings, p. 72.
68 C. S. McArdell, M. Clara, A. C. Alder, A. Göbel, A. Joss, E. Keller, N. Kreuzinger, H. Plüss, H. Siegrist, B. Strenn, Poseidon Symposium Braunschweig, 4–5 November **2003**, proceedings, p. 24.
69 J. Qiting, Z. Xiheng, *Water Treat.* **1988**, 3, 285–291.
70 T. Kiffmeyer, *Minimisation of human drug input by oxidative treatment of toilet effluents from hospital wards.* European Conference on Human and Veterinary

Pharmaceuticals in the Environment. Lyon, 14–16 April **2003**, can be found under http://www.envirpharma.org/presentation/speeche/ kiffmeyer.pdf (access 29 June 2005).
71 T. A. Ternes, J. Stuber, N. Herrmann, D. McDowell, A. Ried, M. Kampmann, B. Teiser, *Wat. Res.* **2003**, 37, 1976–1982.
72 W. Blüm, Ch. S. McArdell, E. Hoehn, R. Schabhut, *Organische Spurenstoffe im Grundwasser des Limmattales – Ergebnisse der Untersuchungen der Untersuchungskampagne* 2004, Eawag, Kantonales Labor Zürich, AWEL Amt für Abfall, Wasser, Energie und Luft, Zürich, June **2005**.
73 F. Sacher, F. T. Lange, H. J. Brauch, I. Blankenhorn, *J. Chromatogr. A* **2001**, 938, 199–210.
74 D. Kolpin, E. T. Furlong, M. T. Meyer, E. M. Thurman, S. D. Zaugg, L. B. Barber, et al., *Environ. Sci. Technol.* **2002**, 36, 1202–1211.
75 P. Jacobsen, L. Berglind, *Aquaculture* **1988**, 70, 365–370.
76 R. Coyne, M. Hiney, B. O' Conner, D. Cazabon, P. Smith, *Aquaculture* **1994**, 123, 31–42.
77 L. Migliore, C. Civitareale, G. Brambilla, G. Dojmi di Delupis, *Wat. Res.* **1997**, 31, 1801–1806.
78 H. Hektoen, J. A. Berge, V. Hormazabal, M. Yndestad, *Aquaculture* **1995**, 133, 175–184.
79 J. R. Marengo, R. A. O' Brian, R. R. Velagaleti, J. M. Stamm, *Environ. Toxicol. Chem.* **1997**, 16, 462–471.
80 O. B. Samuelsen, V. Torsvik, A. Ervik, *Sci. Total Environ.* **1992**, 114, 25–36.
81 O. B. Samuelsen, B. T. Lunestad, S. Fjelde, *Aquaculture* **1994**,126, 183–290.
82 P. K. Hansen, B. T. Lunestad, O. B. Samuelsen, *Can. J. Microbiol.* **1992**, 38, 1307–1312.
83 D. G. Capone, D. P. Weston, V. Miller, C. Shoemaker, *Aquaculture* **1996**, 145, 55–75.
84 G. Hamscher, S. Sczesny, H. Höper, H. Nau, *Anal. Chem.* **2002**, 74, 1509–1518.
85 J. V. Holm, K. Rugge, P. L. Bjerg, T. H. Christensen, *Environ. Sci. Technol.* **1995**, 29, 1415–1420.
86 T. A. Ternes, B. Teiser, N. Herrmann, 2003, Poseidon Symposium Braunschweig, 4–5 November **2003**, Proceedings, p. 21.
87 S. Korhonen, N. Lindqvist, J. Ristimäki, V.-P. Perttinä, T. Tuhkanen, Removal of PPCPs in Finnish Waters with high DOC Tagungsband Brauschweig **2003**, S. 33.
88 K. Ohlsen, G. Werner, T. Ternes, W. Ziebuhr, W. Witte, J. Hacker, *Environ. Microbiol.* **2003**, 5, 711–716.
89 L. E. Edquist, K.B. Pedersen In: *Late lesson from early warnings: the precautionary principle 1896–2000.* P. Harremoëes, D. Gee, M. MacGarvin, A. Stirling, J. Keys, B. Wynne, Vas S. Guedes, Environmental Issue Report No. 22, European Environment Agency. Copenhagen. **2001**, 93–109.
90 J. Davison, *Plasmid.* **1999**, 42, 73–91.
91 http://www.epa.gov/nerlesd1/chemistry/ppcp/images/acs/klauskuemmerer.mht
92 S. Webb, In: *Pharmaceuticals in the Environment. Sources, Fate, Effects and Risks.* Kümmerer K, Ed. Springer Heidelberg Berlin. **2004**, 2nd ed.
93 P. F. Lanzky, B. Halling-Sørensen, *Chemosphere* **1997**, 35, 2553–2561.
94 C. G. Daughton, *Environ. Health Perspect.* **2003**, 111, 757–774.
95 P.F. Harrison, J. Lederberg, (Eds.) *Antimicrobial Resistance, Issues and Options.* Washington DC, National Academic Press, **1998**.
96 C.G. Daughton, *In: Pharmaceuticals in the Environment.* K. Kümmerer (Ed.) Springer. **2004**, 2nd ed.
97 B. Lemmer, *Chronopharmakologie. Tagesrhythmen und Arzneimittelwirkung.* Wissenschaftliche Verlagsgesellschaft Stuttgart, **2004**, 3rd ed.
98 H. L. Boreen, X. A. Arnold, K. McNeill, Photochemical fate of sulfa drugs in the aquatic environment. Sulfa drugs containing five-membered heterocyclic groups. *Environ. Sci. Technol.*, 2004, 38, 3933–3940.

4
Iodinated X-ray Contrast Media

Anke Putschew and Martin Jekel

4.1
Introduction

2,4,6-Triiodinated benzoic acid derivatives are used as X-ray contrast media and are applied in large quantities in hospitals and X-ray practices. Figure 4.1 shows the structures of some selected iodinated X-ray contrast media (ICM). The triiodinated benzene derivatives are applied to enhance the contrast between vessels or organs which otherwise could not be examined by X-ray.

The iodine atoms bound in the molecules are responsible for the absorption of X-rays. The human and biological properties are controlled by the side-chains in positions 1, 3, and 5, and, as required by the application, the compounds are designed to be very polar and persistent. The high polarity guarantees that the com-

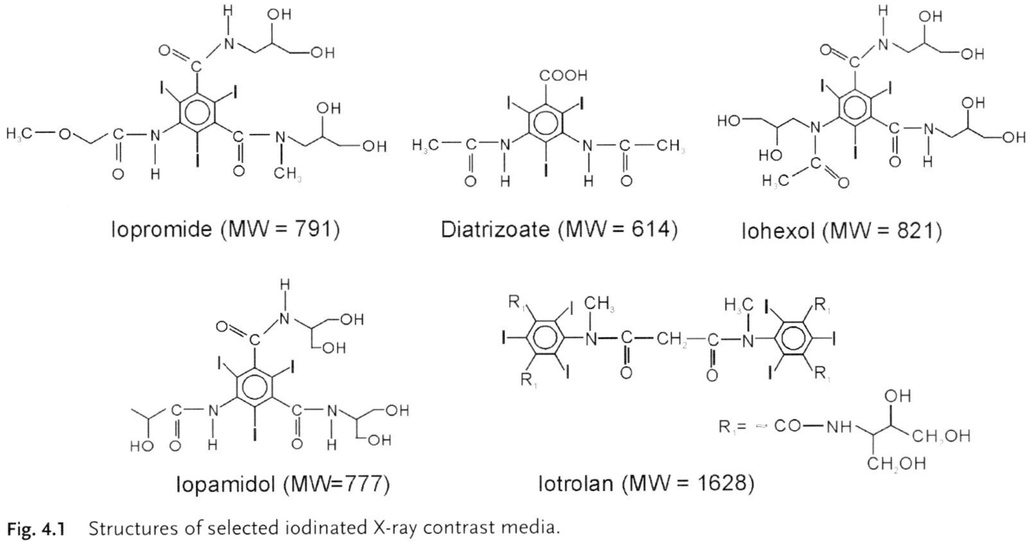

Fig. 4.1 Structures of selected iodinated X-ray contrast media.

Organic Pollutants in the Water Cycle. T. Reemtsma and M. Jekel (Eds.).

ISBN: 3-527-31297-8

pounds are excreted within a few hours after application. The stability is required to avoid interactions after application. Both properties guarantee rapid excretion of the unmetabolized ICM via urine within 24 h after application [1].

ICM can be divided up into ionic (e.g., Diatrizoate) and non-ionic (e.g., Iopromide, Iohexol) compounds. Non-ionic contrast media are applied, e.g., for angiography and urography. Of all applied ICM, 90% are non-ionic [2], and for one examination more than 100 g ICM is applied [1]. The world-wide consumption of the iodinated compounds is about 3500 t year^{-1} [3]. In Germany, 360 t year^{-1} ICM are applied; the consumption of Iopromide, the most prominent X-ray contrast media of this class, is estimated to be 130 t year^{-1} in Germany (personal communication, Schering AG, Berlin). Other data can be found in the BLAC study [4]. Here it is reported, based on sales data, that the consumption of Iopromide in Germany is around 65 t year^{-1}. It is not easy to get production and sales data, but the data available clearly indicate that ICM are used in large quantities for diagnostic purposes. For comparison, 86 t Diclofenac were sold in Germany in 2001 [4].

A risk assessment performed for the widely used ICM Iopromide revealed no toxic effects in short-term toxicity tests with bacteria, crustaceans, and fish for 10 g L^{-1} Iopromide, the highest concentration tested [5]. Even a chronic test with *D. magna* showed no effect for concentrations up to 1 g L^{-1} Iopromide [5]. In the same study, it was shown that Iopromide is not readily biodegradable, which is confirmed by many other studies [2, 6, 7, 8]. The predicted environmental concentration (PEC) of Iopromide is calculated to be 2 µg L^{-1}, and the ratio between the PEC and the predicted no-effect concentration (PNEC) is PEC/PNEC <0.0002. Thus, no environmental risk is to be expected from Iopromide [5]. ICM are expected to remain in the water phase because of the high polarity, as shown by the low log P_{OW} values: Diatrizoate <–1 (butanol/water), Iohexol –1.22 (estimated), Iotrolan –3.65, and Iopromide –2.35 [5].

**4.2
Source**

ICM are applied in hospitals and X-ray practices and are excreted unaltered within 24 h in the patients urine. Based on an applied dose of 100 g of ICM per patient, the concentration in urine should be around 50 g L^{-1} in 24 h urine if a volume of 2 L of urine is excreted. The maximum iodine concentration in urine is found even within the first hour after application and ranges between 20 and 70 g L^{-1} [9].

The sum parameter adsorbable organic halogens (AOX) is widely used in wastewater analysis as well as in environmental analysis for the quantification of organically bound halogens. The conventional AOX analysis combines carbon adsorption and combustion, followed by microcoulometric detection of the produced halides. As early as 1996 it was recognized that hospital wastewaters are characterized by high AOX contaminations (0.1–0.9 mg L^{-1}), most of the AOX being generated by ICM (0.1–0.5 mg L^{-1}, [7]). The mean AOX concentration of hospital wastewater is reported to be 0.41 mg L^{-1} AOX, which is equal to ~1.47 mg L^{-1} adsorbable or-

Tab. 4.1 Hospital wastewater flow, AOI and AOI load.

Date	Flow rate (m³)	AOI (µg L^{-1})	AOI load (g d^{-1})
Tue	428.4	346.0	148.2
Wed	1143.5	34.2	39.1
Thu	425.7	377.4	160.7
Fri	348.7	629.7	219.6
Sat	446.3	1102.1	491.9
Sun	137.5	44.9	6.2
Mon	570.2	24.6	14.0

ganic iodine (AOI) [2] assuming that the AOX is formed solely by iodinated compounds. In the conventional AOX analysis with microcoulometric detection it is not possible to distinguish between organic chlorine, bromine and iodine. A method permitting this distinction was developed by Oleksy-Frenzel et al. [10, 11], where the combustion gas containing HX is trapped and the halides Cl$^-$, Br$^-$, and I$^-$ determined by ion chromatography. The differential AOX analysis was used for the monitoring of ICM in wastewater of a small hospital (300 beds, Berlin, unpublished). Over one week, the applied amount of ICM was documented and 24-h mixed hospital wastewater samples were taken and analyzed for iodinated organic compounds (as AOI, differentiation of the AOX into AOCl, AOBr, and AOI [10, 11]). The wastewater flow was monitored, and thus it was possible to calculate the discharged AOI load, which can be compared with the applied amounts of ICM. Within one week, 60.4 g Diatrizoate, 245.0 g Iomeprol, 1464.0 g Iopromide, and 191.0 g Ioversol were applied. The applied ICM amounted to 980 g adsorbable organic iodine (AOI). Based on the measured data (Table 4.1), the hospital has discharged 1080 g AOI within one week.

The difference between the applied AOI as ICM and the amount of AOI detected in the wastewater is 11%. For the deviation, analytical errors as well as documentation errors of applied ICM can be responsible, or other iodinated organic compounds are discharged. But nevertheless it could be shown that the AOI of hospital wastewater is mainly formed by ICM. With the patients' urine, the ICM are transported via the public sewer system to municipal wastewater treatment plants.

4.3
Wastewater Treatment

Figure 4.2 shows the AOI of the influent and effluent of a wastewater treatment plant (WWTP). First of all, it can be seen that the AOI cannot be reduced by wastewater treatment to a significant extent. The reduction rate is only around 20%. The high stability of the compounds or at least the iodinated aromatic ring is most probably caused by the three iodine atoms bound in the 2, 4, and 6 positions.

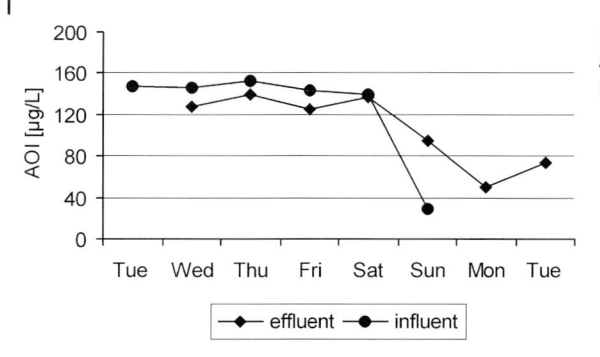

Fig. 4.2 Influent and effluent AOI of a wastewater treatment plant (24-h mixed samples).

The AOI influent concentration varies between 140 and 153 µg L^{-1} during the working days; lower concentrations are detected at the weekend, because the ICM are only applied during working days. The AOI of the effluent is around 130 µg L^{-1} during the working days, and thus high amounts of AOI and ICM are discharged into the aquatic environment. Similar results, including the profile over the week, are published by Oleksy-Frenzel et al. [10, 11]. The persistence of the iodinated compounds during wastewater treatment is also documented by single-compound analysis [12, 13]. In Table 4.2, corresponding influent and effluent concentrations of ICM of a German municipal WWTP are given [12]. The influent and effluent concentrations, varying between 0.14 and 8.1 µg L^{-1}, depending on the ICM, are similar, and thus the WWTP is not able to remove these compounds.

Table 4.3 gives an overview of ICM concentrations of German municipal WWTP effluent. The mean values are in the lower µg L^{-1} range, but maximum concentrations of up to 15 µg L^{-1} were detected.

A WWTP effluent analyzed by Hirsch et al. [14] was contaminated with 3.07 µg L^{-1} Iopromide, 1.14 µg L^{-1} Diatrizoate and 0.59 µg L^{-1} of Iopamidol (grab sample). In an effluent of a WWTP receiving a high load of hospital wastewater, high concentrations of 20 µg L^{-1} Iopromide and 13 µg L^{-1} Diatrizoate were detected (24-h mixed samples [15, 16]). The general properties of the ICM as well as the data shown indicate that ICM cannot be removed by activated-sludge municipal waste-

Tab. 4.2 Concentration of ICM in the influent and final effluent of a municipal WWTP, daily composite samples (flow proportional) [12].

	Mean concentration (µg L^{-1})	
	Influent	Effluent
Iopamidol	4.3 ± 0.9	4.7 ± 1.0
Diatrizoate	3.3 ± 0.7	4.1 ± 0.8
Iothalamic acid	0.18 ± 0.1	0.14 ± 0.1
Ioxithalamic acid	0.17 ± 0.1	0.16 ± 0.1
Iomeprol	1.6 ± 0.4	1.3 ± 0.3
Iopromide	7.5 ± 1.5	8.1 ± 1.6

Tab. 4.3 Concentration of ICM in the effluent of different municipal WWTP, daily composite samples (time proportional) [12]; LOQ = 0.05 µg L.

	Number of WWTP	Median ($\mu g\ L^{-1}$)	90-Percentile ($\mu g\ L^{-1}$)	Maximum ($\mu g\ L^{-1}$)
Iopamidol	25	0.66	8.0	15
Diatrizoate	24	0.75	4.4	11
Iothalamic acid	21	0.25	1.5	8.7
Ioxithalamic acid	14	<LOQ	0.12	0.64
Iomeprol	14	<LOQ	0.10	0.21
Iopromide	12	0.37	2.8	3.8

water treatment to an adequate extent. Up to now it has not been possible to identify or detect transformation products in WWTP effluents, but, based on the AOI, data, a partial deiodination may be possible. The influent and effluent concentrations of the ICM can be classified as very high. Effluent concentrations of drugs like Diclofenac (0.81 µg L^{-1} median, max 2.1µg L^{-1}) or Ibuprofen (0.37 µg L^{-1} median, 3.4 µg L^{-1} max) are generally lower [17]. The complexing agent EDTA is one of the rare compounds showing higher effluent concentrations (10–1000 µg L^{-1}) than ICM [18].

4.4
Receiving Water

The ICM are released into the aquatic environment with the WWTP effluents. In Table 4.4, surface water concentrations of selected ICM are summarized. The concentrations given for Berlin [19] are mean values over nearly one year, the others being single measurements. Compared to other regions of Germany, slightly higher concentrations of ICM can be expected in surface waters of Berlin, because wastewater discharges become less diluted in surface waters than elsewhere.

The behavior of the iodinated compounds in the environment was studied in a partially closed water cycle in Berlin (Fig. 4.3 [15, 16, 19]). In this system, the effluent of a tertiary WWTP that receives hospital wastewater is directed to a lake (Tegeler See) via a receiving channel. This lake water, after bank filtration over several weeks, becomes a part of the raw water used for drinking water production.

Tab. 4.4 Concentration of ICM in surface water.

Compound	Central Germany rivers and creeks [12]	River Rhine [14]	Surface water (Tegeler See, Berlin) [19]
Iopromide [µg L^{-1}]	0.01–0.9	0.15	0.86
Diatrizoate [µg L^{-1}]	0.08–100	0.14	0.96
Iopamidol [µg L^{-1}]	0.17–2.8	0.18	0.29

Tab. 4.5 Concentration of ICM and AOI in grab samples of the studied site (Fig. 4.3)

Sample	Diatrizoate ($\mu g\ L^{-1}$)	Iopromide ($\mu g\ L^{-1}$)	Iohexol ($\mu g\ L^{-1}$)	AOI ($\mu g\ L^{-1}$)	% AOI due to identified ICM
Effluent[a]	14	22	7	57.4	38
Receiving channel	7.2	8.5	1.5	32.9	28
Tegeler See	1.2	2.0	0.5	12.9	20

[a] 24-h mixed sample [15, 16]

In Table 4.5, the concentration of ICM and the AOI are summarized. The ICM concentration and the AOI of the WWTP effluent are very high. Within the water cycle, the concentrations of the ICM are reduced, but still remain elevated. In the receiving channel, the concentrations vary between 1.5 and 8.5 $\mu g\ L^{-1}$, and in the surface water Tegeler See between 0.5 and 2.0 $\mu g\ L^{-1}$, depending on the compound. The concentration decrease is mainly a result of dilution, but a transformation of ICM to other iodinated compounds may even occur, because the AOI is less reduced than the concentrations of the individual ICM. Thus, the contribution of the known ICM to the AOI decreases (see Table 4.5). However up to now it has not been possible to identify the products. This finding is in agreement with results of a laboratory study in which the transformation of Iopromide was determined in a sediment/water system [20] but the transformation products could not be identified.

In general it can be concluded that ICM occur in all surface waters which are influenced by WWTP effluents with concentrations in the upper ng L^{-1} or lower $\mu g\ L^{-1}$ range. To our knowledge up to now, data on the occurrence of ICM in other European Countries or the United States have not yet been published.

4.5
Groundwater/Exposed Groundwater

Leaching experiments with different soils (grain size distribution, pH, C_{org} content) indicate that groundwater contamination with ICM is very likely [21]. Under the experimental conditions, Iopromide was very mobile, which is in agreement with its high polarity (see Introduction). Correspondingly, ICM have been found in groundwaters in different regions of Germany. In Baden-Württemberg (southwest Germany) ICM were analyzed in 105 groundwater test points, whereby 25 are classified as exposed wells [12]. Iopamidol was found in 5 and Diatrizoate in 21 test points. Iopromide and Iomeprol were not detected. The maximum concentration of Iopamidol and Diatrizoate was 0.3 $\mu g\ L^{-1}$ and 1.1 $\mu g\ L^{-1}$, respectively. The study gives no information about the concentrations found in the groundwater that was not exposed. All other data found in literature are related to exposed groundwater. In mid-Germany, exposed groundwater is contaminated with 0.16 $\mu g\ L^{-1}$ Iopamidol and 0.03 $\mu g\ L^{-1}$ Diatrizoate (median values, $n = 17/10$), the concentrations of all other quantified ICM (Iopromide, Iothalamic acid, and Ioxithalamic acid) were below the limit of quantification of 0.01 µg L^{-1} [12].

In Berlin, groundwater concentrations were determined in connection with a study of the behavior during bank filtration [19]. At the lake Tegeler See a waterworks is located using bank filtered lake water for the production of drinking water. The quality of the groundwater was examined over one year via monitoring wells (see Fig. 4.3). Table 4.6 summarizes the average concentration of selected ICM and the AOI of the monitoring wells and of the lake water.

ICM were found at concentrations in the middle to lower ng L^{-1} range. The data in Table 4.6 indicate that the concentration of the ICM and the AOI is reduced during bank filtration. The main reduction is observed on the way between the lake and the first monitoring well 3311 under anoxic conditions. The data found here and in another study [23] indicate that ICM are (partly) deiodinated under anoxic conditions, but, besides transformation, dilution is responsible for the reduced concentrations (for details see [19]). Some unknown iodinated organic compounds could be detected in groundwater samples by selective detection of iodinated organic compounds (Fig. 4.4), but because of the low concentrations identification was not possible.

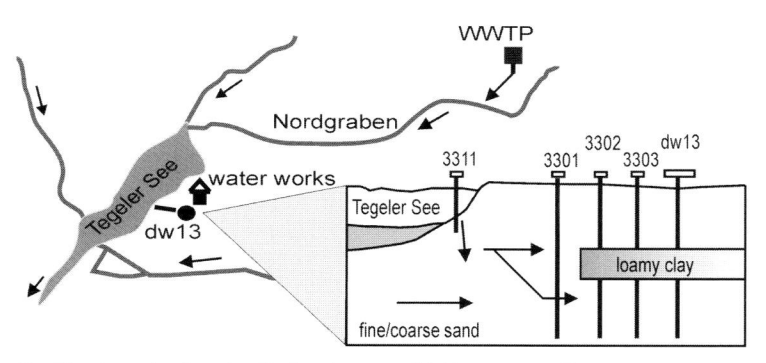

Fig. 4.3 Investigation site, Berlin, dw = raw drinking water well.

Tab. 4.6 Concentrations (in µg L^{-1}) of ICM along a bank filtration site (average values of grab samples [19]); dw = raw drinking water well.

Distance from lake: µg L^{-1}	lake	0 m 3311	40 m 3301	60 m 3302	80 m 3303	100 m dw
AOI (n=11)	9.6	6.9	3.6	3.8	4.2	2.3
Iopromide	0.86 (n=11)	0.22 (n=7)	0.04 (n=7)	0.07 (n=8)	0.04 (n=5)	0.02 (n=8)
Iopamidol	0.68 (n=8)	0.25 (n=5)	0.26 (n=4)	0.12 (n=4)	0.17 (n=3)	0.17 (n=5)
Iohexol	0.30 (n=11)	0.03 (n=6)	0.03 (n=3)	0.05 (n=2)	0.03 (n=2)	0.03 (n=3)
Diatrizoate	0.96 (n=11)	0.46 n=6)	0.29 (n=7)	0.61 (n=8)	0.30 (n=5)	0.20 (n=9)

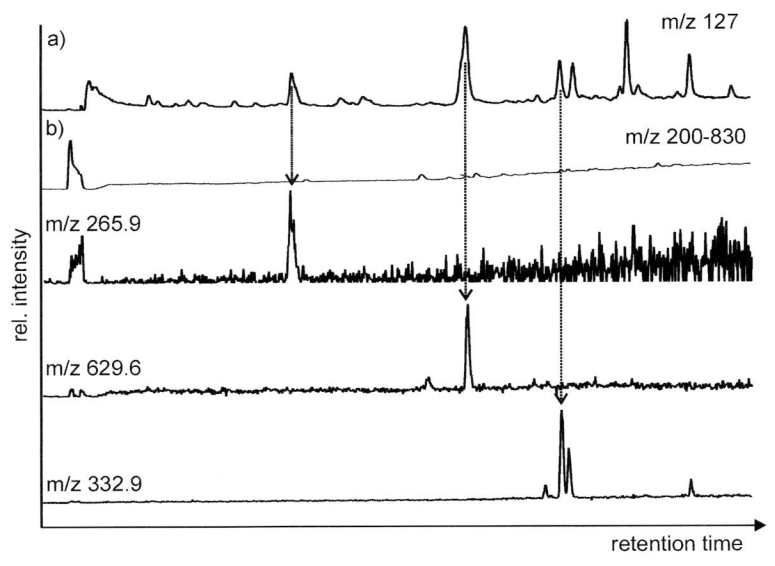

Fig. 4.4 Chromatograms of a groundwater sample. (a) selective detection of iodinated organic compounds, for details see [24]; (b) TIC of a scan and (below) mass chromatograms, most probably representing the molecular ions of unknown iodinated organic compounds.

4.6
Treatment

As outlined in Chapter 3, conventional municipal wastewater treatment is inappropriate to remove the ICM. For this reason, the potential of ozonation and advanced oxidation processes for the oxidation of ICM and other pharmaceuticals in drinking water [25] and wastewater has been studied [26, 27]. The removal efficiencies are in general very low for the ionic contrast media (Diatrizoate) and higher for the non-ionic compounds, e.g., Iopromide. With a dose of 5 mg L^{-1} O_3, Diatrizoate could not be removed; a dose of 15 mg L^{-1} leads to 14% removal of Diatrizoate [27]. For Iopromide the oxidation efficiency is higher, 34% of Iopromide is removed with an O_3 dose of 5 mg L^{-1}, and 84% with a dose of 15 mg L^{-1}. The treatment with O_3/H_2O_2 or O_3/UV can increase the removal of ICM by 10–20% [27]. Oxidation products were not investigated in the studies cited above. In case of ozonation, the concentration of the iodinated compounds can be reduced, but the AOI remains almost constant [28], and several iodinated transformation products with low concentrations were detected via selective detection of organically bound iodine [24, 29]. Photocatalytic degradation of ICM has also been investigated. With TiO_2 as catalyst, a fast transformation but not a complete deiodination of the ICM occurs [30, 31]. By irradiation of highly contaminated wastewater, e.g., hospital wastewater, deiodination of the ICM is achieved, and the released iodide is transformed to iodine by oxidation, and then can be recovered. The efficiency of the reaction depends strongly on the energy input [9, 32]. For the treatment of contaminated water at the production site of ICM, nanofiltration can be used [33] to minimize the AOX load and thus the disposal costs for discharging AOX (legal limit: 1 ppm). Ultrafiltration is not suitable for the retention of ICM [34].

A further treatment of ICM has just recently been published and is suited for the treatment of highly contaminated wastewater, like hospital wastewater and urine [35]. The reductive dehalogenation by zero-valent iron, which is used for the treatment of contaminated groundwater, is appropriate for the treatment of wastewaters with a high ICM load, like hospital wastewater or urine. The treatment of 1.6 g L^{-1} Iopromide in urine with 50 g L^{-1} iron at pH 2 showed an Iopromide removal of 43% within ca. 40 min, and after 24 h the Iopromide was completely removed (Fig. 4.5).

The Iopromide was not completely deiodinated, as only 63% of the theoretical amount of iodide was detected in the solution. Figure 4.6 shows chromatograms of the reaction solution obtained after 30-min and 24-h treatments. After 30 min, Iopromide is still detectable, the main transformation product being Iopromide that has lost one iodine atom. Minor products are Iopromide minus 2 and 3 iodine atoms. After 24 h, Iopromide is not detectable and the amount of the single deiodinated compound has decreased, while the amounts of the doubly and triply deiodinated Iopromides have increased.

This reaction will have to be examined in more detail, but it appears to be a promising, less energy-consuming treatment than the advanced oxidation procedure or membrane filtration. In further experiments the behavior of the partly deiodinated transformations products will be investigated, e.g., in biodegradation and ozonation.

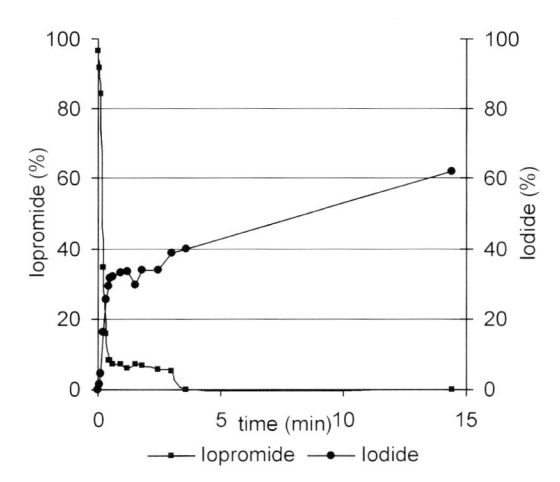

Fig. 4.5 Iopromide (1.6 g L^{-1}) in urine treated with 50 g L^{-1} iron at pH 2.

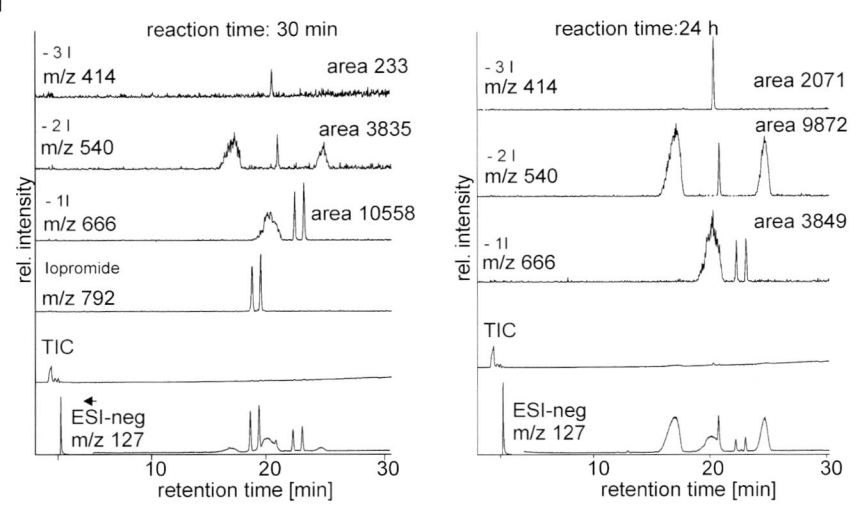

Fig. 4.6 Chromatograms (ESI-neg, m/z 127: iodinated organic compounds; ESI-pos, m/z 200–900: TIC) and mass chromatograms of (partly) deiodinated Iopromide in urine treated with zero-valent iron.

4.7
Summary

Iodinated X-ray contrast media (ICM) which are applied in large quantities (360 t year^{-1} in Germany) in hospitals and X-ray practices are excreted unaltered within 24 h after application. A mass balance showed that over one week a small hospital can discharge about 1 kg of organically bound iodine, which is equal to about of 500 g of X-ray contrast media.

With the patients' urine, the ICM are transported to the municipal wastewater treatment plants, which are not able to remove the compounds. Thus, the iodinated compounds are released into the aquatic environment along with the wastewater effluent. The ICM concentration of WWTP effluents, with values up to 20 µg L^{-1}, is higher than the effluent concentration of, e.g., Diclofenac or Ibuprofen.

The concentration of the ICM in surface waters is in the upper ng L^{-1} range. In exposed groundwater, contrast media could be found with concentrations in the middle to lower ng L^{-1} range. As indicated by the widespread occurrence of the X-ray contrast media, the compounds are very polar and persistent. The occurrence and behavior in surface water indicates that at least fractions of the compounds are transformed into products which are most probably still iodinated. During bank filtration, decreasing concentrations are recognized. This is mainly a result of dilution, whereby under anoxic condition a (partial) deiodination is possible. Several unknown iodinated organic compounds could be detected in groundwater samples by selective detection of organically bound iodine, but because of the low concentrations of these compounds identification was not possible.

The usual wastewater treatment techniques are inappropriate for the removal of ICM, and therefore the potential of ozonation and advanced oxidation processes for the oxidation of ICM and other pharmaceuticals in drinking water and wastewater have been studied. The removal efficiencies are in general poor for the ionic contrast media and higher for the non-ionic compounds. With an ozone dose of 15 mg L^{-1}, the concentration of Diatrizoate and Iopromide can be reduced by 14% and 84%, respectively. Treatment with O_3/H_2O_2 or O_3/UV can increase the removal of ICM by 10–20%. In the case of ozonation, the concentration of the iodinated compounds can be reduced, but the AOI remains almost constant. Photocatalytic degradation of ICM has also been investigated. With TiO_2 as catalyst, a fast transformation but not a complete deiodination of the ICM occurs. On irradiation of highly contaminated wastewater, e.g., hospital wastewater, deiodination of the ICM is recognized, and on oxidation the released iodide is transformed into iodine, which can then be recovered, but the efficiency of the reaction depends strongly on the energy input. Nanofiltration is a suitable method for the AOX reduction of contaminated water at the production site. A reductive dehalogenation of ICM by zero-valent iron, which is used for the treatment of contaminated groundwater, is appropriate for the treatment of highly ICM-contaminated wastewater, like hospital wastewater or urine. The reaction will have to be examined in more detail, but seems to be a promising, less energy-consuming treatment than AOP or photocatalytic processes.

References

1 Speck U., Hübner-Steiner H. *Pharmakologie und Toxikologie*, Eds. Oberdisse, Hackenthal, Kuschinsky, 2nd Ed., Springer-Verlag Berlin, Heidelberg, New York **1999**, 597–601.

2 Hündesrügge T. *Krankenhauspharmazie* **1998**, *19*, 245–248.

3 Sprehe M., Geissen S.U. *Halogenorganische Verbindungen*, Herausgeber ATV-DVWK, ATV-DVWK Schriftenreihe 18, **2000**, 257–268.

4 BLAC: Bund/Länderausschuss für Chemikalien, Arzneimittel in der Umwelt Auswertung der Untersuchungsergebnisse, Hamburg, November 2003, Herausgeber: Freie und Hansestadt Hamburg Behörde für Umwelt und Gesundheit Institut für Hygiene und Umwelt, www.hu.hamburg.de.

5 Steger-Hartmann T., Länge R., Schweinfurth H. *Ecotoxicol. Environ. Saf.* **1999**, 42, 274–281.

6 Steger-Hartmann T., Länge R., Schweinfurth H.,Tschampel M., Rehmann I. *Water Res*. **2002**, 36, 266–274.

7 Gartiser S., Brinker L., Erbe T., Kümmerer K. *Acta Hydrochim. Hydrobiol.* **1996**, 24, 90–97.

8 Kalsch W. *Sci. Total Environ*. **1999**, 255, 14 353.

9 Sprehe M. *CUTEC-Schriftenreihe* 44, **1999**.

10 Oleksy-Frenzel J., Wischnack S., Jekel M. *Vom Wasser* **1995**, 85, 59–67.

11 Oleksy-Frenzel J., Wischnack S., Jekel M. *Fresenius J. Anal. Chem.* **2000**, 366, 89–94.

12 Ternes T. A., Hirsch R. *Environ. Sci. Technol*. **2000**, 34, 2741–2748.

13 Carballa M., Omil F., Lema J.M.J.M., Liompart M., Garcia-Jares C., Rodrigues I., Gomez M., Ternes T. *Water. Res.* **2004**, 38, 2918–2926.

14 Hirsch R., Ternes T.A., Lindart A., Haberer K., Wilken R.-D. *Fresenius J. Anal. Chem.* **2000**, 366, 835–841.

15 Putschew A., Jekel M. *Vom Wasser* **2001**, 97, 10 314.

16 Putschew A., Schittko S., Jekel M. *J. Chromatogr. A* **2001**, 930, 12 734.

17 Ternes A. T. *Water Res.* **1998**, 32, 3245–3260.
18 Knepper T.P., Werner A., Bogenschütz G. *J. Chromatogr. A* **2005**, 1085, 240–246.
19 Schittko S., Putschew A., Jekel M. *Water Sci. Technol.* **2004**, 50 (5), 261–268.
20 Löffler D., Römbke J., Meller M., Ternes T. *Environ. Sci. Technol.* **2005**, 39, 5209–5218.
21 Oppel J., Broll G., Löffler D., Meller M., Römbke J., Ternes T. *Sci. Total Environ.* **2004**, 328, 265–273.
22 Sacher F., Gabriel S., Metzinger M., Strez A., Wenz W., Lange F.T., Brauch H.-J., Blankenhorn I. *Vom Wasser* **2002**, 99, 18395.
23 Drewes E.J., Fox P., Jekel M. *J. Environ. Sci. Health* **2001**, A36 (9), 1633645.
24 Putschew A., Jekel M. *Rapid Commun. Mass Spectrom.* **2003**, 17(20), 2279–2282.
25 Huber M. M., Canonica S., Park G.Y., Gunten von U. *Environ. Sci. Technol.* **2003**, 37, 1016024.
26 Huber M. M., Göbel A., Joss A., Hermann N., Löffler D., Mcardell C.S., Ried A., Siegrist H., Ternes T.A., Von Gunten U. *Environ. Sci. Technol.* **2005**, 39, 4290–4299.
27 Ternes T. A., Stüber J., Herrmann N., McDowell D., Ried A., Kampmann M., Teiser B. *Water Res.* **2003**, 37, 1 976 982.
28 Schumacher J., Pi Y.Z., Jekel M. Cutec-Serial Publication No. 57: 3rd International Conference on Oxidation Technologies for Water and Wastewater Treatment. 18.–22. May 2003, Goslar, Germany.
29 Putschew A., Jekel M. Proceedings: Jahrestagung der Wasserchemischen Gesellschaft in der GDCh. 26.–28. Mai 2003, Stade, Seite 15 457. ISBN 3–936028-09-5.
30 Doll T.E., Frimmel F.H. *Water Res.* **2004**, 38, 955–964.
31 Doll T.E., Frimmel F.H. *Water Res.* **2005**, 39, 847–854.
32 Fitzke B., Sprehe M., Geissen S. U., Vogelpohl A. Preprints Colloquium Produktionsintegrierte Wasser-/Abwassertechnik, Bremen 6.–7.9.2000, Institut für Umwelttechnik-Universität Bremen, September 2000.
33 Drews A., Klahm T., Renk B., Saygili M., Baumgarten G., Kraume M. *Desalination* **2002**, 159 (2), 119–129.
34 Baus C. Proceedings Aktuelle Themen bei der Trinkwassergewinnung, Herausgeber DVGW-Technologiezentrum Wasser (TZW), 7. Kolloquium, 10. Dezember, 2002.
35 Putschew A., Soliz Tellez A., Jekel M. Proceedings Jahrestagung der Wasserchemischen Gesellschaft in der GDCh. 02.–04. Mai 2005, Bad Mergentheim, Seite 70–74. ISBN 3-936028-29-X.

5
Veterinary Pharmaceuticals

Gerd Hamscher

5.1
Introduction

In recent years, there has been a continuously growing interest in the occurrence, fate, and possible effects of human and veterinary drug residues in the environment [1–8]. Studies with a special focus on drugs used in human medicine have shown that these compounds reach surface waters mainly via the release of effluent from sewage treatment plants. Today, up to 80 compounds have been identified and quantified in concentrations between the low range of nanograms up to micrograms per liter [9]. Studies performed in Canada, Denmark, Germany, the United Kingdom, and the United States reveal that these agents represent a new class of organic environmental contaminants worldwide [2–5, 10, 11]. There is concern about the effects of the entry of these compounds into the environment, including the possibility of the spread of antibiotic resistance [12–14] and/or effects on the endocrine system due to the ability of some of these compounds to act as hormones [2, 3].

Veterinary pharmaceuticals are in use in large amounts worldwide. Unfortunately, there are only limited data with a high uncertainty available on the sale and usage of these compounds in Denmark, Germany, the United Kingdom (UK), and the United States of America (USA). However, the European Federation of Animal Health estimates that approximately 8500 tons of antibiotics were used in human medicine and 4700 tons in veterinary medicine in the European Union (including Switzerland) in 1999 [15]. Tetracyclines and sulfonamides represent the major classes of veterinary antibiotics prescribed most intensively worldwide [3–5].

Veterinary pharmaceuticals have been in use for many years as feed additives, and for prophylactic, metaphylactic, and therapeutic purposes. Although the topic has been intensively discussed in the scientific community during the last couple of years, there is still a lack of knowledge, especially about the behavior and fate of veterinary pharmaceuticals in the environment under field conditions [5].

For decades, liquid manure from livestock farming and sewage sludge from wastewater treatment plants have been applied to agricultural fields as a sustain-

Organic Pollutants in the Water Cycle. T. Reemtsma and M. Jekel (Eds.).
Copyright © 2006 WILEY-VCH Verlag GmbH & Co. KGaA, Weinheim
ISBN: 3-527-31297-8

able principle of nutrient recycling. With the rapidly increasing knowledge about pharmaceuticals entering the environment via liquid manure and thus probably contaminating our feed, food, and groundwater resources, there is increasing concern about the potential risks associated with this common practice. The first step in this process is to generate data on the environmental exposure levels and to investigate the fate and behavior of these compounds in relevant environmental compartments. Further in-depth studies on this topic have recently been completed [16–31] or are currently under way to obtain substantial data for a proper risk assessment of veterinary pharmaceuticals in soil and the aquatic environment. Furthermore, laboratory studies quantifying the sorption coefficients of antibiotics in various soils should allow an assessment of their leaching potential to the aquatic environment [5, 32].

5.2
Substance Classes

Veterinary pharmaceuticals have been widely used in the treatment and protection of farm animals for many years, and these compounds are also used to treat pet or companion animals [33]. However, the latter applications are probably of minor importance regarding the environmental impact of veterinary pharmaceuticals.

This whole book focuses on polar organic compounds, primarily those with a log K_{ow} < 3. This classification covers the most frequently used veterinary pharmaceuticals worldwide. Substance classes which will be discussed in the paper are mainly antibiotics:

- aminoglycosides (gentamicin, kanamycin, neomycin B, streptomycin),
- β-lactams (amoxicillin, ampicillin, benzylpenicillin, cefazolin),
- macrolides (oleandomycin, tylosin),
- quinolones and fluoroquinolones (oxolinic acid, enrofloxacin, sarafloxacin),
- sulfonamides (sulfachloropyridazine, sulfadiazine, sulfadimethoxine, sulfamethazine, sulfathiazole),
- tetracyclines (chlortetracycline, doxycycline, oxytetracycline, tetracycline),
- various antibiotics (chloramphenicol, lincomycin, trimethoprim).

Other important substance classes frequently used in veterinary medicine are ecto- and endoparasiticides. These substances may be released into the environment after topical application. Investigations performed in the UK have revealed that diffuse and point-source pollution can occur, e.g., when sheep in particular are treated with dipping solutions. Most of the drugs appplied (avermectin, chlorfenvinphos, coumaphos, cypermethrin, diazinion, fenchlorphos, flumethrin, ivermectin, propetamphos) have log K_{ow} values > 3 and are therfore not discussed in this paper. For a recently published comprehensive review of these veterinary pharmaceuticals and their occurrence in the environment, please refer to Boxall et al. [8].

5.2.1
Aminoglycosides

Since the isolation of streptomycin in 1944, aminoglycosides have been used as highly effective antibiotic agents, especially against gram-negative bacteria. In recent years, these compounds have been applied less frequently in human medicine because of severe side effects and the development of well-tolerated β-lactam antibiotics. In veterinary medicine, gentamicin in particular is used frequently for the treatment of calves, pigs, and turkeys. All macrolides have a comparable mode of action: they inhibit protein synthesis by binding to the 50S ribosomal subunit [34].

Aminoglycosides are highly polar, water-soluble compounds. They are stable in aqueous solutions in a pH range from 2.2. to 10. Several primary amino groups are responsible for the weakly basic reaction of this substance class (pKa between 7.2 and 8.8). The physicochemical properties of gentamicin, kanamycin, neomycin, and streptomycin are summarized in Table 5.1. The molecular structures of gentamicin C_1 and neomycin B are shown in Fig. 5.1.

Because of their behavior as highly polar polycations, aminoglycosides are poorly resorbed in the gastrointestinal tract. They are not metabolized to any extent and are excreted in large quantities via the feces [34].

Tab. 5.1 CAS-Nos., molecular weights, and some physical-chemical properties of aminoglycosides used in veterinary medicine.

Compound	Gentamicin	Kanamycin	Neomycin B	Streptomycin
CAS#	1405-41-0	25389-94-0	1404-04-2	57-92-1
MW	consists of three closely related components C_1: 477.60 C_2: 463.57 C_{1a}: 449.55	consists of three closely related components A: 486.51 B: 483.51 C: 484.50	614.64	581.58
log K_{ow}	−1.88 (Gentamicin C_1)	− 6.70 (Kanamycin C)	CLOGP: −1.01	−7.53
S	freely soluble	1 g L^{-1}	250 g L^{-1}	1 g L^{-1}; >20 g L^{-1} (as trihydrochloride)
pK_a	7.2–8.8	7.2–8.8	7.2–8.8	7.2–8.8
References	34–37	34–37	34–38	34–37

CAS# = chemical abstract services registry number; MW = molecular weight, log K_{ow} = logarithm of the octanol/water partition coefficient; CGLOP = TerraQSAR-LOGP™ computed octanol/water partition coefficient; S: solubility in water; pK_a = the negative logarithm of the acid dissociation constant, K_a.

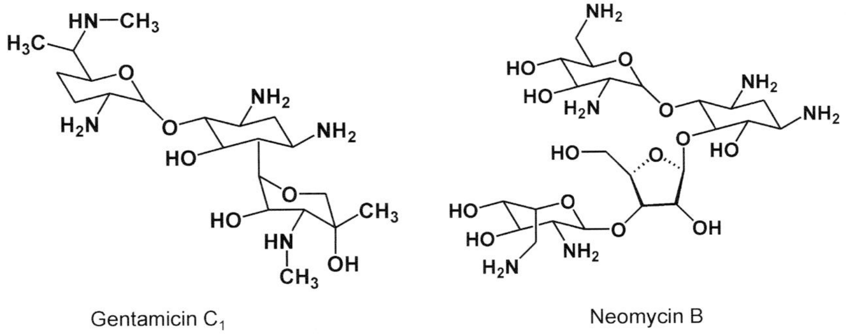

Gentamicin C₁ structure and Neomycin B structure

Fig. 5.1 Molecular structures of gentamicin and neomycin B, which are frequently used in veterinary medicine.

5.2.2
β-Lactam Antibiotics

Alexander Fleming's 1929 discovery of the bactericidal effects of penicillin isolated from *penicillium notatum* stimulated intensive research in the field of antibiotics. Although up to 50 different substances are now used in human and veterinary medicines, benzylpenicillin is still the drug of choice for many indications. Penicillins possess broad antimicrobial activity against both gram-positive and gram-negative organisms by their inhibition of cell wall synthesis. All penicillins have a carboxylic group (pK_a = 2–4). Derivatives of penicillins (e.g., calcium and potassium salts) are freely soluble in water under mildly acidic conditions [35], but are very sensitive to light and temperature [34].

Cephalosphorins represent another family of β-lactam antibiotics commonly used for the treatment of infections caused by gram-positive and gram-negative bacilli. They are produced semisynthetically and are highly polar, water soluble drugs.

Penicillins and cephalosporins are widely used in the treatment of various farm animals because of their very low toxicity and their wide availibilty. The pharmaco-

Tab. 5.2 CAS-Nos., molecular weights, and some physical-chemical properties of β-lactams used in veterinary medicine.

Compound	Amoxicillin	Ampicillin	Benzylpenicillin (Penicillin G)	Cefazolin
CAS#	26787-90-3	69-53-4	61-33-6	25953-19-9
MW	365.41	349.41	334.40	454.51
log K_{ow}	0.87	1.35	1.83	–0.58
S	4 g L^{-1}	10 g L^{-1}	0.21 g L^{-1}	0.21 g L^{-1}
pK_a:	2–4	2–4	2.74	2–4
References	34–37	34–37	35–37	34–37

CAS# = chemical abstract services registry number; MW = molecular weight; log K_{ow} = logarithm of the octanol/water partition coefficient; S = solubility in water; pK_a = the negative logarithm of the acid dissociation constant, K_a.

logical behavior of β-lactam antibiotics is quite different and depends on the properties of the side chain modifications in the molecules. In general, after parenteral application, most of the compounds are excreted renally [34]. The physicochemical properties of amoxicillin, ampicillin, benzylpenicillin (penicillin G), and cefazolin are summarized in Table 5.2. The molecular structures of one representative of penicillins and cephalosphorins are shown in Fig. 5.2.

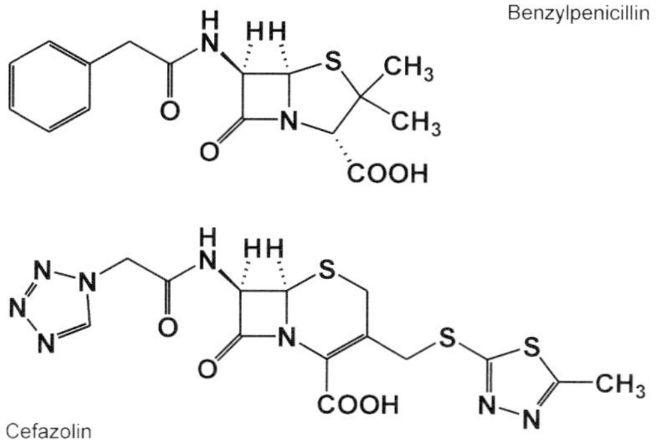

Fig. 5.2 Molecular structures of β-lactam antibiotics used in veterinary medicine: benzylpenicillin (= penicillin G) and cefazolin.

5.2.3
Macrolides

Macrolide antibiotics are lipophilic drugs with high molecular weights. They have a macrocyclic ring structure containing 14–16 carbonlactones. This ring system is coupled to neutral or amino sugars giving the molecules a basic character [34]. The physicochemical properties of oleandomycin and tylosin, which are important in veterinary medicine, are summarized in Table 5.3. The molecular structure of tylosin is shown in Fig. 5.3. Erythromycin, which is frequently used in human and veterinary medicine, has a log $K_{ow} > 3$ and is therefore not considered in the current publication.

Tab. 5.3 CAS-Nos., molecular weights, and some physical-chemical properties of macrolides used in veterinary medicine.

Compound	Oleandomycin	Tylosin
CAS#	3922-90-5	1401-69-0
MW	687.88	917.14
log K_{ow}	1.69	1.63
S	15.5 mg L^{-1}	0.5 mg L^{-1}
pK_a	8.84	7.73
References	37	37

CAS# = chemical abstract services registry number; MW = molecular weight, log K_{ow} = logarithm of the octanol/water partition coefficient; S = solubility in water; pK_a = the negative logarithm of the acid dissociation constant, K_a.

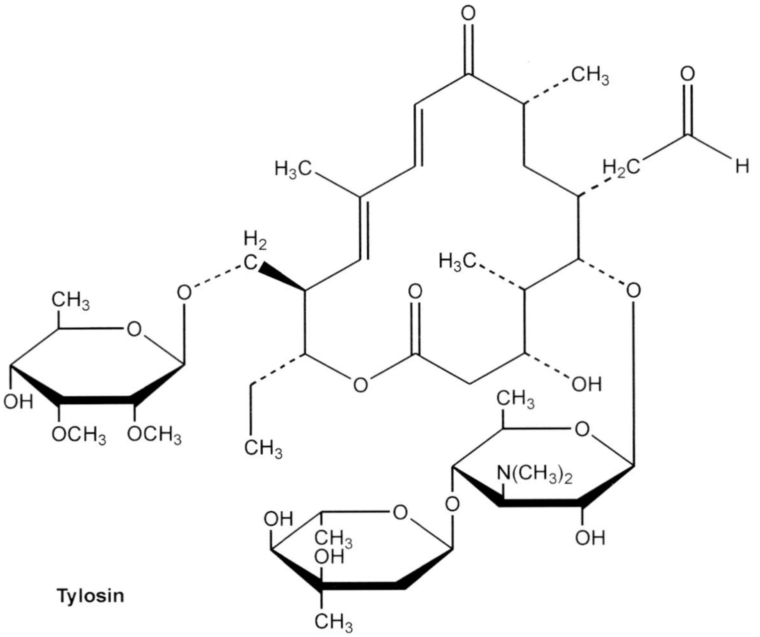

Fig. 5.3 Molecular structure of tylosin, a well-known macrolide frequently used in veterinary medicine.

Macrolides possess antimicrobial activity against gram-positive organisms and – in rare cases – against gram-negative ones. They are nevertheless useful against mycoplasms, which have no cell wall. Macrolides inhibit protein synthesis by binding to the bacterial 50S ribosome, thus inhibiting polypeptide translation [34].

Salts (tartrates, phosphates, lauryl sulfates) and esters (estolate, adipate, ethyl succinate) of the macrolides are synthesized to enhance their bioavailability and water solubility. After hydrolytic or enzymatic cleavage of the derivatives, the free bases are resorbed, and accumulate preferably in tissue. However, the oral bioavailability of macrolides depends on individual differences and on feed consumption, and is in the approximate range between 9% and 45% [34].

5.2.4
Quinolones and Fluoroquinolones

Quinolones are synthetic chemotherapeutic agents used in human and veterinary medicine. They are polar compounds and highly active against a broad spectrum of gram-negative bacteria and mycoplasma. With the synthesis of the chemically more stable fluoroquinolones it became possible also to treat gram-positive bacterial infections. These agents interfere with nucleic acid synthesis by inhibiting the enzyme DNA gyrase, thus preventing DNA replication, gene transcription, and finally cell division. However, quinolones and fluoroquinolones are bactericidal agents, and other potential, as yet unknown, modes of action must still be considered [34].

Enrofloxacin and sarafloxacin are of increasing importance in veterinary medicine, and their physicochemical properties are summarized together with that of oxolinic acid in Table 5.4. The molecular structures of these compounds are shown in Fig. 5.4.

Oxolinic acid has been frequently used to treat fish diseases, wheras enrofloxacin and sarafloxacin are applied to treat pig and poultry infections. In general, after oral or parenteral application, most of these compounds are excreted renally (approximately 80%) and via the feces (20%). It is important to note that, in the case of enrofloxacin, a microbiologically active metabolite, ciprofloxacin, is generated via dealkylation of the parent drug. Ciprofloxacin itself is used exclusively in human medicine [34].

Tab. 5.4 CAS-Nos., molecular weights, and some physical-chemical properties of quinolones and fluoroquinolones used in veterinary medicine.

Compound	Enrofloxacin	Oxolinic acid	Sarafloxacin
CAS#	93106-60-6	14698-29-4	98105-99-8
MW	359.40	261.24	385.37
log K_{ow}	0.7	0.94	1.07
S	3.4 g L^{-1}	S: 3.2 mg L^{-1}	0.1 g L^{-1}
pK_a	p$K_{a,1}$: 6.27	pK_a: 6.87	p$K_{a,1}$: 6.0
	p$K_{a,2}$: ca. 8.3		p$K_{a,2}$: 8.6
References	32, 37	32, 37	32, 37

CAS# = chemical abstract services registry number; MW = molecular weight, log K_{ow} = logarithm of the octanol/water partition coefficient; S: solubility in water; pK_a = the negative logarithm of the acid dissociation constant, K_a.

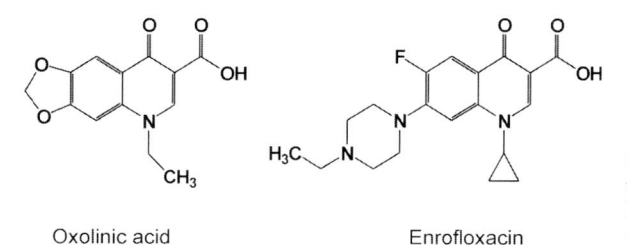

Fig. 5.4 Molecular structures of important quinolones used in veterinary medicine.

5.2.5
Sulfonamides

In 1935, Domagk discovered a sulfonamide-containing dye, Prontosol, which showed antibacterial activity. Since then, thousands of sulfonamide derivatives have been synthesized, but only a couple of them have been widely used in both human and veterinary medicine. Sulfonamides have bacteriostatic properties due to their interference with the synthesis of folic acid in susceptible organisms. These agents inhibit dihydropteroase because of the substrate analogy of sulfonamides with paraaminobenzoic acid. This effect can be potentiated when trimethoprim is coadministered together with sulfonamides. Therefore, sulfonamides are commonly used for the treatment of infections caused by gram-positve and gram-negative bacteria, chlamydias, and protozoa (coccidia, toxoplasms) [34].

Sulfonamides are only poorly water soluble. Therefore, highly water-soluble sodium salts are used for therapeutic purposes. Sulfonamides show an amphoteric behavior characterized by two different pK_a values: they behave as weak acids and form salts in strong acids as well as in strong bases [34]. The physicochemical properties of sulfadiazine, sulfadimethoxine, sulfamethazine, sulfathiazole, and sulfachloropyradizine important in veterinary medicine worldwide are summarized in Table 5.5. The molecular structures of these compounds are shown in Fig. 5.5.

The oral bioavailabilty of sulfonamide sodium salts is generally high and is independent of food consumption. Sulfonamides are frequently used in pig and poultry production and can be metabolized extensively. In many species, including humans and pigs, the main metabolic pathway is the formation of acetyl derivatives. Sulfonamides or their metabolites are mainly excreted renally [34].

Tab. 5.5 CAS-Nos., molecular weights, and some physical-chemical properties of sulfonamides used in veterinary medicine.

Compound	Sulfa-diazine	Sulfa-dimethoxine	Sulfa-methazine	Sulfa-thiazole	Sulfachlor-pyradizine
CAS#	68-35-9	122-11-2	57-68-1	72-14-0	80-32-0
MW	250.28	310.33	278.33	255.32	284.73
log K_{ow}	−0.09	1.63	0.89	0.05	0.31
S	0.077 g L^{-1}	0.343 g L^{-1}	1.5 g L^{-1}	0.6 g L^{-1}	7 g L^{-1}
pK_a	$pK_{a,1}$: 2.1; $pK_{a,2}$: 6.4	$pK_{a,1}$, $pK_{a,2}$: comparable to other sulfonamides	$pK_{a,1}$: 2.65; $pK_{a,2}$: 7.65	$pK_{a,1}$: 2.00; $pK_{a,2}$: 7.24	$pK_{a,1}$: 1.76; $pK_{a,2}$: 5.71
References	32, 37, 39	32, 37	32, 37	32, 37	32, 37

CAS# = chemical abstract services registry number; MW = molecular weight, log K_{ow} = logarithm of the octanol/water partition coefficient; S = solubility in water; pK_a = the negative logarithm of the acid dissociation constant, K_a.

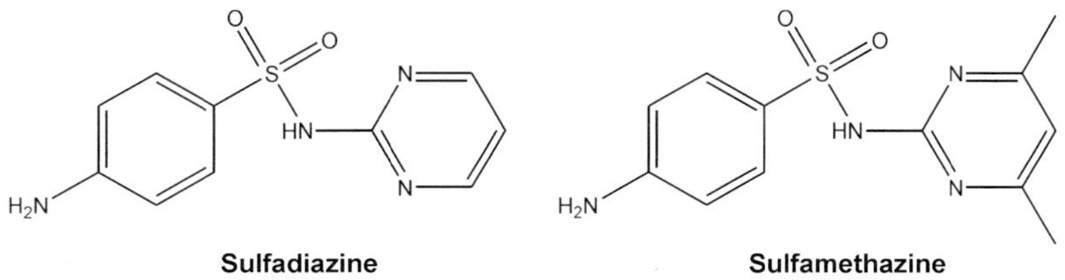

Sulfadiazine   **Sulfamethazine**

Fig. 5.5 Molecular structures of two frequently used sulfonamides used in veterinary medicine.

5.2.6
Tetracyclines

Tetracyclines have been the most important class of antibiotics in animal agriculture for the last four decades. Because of their broad-spectrum effectiveness, tetracyclines are commonly used for the treatment of infections caused by gram-positive and gram-negative bacteria, mycoplasms, and large viruses. Tetracyclines inhibit protein synthesis by preventing the association of aminoacyl-tRNA with the bacterial ribosome.

Tetracyclines consist of a hydronaphthacene skeleton containing four fused rings; they are only poorly water soluble. Therefore, highly water-soluble hydrochlorides have been produced semisynthetically and are generally used for therapeutic purposes. Tetracyclines are frequently used worldwide in aquaculture (mainly oxytetracycline), pig (chlortetracycline, doxycycline, and tetracycline) and poultry production (mainly chlortetracycline) [34].

Tetracyclines show an amphoteric behavior characterized by three different pK_a values. They are ionized throughout the pH range between 3 and 8. At a pH of around 3 the molecule exists in the cationic form, as a zwitterion between a pH of 3.5 and 7.5, and above that as an anoin. Tetracycline solutions are quite stable at neutral and weakly acidic pH values. At alkaline pH the compounds are comparably unstable, and biologically inactive isomers are formed. At an acidic pH below 3 anhydro-tetracyclines are formed, which also show limited bioactivity in comparison to the parent drug [34].

The physicochemical properties of chlortetracycline, doxycycline, oxytetracycline, and tetracycline are summarized in Table 5.6. The molecular structures of three tetracyclines, which are most commonly applied to food-producing animals, are shown in Fig. 5.6.

Tab. 5.6 CAS-Nos., molecular weights, and some physical-chemical properties of tetracyclines used in veterinary medicine.

Compound	Chlortetracycline	Doxycycline	Oxytetracycline	Tetracycline
CAS#	57-62-5	564-25-0	79-57-2	60-54-8
MW	478.89	444.45	460.44	444.45
log K_{ow}	– 0.62	– 0.02	– 0.90	– 1.30
S	0.63 g L^{-1}	0.63 g L^{-1}	0.31 g L^{-1}	0.23 g L^{-1}
pK_a	$pK_{a,1}$: 3.30; $pK_{a,2}$: 7.44; $pK_{a,3}$: 9.27	$pK_{a,1}$; $pK_{a,2}$; $pK_{a,3}$: comparable to other tetracyclines	$pK_{a,1}$: 3.27; $pK_{a,2}$: 7.32; $pK_{a,3}$: 9.11	$pK_{a,1}$: 3.30; $pK_{a,2}$: 7.68; $pK_{a,3}$: 9.69
References	37, 40	37	32, 37	32, 37

CAS# = chemical abstract services registry number; MW = molecular weight, log K_{ow} = logarithm of the octanol/water partition coefficient; S = solubility in water; pK_a = the negative logarithm of the acid dissociation constant, K_a.

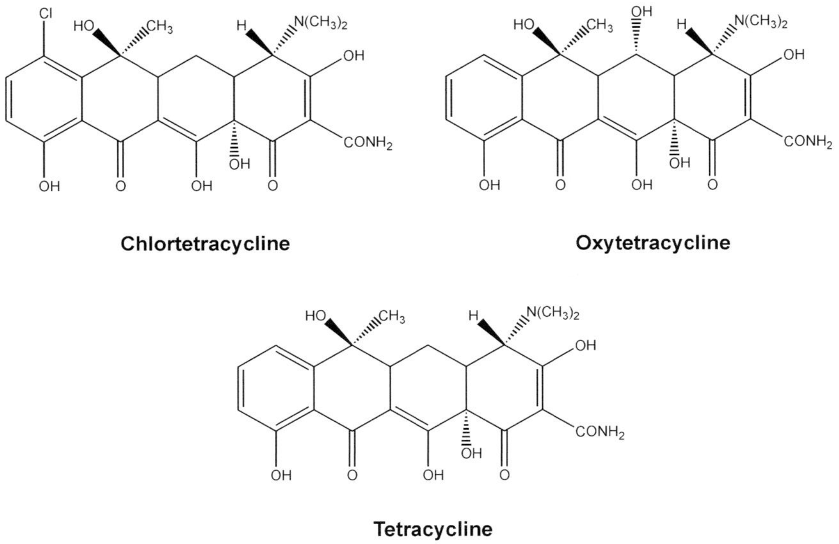

Fig. 5.6 Structures of the most important tetracyclines used in veterinary medicine worldwide.

The oral bioavailabilty of tetracyclines is strongly influenced by feed composition. Together with divalent metallic ions (e.g., calcium and magnesium), tetracyclines form stable complexes. Tetracylines are not metabolized to any extent and are excreted in large quantities renally and via the feces [34].

5.2.7
Various Antibiotics

Beside the antibiotics mentioned above, some other compounds may have an environmental impact because of their usage patterns. The physicochemical properties of chloramphenicol, lincomycin, and trimethoprim are summarized in Table 5.7. The molecular structures of these pharmaceuticals, which are used in both veterinary and human medicine, are shown in Fig. 5.7.

Chloramphenicol is a highly effective broad-spectrum bacteriostatic compound active against aerobic and anaerobic gram-negative and gram-positive bacteria, mycoplasma, and chlamydia. The drug inhibts the translation on the 50S ribosomal subunit at the peptidyltransferase step, thus inhibiting elongation. Because of reported severe side effects in humans (aplastic anaemia) and its genotoxic properties it was totally banned for food-producing animals within the European Union and the USA in 1994 [34].

Lincomycin belongs to the family of lincosamide antibiotics and has similar applications as the macrolides. The drug is frequently used in pig and poultry production.

Tab. 5.7 CAS-Nos., molecular weights, and some physical-chemical properties of various other antibiotics used in veterinary and human medicine. Please note: Chloramphenicol is not approved for use in food-producing animals, whereas administration to pets is allowed.

Compound	Chloramphenicol	Lincomycin	Trimethoprim
CAS#	56-75-7	154-21-2	738-70-5
MW	323.13	406.55	290.32
log K_{ow}	1.14	0.56	0.91
S	2.5 g L^{-1}	0.9 g L^{-1}	0.4 g L^{-1}
pK_a	pK_a: 5.5	pK_a: 7.60	pK_a: 7.12
References	35, 37	35, 37	37

CAS# = chemical abstract services registry number; MW = molecular weight, log K_{ow} = logarithm of the octanol/water partition coefficient; S = solubility in water; pK_a = the negative logarithm of the acid dissociation constant, K_a.

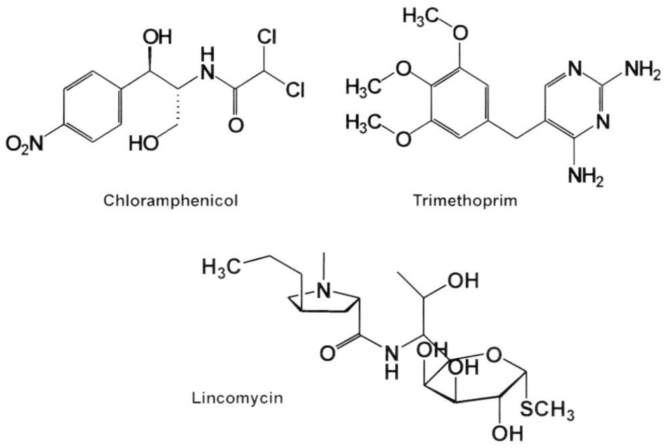

Fig. 5.7 Molecular structures of various antibiotics used in veterinary and human medicine. Please note: Chloramphenicol is not approved for use in food producing animals, whereas administration to pets is allowed.

Trimethoprim is a diaminopyrimidine, which are well-known inhibitors of dihydrofolate reductase. This effect can be potentiated when sulfonamides are coadministered together with trimethoprim. Although the effectiveness of this combination in farm animals is limited because of an increasing antimicrobial resistance against sulfonamides, it is frequently used in pig production [34].

5.3
Pathways to the Environment

Most antibiotics (e.g., amoxicillin, aminoglycosides, tetracyclines) are only poorly metabolized after their administration to humans or animals. Therefore, antibiotics and, in some cases, also their metabolites, are excreted in substantial amounts and can enter the environment via several pathways. Furthermore, in the case of sulfonamides, N^4-acetyl metabolites of sulfamethazine and sulfmethoxazole are degraded in liquid manure [16] or in wastewater treatment plants [39] thus liberating the parent drugs.

Today, there are at least three well-established pathways along which veterinary pharmaceuticals can enter the soil compartment (see also Fig. 5.8): they reach sediments after direct application of antibiotics in aquaculture, or they reach soil directly via the excrement of animals kept outdoors or via liquid manure applied as fertilizer.

Very recently, it was shown that dust-bound antibiotics (including tylosin, various tetracyclines, sulfamethazine, and chloramphenicol) occur in the mg kg^{-1} range and may enter the environment via exhaust air [42]. The inhalation of dust particles may be of major importance as a potential health hazard, and only further investigations of this heretofore unsuspected exposure route will make it possible to assess the contribution of this pathway to environmental contamination. On the other hand, contamination of water resources via this pathway seems to be of minor importance.

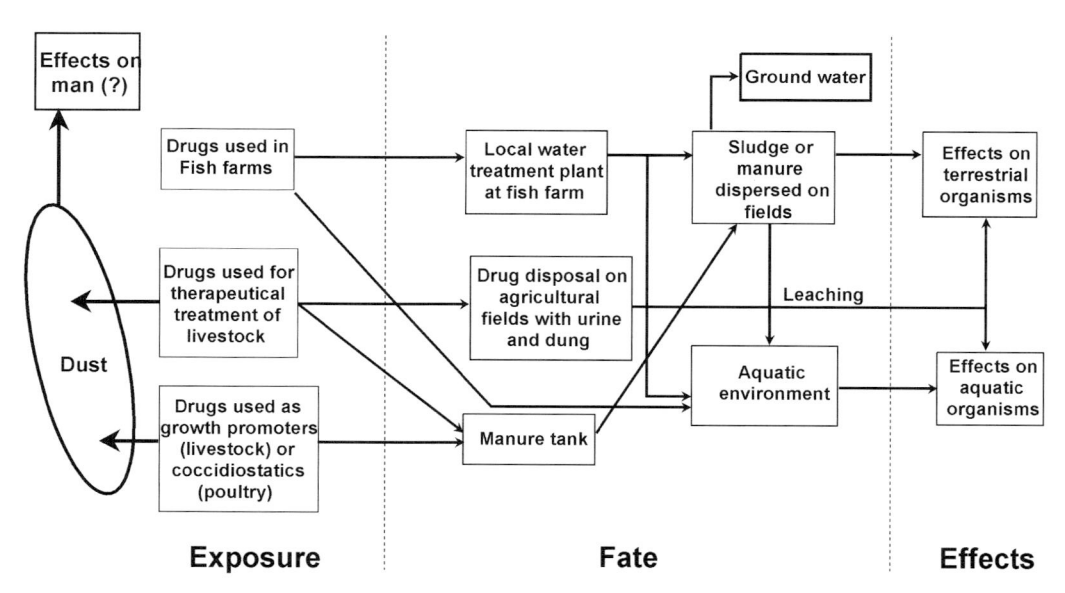

Fig. 5.8 Anticipated exposure routes of veterinary drugs into the environment (modified according to Refs. [41, 42]).

5.3.1
Liquid Manure

Recently it was shown that tetracycline and sulfamethazine are present in liquid manure at concentrations of up to 66 mg L^{-1} and 40 mg L^{-1}, respectively, after application of these drugs in recommended dosages [43–45]. Reported concentrations of various antibiotics in liquid manure samples are summarized in Table 5.8.

A comparable spectrum of substances was found in studies in the USA, where manure samples and hog lagoons were screened for various antibiotics including macrolides [46–49]. However, the absolute concentrations in the samples from hog lagoons were significantly lower than those found in European studies. One explanation for this difference may be the fact that all wastewater from animal feeding operations in the United States is collected in such lagoon systems, resulting in a dilution of the pharmaceuticals. In Europe, liquid manure is collected usually in manure tanks (see Fig. 5.8).

Furthermore, Eichhorn and Aga [50] identified degradation and photooxygenation products of chlortetracycline in hog lagoons. The microbiologically inactive isochlortetracycline, which is formed under alkaline conditions, was found in concentrations from 1.9 mg kg^{-1} to 15.8 mg kg^{-1}. The so called photooxygenation product epi-M510 was detected in concentrations from 0.05 mg kg^{-1} to 0.30 mg kg^{-1}. However, the environmental impact of these degradation products remains unclear and will have to be investigated in further studies [50].

Tab. 5.8 Maximum concentrations (max. conc.) of various antibiotics detected in liquid manure samples originating from animal feeding operations for pigs or calves.

Pharmaceuticals	Samples [n]	Max. conc. [mg kg^{-1}]	References
Tetracyclines			
Chlortetracycline	7	2.7	21
Doxycycline	3	0.04	49
Oxytetracycline	1	19	22
Tetracycline	181	66	43
Sulfonamides			
Sulfadiazine	6	1.1	21
Sulfadimethoxine	8	0.003	48
Sulfamethazine	?	40	43
N^4-Acetyl-Sulfamethazine	6	2.6	51
Sulfathiazole	6	12.4	51
Macrolides			
Tylosin	7	<0.02	21
Others			
Lincomycin	8	0.24	48
Trimethoprim	6	<0.1	48, 51

5.3.2
Soil Fertilization

Because of the persistence of various veterinary pharmaceuticals in liquid manure, these compounds reach soil after the fertilization process, and have already been shown to persist in this environmental compartment. Maximum concentrations of various antibiotics detected in soil are summarized in Table 5.9.

Recently, tetracyclines were detected on farmed land at concentrations of up to 300 µg kg^{-1} soil, demonstrating that this class of antibiotics in use world-wide is only slowly degraded under field conditions and may accumulate in soil after repeated fertilization with liquid manure from intensive pig farming. Another interesting result of this study was the high concentrations of antibiotics found in dried liquid manure aggregates at the soil surface, where 0.35 mg kg^{-1} of tetracycline and 1.44 mg kg^{-1} of chlortetracycline were detected. These aggregates can remain intact after incorporation into the soil and create spots with high antibiotic concentration in the soil matrix [19].

Furthermore, these field studies gave no evidence of leaching of these compounds into deeper soil segments or into groundwater because of the strong sorption of the drugs in topsoil [18, 19, 21]. Field studies recently performed in Italy and the USA confirmed the persistence of tetracyclines in soil after the application of liquid manure [22, 49]. Preliminary results of a screening study performed in 14 different soil-monitoring plots in Germany showed that the only detectable sulfonamide in soil was sulfamethazine, which occurred in four areas at a maximum concentration of 11 µg kg^{-1} [53]. Six other sulfonamides including sulfadiazine, sulfathiazole, sulfamerazine, sulfamethoxypyridazine, sulfamethazine, sulfamethoxazole, and sulfadimethoxine could not be found above the detection limit of 1 µg kg^{-1} [21]. A follow-up study performed from June 2002 to May 2004 monitor-

Tab. 5.9 Maximum concentrations (max. conc.) of tetracycline (including their epimers) and sulfonamide antibiotics detected in farmed soil under conventional fertilization with liquid manure.

Pharmaceuticals	Number of sites [n]	Max. conc. [µg/kg]	References
Tetracyclines			
Chlortetracycline	14	93	52
Oxytetracycline	14	27	52
Tetracycline	14	443	52
Sulfonamides			
Sulfadiazine	14	<5 (LOQ)	52
Sulfamethazine	14	11	21
Macrolides			
Tylosin	14	<1 (LOD)	21, 52

LOQ = limit of quantification, LOD = limit of determination.

ing the same agricultural fields representative of northwestern Germany showed that as many as one in three samples could be contaminated with residues at concentrations of more than 100 µg kg^{-1}. In that study, sulfadiazine was detected for the first time in an agricultural field regularly fertilized with liquid manure. Again, sulfamethazine was present only at concentrations of a few µg kg^{-1} soil [52].

A detailed investigation of a soil-monitoring plot which had been previously shown to be highly exposed to tetracyclines [19] revealed the different behavior of tetracyclines and sulfonamides under field conditions: tetracycline occurred in average amounts of more than 150 µg kg^{-1} soil, whereas sulfamethazine was detectable in the plough layer at concentrations approximately two orders of magnitude lower. Although sulfamethazine was apparently present at very low concentrations in soil, it was found repeatedly in shallow groundwater sampled with suction probes at 1.4 m below the soil surface at a maximum concentration of 0.24 µg L^{-1}. This provided direct evidence of the continuous leaching of a veterinary drug from soil into groundwater under field conditions, as the drug was simultaneously detected in shallow groundwater, soil, and the liquid manure applied. One explanation for this observation may be the different sorption coefficients of these substances in soil indicating among other things their different mobilities in that ecosystem [54].

Boxall et al. reported in 2002 [20] on differing behavior of sulfachloropyridazine in two soil systems. Although the sorption coefficients of the compound suggested a high mobility in the aquatic environment, the predicted behavior was confirmed only in a study carried out on a clay soil. However, leaching studies at a sandy site indicated only a low potential for this substance to reach groundwater.

Kay et al. [28] performed three soil-leaching studies in clay soils with artificial macropores in order to investigate the environmental fate of oxytetracycline, sulfachloropyridazine, and tylosin, which are highly important veterinary pharmaceuticals in the UK. Whereas sulfachloropyridazine and oxytetracycline were found in concentrations of > 200 µg L^{-1} in the leachate, no tylosin was detected either in the soil or in the leachate. In another set of experiments it was clearly shown that pre-tillage can substantially reduce the leaching of antibiotic residues through macroporous clay soils [28].

5.3.3
Aquaculture

After the application of antibiotics in aquaculture, the drugs can settle into the sediments (directly, via fish feed, or via fish excrement). In this particular form of application, substantial amounts of the drugs are neither metabolized nor degraded. Thus, oxytetracycline, a drug frequently used in aquaculture world-wide, has been detected at a very high concentration (285 mg kg^{-1}) in the sediment of a fish farm [55]. Lower concentrations of this antibiotic have also been reported in marine sediments by other groups [56–58], indicating the widespread exposure of this environmental compartment to oxytetracycline from aquaculture.

5.4
Occurrence in Wastewater Treatment Plants

As mentioned in Chapter 3.2, most of the veterinary pharmaceuticals used in intensive livestock farming reach the soil via the spreading of liquid manure or via the sediments after direct use in aquaculture. It is unlikely that significant concentrations of veterinary pharmaceuticals will occur in municipal wastewater or in sewage sludge.

However, veterinary pharmaceuticals may occur in municipal wastewater after use in private practices for the treatments of pets. Although the significance of this pathway is difficult to assess, it seems to be of minor importance.

Very few studies have reported concentrations of veterinary pharmaceuticals in the influent and effluent of wastewater treatment plants (WWTPs, see Table 5.10). Yang and Carlson [59] investigated influents and effluents of a WWTP in an agricultural area in northern Colorado, USA. The emissions from the WWTP were consistent with the amounts of antibiotics measured in the river into which the effluent was finally directed. Most of the compounds were removed completely during the wastewater treatment process. However, tetracycline and traces of democlocycline and doxycycline did appear in the effluent. In another study, Batt and Aga demonstrated the occurrence of tetracycline and enrofloxacin at the sub-µg L^{-1} level in the effluent of a local WWTP in Holland, NY, USA [60]. Recently Göbel et al. showed that sulfamethazine is only partly degraded in a municipal WWTP and can reach surface waters [39]. These preliminary studies show that two of the most

Tab. 5.10 Maximum concentrations of veterinary pharmaceuticals dectected in the influent and effluent of waste water treatment plants.

Pharmaceuticals	Max. concentration (µg L^{-1})		References
	Influent	Effluent	
Tetracyclines			
Chlortetracycline	0.26	<0.05	59
Democlocycline	1.14	0.09	59
Doxycycline	*	0.05	11
	0.22	0.09	59
Oxytetracycline	0.64	<0.05	59
Tetracycline	*	0.98	11
	0.98	0.16	59
	*	0.62	60
Others			
Chloramphenicol	*	0.56	63
Enrofloxacin	*	0.10	60
Sulfamethazine	*	0.363	11
	0.04	0.02	39
	0.68	<0.05	59

* = not investigated

important veterinary pharmaceuticals, tetracycline and sulfamethazine, may reach surface waters via WWTPs. However, in the case of tetracycline and chloramphenicol, contamination may also originate from prescriptions in human medicine [59] (see also Chapter 3).

There is no current study available on the concentrations of veterinary pharmaceuticals in sewage sludge, but they should behave similarly to human pharmaceuticals. It therefore may be assumed that tetracyclines and fluoroquinolones strongly sorb to sludge, as has already been shown for these substances in soil [18, 19, 61, 62].

5.5
Surface Waters

Veterinary pharmaceuticals may reach surface waters after runoff from manured soils [1]. This can occur when liquid manure is not ploughed into the soil immediately after application (Fig. 5.8) [27].

Kay et al. [27] recently investigated the transport of three veterinary antibiotics (oxytetracycline, sulfachloropyridazine, and tylosin) in overland flow. The runoff was generated following the application of liquid manure to arable land prior to incorporation. Subsequently, oxytetracycline and sulfachloropyridazine were found in runoff at concentrations of 71 and 703 $\mu g\ L^{-1}$, respectively. Thus, in this study, overland flow was identified as one important route by which veterinary pharmaceuticals may reach surface water. Tylosin was not detected at all, and the authors concluded that this drug is degraded during the storage of liquid manure prior to application. This is in good agreement with recently published studies by other groups regarding the behavior of tylosin in liquid manure [64–68] and the negative findings for tylosin in soil [18, 19, 21, 52].

Kreuzig et al. [69] performed test-plot studies on runoff of sulfonamides from manured soils after sprinkler irrigation. They found a comparatively low runoff risk for sulfadiazine, sulfamethazine, and sulfamethoxazole from agricultural fields, even though they simulated worst-case conditions. The runoff risk was considered higher for grassland, especially when the manure application was directly followed by a heavy rainfall.

However, veterinary pharmaceuticals have only rarely been detected in surface waters (oxytetracycline, sulfamethazine, tylosin, and in one study chloramphenicol and enrofloxacin, see Table 5.11) in the lower $\mu g\ L^{-1}$ range. Comparison of these figures with effective concentrations (EC50) or maximum inhibitory concentrations of each medicine with respect to standard test organisms indicates that veterinary pharmaceuticals may not pose much of an acute risk to the aquatic environment. However, the occurrence of antibiotics in the aquatic environment in subtherapeutic concentrations may induce more subtle effects on water micro-organisms, e.g., the development of antibiotic resistance [2, 7, 8, 31] (see also Chapter 3). One should keep in mind that it is still unclear at what threshold concentration of a certain antibiotic in water a shift toward an increase in the percentage of resistant strains can be expected. Studies in humans revealed that the development of resistant strains in the intestinal microflora occurs at doses between 10 and 100 mg/day, which is far away from any concentration of an antibiotic measured in the aquatic environment. Another finding was that resistant organisms already existed before treatment and that the resistance was transitory [70].

It should be noted that lincomycin, tetracycline, and especially chloramphenicol contamination may also originate from prescriptions in human medicine [71]. Because of the genotoxicity of chloramphenicol and three metabolites thereof (nitroso-chloramphenicol, dehydro-chloramphenicol, and dehydro-chloramphenicol-base) in several *in vitro* and *in vivo* test systems, it was not possible to confirm an acceptable daily intake [72]. Therefore, the European Union and the United States completely prohibited its use in livestock farming within their territories in 1994. Besides that, chloramphenicol is still used to treat pets, and is also used in human medicine to treat, e.g., severe eye infections.

Tab. 5.11 Maximum concentrations of veterinary pharmaceuticals dectected in surface water monitoring studies.

Pharmaceuticals	Max. concentration ($\mu g\ L^{-1}$)	References
Tetracyclines		
Chlortetracycline	0.11–0.69	10, 71, 73
Doxycycline	0.07	67
Oxytetracycline	0.15–1.34	10, 71, 73
Tetracycline	0.11	10
Sulfonamides		
Sulfadimethoxine	0.08	71
Sulfamethazine	0.06–0.22	10, 71
Others		
Chloramphenicol	0.06	63
Lincomycin	0.73	10
Tylosin	0.28	10

5.6
Groundwater

Veterinary pharmaceuticals have only rarely been detected in groundwater, mainly because of their sorptive behavior in soil. Several studies have been performed worldwide with a focus on tetracyclines and sulfonamides (see Table 5.12).

After veterinary pharmaceuticals have entered the soil, sorption, transport, and degradation processes may occur. The sorption behaviour of most of the compounds studied is not appropriately characterized by their hydrophobic properties. Other mechanisms must also be taken into account, such as ion exchange, cation bridging at clay surfaces, surface complexation, and hydrogen bonding [5–8, 32]. Therefore, leaching of veterinary pharmaceuticals into groundwater may be prevented by various soil interactions, which are currently not fully understood.

Tab. 5.12 Maximum concentrations of veterinary pharmaceuticals detected in groundwater monitoring studies.

Pharmaceuticals	Max. concentration ($\mu g\ L^{-1}$)	References
Sulfamethazine	0.16–0.24	10, 54, 63
Tetracycline	0.13	48, 52, 75

However, sulfamethazine in particular has been recognized in several studies as a compound which represents a risk of groundwater contamination [10, 54, 63]. Leaching of tetracyclines seems to be more unlikely because of the stronger sorption of these pharmaceuticals to topsoil in comparison to sulfonamides, and they are, thus, not usually detectable in groundwater [18, 19, 63, 73]. Hotspot studies performed in the United States in the vicinity of large-scale swine and poultry feeding operations [48] and in an area in Germany with very high tetracycline concentrations in soils over several years revealed that traces of tetracyclines occur in pore water [74] and may also reach shallow groundwater [48, 52, 75].

5.7
Water Treatment

Veterinary pharmaceuticals have – to the best of my knowledge – never been detected in raw waters used for the production of drinking water. However, several human pharmaceuticals have already reached our groundwater resources but have rarely been detected in trace levels in drinking water samples. As a consequence of this, water treatment techniques (e.g., ozonation, membrane filtration) have been intensively studied during the last years. It was shown that pharmaceutically active compounds can be efficiently removed from drinking water or even from municipal sewage effluents [9] (see also Chapter 2). Although veterinary pharmaceuticals were not included in these investigations, one may expect similar chemical or physical behavior of these drugs in the different water treatment processes used for drinking water production.

5.8
Summary

The use of large amounts of veterinary pharmaceuticals in intensive livestock farming has led primarily to residues of tetracycline and sulfonamide antibiotics in liquid manure or hog lagoons and in topsoil. Trace concentrations, mainly of sulfamethazine and various tetracyclines, have on rare occasions been detected in surface water and in shallow groundwater samples in agricultural areas. Therefore, it seems highly unlikely today that veterinary drug residues could reach drinking water resources in significant concentrations. Moreover, veterinary pharmaceuticals should behave comparably to human pharmaceuticals during water treatment processes because of their comparable molecular structures.

Acknowledgments

I am obliged to Dr. Judith McAlister-Hermann for careful editing of the manuscript. Many thanks also to Dr. Heike T. Pawelzick and Dr. Daniel Eikel for preparing the molecular structures.

References

1 B. Halling-Sørensen, S. Nors Nielsen, P. F. Lanzky, F. Ingerslev, H. C. Holten Lützhoft, S. E. Jørgensen, *Chemosphere* **1998**, 36, 357–393.
2 C. G. Daughton, T. A. Ternes, *Environ. Health Perspect.* **1998**, 107, 907–938.
3 K. Kümmerer, ed., *Pharmaceuticals in the environment: Sources, fate, effects, and risks*, 2nd ed. **2004**, Springer, Berlin, Heidelberg, Germany.
4 A. B. Boxall, L. A. Fogg, P. Kay, P. A. Blackwell, E. J. Pemberton, A. Croxford, *Toxicol. Lett.* **2003**, 142, 207–218.
5 S. Thiele-Bruhn, *J. Plant Nutr. Soil Sci.* **2003**, 166, 145–167.
6 M. S. Diaz-Cruz, M. J. L. de Alda, D. Barcelo, *Trends Anal. Chem.* **2003**, 22, 340–351.
7 A. B. Boxall, D. W. Kolpin, B. Halling-Sørensen, J. Tolls, *Environ. Sci. Technol.* **2003**, 37, 286A–294A.
8 A. B. Boxall, L. A. Fogg, P. A. Blackwell, P. Kay, E. J. Pemberton, A. Croxford, *Rev. Environ. Contam. Toxicol.* **2004**, 180, 1–91.
9 T. Heberer, *Toxicol. Lett.* **2002**, 131, 5–17.
10 D. W. Kolpin, E. T. Furlong, M. T. Meyer, E. M. Thurman, S. D. Zaugg, L. B. Barber, H. T. Buxton, *Environ. Sci. Technol.* **2002**, 36, 1202–1211.
11 X. S. Miao, F. Bishay, M. Chen, C. D. Metcalfe, *Environ. Sci. Technol.* **2004**, 38, 3533–3541.
12 W. Witte, *Science* **1998**, 279, 996–997.
13 V. C. Nwosu, *Res. Microbiol.* **2001**, 152, 421–430.
14 K. Kümmerer, *J. Antimicrob. Chemother.* **2003**, 52, 5–7.
15 FEDESA, European Federation of Animal Health, *Deutsches Tierärzteblatt* **2001**, 8, 841.
16 J. P. Langhammer, F. Führ, H. Büning-Pfaue, *Lebensmittelchem. Gerichtl. Chem.* **1990**, 44, 93.
17 J. Gavalchin, S. E. Katz, *J. AOAC Intern.* **1994**, 77, 481–485.
18 G. Hamscher, S. Sczesny, A. Abu-Qare, H. Höper, H. Nau, *Deut. Tierärztl. Woch.* **2000**, 107, 332–334.
19 G. Hamscher, S. Sczesny, H. Höper, H. Nau, *Anal. Chem.* **2002**, 74, 1509–1518.
20 A. B. A. Boxall, P. Blackwell, R. Cavallo, P. Kay, J. Tolls, *Toxicol. Lett.* **2002**, 131, 19–28.
21 H. Höper, J. Kues, H. Nau, G. Hamscher, *Bodenschutz* **2002**, 7, 141–148.
22 M. De Liguoro, V. Cibin, F. Capolongo, B. Halling-Sørensen, C. Montesissa, *Chemosphere* **2003**, 52, 203–212.
23 M. P. Schlüsener, K. Bester, M. Spiteller, *J. Chromatogr. A* **2003**, 1003, 21–28.
24 S. Thiele-Bruhn, T. Seibicke, H. R. Schulten, P. Leinweber, *J. Environ. Qual.* **2004**, 33, 1331–1342.
25 P. Kay, P. A. Blackwell, A. B. Boxall, *Environ. Toxicol. Chem.* **2004**, 23, 1136–1144.
26 T. Soeborg, F. Ingerslev, B. Halling-Sorensen, *Chemosphere* **2004**, 57, 1515–1524.
27 P. Kay, P. A. Blackwell, A. B. Boxall, *Chemosphere* **2005**, 59, 951–959.
28 P. Kay, P. A. Blackwell, A. B. Boxall, *Environ. Pollut.* **2005**, 134, 333–341.

29 B. Halling-Sørensen, A. M. Jacobsen, J. Jensen, G. Sengelov, E. Vaclavik, F. Ingerslev, *Environ. Toxicol. Chem.* **2005**, 24, 802–810.

30 P. Kay, P. A. Blackwell, A. B. Boxall, *Chemosphere* **2005**, 60, 497–507.

31 H. Sanderson, F. Ingerslev, R. A. Brain, B. Halling-Sørensen, J. K. Bestari, C. J. Wilson, D. J. Johnson, K. R. Solomon, *Chemosphere* **2005**, 60, 619–629.

32 J. Tolls, *Environ. Sci. Technol.* **2001**, 35, 3397–3406.

33 H. H. Frey, W. Löscher, ed., *Lehrbuch der Pharmakologie und Toxikologie*, 2nd ed. **2002**, Enke, Stuttgart, Germany.

34 R. Kroker, R. Scherkl, F. R. Ungemach, *Chemotherapie bakterieller Infektionen*, in: Lehrbuch der Pharmakologie und Toxikologie, 2nd ed. **2002**, Enke, Stuttgart, Germany, 353–389.

35 Budavari S, ed., *The Merck index: an encyclopedia of chemicals, drugs, and biologicals.* 12th ed. **1996**, Merck and Co., Inc., Whitehouse Station, N.Y., U.S.A.

36 J. Falbe, M. Regitz, ed., *Römpp Chemie Lexikon.* 9th ed. **1995**, Thieme, Stuttgart, New York, Germany.

37 National Institute of Health, U.S. National library of medicine: Toxnet – Toxicology data network **1999**, can be found under http//toxnet.nlm.nih.gov.

38 Terrabase Inc., TerraQSAR-LOGPTM computed octanol/water partition coefficients (CLOGPs) of antibiotics and antibacterials **2005**, can be found under http://www.terrabase-inc.com/antibiots-antibacts.html.

39 A. Gobel, C. S. McArdell, M. J. Suter, W. Giger, *Anal. Chem.* **2004**, 76, 4756–4764.

40 A. M. Jacobsen, B. Halling-Sørensen, F. Ingerslev, S. H. Hansen, *J. Chromatogr. A* **2004**, 1038, 157–170.

41 S. E. Jørgensen, B. Halling-Sørensen, *Chemosphere* **2000**, 40, 691–699.

42 G. Hamscher, H. T. Pawelzick, S. Sczesny, H. Nau, J. Hartung, *Environ. Health Perspect.* **2003**, 111, 1590–1594.

43 K. Berger, B. Pertersen, H. Büning-Pfaue, *Arch. Lebensmittelhyg.* **1986**, 37, 99–102.

44 J. P. Langhammer, H. Büning-Pfaue, J. Winkelmann, E. Körner, *Tierärztl. Umschau* **1988**, 43, 375–382.

45 C. Winckler, A. Grafe. *J. Soils Sediments* **2001**, 1, 66–70.

46 M. T. Meyer, J. E. Bumgarner, J. L. Varns, J. V. Daughtridge, E. M. Thurman, K. A. Hostetler, *Sci. Total Environ.* **2000**, 248, 181–187.

47 J. Zhu, D. D. Snow, D. A. Cassada, S. J. Monson, R. F. Spalding, *J. Chromatogr. A* **2001**, 928, 177–186.

48 E. R. Campagnolo, K. R. Johnson, A. Karpati, C. S. Rubin, D. W. Kolpin, M. T. Meyer, J. E. Esteban, R. W. Currier, K. Smith, K. M. Thu, M. McGeehin, *Sci. Total Environ.* **2002**, 299, 89–95.

49 D. S. Aga, R. Goldfish, P. Kulshrestha, *Analyst* **2003**, 128, 658–662.

50 P. Eichhorn, D. S. Aga, *Anal. Chem.* **2004**, 76, 6002–6011.

51 M. Y. Haller, S. R. Müller, C. S. McArdell, A. C. Alder, M. J. F. Suter, *J. Chromatogr. A* **2002**, 952, 111–120.

52 H. T. Pawelzick, *PhD thesis*, University of Veterinary Medicine Hannover, Foundation, and University of Hannover, Germany, **2005**.

53 B. Kleefisch, J. Kues J., *Arbeitshefte Boden* **1997**, 2, 1–122.

54 G. Hamscher, H. T. Pawelzick, H. Höper, H. Nau, *Environ. Toxicol. Chem.* **2005**, 24, 861–868.

55 O. B. Samuelsen, V. Torsvik, A. Ervik, *Sci. Total Environ.* **1992**, 114, 25–36.

56 P. Jacobsen, L. Berglind, *Aquaculture* **1994**, 70, 365–370.

57 D. G. Capone, D. P. Weston, V. Miller, C. Shoemaker, *Aquaculture* **1996**, 145, 89–95.

58 G. M. Lalumera, D. Calamari, P. Galli, S. Castiglioni, G. Crosa, R. Fanelli, *Chemosphere* **2004**, 54, 661–668.

59 S. Yang, K. Carlson, *Water Res.* **2003**, 37, 4645–4656.

60 A. L. Batt, D. S. Aga, *Anal. Chem.* **2005**, 77, 2940–2947.

61 M. Rabolle, N. H. Spliid, *Chemosphere* **2000**, 40, 715–722.

62 E. M. Golet, A. Strehler, A. C. Alder, W. Giger, *Anal. Chem.* **2002**, 74, 5455–5462.

63 R. Hirsch, T. Ternes, K. Haberer, K. L. Kratz, *Sci. Total Environ.* **1999**, 225, 109–118.

64 M. L. Loke, F. Ingerslev, B. Halling-Sørensen, J. Tjornelund, *Chemosphere* **2000**, 40, 759–765.
65 F. Ingerslev, B. Halling-Sørensen, *Ecotoxicol. Environ. Saf.* **2001**, 48, 311–320.
66 M. L. Loke, J. Tjornelund, B. Halling-Sørensen, *Chemosphere* **2002**, 48, 351–361.
67 A. C. Kolz, S. K. Ong, T. B. Moorman, *Chemosphere* **2005**, 60, 284–289.
68 A. C. Kolz, T. B. Moorman, S. K. Ong, K. D. Scoggin, E. A. Douglass, *Water Environ. Res.* **2005**, 77, 49–56.
69 R. Kreuzig, S. Holtge, J. Brunotte, N. Berenzen, J. Wogram, R. Schulz, *Environ. Toxicol. Chem.* **2005**, 24, 777–781.
70 JECFA Joint Expert Committee on Food Additives, Toxicological evaluation of certain veterinary drug residues in food. WHO Food Additives Series 41, No. 920. Geneva:World Health Organization, **1998**.
71 S. Yang, J. Cha, K. Carlson, *Rapid Commun. Mass Spectrom.* **2004**, 18, 2131–2145.
72 JECFA Joint FAO/WHO Expert Committee on Food Additives, Toxicological evaluation of certain veterinary drug residues in food, WHO Food Additives Series 33; Geneva: World Health Organization, **1994**.
73 M. E. Lindsey, T. M. Meyer, E. M. Thurman, *Anal. Chem.* **2001**, 73, 4640–4646.
74 G. Hamscher, A. Abu-Quare, S. Sczesny, H. Höper, H. Nau, *Determination of tetracyclines and tylosin in water and soil samples from agricultural areas in lower saxony.* In: Proceedings of the Euro-residue IV conference, Ed.: L. A. Ginkel and A. Ruiter, Veldhoven **2000**, The Netherlands, 08.–10.05.2000, ISBN 90-804925-1-5, 522–526.
75 G. Hamscher, H. T. Pawelzick, H. Höper, H. Nau, *Tierarzneimittel in Böden – eine Grundwassergefährdung?* In: UBA-Texte 29/05 "Arzneimittel in der Umwelt – Zu Risiken und Nebenwirkungen fragen Sie das Umweltbundesamt", Editor: Umweltbundesamt Berlin **2005**, ISSN 0722-186X, 175–184.

6
Polar Herbicides and Metabolites

Rita Fobbe, Birgit Kuhlmann, Jürgen Nolte, Gudrun Preuß, Christian Skark, and Ninette Zullei-Seibert

6.1
General

Plant protection serves very diverse needs, and therefore a very wide variety of types of plant protection agents have been developed. The herbicides are quantitatively the most important class, comprising more than 50% of all pesticides used. The other main classes of pesticides are the insecticides and fungicides, followed by the acaricides, nematicides, molluscicides, and rodenticides.

Herbicides are chemical agents specially designed for weed control. The extensive use of herbicides is a precondition for total mechanization in the cultivation of all kind of vegetables, beets, cereals, potatoes, corn, and cotton. For example, United States farmers spend about $5.5 billion annually on herbicides.

But in addition to their use in agriculture, herbicides are used extensively for the prevention of uncontrolled weed growth in urban and industrial areas, at roadsides, on railroad embankments and canal banks, and on buildings where undesired plants might cause damage.

To reduce their environmental impact, modern herbicides are expected to decompose within a relatively short time after application without forming persistent metabolites or degradation products.

As this book is devoted to polar organic pollutants (log K_{ow} < 3), some well-known groups of herbicides are not included in this review because most of their members possess a higher value of this parameter: amides, anilides, arylalanines, chloroacetanilides, carbanilates, cyclohexene oximes, dinitroanilines, and dinitrophenylethers. Further disregarded groups are the aryloxyphenoxypropionic herbicides and also the phenylendiamine, phenylpyrazolyl ketone, pyridazine, carbamate, and phenylenediamine herbicides. Fundamental information on pesticides included in this chapter is based on comprehensive textbooks on pesticide chemistry and application [1–5].

Organic Pollutants in the Water Cycle. T. Reemtsma and M. Jekel (Eds.).
Copyright © 2006 WILEY-VCH Verlag GmbH & Co. KGaA, Weinheim
ISBN: 3-527-31297-8

6.1.1
History

In about 1874, sulphuric acid and its salts were applied as the first inorganic herbicides, acting by altering pH of soil. The first organic herbicide was 2-methyl-4,6-dinitrophenol (DNOC), which was developed in 1892 and initially used as an insecticide. It has been applied as a herbicide since 1934 and today is the longest used organic herbicide. However, it took until the end of Second World War for the boom in pesticide development to start. 2,4-Dichlorophenoxyacetic acid (2,4-D) was the first herbicide to be widely used against broadleaf plants. As a low-cost product 2,4-D is mostly applied as the amine salt or in the ester form. With the introduction of atrazine into the market in the 1970s, groundwater contamination became a real problem, especially as atrazine has relatively high stability, a very unfavorable property of a pesticide. Ten years later, glyphosate was introduced as a nonselective agent. Meanwhile, following the parallel development of seeds resistant to glyphosate, this chemical has become a major herbicide in selective weed control.

Detailed information on some individual agents is provided in the following section. Further research is also focused on so-called safeners. These are additives designed specifically for the exclusive protection of the cultivated crop against the applied herbicide [6].

6.1.2
Classification and Application

6.1.2.1 Classification

Herbicides can be classified under different aspects such as selectivity, chemical structure, the mode and mechanism of the biological action, and the types of plants to be controlled.

The simplest distinction is that between selective and nonselective herbicides, and both classes can be applied when the mode of biological action does not affect the cultivated plant. Proper timing and dosage are decisive for an optimal effect.

Mode of biological action is another basis of classification, distinguishing between contact herbicides, systemic herbicides, and soil-borne herbicides. As the name indicates, contact herbicides act directly on the plant's surface at the point where the pesticide spray is deposited, and they are therefore the fastest acting agents, destroying the plant tissue. Systemic herbicides, also called translocated herbicides, are applied to the plant part above the ground surface and are translocated through the total plant; in this way, they destroy more plant tissue than contact herbicides. Their outstanding advantage is their great success when used against perennials. Systemic or translocated pesticides may also be adsorbed by the roots. Herbicides acting via soil can only be adsorbed by the roots and act similarly to systemic herbicides.

Another classification is based on the time of application. A pre-emergent herbicide is active before germination or in the early phase of growth of the weeds; post-emergent applications are done after the crop or weed emerges from the soil.

Proper timing depends on the type of chemical family of the herbicide, its persistence, the crop type and its tolerance to the herbicide, the weed type, the climate, and the soil type.

Other herbicides can be classified according the way they inhibit the mechanism of some biological cell processes, in particular ACCase, the ALS and EPSPs enzymes, the photosystem II, and the auxin generation.

The ACCase and the ALS enzyme regulate cell membrane production in the meristem. Whereas the ALS enzymes of both grasses and dicotyledonous plants are sensitive to herbicides, only the ACCase of grasses is sensitive to such herbicides. The hindered EPSPs enzyme is necessary for the synthesis of the amino acids tryptophan, phenylanaline, and tyrosine, and the photosystem II manages the ion flow of $NADPH_2^+$ in photosynthesis. The hormone system of the plants regulating the growth can be influenced effectively by application of herbicides containing auxins, a group of natural and especially synthetic hormones.

6.1.2.2 Application

Herbicides are used wherever the growth of undesired weeds has to be inhibited, and are used extensively in North America, Western Europe, Japan, and Australia. Consumption per km^2 can vary widely, as is illustrated for Europe in Fig. 6.1 and Table 6.1 [7, 8]. The application rate may vary depending on the crop cultivated, and the form of application can be by band, broadcast, and spot treatments, and by directed spraying. Band application treats a strip along or in a plant row, broadcast application covers the total cultivation area, and spot treatment is applied to small areas of weeds. Directed sprays are applied to selected weeds or to the soil, contact with the crop being undesired. A modern technique is application over-the-top to both the crop and the weeds. This technique is only indicated shortly after germination when the crops are naturally tolerant to specific herbicides.

The herbicides are mostly applied as aqueous sprays by appropriate equipment from a vehicle. In large areas or topographically adverse situations, airplanes and helicopters are also used for fumigation. In order to achieve high application efficiency, auxiliary substances are mixed with the herbicides. These so-called formulation additives ensure optimal distribution in the spray and ultimately at the plant surface.

The deployed concentrations of the active ingredients, even of the same herbicide, vary strongly and depend mostly on the effectiveness of the agent, the time of application, the field conditions, and the technical way of spraying. For example, some amounts are presented in the literature for diuron: in the UK 6.7 kg ha^{-1}, in France 1.8 kg ha^{-1}, and in Germany 0.6–4.8 kg ha^{-1} are applied in pre-emergence mode and 10–30 kg ha^{-1} as a total herbicide [9, 10]. Another example is the varying application amount for atrazine: about 1 kg ha^{-1} is applied as a selective herbicide in corn cultivation, but approximately 9 kg ha^{-1} as a total herbicide. After contamination of the aquatic environment was observed, maximum application rates were limited to 1 kg ha^{-1} and the use of atrazine was even banned in many European countries [11, 12].

6 Polar Herbicides and Metabolites

Tab. 6.1 Applied active agents of herbicides and the respective area under cultivation in 14 European countries [km^2] in relation to the national territory [%] (1999) [8].

Country	Herbicides [t year^{-1}]	Area under cultivation [km^2]/(%)
Belgium	1846.58	7558/(25.0)
Denmark	864.89	23711/(55.9)
Germany	12035.26	120587/(34.5)
Greece	661.72	40012/(30.6)
Spain	5439.64	190625/(38.2)
France	27941.04	193208/(35.4)
Ireland	280.00	13454/(19.5)
Italy	4594.08	109728/(37.3)
Netherlands	1595.12	9338/(27.6)
Austria	1057.94	14794/(17.9)
Portugal	1273.24	26031/(28.3)
Finland	518.72	21352/(7.0)
Sweden	611.67	27944/(6.8)
Great Britain	5763.68	61823/(25.6)

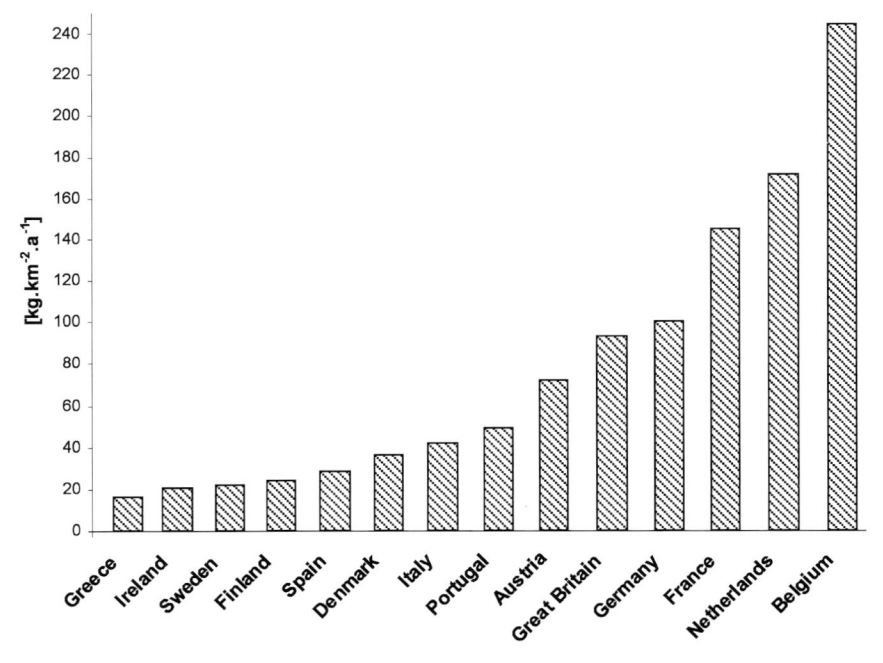

Fig. 6.1 Herbicides per cultivated area of 14 European countries in 1999.

6.1.3
Herbicide Classes Considered

6.1.3.1 Phenoxycarboxylic Acids

Phenoxyacetic acid herbicides as well as phenoxypropionic and phenoxybutyric acid herbicides belong to this class. 2,4-D and MCPA are phenoxyacetic acids, 2,4-DP (dichlorprop), 2,4,5-TP (fenoprop) and MCPP (mecoprop) are phenoxypropionic acids, whereas 3,4-DB and MCPB are phenoxybutyric acid herbicides.

Apart from benzoic acids and pyridines, phenoxyacetic acids are the most important growth regulators. Growth regulators disrupt the hormone balance and protein synthesis in a plant in different ways [13].

In 1942, the first compound of this group, 2,4-D, was introduced into the market, and the group as a whole was found to have herbicidal properties in 1944 after studies in the United States and the UK. They are usually formulated as emulsions in oil or as water-soluble amines or salts. The water solubility of 2,4-D in natural waters ranges from 20 g L^{-1} (pH 5) to 34.2 g L^{-1} (pH 9) and is the highest in this group [14].

(2,4,5-Trichlorophenoxy)acetic acid (2,4,5-T) should be mentioned even though its log K_{ow} > 4.9, because in the 1970s 2,4,5-T together with 2,4-D was the main agent in the defoliant "Agent-Orange". Agent Orange was discredited because of its content of dioxin impurities caused by a side reaction in synthesizing the precursor 2,4,5-trichlorophenol. Nowadays 2,4,5-T plays only a secondary role as a herbicide.

During degradation of the phenoxycarboxylic acids, hydrolysis occurs, the respective substituted phenols being the main metabolites, which are considered to be persistent because of the number of chlorine atoms.

Tab. 6.2 Representative phenoxycarboxylic acids.

	R_1	R_2	R_3
MCPA [(4-chloro-2-methylphenoxy)acetic acid]	CH_3	H	CH_2
MCPP (Mecoprop) [2-(4-chloro-2-methylphenoxy)propionic acid]	CH_3	H	$CHCH_3$
2,4-D [(2,4-dichlorophenoxy)acetic acid]	Cl	H	CH_2
2,4-DP (Dichlorprop) [2-(2,4-dichlorophenoxy)propionic acid]	Cl	H	$CHCH_3$
2,4,5-TP (Fenoprop) [(2,4,5-trichlorophenoxy)propionic acid]	Cl	Cl	$CHCH_3$

6.1.3.2 Triazines

Chlorotriazines and methylthiotriazines are differentiated. Chlorotriazines are for example atrazine, cyanazine, sebutylazine, and simazine, whereas ametryne, desmetryne, prometryne, and simetryne are methylthiotriazines. Hexazinone, metamitron, and metribuzin are triazinones.

The efficiency of triazines increases with the introduction of chloro-, methoxy- or alkylthiogroups into the molecule. Triazines are very diverse with respect to their target spectrum, selectivity, and duration of effect. Their main effect is based on the inhibition of photosynthesis, similarly to the ureas. Triazines are mostly applied in tropical areas, in Europe and in North America, especially in the cultivation of crops and corn.

Today the economically most important compounds are the 1,3,5-triazine derivatives, which were discovered in 1955. In the United States, atrazine has been the most widely used herbicide since the 1960s. In the early 1990s the global consumption of herbicides amounted to 2.5 million t year^{-1}, of which 36 000 t year^{-1} were due to atrazine used in the United States [12, 15].

In parts of Europe, triazines constitute 27% of all herbicides used [16]. Also in China, triazines are the most commonly applied herbicides [17].

Tab. 6.3 Representative triazines.

	R_1	R_2	R_3
Atrazine [1,3,5-triazine-2,4-diamine-6-chloro-N-ethyl-N'-(1-methylethyl)]	NHCH(CH$_3$)$_2$	Cl	NHCH$_2$CH$_3$
Simazine [1,3,5-triazine-2,4-diamine-6-chloro-N,N'-diethyl]	NHCH$_2$CH$_3$	Cl	NHCH$_2$CH$_3$
Ametryne [1,3,5-triazine-2,4-diamine-6-methylthio-N-ethyl-N'-1-methylethyl]	NHC$_2$H$_5$	SCH$_3$	NHC(CH$_3$)$_2$
Terbutylazine [4-*tert*-butylamino-2-chloro-6-ethylamino-1,3,5-triazine]	NHC(CH$_3$)$_3$	Cl	NHCH$_2$CH$_3$
Cyanazine [2-(4-chloro-6-ethylamino-1,3,5-triazine-2-ylamino)-2-methylpropionitrile]	NHC(CH$_3$)$_2$CN	Cl	NHCH$_2$CH$_3$

A significant reduction in application started in the 1990s. Atrazine was the preferred herbicide in corn cultivation until it became prohibited in Germany in 1991 [18]. Also in Austria, atrazine-containing products have been forbidden since May of 1995 [19]. In Switzerland, the consumption was reduced from 105 to 35 t year^{-1} within 10 years [12]. In the 1990s, a ten-year study in Southern Germany demonstrated that atrazine followed by desethylatrazine, desethylsimazine and simazine were the most frequently detected compounds. In 1999, 8 years after the ban on atrazine application in Germany, concentrations exceeding the precautionary limit value of 0.1 µg L^{-1} could still be found in groundwater and surface water. Meanwhile, atrazine and simazine are being replaced by less polar substances such as terbutylazine and terbutryne, which are the only s-triazines presently permitted in Germany [11]. It has been shown that triazines accumulate in soil by forming non-extractable residues of about 10–30% [20]. Thus, the real amount of triazines in soils is significantly underestimated.

The main three chlorinated metabolites of atrazine are desethylatrazine (DEA; 2-amino-4-chloro-6-isopropylamino-s-triazine), desisopropylatrazine (DIA; 2-amino-4-chloro-6-ethylamino-s-triazine), and diaminoatrazine (2-chloro-4,6-diamino-s-triazine). In the United States, DEA and DIA are two of the six most frequently found metabolites in groundwater [21, 22]. The degradation rates of atrazine are strongly affected by the pH value, the surface area, and the type of manganese oxides in the soil [18].

Photodegradation of triazines plays an important role, depending on the position of the chlorine atoms. The unsaturated form of triazines degrades faster than the saturated form [23]. Although irradiation contributes to degradation and leads to decomposition within 5 days, triazines are said to be persistent, because they are detectable in soil over 70 days [15].

6.1.3.3 Aromatic Acid Herbicides

After the discovery of their growth-promoting effect it was found that synthetic auxins can be used as weed killers, and, besides the phenoxy acids discussed above, other aromatic acids were developed as selective herbicides. Aromatic carboxylic

Tab. 6.4 Representative benzoic acids.

	R
Dicamba [3,6-dichloro-2-methoxybenzoic acid]	OCH$_3$
2,3,6-TBA [2,3,6-trichlorobenzoic acid]	Cl

Tab. 6.5 Representative picolinic acids.

	R_1	R_2
Clopyralid [3,6-dichloropyridine-2-carboxylic acid]	H	H
Picloram [4-amino-3,5,6-trichloro-2-pyridinecarboxylic acid]	NH_2	Cl

acid herbicides comprise various subgroups, for example the benzoic acids, including chloramben, 3,6-dichloro-2-methoxybenzoic acid (dicamba), 2,3,6-trichlorobenzoic acid (2,3,6-TBA), and tricamba. Further representatives are the picolinic acids aminopyralid, clopyralid, and picloram, and the quinolinecarboxylic acids quinclorac and quinmerac. Already 1954, 2,3,6-TBA was introduced into the market by DuPont, and dicamba has been available since 1965. In these years, products based on picolinic acid (e.g., picloram, Dow Chemicals) were also added. In Germany 2,3,6-TBA has been forbidden since 1979.

In the class of aromatic acids, the benzoic acid derivatives, in particular dicamba, are the most important. Analogously to the phenoxycarboxylic acids, they act as growth regulators after uptake by leaves and roots. Some plants are able to metabolize dicamba. It is in use in forestry, viniculture, grassland, and nonagricultural areas. Weed control takes place by ground or aerial broadcast, band, basal bark, cut surface, or spot treatment. Application of such amounts as are customary in agriculture does not cause any lasting damage to soil microflora. In contrast, noticeable effects are the consequence using these agents as total herbicides.

The application rate for the active agent of dicamba is 0.3–9.0 kg ha^{-1} and 0.3–3.3 kg ha^{-1} for picloram. While the degradation of dicamba occurs relative quickly (within a maximum of 40 days), picloram is much more stable and may persist for 1 year because of strong sorption to soil organic matter.

6.1.3.4 Benzonitriles

Of the group of benzonitriles, dichlobenil and the hydroxybenzonitriles bromoxynil, ioxynil, and chloroxynil are the most important.

In 1959, the herbicide effect of halogen-substituted 4-hydroxybenzonitriles was observed. Especially the 3,5-dihalogenated compounds worked effectively. Their biological activity is the highest in presence of substituted iodine, decreases with bromine, and is relatively small with substituted chlorine. They act both by interruption of oxidative phosphorylation and by inhibition of photosynthesis (Hill reaction). These agents are used as selective post-emergence agents, mainly in the cultivation of grain. A further development led to the less toxic bromofenoxim,

Tab. 6.6 Representative benzonitriles.

	R_1	R_2	R_3	R_4	R_5
Dichlobenil [2,6-dichlorobenzonitrile]	Cl	H	H	H	Cl
Bromoxynil [3,5-dibromo-4-hydroxybenzonitrile]	H	Br	OH	Br	H
Ioxynil [3,5-diiodo-4-hydroxybenzonitrile]	H	I	OH	I	H

although after a short time this hydrolyzes to the more toxic bromoxynil and 2,4-dinitrophenol [24, 25].

6.1.3.5 Dinitroalkylphenol Herbicides

Besides DNOC and dinoseb (Fig. 6.7), further dinitrophenol herbicides such as dinoprop and dinoterb have gained in importance since 1945.

Based on their extremely acidic character, these agents are typical contact herbicides, etching the surface of the leaves. Plants having a natural wax layer are protected against this. The preferred mode of action is the interruption of oxidative phosphorylation. They are still present in the environment, but in recent years have become less important.

Tab. 6.7 Representative dinitroalkylphenol herbicides.

	R
DNOC [2-methyl-4,6-dinitrophenol]	CH_3
Dinoseb [2-(1-methylpropyl)-4,6-dinitrophenol]	$CH(CH_3)C_2H_5$

6.1.3.6 Phosphorus Compounds

A further class of herbicides is the organophosphorus agents such as fosamine, glyphosate, and glufosinate.

In the end of the 1950s, phosphorus herbicides were developed as harvest aids. In practice they serve as defoliating agents, for example in cotton cultivation. Tris[2-(2,4-dichlorophenoxy)ethyl]phosphite (2,4-DEP) was one of the first selective herbicides among phosphorus-based compounds. It is used against dicotyledonous plants in grass. Today, predominantly glyphosate and glufosinate are in use. O-Methyl-O-2-nitro-p-tolyl isopropylphosphoramidothioate (amiprofos-methyl), developed in Japan, is of significance. It is a highly efficient mitosis inhibitor and is applied against dicotyledonous weeds and annual grasses. Amiprofos-methyl is a selective herbicide and often applied in market gardening as well as in dry cultivation of rice as a pre-emergence herbicide.

Glyphosate is a broad-spectrum, nonselective herbicide, which is active on most green plants and exclusively acts via the above-ground biomass, the leaves [26]. It is a pale yellowish liquid of low purity, completely soluble in water and polar solvents. It has been in frequent use in agriculture since 1971 and is known as Roundup or Touchdown. The application amount of 0.5–4 kg ha^{-1} depends on the type of weed [27]. It is considered to be environmentally friendly and nontoxic for humans. Glyphosate acts by inhibition of a metabolism enzyme needed for the EPSP synthesis [28]. The similarity to natural amino acids is the cause for its high efficiency. Surface-active agents increase the effect. It is formulated as the isopropylamine salt [29]. Since about 2000, glyphosate has been the most sold and most expanding herbicide worldwide.

The mobility of glyphosate is restricted by strong adsorption in soils, where glyphosate competes with inorganic phosphate for soil binding sites [30]. On the other hand, because of its enormous water solubility (12 g L^{-1}), glyphosate is transported into the subsoil if there are no sorption sites available. Hydrolysis of glyphosate occurs with a half-life of 12 to 60 days [31], but it is relative stable when directly bound to particles. After biological decomposition it can only be analyzed by its main metabolite aminomethylphosphonic acid (AMPA). AMPA is more persistent than its parent compound; in different soils in the United States its half-life was

Tab. 6.8 Representative phosphorus compounds.

Structure: R$_1$–P(=O)(OH)–CH$_2$–R$_2$–CH(R$_3$)–COOH	R_1	R_2	R_3
Glufosinate [L-2-amino-4-[hydroxy-(methyl)-phosphinyl]-butanoic acid]	CH$_3$	CH$_2$	NH$_2$
Glyphosate [N-(phosphonomethyl)glycine]	OH	NH	H

determined to be between 119 and 958 days [31]. In water, the long-lasting high concentration of AMPA, exceeding 1µg L^{-1}, does not correlate with the application amount of glyphosate. A possible explanation is that AMPA results also from the decomposition of phosphonic acids and complexing agents of detergents [32, 33, 34].

In embankments, glyphosate has replaced the previously used phenoxyacetic acids and diuron. Present annual consumption in the USA is about 22 million kg and in general 2.4 kg ha^{-1} is used [26, 31]. The application of the nonselective glyphosate was not suitable in soybean, colza, or corn cultures until the introduction of genetically modified plants and the modified herbicide 2-amino-4-[hydroxy(methyl)phosphinoyl]butyrate (glufosinate) with the trade name Liberty (in Germany also Basta) [35].

Glufosinate was developed in 1984 and is only suitable for controlling weeds among genetically modified plants. In the form of the ammonium salt it is also used in fruit- and wine-growing. It acts as a toxic agent for nearly all plants by inhibition of the glutamine synthesis, but it can be used in association with the herbicide-resistant system LibertyLink.

Degradation quickly takes place, with a half-life in soil of 3 to 20 days, dependent on climatic conditions. Enrichment of the herbicide or any metabolites has not been detected so far.

6.1.3.7 Ureas

The negative effect of ureas on plant growth was first observed in 1946 but without a direct consequence for agricultural practice. Only since the introduction of the phenylureas monuron and diuron as herbicides the urea group has developed into one of the commercially most interesting groups of weed-regulating products. The tri-substituted urea derivatives show the highest herbicidal activity.

In addition to the phenyl- and sulfonylureas mentioned below, thiadiazolylureas are an important subgroup of ureas which includes ethidimuron and tebuthiuron as members. Benzthiazuron and methabenzthiazuron also belong to this class. By variation of the substitution of the urea, a wide variation in the effect, selectivity, and behavior in the environment is obtainable. They can be absorbed via leaves and soil. Most of them are phenyl-substituted.

Phenylureas

Representatives of the phenylurea herbicides include buturon, chlorotoluron, dimefuron, diuron, fluometuron, isoproturon, linuron, methiuron, metobromuron, metoxuron, monolinuron, monuron, and neburon.

The mode of action for all phenylureas is the inhibition of photosynthesis by blocking the electron transport in the photosystem II.

In 1951, monuron was introduced as the first herbicide of this class. Soon after, diuron came onto the market, and it is still today one of the most important herbicides of the substituted urea type. Known as DCMU, it was described in basic research on plant physiology. It is applied both as a total herbicide and at a lower rate

Tab. 6.9 Representative phenylureas.

	R_1	R_2	R_3
Isoproturon [N-(4-isopropylphenyl)-N',N'-dimethylurea]	$(CH_3)_2CH$	H	CH_3
Diuron [N-(3,4-dichlorophenyl)-N',N'-dimethylurea]	Cl	Cl	CH_3
Monuron [N-(4-chlorophenyl)-N',N'-dimethylurea]	Cl	H	CH_3
Chlorotoluron [N-(3-chloro-4-methylphenyl)-N',N'-dimethylurea]	CH_3	Cl	CH_3
Linuron [N-(3,4-dichlorophenyl)-N'-methoxy-N'-methylurea]	Cl	Cl	OCH_3

as a selective herbicide. The application amount of diuron varies between 0.7 and 1.5 g m^{-2}, depending on its formulation. Because of its low selectivity and high persistence, application in tillage is not common. The metabolism of diuron leads to 3,4-dichlorophenyl derivatives, whereas linuron degrades to dichloroaniline and dimethylhydroxyamine. As well as diuron, isoproturon is of considerable significance and is widely used against weeds in grain cultures.

Sulfonylureas
The sulfonylurea herbicides constitute an important subgroup of the phenylureas and can be divided into pyrimidinylsulfonylurea and triazinylsulfonylurea herbicides. Amidosulfuron, bensulfuron, ethoxysulfuron, flupyrsulfuron, nicosulfuron, primisulfuron, rimsulfuron, and sulfometuron are pyrimidinylsulfonylureas, whereas chlorsulfuron, ethametsulfuron, iodosulfuron, metsulfuron, prosulfuron, thifensulfuron, triasulfuron, and tribenuron belong to the triazinylsulfonylureas. Sulfonylureas are characterized by their low application rates, in the lower grams per hectare range. The mode of action differs from the other ureas; it works by inhibition of the acetolactate synthesis. Hydrolysis occurs more quickly in an acidic ambience than in a basic one. They have a good selectivity against a wide range of weeds in cereal crops at application rates of 4–60 g ha^{-1}. For noncrop industrial use this rate is much higher, at 175 g ha^{-1}. Sulfonylurea herbicides can be adsorbed by foliage and roots and are transported in the xylem and phloem [36].

The first sulfonylurea herbicide applied in row crops was chlorsulfuron, which was introduced into the market in 1981. It was quickly accepted in cereal protection: in 1995 more than 6 million hectares was treated in both pre- and post-emer-

Tab. 6.10 Representative sulfonylureas.

	R_1	R_2	R_3
Metsulfuron-methyl [methyl-2-(4-methoxy-6-methyl-1,3,5-triazine-2-ylcarbamoylsulfamoyl)benzoate]	Ph-CO$_2$CH$_3$	T-CH$_3$-OCH$_3$	H
Tribenuron-methyl [methyl-2-(4-methoxy-6-methyl-1,3,5-triazine-2-yl(methyl)-carbamoylsulfamoyl)benzoate]	Ph-CO$_2$CH$_3$	T-CH$_3$-OCH$_3$	CH$_3$
Rimsulfuron [1-(4,6-dimethoxypyrimidine-2-yl)-3-(3-ethylsulfonyl-2-pyridylsulfonyl)urea]	P-SO$_2$C$_2$H$_5$	Py-OCH$_3$-OCH$_3$	H
Amidosulfuron [1-(4,6-dimethoxypyrimidine-2-yl)-3-mesyl(methyl)sulfamoylurea]	NCH$_3$SO$_2$CH$_3$	Py-OCH$_3$-OCH$_3$	H

T = triazine; Ph = phenyl; P = pyridine; Py = pyrimidine

gent application worldwide [37]. Chlorsulfuron hydrolyzes at pH 5 with a half-life of 24 d, and this time increases to more than 1 year in a pH range of 7 to 9. Similar data are observed for aerobic and anaerobic decomposition. Hydrolysis of chlorsulfuron leads to dihydroxytriazine. A similar degradation tendency is noticed for metsulfuron-methyl. The half-life is 5 d in acid soil, 69 d in basic soil, and 139 d in sterile soil [38].

6.1.3.8 Nitrogenous Compounds

Pyridines are nitrogen-containing compounds, and their derivatives include the herbicides aminopyralid, clopyralid, fluroxypyr, pyriclor, thiazopyr, and triclopyr.

Further nitrogenous herbicides are pyrazolylphenyl, pyridazinone, and quaternary ammonium compounds. Fluazolate is a pyrazolylphenyl, and chloridazon and norflurazon are pyridazinone herbicides. Chlormequat, diquat, and paraquat represent the quaternary ammonium herbicides.

In the mid-1960s, pyridine herbicides were developed. Although the efficiency of pyridines is described in numerous patent specifications, only pyrichlor gained importance. It finds application as a total herbicide on industrial sites and at the roadside. A further product is thiazopyr, applied against weeds in the cultivation of cotton, peanuts, soybeans, fruit, and wine. Of the quaternary ammonium herbicides, paraquat and diquat (also deiquat), both bipyridylium salts, should be mentioned. They act as contact herbicides of fast effectiveness against dicotyledonous plants, solely affecting the above-ground parts. Diquat is predominantly used in potato growing to destroy haulm previous to harvest. Paraquat acts as a total herbicide, but without permanent effect.

6.2
Entry into the Environment

Figure 6.2 provides a scheme of the various application sites and the different subsequent pathways of herbicide entry into the aquatic environment. The main release both in terms of amount and frequency occurs in agricultural and horticultural use. But application can also take place at nonagricultural areas, on pavements, at industrial sites, in residential areas, or on railway tracks or roads. Some unintentional dissipation may occur via rainwater and from landfills with herbicide package deposits or other related wastes.

Contamination of surface water is mainly due to surface runoff and effluent from wastewater treatment plants (WWTP). Groundwater is primarily contaminated by release of herbicides from topsoil and further transport through the vadose zone to the aquifer. Details are given in the individual chapters.

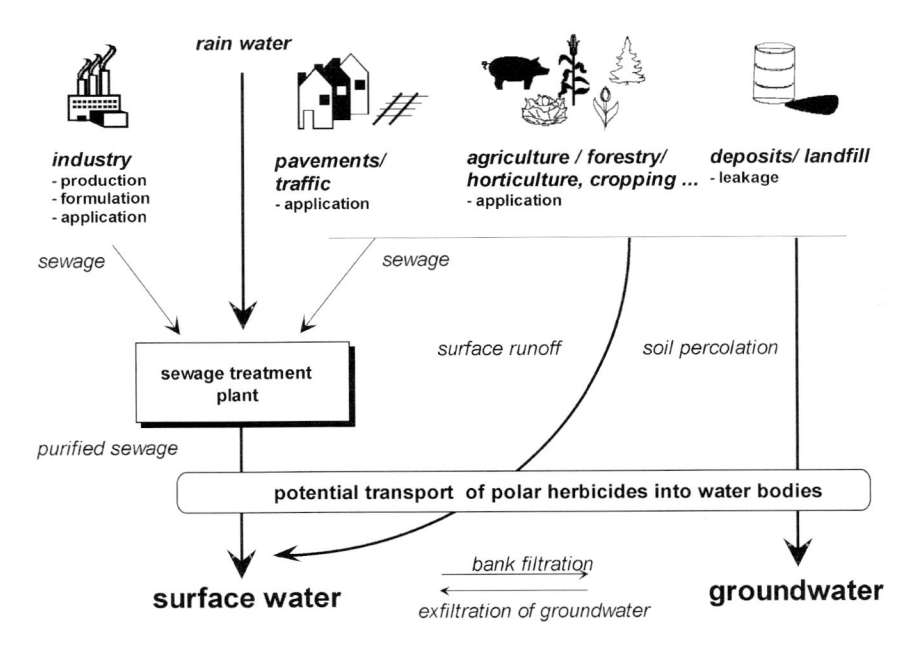

Fig. 6.2 Possible pathways for the entry of herbicides into water.

6.3
Polar Herbicides in Wastewater

By cleaning pesticide spraying equipment on the farmyard instead of washing it at the field edge, highly concentrated solutions of pesticides may enter the sewerage system if the farmyard is connected to it [39]. The same will happen when herbicides are applied at sealed surfaces in residential areas. In this way, municipal wastewater may be significantly loaded with herbicides during the period of herbicide application.

In municipal wastewater treatment plants (WWTP), the polar herbicides are not adequately removed. Therefore, in WWTP effluents, high concentrations of different herbicides can be observed, which contribute to the surface water contamination.

A concentration decrease in WWTP is often due to sorption onto suspended solids rather than to biodegradation. The relative importance of sorption as compared to biodegradation can increase when [40, 41]

- the residence time in the WWTP is too short for implementing an efficient degradation process;

- there is a high nutrient level for microorganisms, which hinders the adaptation of bacteria leading to degradation of a specific herbicide;

- even microorganisms which are specialized in the biodegradation of herbicides will find an alternative, more easily accessible nutrient source in the activated sludge treatment system.

For polar herbicides, removal by sorption may be negligible, resulting in considerable herbicide concentration in WWTP effluents. The contribution of WWTP to the pesticide load in surface waters can vary over a wide range. For herbicides in agricultural use, it depends on the degree of drained farmyards, the actual practice in cleaning of the spraying equipment in the catchment area, and on the season. Concentration peaks of agriculturally used herbicides during dry weather periods are an indication of inputs from cleaning. In a Swiss study, the contribution of WWTP effluents to the input of agricultural herbicides (atrazine, isoproturon, metolachlor) into a lake was determined to be almost 20%. When every farmyard is connected to the sewerage system, this contribution can reach more than 90% [42, 43]. In certain regions, herbicide contamination of surface water may be explained in this way, whereas in other regions diffuse sources such as surface run-off or exfiltration from groundwater may prevail [44, 45]. For nonselective herbicides in nonagricultural use (e.g., diuron), the path from impervious surfaces via WWTP or directly into surface water is the major entry path.

Apart from its application in agriculture, the phenoxy acid herbicide MCPP is also used as a root protection agent in bituminous roof sealing membranes (e.g., in Switzerland). The racemic ratio for the product used for this purpose is 1:1 between R- and S-MCPP, whereas in other application only R-MCPP is used. If an MCPP enantiomeric ratio of 1 is observed in WWTP outlets, this is a strong indication of MCPP leaching from roof sealing [42, 46, 47].

6.4
Polar Herbicides in Surface Water

Regarding the literature on the occurrence of herbicides in the aquatic environment, much data are available from the 1980s and the early 1990s, but less data have been published within recent years. Because the pesticide pollution of surface waters strongly depends on the actual use profile, data before 1995 are rarely included in this review. Reviews on the situation before 1995 can be found in [48, 49].

6.4.1
Pathways to Surface Waters

Herbicides may enter surface water via different pathways, of which run-off from treated fields is the most important. Entry from treated fields can happen directly as well as after a short passage through soil. The amount of herbicide run-off depends on the soil type, the topography, the weather conditions, the types of plants, and of course on the technique, quantity, and time of herbicide application. Very high concentrations are found in the run-off

- shortly after application,
- after application of large amounts,
- after the first strong rainfall,
- from slopes,
- when herbicides of high water solubility are used, which is the case with polar herbicides.

While input with the run-off leads to short, but high concentration peaks, entry of herbicides via field drains leads to a constant pollution at relatively low concentrations. Depending on the weather, herbicides may also evaporate from treated fields and may enter surface water via wet or dry deposition.

When looking at the pathways described above, it is obvious that the type of herbicide entering the aquatic environment depends on the actual agricultural practice. Most farmers use the products recommended by agricultural advisory boards. Practice may vary from time to time because of changes in herbicide registration or because of the introduction of new active ingredients, but there are some substances like the phenoxyacetic acids that have already been used for decades. Therefore, most of the herbicides actually used in an area can be found in the surface water. Even when a certain herbicide is not used in an area or is prohibited in a country, traces of this compound may occur in surface water when sewage of a local herbicide producer is discharged. Furthermore farmers sometimes use pesticides that are no longer registered. Besides regular agricultural practice, a source of herbicides in surface water can be their irregular use, for example, directly beside rivers, or the incorrect disposal of remnants of applied solutions.

Moreover, nonselective herbicides are not only used in agriculture but also in domestic, municipal, and industrial environments. For example, the occurrence of diuron in surface waters in Germany is strongly connected with its nonagricultural use.

6.4.2
Occurrence in Surface Water

Herbicides are routinely monitored by many waterworks. Of the group of polar herbicides, triazines, phenoxyacetic acids, and phenylureas are common subjects of monitoring programs. When surface water is analyzed, the most frequent positive findings are atrazine, simazine, and terbutylazine and their metabolites, together with MCPP, 2,4-D, diuron, and isoproturon. Because of analytical difficulties, glyphosate is not a common part of the monitoring programs, but when it is monitored it is detected in almost every sample. As mentioned above, the link between actual herbicide usage and occurrence in surface water is very close. This can clearly be shown for herbicides in the river Ruhr (Germany) between 1998 and 2003. Figure 6.3 presents the percentage of positive findings in all samples taken during the regular monitoring program [50].

In 1998, atrazine, despite its ban in 1991, was still found in Germany in about 15% of all samples, while there are only a few findings in 2002 and 2003. A similar trend was found for other triazine herbicides in the Ruhr (simazine and terbutylazine).

Diuron, the most important herbicide used in the nonagricultural sector, is present in 18–25% of all samples, with quite constant findings. MCPP is the most frequently detected phenoxyacetic acid (up to 15% positive findings), but it shows

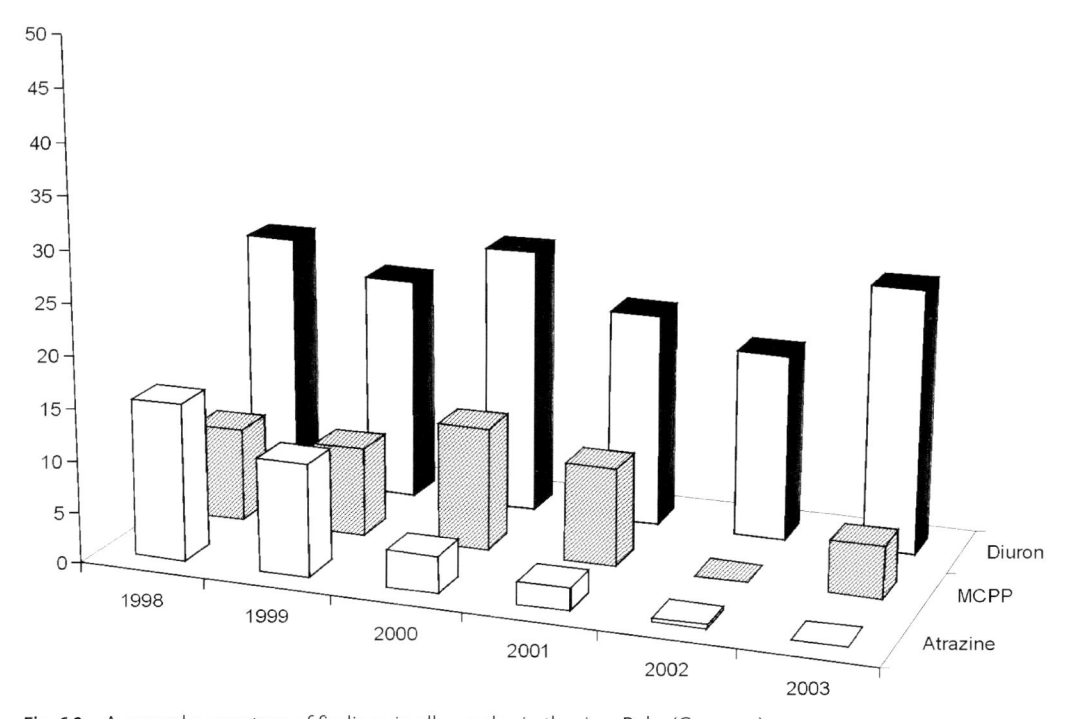

Fig. 6.3 Averaged percentage of findings in all samples in the river Ruhr (Germany).

high variability without a clear trend. Another pesticide frequently found in the river Ruhr is glyphosate, a commonly used nonselective herbicide. In contrast to diuron, which seems to be present over the whole year, glyphosate occurs as shock loads. It was investigated in some research programs within the river Rhine area and in some tributaries of the Ruhr in detail. In the rivers Neckar, Rhine, Main, Danube and two tributaries to Lake Constance, it was found in concentrations up to 0.1 µg L^{-1}. The concentration of its metabolite AMPA tends to be 5 to 20 times higher than the concentration of glyphosate itself [51, 52].

A summary of the monitoring done by the German Federal States (LAWA) up to 1998 also showed the important contribution of some polar herbicide groups (triazines, phenoxyacetic acids, and phenylurea derivatives) to the contamination of surface waters [44]. This is also valid for the monitoring from 2001 to 2003 [53]. The river Rhine near Cologne is regularly analyzed for about 25 different polar herbicides. In 2002, atrazine, bentazone, chlorotoluron, dichlorprop, diuron, isoproturon, and MCPP showed at least one positive result over the year [54].

A monitoring program in the Dutch part of the river Rhine confirmed that the glyphosate concentrations are increasing from 1995 to 2003. Other polar herbicides detected in the river Rhine are shown in Table 6.11. Chlorotoluron, linuron, metobromuron, methabenzthiazuron, metoxuron, monolinuron, desethylatrazine, metamitron, propazine, simazine, and terbutylazine could not be detected (limit of quantification between 0.03 and 0.1 µg L^{-1} depending on the substance) [55].

The river Lek near Nieuwegen (The Netherlands) is regularly monitored, and positive findings are reported for bentazone, 2,4-D, MCPA, MCPP, chlorotoluron, diuron, isoproturon, dinoterb, and terbutryne [55]. In the Dutch part of the river Meuse, glyphosate exceeds the level of 0.1 µg L^{-1} from time to time, while its metabolite AMPA is almost continuously present in concentrations above this level [56].

A national survey of pollutants in surface waters in Austria (1996–1998) included some triazine and phenoxyacetic acids. Atrazine was found in about 5% of all samples, and simazine in 0.5%. 2,4-D, MCPA, MCPP, and MCPB were found in less than 1% of the samples [57]. The relatively few findings of atrazine in Austrian surface waters in this study reflects the banning of atrazine in 1994. A countrywide overview from 1991 to 1995 showed about 27% positive findings for atrazine and 24% for desethylatrazine [19]. Glyphosate is one of the widely used herbicides in Austria. A national survey, with 345 samples taken from 42 surface waters from

Tab. 6.11 Polar herbicides detected in the Rhine at Lobith (The Netherlands) in 2003 [56].

Herbicide	Max. concentration (µg L^{-1})	Positive samples/samples
Glyphosate	0.17	11/12
Diuron	0.08	6/12
Isoproturon	0.07	4/12
Atrazine	0.07	3/12
Terbutryne	0.06	1/12

June 2001 to June 2002, revealed glyphosate concentrations >0.1 µg L^{-1} in four samples. AMPA was analyzed in 90 samples with a concentration >0.1 µg L^{-1}, but it is doubtful if this only relates to the degradation of glyphosate, because other anthropogenic compounds are also metabolized via AMPA [58].

In a study from Switzerland, about 75% of herbicides without use in agriculture were shown entering surface waters via WWTPs [42].

In a river in Greece, which is located in an agricultural area, atrazine, desethylatrazine, and simazine are detected over the whole year, with peak concentrations of 0.174 µg L^{-1} (atrazine), 0.121 µg L^{-1} (desethylatrazine), and 0.07 µg L^{-1} (simazine) [59]. In another area of Greece, the above-mentioned herbicides were detected frequently in surface waters [16].

Triazine findings prevailed also in the Charente River in France. Of all herbicides detected, 90% belong to this group, including 60% atrazine and terbutylazine metabolites and simazine [60]. Another French study presented positive findings for triazines (atrazine and metabolites, simazine) and phenylureas (chlorotoluron, isoproturon, diuron, linuron, neburon) [61].

In Portuguese rivers, atrazine and simazine concentrations revealed clear seasonal and spatial variations. Their concentration was below the detection limit (of about 10 ng L^{-1}) in winter and in the upper parts of the rivers, where no agriculture is found. During summer, tributaries of the river Tejo exhibited atrazine and simazine concentrations up to 0.63 and 0.29 µg L^{-1}, respectively [62].

In Great Britain, pesticides are regularly monitored in surface waters by the Environment Agency. From 1998 to 2003, 9 polar herbicides exceeded the maximum contamination level (MCL) of 0.1 µg L^{-1} according to the EU Drinking Water Directive, most frequently MCPP, isoproturon, MCPA, diuron, 2,4-D, chlorotoluron, dichlorprop, simazine, and atrazine [63].

The linkage between the actual use of polar herbicides and their occurrence in surface water is also reported by the United States Geological Survey. They analyzed the Mississippi River at Baton Rouge from 1991 to 1997, taking into account the herbicides use in the Mississippi River Basin. The most frequently used herbicides were triazines, especially atrazine and cyanazine, and some chloroacetanilides such as alachlor and metolachlor. The total herbicide concentration in the Mississippi River varied seasonally, with highest concentrations from May to August, but did not exceed 10 µg L^{-1}. No single compound was found above 5 µg L^{-1}, while smaller tributaries showed higher concentration levels. Regarding the triazines, atrazine, desethylatrazine, desisopropylatrazine, and cyanazine were found in more than 50% of the samples. Cyanazine amide, simazine, metribuzin, and prometon were observed as well, but in minor amounts. Ametryne, prometryne, propazine, and terbutryne were detected in less than 2% of all samples [64]. In the Mississippi Delta, 1995 samples from three streams were analyzed for selected cotton and rice herbicides. Four out of the eight herbicides and metabolites that exceeded the concentration of 5 µg L^{-1} throughout most of the growing season belong to the polar herbicides fluometuron, cyanazine, atrazine, and prometryn [65]. In 97% of all samples taken from the Mississippi River Basin between March 1999 and May 2001, atrazine was detected [66]. A further study by the United States Geological Survey assessed the quality of the main streams in the United States, based on 1600 samples from 65 sites. More than 95% of all samples contained at least one herbicide. The most frequently detected polar herbicides in agricultural areas were atrazine (about 70% of all samples) and its metabolites (20%) and cyanazine (5%), while in urban areas simazine (80%), 2,4-D (15%), and diuron (10%) were repeatedly detected. Stream concentrations of atrazine and cyanazine frequently exceeded the aquatic life criteria established by the Environmental Protection Agency (EPA) [67]. The prominent role which the triazines, its metabolites, and diuron play in the contamination of United States surface waters is also confirmed by other investigations [68, 69]. Concentrations above 0.1 µg L^{-1} were found for 2,4-D, dicamba, bromacil, diuron, and sulfometuron in a river in south-eastern New York State from 2000 to 2001 [70].

In a Canadian investigation, residues of atrazine, the second most applied herbicide in Ontario, were found in a small agricultural river at a low µg L^{-1} concentration level in nearly 90% of the samples; the concentration in run-off from treated fields was up to 14 µg L^{-1} [71].

Triazine herbicides were analyzed in a river located in Eastern China as well. Atrazine, simazine, propazine, and terbutylazine were detected, with especially high values during the growing season. In March, the atrazine concentration was approximately 0.18 µg L^{-1}; in June, after application, 1.6 µg L^{-1} was found [72].

6.5
Polar Herbicides in Groundwater

6.5.1
Pathways to Groundwater

Herbicide transport with the seepage water to soil horizons below the root zone has often been described. The applied amount, application technique, and hydraulic conditions during and after the application are the main factors influencing the extent of herbicide transport from topsoil to subsoil. During that transport, sorption and degradation may occur [73, 74].

The main sorbents in soils are its organic matter fraction and clay minerals. In the topsoil and the upper part of the unsaturated zone, organic carbon content exceeds 1% in general, while it decreases in subsoil, sediments of porous aquifers, and in bedrock aquifers. Consequently, the importance of herbicide sorption decreases during transport from topsoil to deeper soil layers. Particularly for polar substances, the interaction with clay minerals is of great importance. Additionally, the pH influences sorption of polar herbicides with acidic or basic moieties.

Because of the high microbial activity and diversity in topsoils, herbicides may be degraded very fast after application. However, microbial activity declines by several orders of magnitude toward deeper soil layers and aquifer sediments, while microbial diversity also decreases. Hence degradation rates of herbicides decrease with increasing depths, and abiotic processes like hydrolysis become more important. For example, the half-life of atrazine can increase from 60 d in topsoil to more than 10 years at aquifer conditions [75, 76].

Interactions between surface water and groundwater can influence the herbicide concentration in both aquatic compartments. Pesticides in surface water may enter groundwater bodies via bank filtration. In general, groundwater is discharged to receiving water and may contribute to pesticide load in the surface water. This process is of particular importance during base flow conditions in the streams [45, 77–81]. Nonequilibrium processes concerning hydraulic movement of seepage water by preferential flow phenomena or sorption kinetics may influence the degree of herbicide retardation during the soil passage substantially [77, 81–87]. Particularly in drained farmland preferential flow paths may generate a by-pass flow to shortcut the soil passage [88].

The amount of pesticides leached to the underground can reach up to 5% of the applied mass in the worst case when heavy rainfall events accompany the application. A herbicide application during dry weather conditions may also result in leaching, but this ranges from <0.1% to 1% of the applied amount [74]. Once leached to groundwater, herbicides pass through the aquifer like other solutes, and only limited retardation compared to the water movement was observed [73, 89].

6.5.2
Occurrence in Groundwater

In the United States, during the pesticide monitoring undertaken by the National Water Quality Assessment (NAWQA), groundwater samples were taken from 1034 sites in shallow aquifers, representing 20 major hydrological basins in agricultural and urban settings across the United States. These samples were examined for the residues of 46 pesticide compounds during the period 1993 – 1995 [90]. Pesticides were commonly detected in more than half of all sampling points. The most frequently detected substances were atrazine, desethylatrazine, simazine, metolachlor, and prometon, which occurred both at agricultural and urban sites. Concentrations were below 1 µg L^{-1} in almost 95% of all positive results. The enlargement of that monitoring campaign with respect to time, sites investigated, and compounds did not change these preliminary results in a relevant way [67, 91–94]. Sulfonylurea, sulfonamide, and imidazolinone herbicides were rarely investigated in groundwater. In one study in the Midwestern United States, nicosulfuron, imazethapyr, and flumetsulam were observed in groundwater at maximum concentrations of 0.016 µg L^{-1}, 0.059 µg L^{-1}, and 0.035 µg L^{-1} respectively [95].

In various countries of the European Union, pesticides have been monitored in groundwater for many years, because a maximum contamination level (MCL) for drinking water of 0.1 µg L^{-1} for a single active ingredient or its metabolite was established in 1980. The results of early groundwater investigations (up to 1995) in several countries have been compiled and assessed in various reviews [48, 96, 97]. Analytical procedures focused on triazines and phenylurea herbicides, but also included phenoxy acids. Atrazine was the most frequently analyzed and detected compound in groundwater of European countries, followed by one of its metabolites, desethylatrazine (DEA); apart from these two compounds 2,4-D, MCPP, and simazine were found in 6 out of 8 EU countries with concentrations exceeding 0.1 µg L^{-1}.

In Denmark, shallow groundwater (1.5 to 5 m below surface) was investigated for 46 compounds. Under sandy soil, mainly triazines and their degradation products were frequently found, whereas under clayey soil phenoxy acids occurred additionally, because the oxygen availability in these subsoils is restricted and in clayey subsoils anoxic conditions prevail [98]. These findings were underlined in a Danish monitoring program. However, bentazone, isoproturon, and 2,6-dichlorobenzamide (BAM), a metabolite of dichlobenil, were also commonly found [49, 96]. BAM also seems to persist under anoxic conditions [99]. During the resumption of the monitoring program (1998–2003), triazines and their metabolites as well as BAM were the most abundant compounds. In groundwater abstraction wells, the phenoxy acids MCPP and dichlorprop were also frequently detected [100].

In the survey of groundwater in England and Wales by the Environment Agency, atrazine, bentazone, dichlorprop, diuron, MCPP, and prometryne were observed in more than 3% of the analyzed samples at concentrations above 0.1 µg L^{-1} in at least one of the years from 1998 to 2003 [63].

The groundwater of the major corn-growing areas in Greece was monitored for selected herbicide residues during the period 1996–1997. In 38 wells (out of 80 analyzed) herbicides were detected with concentrations below 0.1 µg L^{-1}. In only 9 wells the concentration exceeded 0.1 µg L^{-1}. The most frequently occurring compounds were atrazine, DEA, alachlor, and metolachlor. Unexpectedly, the group of wells with the statistically significant highest average concentration of herbicides were at depths between 50 and 100 m [101]. These results were confirmed by groundwater analyses of the Imathia region in Greece, where mainly fruit trees and cotton are cultivated [16]. Investigations of shallow groundwater under wheat, corn, cotton, and rice fields in northern Greece also showed the frequent occurrence of atrazine, alachlor, prometryne, and propanil [102, 103]. The maximum concentration of atrazine exceeded 1 µg L^{-1}. For the other herbicides the maximum concentration remained below 0.5 µg L^{-1}.

In the groundwater of a watershed near Barcelona, Spain, during an investigation period between February and August 2002, mainly atrazine, simazine, and diuron were detected, mostly with concentrations below 0.1 µg L^{-1} [104].

In the drinking water abstraction wells situated in catchment areas with intensive agriculture in the south east of the Netherlands, bentazone and atrazine were detected at concentrations up to 0.28 µg L^{-1} and 0.06 µg L^{-1} respectively [105].

Results from groundwater monitoring in Germany were published in a survey by the Länderarbeitsgemeinschaft Wasser (LAWA) [106, 107]. During the period 1990 to 2000 the by far most abundant compounds were atrazine and DEA (Fig. 6.4). Atrazine was applied in large quantities in corn cropping and as a nonselective herbicide for nonagricultural purposes (e.g., weed control at railway tracks and in residential areas), but its use was banned in Germany in 1991. Therefore, decreasing concentrations in groundwater were expected [108]. In fact the frequency of detection and the concentration level in groundwater of the compounds atrazine

6.5 Polar Herbicides in Groundwater

and DEA decreased as a general trend but much more slowly than expected (Fig. 6.5). In a karst aquifer with a relatively small mean groundwater residence time (approx. 14 years), it could take 16 years for the DEA concentration to decrease to a

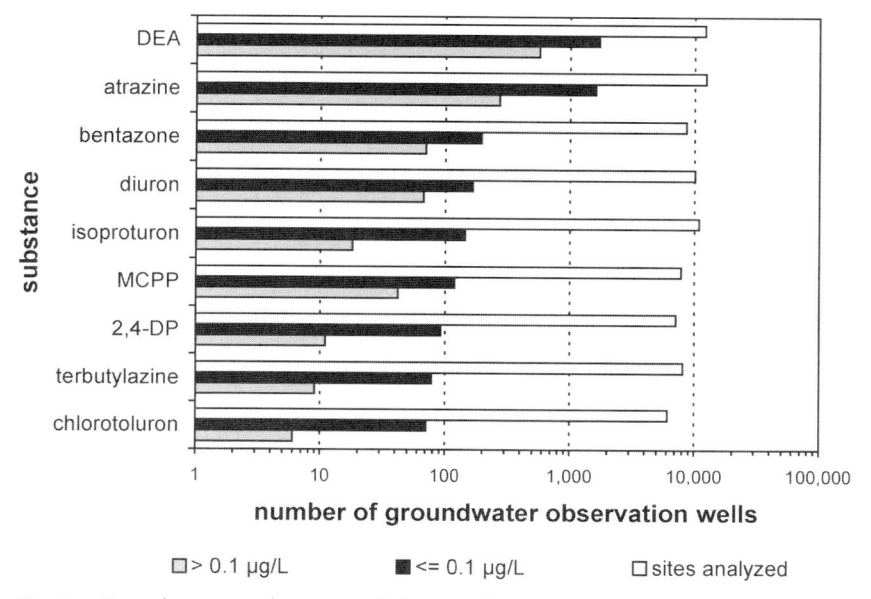

Fig. 6.4 Groundwater sampling sites with frequent detection of selected herbicides and metabolites in Germany during the period 1996–2000.

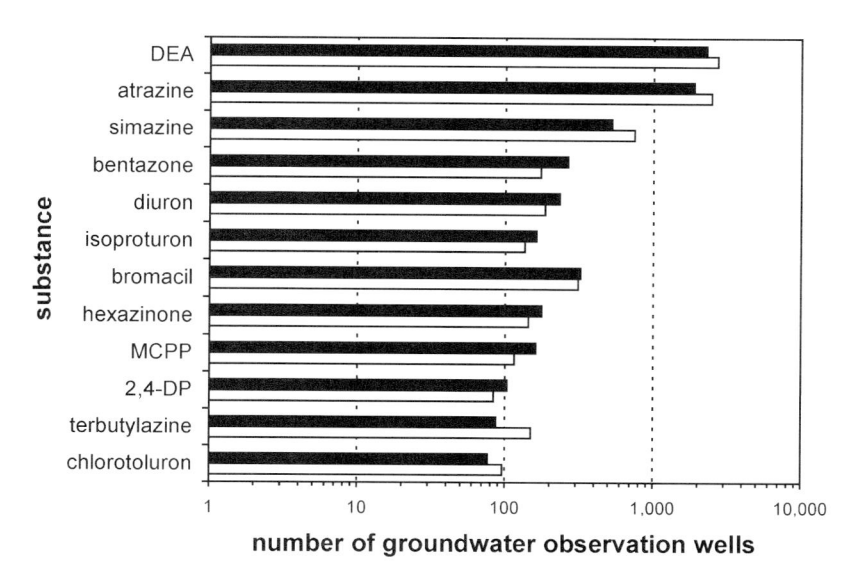

Fig. 6.5 Groundwater sampling sites with frequent detection of selected herbicides and metabolites in Germany; positive results of the periods 1990–1995 (white) and 1996–2000 (black).

quarter of its original value so that the groundwater would comply with the drinking water MCL [109].

In the LAWA survey, the agricultural herbicides bentazone, isoproturon, chlorotoluron, terbutylazine, MCPP, and dichlorprop were also observed at concentrations above 0.1 µg L^{-1} in various sampling sites [106, 107].

The use of nonselective herbicides for weed control apart from agriculture results in remarkable pollution of groundwater in Germany. The nonselective active ingredients diuron, bromacil, hexazinone, and the dichlobenil metabolite 2,6-dichlorobenzamide deserve particular attention. Although in Germany these herbicides are applied in much smaller amounts than herbicides used in agriculture and relative restricted areas (railway tracks, residential areas), they were found in groundwater samples quite abundantly. At railway tracks, the herbicides are sprayed at the ballast, which does not offer similar sorption sites and microbial biocoenosis like a topsoil. Therefore, the retardation as well as the biodegradation of herbicides at railway tracks is diminished compared to field application. In residential areas, private house owners may infringe legal regulations by applying herbicides on paved areas. Because of this use, the herbicides may infiltrate into the pavement gaps without passing topsoil.

The dominance of atrazine and DEA in groundwater analysis results was also well documented by the groundwater investigations of the German waterworks [110, 111]. However, not all herbicides that are applied were included in these programs. Thus, a possible groundwater contamination may remain unrevealed. This is particularly true for glyphosate, which was – together with isoproturon – one of the most widely spread herbicides in Germany but is not regularly analyzed in the water quality control of waterworks. Thus, few reports on the occurrence of herbicides in groundwater include glyphosate, and reports on positive results are rare [58, 112–114].

Karst aquifers can be assumed to be highly vulnerable, whereas their hydraulic conditions, with sinkholes and networks of conduits, are very complex and vary from site to site. Therefore, the occurrence of herbicides is to be expected if they are used in the karst area, but the flow path and the points of resurgence are hard to predict. Reported groundwater occurrences of herbicides include the active ingredients mentioned above [109, 115, 116].

6.5.3
Metabolites

Degradation of herbicides may yield metabolites which are more polar and therefore more mobile in groundwater than the parent compounds. For example, DEA and desisopropylatrazine show lower retardation factors than atrazine in field studies [89]. As well as the dealkylated metabolites, the hydroxyl degradation products of triazines are of great relevance to the assessment of the fate of these herbicides in aquatic environment [117, 118]. The relevance of various degradation products as groundwater contaminants has already been shown by various studies on cyanazine [22, 92, 119]. The metabolites may contribute to a total pesticide concentration of between 55 and 99% of the applied amount according to a study in Iowa [115]. The most frequent detections and the highest concentrations were found in bedrock aquifers of alluvial and karst regions.

Studies in the United Kingdom (UK) showed that groundwater can be contaminated by Diuron and its three main metabolites 3-(3,4-dichlorophenyl)-1-methylurea (DCPMU), 1-(3,4-dichlorophenyl)urea (DCPU), and 3,4-dichloroaniline (DCA) more than 50 days after application [9].

Laboratory studies in the United States demonstrated that the pH value is a very important factor for the degradation of chlorsulfuron in water; a half-life of 24 days at a pH-value of 5 increased to more than one year at a pH value of 7–9. In an aerobic silt loam soil study, degradation rapidly occurs within 20 days, whereas in an anaerobic sediment/water system it was much slower and finally took one year [37].

In a field study in Denmark, the fate of commonly used herbicides was investigated at different sites representing a broad range of conditions including climate, soil, and crops cultivated. Average concentrations of degradation products of glyphosate, metribuzin, rimsulfuron, and terbutylazine exceeding 0.1 µg L^{-1} in the leachate deeper than 1 m below the surface were found. Particularly, the metabolites of metribuzin seemed to persist in groundwater over a period of several years [114].

6.6
Water Treatment

Studies on the behavior of pesticides in water treatment processes refer predominantly to laboratory tests or investigations of semi-technical plants, sometimes under conditions related to waterworks practice. In addition, water supply companies have collected a lot of data on the behavior of pesticides during drinking water purification processes, but have seldom published their results. However, information on the behavior of polar herbicides is needed, because the limited number of reports available that are based on realistic operating conditions or which reproduce practical conditions are already several years old [48, 120, 121].

6.6.1
Activated Carbon

The effectiveness of sorption processes depends fundamentally on the characteristics of the substance itself and on the type of activated carbon selected. Organic background contamination of the untreated water can reduce the sorption capacity by two orders of magnitude [122]. In addition, operational requirements may lead to contact times which are too short to ensure equilibration. As a basic rule, the extent of removal rises as more activated carbon is used. This applies to powdered activated carbon as well as to granulated fixed-bed adsorbers. The specific filter capacity (m^3 kg^{-1}) describes the quantity of water that can pass through a filter before a given final concentration (in the case of Fig. 6.6 the European drinking water

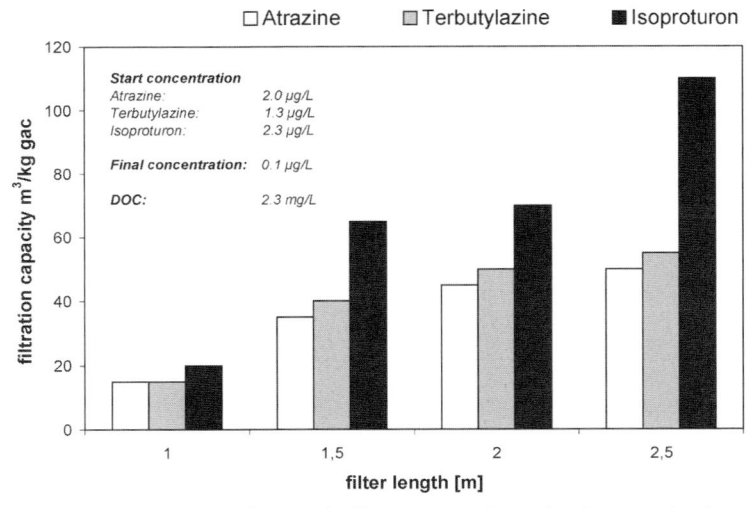

Fig. 6.6 Improvement in the specific filter capacity of granulated activated carbon (gac) for selected herbicides by increasing the filter length.

standard for single pesticides [123] of 0.1 µg L^{-1}) is reached. High specific filter capacities indicate a good and cost-effective performance.

Atrazine, isoproturon, simazine, and terbutylazine lead to markedly reduced filter capacities under given background conditions and thus to increased regeneration costs. In comparison to atrazine, the phenylurea herbicide diuron is strongly adsorbed, whereas the phenoxycarboxylic acid MCPA adsorbs only weakly. No direct relationship between the water solubility and the K_{ow} value could be observed [124]. Nevertheless, the following polar pesticides are well adsorbed by activated carbon: 2,4-D, bentazone, bromacil, dichlorprop, diuron, linuron, MCPP, monuron, terbutryne, triclopyr [125].

6.6.2
Physico-chemical Processes (Flocculation, Rapid Filtration, Sedimentation)

The efficiency of physico-chemical processes like flocculation, filtration, and sedimentation depends on the properties of the respective herbicide. Only herbicides of low water solubility that tend to adsorb on suspended solids will be removed effectively. Generally, these classic methods of water purification have limited potential to reduce polar herbicide contamination of raw waters [122, 126].

6.6.3
Oxidation and Disinfection

Oxidation and disinfection methods have been intensively investigated on the laboratory scale. Information is available on ozonation and hydrogen peroxide treatment, chlorination and chlorine dioxide treatment, potassium permanganate dosage, and ultraviolet irradiation [48]. Ametryne, bentazone, and MCPB were reduced by ozone by between 76 and 100%. In contrast, elimination rates for 2,4-D, atrazine, diuron, linuron, MCPA, monuron, and simazine show large variations. Depending on several influence factors, any individual substance may be reduced by between 0 and 100% [48]. The herbicides chlorotoluron, diuron, isoproturon, MCPA, MCPP, methabenzthiazuron, and metobromuron are easily degradable by ozone (>80 to >99%; pH 2.2 – 8.3, temperature 5–20 °C, ratio $O_3/DOC = 1.0$) [127]. Glyphosate could be completely destroyed by ozone after 5 – 7 min [126]. Highly resistant pesticides require a higher dosage of ozone or combination of ozone with hydrogen peroxide [127]. If ozone is intended to be used for pesticide removal, case-specific conditions, such as the quality of the raw water, must be considered. Preliminary trials at a French waterworks showed that although a combination of ozone and hydrogen peroxide is in principle suitable for reducing atrazine and simazine, an activated carbon stage is still needed because of the overall quality of the treated water [128].

The use of chlorine or chlorine dioxide hardly reduces the concentration of herbicides. Propazine seems to be an exception and, to a lesser extent, atrazine, diuron, and simazine [48]. Other studies on chlorination show no effect on triazine herbicides. Glyphosate requires high chlorine doses (>2 mg L^{-1}), which are not permitted in Europe, to secure 100% removal after 6.5 min [126].

With ultraviolet irradiation only propazine was removed from the 10 pesticides tested, while alachlor, atrazine, cyanazine, fenchlorphos, and neburon were not affected [48]. The suitability of ultraviolet oxidation to remove atrazine concentrations of 0.02 to 1.8 µg L^{-1} from a groundwater was investigated. Relatively large quantities of hydrogen peroxide had to be added, which led to problems as nitrite was produced. In addition, the high energy requirements were disadvantageous [129].

6.6.4
Natural Filtration Processes (Bank Filtration, Slow Sand Filtration, Underground Passage)

Bank filtration or artificial groundwater recharge via gravel and slow sand filtration involve considerably longer contact times than technical rapid filtration processes. These longer contact times support biological degradation processes and the adjustment of sorption equilibria. Additionally, small scale changes in the oxygen environment (aerobic, anaerobic zones) can amplify substance degradation. Pilot plant investigations have led to the conclusion that aerobic conditions during slow sand filtration and underground passage enhance the degradation of phenoxyacetic acids like 2,4-D, 2,4,5-T, MCPA, and MCPP [130, 131]. In comparison, the aerobic degradation of hydroxybenzonitriles like bromoxynil and ioxynil and the sulfonylurea compounds amidosulfuron, metsulfuron-methyl, rimsulfuron, thifensulfuron-methyl and the herbicide bentazone was weak. Some herbicides like bentazone appear to be stable during riverbank filtration regardless of the redox milieu and can be used as conservative tracers, although they tend to degrade readily in the vadose zone [49, 132–134]. Under anaerobic conditions, the phenoxycarboxylic acids, phenylureas, and sulfonylureas are mostly persistent, while degradation of many triazines and bromoxynil was sufficient [131, 135, 136]. Glyphosate and its degradation product AMPA is completely reduced by natural filtration processes [137, 138]. Also waterworks practice in Germany shows a good performance for the removal of polar pesticides like atrazine, simazine, isoproturon, diuron, and MCPP [139].

6.6.5
Membrane Filtration

Membrane techniques such as reverse osmosis and ultra- or nanofiltration can in principle be used for removing pesticides. Their effectiveness heavily depends on the selection of suitable membranes. However, the reduction of the pesticide concentration in the permeate results in a concentration increase in the retentate that can amount to 10% of the treated water volume. This retentate has to be treated separately. The following polar pesticides were sufficiently reduced: bentazone, bromoxynil, dinoseb, propazine, and terbutylazine [140]. Using reverse osmosis with composite thin film, 80–100% retention was obtained for triazines, acetanilides, urea derivatives, and carbamates. Membranes consisting of cellulose acetate and polyamide show inferior performance [126].

Nanofiltration with four different membranes and unknown input concentrations led to a removal of atrazine in the range 80–98%, simazine 63–93%, diuron 43–87%, and bentazone 96–99%. A combination of membranes with activated carbon gave better results [126].

6.6.6
Conclusions – Water Treatment

Information on the behavior of polar herbicides during water treatment is available for only a limited number of relevant compounds. The majority of the studies was performed at concentration levels which do not correspond to practical conditions or were performed on additives above the permitted levels for water disinfection. These studies are of limited use for waterworks practice.

The processes generally considered best suited for the removal of polar herbicides in waterworks practice are the use of activated carbon or natural biological filtration steps like artificial groundwater recharge or bank filtration. It should be noted that in the activated carbon processes, shorter service life and higher costs of regeneration must be accepted if the processes are designed for removal of very polar herbicides and for compliance with the drinking water limit standards for pesticides. In addition, activated carbon filtration cannot be used effectively at locations where the raw water has a high humic acid content. Membrane techniques show sufficient reduction rates for some pesticides.

Combined processes with downstream disinfection are frequently used after treatment of surface water or bank filtrates for drinking water production. It may be assumed as a rule that a combination of the adsorption with the oxidation stages will be adequate for removing all herbicides [141, 142].

6.7
Summary

Polar herbicides are widely used man-made substances. Because of their chemical diversity, different analytical methods are required for monitoring purpose. As a consequence, only a small fraction of all the herbicides in use is analyzed in groundwater, surface water, and drinking water, or in research work. The few comprehensive investigations show as a rule that the occurrence of polar herbicides in water bodies is strictly related to the amount applied as well as to failures in good application practice. Thus, positive findings mostly concern atrazine, simazine, terbutylazine and its metabolites, bentazone, isoproturon, diuron, chlorotoluron, MCPP, 2,4-D or glyphosate, and rarely sulfonylureas. Even if raw water used for drinking water production is contaminated by polar herbicides, the drinking water itself is usually not affected. Natural as well as technical filtration and oxidation steps or combined purification techniques are able to solve the problems. Therefore, in Europe only minor cases of exceeding the drinking water standard of 0.1 µg L^{-1} are reported. Globally, cases of contamination of drinking water always involve areas where the use of herbicides is intensive and the water treatment is inadequate.

References

1 *Pflanzenschutz und Schädlingsbekämpfung* (Ed.: K.H. Büchel), Georg Thieme Verlag, Stuttgart, **1977**.

2 *Herbizide* (Eds.: B. Hock, C. Fedtke, R. R. Schmidt), Georg Thieme Verlag, Stuttgart, **1995**.

3 *Pestizide im Boden* (Ed.: K. H.Domsch), VCH, Weinheim, **1992**.

4 *The Pesticide Manual*, (Ed.: C. D. S. Tomlin), British Crop Protection Council, **1997**.

5 *Wirksubstanzen der Pflanzenschutz- und Schädlingsbekämpfungsmittel*, 3. Auflage (Eds.: Werner Perkow, Hartmut Ploss), Parey Buchverlag, Berlin, **2001**.

6 "Safer durch Safener", can be found at http://www.bayer.de/geschaeftsbericht2002/reportagen/pflanzenschutz1.html.

7 George W. Ware, "*An introduction to herbicides*", can be found at http://ipmworld.umn.edu/chapters/wareherb.htm.

8 "*Applied active agents of herbicides*" can be found at http://www.welt-in-zahlen.de; EDS European Data Service, Federal Statistical Office, Germany, **2004**.

9 Daren C. Goody, P. John Chilton, Ian Harrison, *Sci. Total Environ.* **2002**, 297, 67–83.

10 Céline Tixier, Philippe Bogaerts, Martine Sancelme, Frédérique Bonnemoy, Landoald Twagilimana, Annie Cuer, Jacques Bohatier, Henri Veschambre, *Pest. Manag. Sci.* **2000**, 56, 455–462.

11 W. Tappe, J. Groeneweg, B. Jantsch, *Biodegradation* **2002**, 13, 3–10.

12 S. R. Müller, M. Berg, M. M. Ulrich, R. P. Schwarzenbach, *Environ. Sci. Technol.* **1997**, 31, 2104–2113.

13 "Herbicide Mode of Action and Sugarbeet Injury Symptoms" can be found at http://www.sbreb.org/brochures/herbicide/herbicide.htm.

14 K. A. Hassall, *The Biochemistry and Uses of Pesticides*, VCH, Weinheim, Second Edition **1990**, p. 398–413.

15 S. I. Rupassara, R. A. Larson, G. K. Sims, K.A. Marley, *Bioremediation J.* **2002**, 6 (3), 217–224.

16 T. A. Albanis, D. G. Hela, T. M. Sakellarides, I. K. Konstantinou, *J. Chromatogr. A* **1998**, 823, 59–71.

17 M. Gfrerer, T. Wenzl, X. Quan, B. Platzer, E. Lankmayr, *J. Biochem. Biophys. Meth.* **2002**, 53, 217–228.
18 J. Y. Shin, M. A. Cheney, T. Spiro, *Preprint of the Orlando Symposium of the American Chemical Society, Division of Environmental Chemistry* **2002**, 42 (1), 43–47.
19 A. Chovanec, *Acta Hydrochim. Hydrobiol.* **1997**, 25, 5, 233–241.
20 G. Henkelmann, "Das Verhalten von Pflanzenschutzmitteln im Agrarökosystem" can be found at http://www.lfl.bayern.de/iab/bodenschutz/ 08142/linkurl_0_0_0_6.pdf.
21 J. E. Hanson, D. E. Stoltenberg, B. Lowery, L. K. Binning, *J. Environ. Qual.* **1997**, 26, 829–835.
22 D. W. Kolpin, E. M. Thurman, D. A. Goolsby, *Environ. Sci. Technol.* **1996**, 30, 335–340.
23 D. Barcelo, T. Konstantinova, M. Doytcheva, M. Mondeshky, *Fresenius Environ. Bull.* **2001**, 10, 203–207.
24 B. Graß, H. Mayer, J. Nolte, G. Preuß, N. Zullei-Seibert, *Pestic. Manag. Sci.* **2000**, 56, 49–59.
25 J. Nolte, F. Heimlich, B. Graß, N. Zullei-Seibert, G. Preuß, *Fresenius J. Anal. Chem.* **1995**, 351, 88–91.
26 "Physikalisch-chemische Eigenschaften und Wirkmechanismen von PBSM" can be found at http://www.lua.nrw.de/veroeffentlichungen/ggb97/gb97k2.htm.
27 "Glyphosate" can be found at http://www.dcchem.co.kr/english/product/p_detail/p_detail11.htm.
28 "Glyphosat" can be found at http://www.biosicherheit.de/lexikon/8.lexi.html.
29 "Was ist Glyphosat" can be found at http://www.roundup.de/wirkung/glyphosat.htm.
30 S. M. Carlisle, J. T. Trevors, *Water Air Soil Pollut.* **1988**, 39, 409–420.
31 "Glyphosate Factsheet" can be found at http://www.mindfully.org/Pesticide/Roundup-Glyphosate-Factsheet-Cox.htm.
32 "Trinkwasserverträgliche Unkrautvernichtungsmittel in der Ruhr" can be found at http:// www.ak-wasser.de/notizen/gwguete/pestiz01.htm.
33 R. Reupert, C. Schlett, *gwf Wasser Abwasser* **1997**, 138, 559–563.
34 W. E. Gledhill, T. C. J. Feijtel, *The handbook of environmental chemistry* (Ed. O. Hutzinger), Vol. 3 Part F, Springer, Berlin, Heidelberg, **1992**, pp. 261–285.
35 "Glyphosat und Glufosinat" can be found at http://www.transgen.de/Lexikon.
36 "How Herbicides Work" can be found at http://eesc.orst.edu/agcomwebfile/edmat/EM8785.pdf.
37 H. J. Strek, *Pestic. Sci.* **1998**, 53, 29–51.
38 N. Pons, E. Barriuso, *Pestic. Sci.* **1998**, 53, 311–323.
39 A. Carter, *Pestic. Outlook* **2000**, 11, 149–156.
40 N. C. Meakins, J. M. Bubb, J. N. Lester, *Chemosphere* **1994**, 28, 1611–1622.
41 D. T. Monteith, W. J. Parker, J. P. Bell, H. Melcer, *Water Environ. Res.* **1995**, 67, 964–970.
42 A. C. Gerecke, M. Schärer, H. P. Singer, S. R. Müller, R. Schwarzenbach, M. Sägesser, U. Ochsenbein, G. Popow, *Chemosphere* **2002**, 478, 307–315.
43 H.-G. Frede, P. Fischer, M. Bach, *J. Plant Nutr. Soil Sci.* **1998**, 161, 395–400.
44 M. Bach, A. Huber, H.-G. Frede, V. Mohaupt, N. Zullei-Seibert, *Schätzung der Einträge von Pflanzenschutzmitteln aus der Landwirtschaft in die Oberflächengewässer Deutschlands*, Erich Schmidt, Berlin, **2000**, p. 273.
45 P. J. Squillace, E. M. Thurman, E. T. Furlong, *Water Resour. Res.* **1993**, 29, 1719–1729.
46 T. D. Bucheli, S. R. Müller, A. Voegelin, R. P. Schwarzenbach, *Environ. Sci. Technol.* **1998**, 32, 3465–3471.
47 H. R. Buser, M. D. Müller, *Environ. Sci. Technol.* **1997**, 31, 1960–1967.
48 I. Heinz., A. Flessau, N. Zullei-Seibert, B. Kuhlmann, U. Schulte-Ebbert, M. Michels, J. Simbrey, F. Fleischer, Directive 80/778/EEC; Part III: The Parameter for Pesticides and related products. Final Report for the European Commission - DG XI, Brüssel **1995**.
49 M. Isenbeck-Schröter, M. Kofod, B. König, T. Schramm, E. Bedbur, G. Matheß, *Grundwasser* **1998**, 2, 57–98.
50 AWWR (Arbeitsgemeinschaft der Wasserwerke an der Ruhr), *Ruhrgüteberichte 1998–2003*, Selbstverlag, Essen **1999–2004**.

51 T. Landrieux, R. Fien, F. T. Lange, H.-J. Brauch, *Vom Wasser* **1999**, 92, 297–306.
52 C. Skark; N. Zullei-Seibert, U. Schöttler, C. Schlett, *Intern. J. Environ. Anal. Chem.* **1998**, 70, 93–104.
53 Data from the German Environment Agency can be found at http://www.umweltbundesamt.de.
54 ARW (Arbeitsgemeinschaft der Rheinwasserwerke), *Jahresbericht 2002*, Selbstverlag, Köln **2003**.
55 RIWA (Arbeitsgemeinschaft der Rhein- und Maaswasserwerke), *Jahresbericht 2003*, Selbstverlag, Nieuwegein **2004**.
56 A. D. Bannink, *Water Sci. Technol.* **2004**, 49, 173–181.
57 R. Goodchild, *Belastung der Österreichischen Fliessgewässer mit gefährlichen Stoffen*, Umweltbundesamt, Wien, **1999**, pp. 7–8.
58 Umweltbundesamt, *Erhebung der Wassergüte in Österreich*, Wien, **2003**, p. 174.
59 T. A. Albanis, D. G. Hela, *Intern. J. Environ. Anal. Chem.* **1998**, 70, 105–120.
60 D. Munaron, P. Scribe, J. F. Dubernet, R. Kantin, A. Vanhoutte, A. C. Fillon, C. Bacher, *XII Symposium Pesticide Chemistry* **2003**, 717–726.
61 S. Irace-Guigand, J. J. Aaron, P. Scribe, D. Barcelo, *Chemosphere* **2004**, 55, 973–981.
62 M. J. Cerejeira, P. Viana, S. Batista, T. Pereira, E. Silva, M. J. Valério, A. Silva, M. Ferreira, A. M. Siva-Fernandes, *Water Res.* **2003**, 37, 1055–1063.
63 Data from the website of the Environment Agency (UK) can be found at http://www.environment-agency.gov.uk.
64 G. M. Clark, D. A. Goolsby, *Sci. Total Environ.* **2000**, 248, 101–113.
65 R. H. Coupe, E. M. Thurman, L. R. Zimmerman, *Environ. Sci. Technol.* **1998**, 32.
66 R. A. Robich, R. H. Coupe, E. M. Thurman, *Sci. Total Environ.* **2004**, 321, 189–199.
67 R. J. Gilliom, J. E. Barbash, D. W. Kolpin, S. J. Larson, *Environ. Sci. Technol.* **1999**, 33, 164 A–169 A.
68 E. A. Scribner, W. A. Battaglin, D. A. Goolsby, E. M. Thurman, *Sci. Total Environ.* **2000**, 248, 255–263.
69 E. M. Thurman, K. C. Bastian, T. Mollhagen, *Sci. Total Environ.* **2000**, 248, 189–200.
70 P. J. Phillips, E. W. Bode, *Pest. Manag. Sci.* **2004**, 60, 531–543.
71 J. D. Gaynor, C. S. Tan, C. F. Drury, I. J. van Wesenbeeck, T. W. Welacky, *Water Qual. Res. J. Can.* **1995**, 30 (3), 513–531.
72 M. Gfrerer, T. Wenzl, X. Quan, B. Platzer, E. Lankmayr, *J. Biochem. Bioph. Meth.* **2002**, 53, 217–228.
73 G. Mattheß (ed.) *Transport- und Abbauverhalten von Pflanzenschutzmitteln im Sicker- und Grundwasser*, G. Fischer, Stuttgart, New York, 1997, p. 439.
74 M. Flury, *J. Environ. Qual.* **1996**, 25, 25–45.
75 G. R. Wehtje, R. F. Spalding, O. C. Burnside, S. R. Lowry, J. R. C. Leavitt, *Weed Sci.* **1983**, 31, 610–618.
76 R. F. Spalding, M. E. Exner, D. D. Snow, D. A. Cassada, M. E. Burbach, S. J. Monson, *J. Environ. Qual.* **2003**, 32, 92–99.
77 A. H. Haria, A. C. Johnson, J. P. Bell, C. H. Batchelor, *J. Hydrol.* **1994**, 163, 203–216.
78 H. Y. F. Ng, S. B. Clegg, *Sci. Total Environ.* **1997**, 193, 215–228.
79 I. M. Verstraeten, J. D. Carr, G. V. Steele, E. M. Thurman, M. T. Meyer, D. F. Dormedy, *J. Environ. Qual.* **1999**, 28, 1396–1405.
80 I. M. Verstraeten, E. M. Thurman, M. E. Lindsey, E. C. Lee, R. D. Smith, *J. Hydrol.* **2002**, 266, 190–208.
81 A. C. Johnson, A. H. Haria, C. L. Bhardwaj, C. Völkner, C. H. Batchelor, A. Walker, *J. Hydrol.* **1994**, 163, 217–231.
82 S. Altfelder, T. Streck, J. Richter, *J. Environ. Qual.* **1999**, 28, 1154–1161.
83 B. Lennartz, J. Michaelsen, W. Wichtmann, P. Widmoser, *Soil Sci. Soc. Am. J.* **1999**, 63, 39–47.
84 B. Lennartz, *Geoderma* **1999**, 91, 327–345.
85 S. Altfelder, T. Streck, J. Richter, *J. Environ. Qual.* **2000**, 29, 917–925.
86 S. K. Kamra, B. Lennartz, M. T. Van Genuchten, P. Widmoser, *J. Contam. Hydrol.* **2001**, 48, 189–212.

87 E. Zehe, H. Flühler, *J. Hydrol.* **2001**, 247, 100–115.
88 U. Traub-Eberhard, W. Kördel, W. Klein, *Chemosphere* **1993**, 28, 273–284.
89 S. K. Widmer, R. F. Spalding in *Herbicide metabolites in surface water and groundwater* (Eds.: M. T. Meyer, E. M. Thurman), ACS Symposium Series 630, Washington, DC, **1996**, pp. 271–287.
90 D. W. Kolpin, J. E. Barbash, R. J. Gilliom, *Environ. Sci. Technol.* **1998**, 32, 558–566.
91 P. J. Squillace, J. C. Scott, M. J. Moran, B. T. Nolan, D. W. Kolpin, *Environ. Sci. Technol.* **2002**, 36, 1923–1930.
92 J. E. Barbash, G. P. Thelin, D. W. Kolpin, R. J. Gilliom, US Geological Survey Water resources Investigations Report 98–4245, Sacramento, CA, **1999**, p. 58.
93 D. W. Kolpin, J. D. Martin, can be found at http://ca.water.usgs.gov/pnsp/pestgw/Pest-GW_2001_Text.html, **2003**.
94 J. E. Barbash, G. P. Thelin, D. W. Kolpin, R. J. Gilliom, *J. Environ. Qual.* **2001**, 30, 831–845.
95 W. A. Battaglin, E. T. Furlong, M. R. Burkhardt, C. J. Peter, *Sci. Total Environ.* **2000**, 248, 133.
96 M. Isenbeck-Schröter, E. Bedbur, M. Kofod, B. König, T. Schramm, G. Mattheß, *Berichte aus dem Fachbereich Geowissenschaften der Universität Bremen*, **1997**, Vol. 91, p. 65.
97 M. Fielding, N. Mole, H. Horth, A. Gendebien, P. Van Dijk, *Report CO 4395*, Medmenham UK, **1998**, p. 173.
98 N. H. Spliid, B. Koeppen, *Chemosphere* **1998**, 37, 1307–1316.
99 L. Ludvigsen, Conference proceedings "Non-agricultural use of pesticides", Kopenhaven, **2003**, 15–16.
100 Geological Survey of Denmark and Greenland – GEUS, can be found at http://www.geus.dk/publications/grundvandsovervaagning/grundvandsovervaagning_uk.html, **2004**.
101 A. Papastergiou, E. Papadopoulou-Mourkidou, *Environ. Sci. Technol.* **2001**, 35, 63–69.
102 E. Papadopoulou-Mourkidou, D. G. Karpouzas, J. Patsias, A. Kotopoulou, A. Milothridou, K. Kintzikoglou, P. Vlachou, *Sci. Total Environ.* **2004**, 321, 147–164.
103 E. Papadopoulou-Mourkidou, D. G. Karpouzas, J. Patsias, A. Kotopoulou, A. Milothridou, K. Kintzikoglou, P. Vlachou, *Sci. Total Environ.* **2004**, 321, 127–146.
104 S. Rodriguez-Mozaz, M. J. Lopez de Alda, D. Barcelo, *J. Chromatogr. A* **2004**, 1045, 85–92.
105 I. Gaus, *Hydrogeol. J.* **2000**, 8, 218–229.
106 LAWA Länderarbeitsgemeinschaft Wasser, *Bericht zur Grundwasserbeschaffenheit – Pflanzenschutzmittel*, Berlin, **1997**, p. 92.
107 LAWA Länderarbeitsgemeinschaft Wasser, *Bericht zur Grundwasserbeschaffenheit – Pflanzenschutzmittel*, Düsseldorf, **2004**, p. 20.
108 Z. Akkan, H. Flaig, K. Ballschmiter, *Pflanzenbehandlungs- und Schädlingsbekämpfungsmittel in der Umwelt*, E. Schmidt, Berlin, **2003**, p. 315.
109 F. Haakh, *Wasserwirtschaft* **2004**, 94, 9–14.
110 N. Zullei-Seibert, *Veröffentlichungen des Instituts für Wasserforschung GmbH und der Dortmunder Stadtwerke AG*, Dortmund, **1990**, 39, p. 102.
111 C. Skark, N. Zullei-Seibert, *Veröffentlichungen des Instituts für Wasserforschung GmbH und der Dortmunder Energie- und Wasserversorgung GmbH*, Dortmund, **1999**, 58, p. 105.
112 R. Kozel, *PhD Thesis*, University Neuchâtel (CH), **1992**.
113 N. J. Smith, R. C. Martin, R. G. S. Croix, *B. Environ. Contam. Tox.* **1996**, 57, 759–765.
114 J. Kjaer, P. Olsen, H. C. Barlebo, R. K. Juhler, F. Plauborg, R. Grant, L. Gudmundsson, W. Brüsch, *The Danish pesticide leaching assessment programme*, Kopenhagen **2004**, p. 110.
115 D. W. Kolpin, E. M. Thurman, S. M. Linhart, *Sci. Total Environ.* **2000**, 248, 115–122.
116 G. C. Pasquarell, D. G. Boyer, *J. Environ. Qual.* **1996**, 25, 755–765.
117 R. N. Lerch, E. M. Thurman, E. L. Kruger, *Environ. Sci. Technol.* **1997**, 31, 1539–1546.
118 R. N. Lerch, P. E. Blanchard, E. M. Thurman, *Environ. Sci. Technol.* **1998**, 32, 40–48.

119 D. W. Kolpin, E. M. Thurman, S. M. Linhart, *Environ. Sci. Technol.* **2001**, 35, 1217–1222.
120 G. Hegenberg, *Veröffentlichungen des Instituts für Wasserforschung GmbH und der Dortmunder Stadtwerke AG*, Dortmund, 46, **1993**.
121 A. Grohmann, H. Dizer, *Examination of methods of treatment to remove pesticides from water intended for human consumption*. Report Institut für Wasser-, Boden- und Lufthygiene, Berlin, **1991**.
122 G. Baldauf, *Acta Hydrochim. Hydrobiol.* **1993**, 21 (4), 203–208.
123 B. Haist-Gulde, G. Baldauf, H.-J. Brauch, *IWSA European Spezialized Conferences*, Amsterdam, **1992**, 159–168.
124 M.-C. Gérard, J.-P. Barthélemy, *Biotechnol. Agron. Soc. Environ.* **2003**, 7 (2), 79–85.
125 "Adsorption/Active Carbon" can be found at http://www.lenntech.com/adsorption.htm.
126 US EPA, (FQPA) *Drinking Water Assessments*, **2001**, p. 50 can be found at http://www.epa-gov/pesticides/trac/science/water_treatment.pdf.
127 "Ozone applications Drinking Water" can be found at http://lenntech.com/ozone/ozone-applications-drinking.water.htm.
128 J. Hotelier, G. Bablon, *Colloque phyt'eau, Recueil des communications ecrites*, A. N. P. P., Versailles, **1992**, 163–168.
129 W. Urban, *personal communication*, **1995**.
130 B. Kuhlmann, B. Kaczmarzyk, U. Schöttler, *Int. J. Environ. Anal. Chem.* **1995**, 58, 199–205.
131 G. Preuß, N. Zullei-Seibert, J. Nolte, B. Graß, *Texte Umweltbundesamt*, Berlin, 71, **1998**, p. 125.
132 H.-C. Schöpfer, *PhD. Thesis*, University of Kaiserslautern (FRG), **1995**.
133 L. Meitzler, *Vom Wasser* **1996**, 87, 163–170.
134 W. Kördel, M. Herrchen, R. T. Hamm, *Chemosphere* **1991**, 23, 83–97.
135 P. J. Stuyfzand, *Künstliche Grundwasseranreicherung* (Eds. : KIWA, DVGW), DVGW Schriftenreihe Wasser, Bonn, **1997**, 97, pp.131–146.
136 G. Preuß, U. Schöttler, N. Zullei-Seibert, *Final report DVGW W 01/98*, Institut für Wasserforschung GmbH, Dortmund, **2002**, p. 68.
137 B. Post, H. Korpien, A. Allendorf, F. T. Lange, T. Landrieux, in *ARW 56. Bericht 1999* (Ed.: Arbeitsgemeinschaft der Rheinwasserwerke ARW), Köln, **2000**, pp. 135–155.
138 C. Schlett, M. Schöpel, *Ruhrgütebericht 2003*, AWWR Arbeitsgemeinschaft der Wasserwerke an der Ruhr and Ruhrverband, Essen, **2004**, 133–136.
139 N. Zullei-Seibert, C. Skark, D. Schwarz, U. Gatzemann, *Künstliche Grundwasseranreicherung* (Eds.: KIWA, DVGW), DVGW-Schriftenreihe Wasser, 90, **1997**, 101–114.
140 J. A. M. H. Hofman, T. H. M. Noij, J. C. Schippers, *Recently Identified Pollutants in Water Ressources – Drinking Water Treatment in the Nineties*, IWSA European Spezialized Conferences, Amsterdam, **1992**, 101 – 111.
141 L. M. Puijker, C. G. E. M. van Beek, H. M. J. Janssen, J. C. Kruithof, G. Krügeman, *Report*, The Netherlands Waterworks Testing and Research Institute KIWA, Rijswijk, **1994**.
142 V. Vidard, J. Magnin, *Colloque phyt'eau, Recueil des communications ecrites*, A. N. P. P., Versailles, **1992**, 153–158.

7
Aminopolycarboxylate Complexing Agents

Carsten K. Schmidt and Heinz-Jürgen Brauch

7.1
Introduction

Synthetic organic compounds forming stable, water-soluble complexes with metal ions are technically applied for a wide variety of purposes in large amounts. Chelating agents insulate metal ions from their catalytic action, avoid the formation of low-solubility metal-salt precipitates, and aid in the removal of previously formed deposits. Complexing agents of the aminopolycarboxylate type have been utilized in industrial applications for more than sixty years [1–6]. Important synthetic representatives of this class are ethylenediaminetetraacetic acid (EDTA, $C_{10}H_{16}N_2O_8$, CAS No. 60-00-4, MW 292.25 g mol^{-1}), 1,3-propylenediaminetetraacetic acid (1,3-PDTA, $C_{11}H_{18}N_2O_8$, CAS No. 1939-36-2, MW 306.27), diethylenetriaminepentaacetic acid (DTPA, $C_{14}H_{23}N_3O_{10}$, CAS No. 67-43-6, MW 393.35), nitrilotriacetic acid (NTA, $C_6H_9NO_6$, CAS No. 139-13-9, MW 191.14), β-alaninediacetic acid (β-ADA, $C_7H_{11}NO_6$, CAS No. 6245-75-6, MW 205.17), and methylglycinediacetic acid (MGDA, $C_7H_{11}NO_6$, CAS No. 29578-05-0, MW 205.17). Each aminopolycarboxylic acid features structural elements (ligands) that exhibit electron donor characteristics and that are able to bind polyvalent metal ions in ring structures, thereby forming very stable complexes. Depending on the complexing agent, five- or six-membered rings are formed, the five-membered ones being thermodynamically more stable. Potential ligand atoms of aminopolycarboxylates are the nitrogen atoms (free electron pair) and the deprotonated, negatively charged oxygen atoms of the carboxylic acid groups (Fig. 7.1). The stability of the complexes depends on the number of possible chelate rings (each ring increases the complex stability roughly by a factor of 100) and thus to a great extent on the number of available nitrogen and carboxylic oxygen atoms. Complexes are typically formed in a 1:1 stoichiometry, with 1 mol aminopolycarboxylate chelating 1 mol metal cations. Aminopolycarboxylate complexes enclose the metal cation that is coordinatively bound to nitrogen and oxygen atoms in their center. The spatial (three-dimensional) structure of most aminopolycarboxylate complexes can be described in a first approximation as an octahedron [1, 7, 8].

Organic Pollutants in the Water Cycle. T. Reemtsma and M. Jekel (Eds.).
Copyright © 2006 WILEY-VCH Verlag GmbH & Co. KGaA, Weinheim
ISBN: 3-527-31297-8

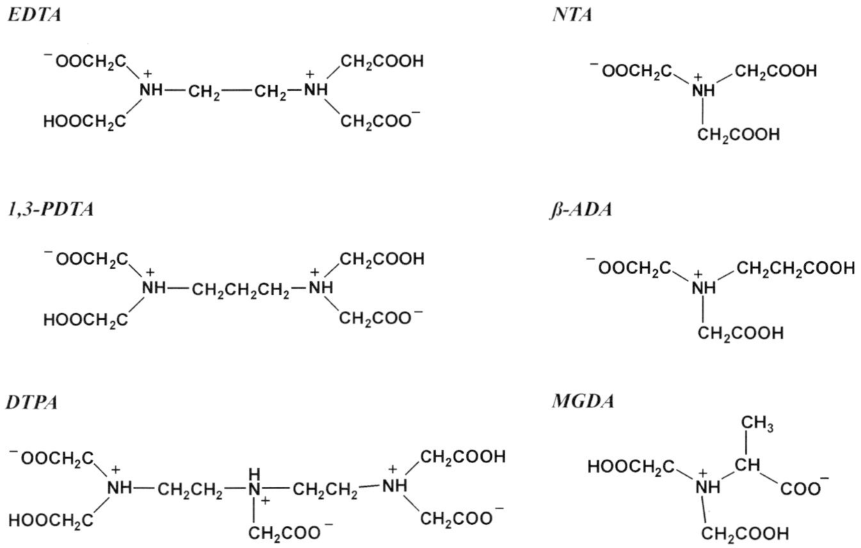

7 Aminopolycarboxylate Complexing Agents

Fig. 7.1 Structural formulae of important aminopolycarboxylates.

Free aminopolycarboxylic acids typically form white crystalline powders that are only slightly soluble in water (e.g., water solubility of EDTA: 0.5 g L^{-1} at 20 °C). The water solubility of aminopolycarboxylates increases with increasing neutralization (water solubilities at 20 °C, Na$_2$EDTA: 100 g L^{-1}, Na$_4$EDTA: 1000 g L^{-1}), making the more soluble alkali salts particularly important for practical application. Also, aminopolycarboxylate–metal complexes are usually highly water soluble, because they are mostly negatively charged, with their hydrophilic oxygen atoms in the complex being directed outwards. Aminopolycarboxylic acids are resistant toward air and water and are attacked neither by strong acids nor by strong bases. Decomposition can take place, however, in the presence of strong oxidizing agents. Melting points of the free aminopolycarboxylic acids are in the range 200–250 °C; however, at temperatures above 150 °C, partial decomposition of the molecules is initiated by decarboxylation reactions [1, 4, 9].

The dissociation behavior of aminopolycarboxylic acids gives rise to a strong pH-dependency of complex formation, i.e. with an increasing degree of dissociation in alkaline solutions more stable complexes are formed. However, depending on the complexed metal ion, this pH effect may be counterbalanced by the formation of low-solubility metal hydroxides under alkaline conditions. The respective complex stability is further determined by competing reactions (e.g., with non-anthropogenic complexing agents) and other influences like temperature and ionic strength. Direct comparison of aminopolycarboxylates with each other shows that their structural differences typically cause different physico-chemical properties and complex formation tendencies. The particular physico-chemical properties of single aminopolycarboxylic acids govern on the one hand options for their industrial utilization, but on the other hand also their environmental fate [1]. EDTA forms highly stable chelate complexes with almost every polyvalent metal cation. It complexes Ca^{2+} and Mg^{2+} above pH 4, the most common metal cations over the entire pH-range, and Fe^{3+} below pH 8. Even though Fe(III)-EDTA represents a highly stable complex, its stability is not sufficient to prevent the precipitation of Fe^{3+} as Fe(III)-hydroxide above pH 8–9. Complexes of Fe^{2+}, however, are stable in the range of pH 1 to 12.5. Under optimal conditions, 1 g of Na_4EDTA binds 105 mg Ca^{2+}, 64 mg Mg^{2+}, 167 mg Cu^{2+}, or 144 mg Fe^{2+}/Fe^{3+} [1, 4, 9, 10]. 1,3-PDTA is a weaker chelating agent than EDTA, and metal-1,3-PDTA complexes are typically less stable than corresponding metal-EDTA complexes [1, 7, 8]. Compared to EDTA, DTPA forms more stable complexes, in particular with metal cations providing a coordination number of 6 or higher, and is thus advantageous for complexation of large metal ions, in particular trivalent cations of rare earth elements and tetravalent cations of thorium and zirconium. DTPA holds the largest complex formation power of all industrially utilized aminopolycarboxylates and can be used for complexation of polyvalent metal cations over almost the entire pH range. For example, DTPA forms stable complexes with Fe^{3+} even at alkaline pH values between 8 and 10 [1, 8, 11]. The stability of a metal-NTA complex, however, is typically several orders of magnitude lower than that of the corresponding EDTA complex. Consequently, the pH-range for maximum effectiveness is also smaller. However, 1:1 complexes with Ca^{2+} and Mg^{2+} are still sufficiently stable to utilize NTA for water softening [1, 3, 7]. Compared to NTA, β-ADA complexes with alkaline earth metals are slightly less stable, while complexes with transition metals provide similar complex stabilities [1, 12]. Metal-MGDA complexes show typically higher stabilities than corresponding metal-NTA chelates. With respect to complex stabilities, MGDA can be classified between NTA and EDTA [1, 12].

7.2
Applications and Consumption

Production processes for aminopolycarboxylates are based on the conversion of the corresponding (poly)amines (>N-H) into cyanomethyl derivatives (>N-CH_2-CN) that are subsequently hydrolyzed to the carboxylic acid (>N-CH_2-COOH). Initially formed sodium salts can be transformed into salts with a lower sodium content or into the free acid by pH variation using mineral acids [1].

In principle, aminopolycarboxylates are used to adjust concentrations of free metal ions in solutions, to prevent trouble that is caused by metal ions, and to solubilize difficultly soluble precipitates of inorganic salts. These properties are advantageous and useful in a considerable number of applications [1, 4, 9, 10].

Detergents
Calcium and magnesium ions form with soap anions (alkylcarboxylates, classical anionic surfactants) low-solubility precipitates (lime soaps) that deposit on clothes

and compromise their usage properties. Since lime soap possesses no surfactant properties, the cleansing action of the detergent is correspondingly reduced. In contrast, the cleansing action of modern synthetic surfactants (e.g., linear alkylbenzene sulfonates or cationic tensides) is highly insensitive to calcium and magnesium ions. However, high levels of hardness builders also cause troublesome deposits on the textiles to be cleaned and on water-sided parts of the washing machine (in particular heating rods). Furthermore, metal ions intensify the adhesion of dirt to surfaces and the cohesion of dirt particles to each other. To reduce these problems, detergents typically contain some type of complexing agent (e.g. aminopolycarboxylates). Nevertheless, colored stains on textiles of technical, botanical, or animal origin, such as spots of ink, fruits, vegetables, coffee, or red wine, are not or only incompletely removable by treatment with surfactants. For removal of these stains, bleaching agents that decompose the colorings by oxidation are added to the detergents. Often, sodium perborate is used for this purpose, and aminopolycarboxylates prevent its metal-catalyzed decomposition. At the same time, aminopolycarboxylates also hinder the metal-catalyzed oxidation of other components such as enzymes, perfume oils, and optical brighteners. The use of aminopolycarboxylates in detergents was primarily necessitated by the abandonment of phosphate-containing complexing agents because of their contribution to the eutrophication of surface waters.

Cleaners
Aminopolycarboxylates are also added to several cleaners to mask hardness builders and heavy metal ions. In sanitary cleaners, aminopolycarboxylates enhance the antibacterial effect, in particular the effect on Gram-negative bacteria, because aminopolycarboxylates sensitize these resistant microorganisms by destruction of their outer cell wall. A major proportion of aminopolycarboxylate consumption by cleaners goes to industrial cleaners used in the beverage industry (brewing, milk, refreshment, fruit juice, mineral water industries, and wine processing). They are used for bottle cleaning in the bottle return system, facilitate the solution of calcareous precipitations, and prevent residues on the bottle surface by improving the run-off of the cleaning agent. In milk-processing factories, aminopolycarboxylates remove the solid, low-solubility coverings of milk stone (consisting of denatured proteins, lipids, milk sugar, and calcium phosphate).

Soap Production
In solid and liquid soaps, aminopolycarboxylates inhibit color changes, rancidity, and calcareous precipitations by masking calcium, magnesium, iron, and other metal ions.

Water Treatment
Aminopolycarboxylates are used for scale control in boilers and cooling towers by cleaning scale deposits from internal surfaces and by softening of the incoming

boiler or cooling feedwater. Because of their neutral or weakly alkaline character, the aminopolycarboxylate additives are non-corrosive.

Textile Industry
In the textile industry, aminopolycarboxylates are used to remove trace metal impurities that would otherwise compromise several high-grade finishing processes such as crosslinking of cellulose molecules (to produce easy-care fabrics), scouring, bleaching, and dyeing of natural and synthetic fibers.

Food Industry
Aminopolycarboxylates are added to foodstuffs (e.g., mayonnaise and salad dressings) as preservatives, preventing heavy-metal-catalyzed oxidation of unsaturated compounds and thus formation of decomposition products with a rancid taste and odor. Furthermore, possible color changes due to iron or other metal ions are avoided (canned vegetables).

Cosmetics Industry
Also in the cosmetics industry, aminopolycarboxylates serve as masking agents for heavy-metal ions for the stabilization of greasy emulsions, creams, lotions, and scents. In this way, precipitations caused by hardness builders are also prevented. Aminopolycarboxylate additives also reduce the danger of allergic reaction to nickel or chromium and act as preservative enhancers.

Pharmaceutical Industry
Aminopolycarboxylates are also approved additives for pharmaceuticals to prevent decomposition of active and auxiliary compounds. In contact lens preparations, aminopolycarboxylates (in particular EDTA) facilitate the removal of inorganic deposits.

Pulp and Paper Industry
Aminopolycarboxylates are used in non-chlorine bleaching processes of the pulp and paper industry in order to complex heavy metals like manganese that would otherwise decompose the chlorine substitute, namely hydrogen peroxide.

Medicine
In medical treatment, aminopolycarboxylates are used as antidotes for heavy metal poisoning and radioactive contamination. Complexation of the toxic metal ion leads to accelerated excretion from the body. In units of stored blood, K_2EDTA prevents clotting. The DTPA complex with gadolinium is used as a contrast agent for carcinoma diagnosis based on nuclear magnetic resonance imaging.

Rubber Industry

In the manufacture of styrene-butadiene rubber, aminopolycarboxylates are applied in order to remove the rubber poisons copper and manganese that would catalyze premature deterioration of the rubber. Aminopolycarboxylates are also constituents of the water-soluble radical initiator system $Fe(II)/H_2O_2$, being used to adjust the concentration of active iron cations in the polymerization process.

Photographic Industry

In the photographic industry, aminopolycarboxylates prevent, by complexation of calcium and magnesium in process solutions, precipitations onto developing equipment and photomaterial. During color film development, aminopolycarboxylate complexes with Fe(III) (e.g., $Fe(III)$-NH_4-EDTA) are used as oxidizing agents for metallic silver in bleach fixation baths.

Galvanotechnics

Another field of aminopolycarboxylate application is electroplating. Here, aminopolycarboxylates are particularly used in the area of chemical (currentless) metal plating. In circuit board fabrication, for example, EDTA acts as a sort of metallic ion buffer, controlling the level of free Cu^{2+}-ions and thus the precipitation rate of copper, thereby improving the quality of the copper layer formed. In other fields, such as nickel and gold plating, aminopolycarboxylates are applied to selectively bind interfering metal ions and to prevent their co-precipitation.

Metalworking

During the processing and surface refinement of iron, steel, zinc, aluminum, brass, and other metals, purifying and degreasing agents are applied that contain aminopolycarboxylates. The complexing agents prevent the precipitation of metal soap, metal hydroxides, and water hardness-derived deposits, which can cause process disruptions by sedimentation and incrustation in containers, pipes, and nozzles, and on material surfaces.

Oil Production

A major problem in oil field operations is the occurrence of scale both downhole and in pipelines. For scale removal, aminopolycarboxylates (in particular EDTA) are applied in a discontinuous cleaning process.

Agriculture

In agricultural applications, aminopolycarboxylates are used to improve the plants' uptake by plants of micronutrients, such as iron, copper, manganese, molybdenum, and zinc, that are needed to correct trace metals deficiencies. In this way, the chelated metal ions cannot be fixed as biologically inactive insoluble salts in the soil and are transported more easily to the plants' roots.

Fuel Gas Cleaning
Aminopolycarboxylates are used during fuel gas cleaning (e.g., fume desulfuration at coal power plants and waste incineration plants) as a stabilizer of the Na_2SO_3 solution that is used for adsorption of SO_2 (Wellmann-Lord process).

Nuclear Industry
Aminopolycarboxylates are used in the nuclear industry for decontamination of reactors and equipment because they form water-soluble complexes with many radionuclides.

Other
Further fields of aminopolycarboxylate application are to be found in the lacquer industry (water-based paints), in offset printing, leather tanning (chrome tanning), remediation of soils contaminated with metals, and in lubricants and impregnating agents.

Figure 7.2 summarizes the different fields of application for each individual aminopolycarboxylate in detail. The dominating role of EDTA with its diverse application spectrum is obvious, but also DTPA and NTA are used in several industries. NTA is used predominantly in detergents and cleaners (about 70% of Europe-wide

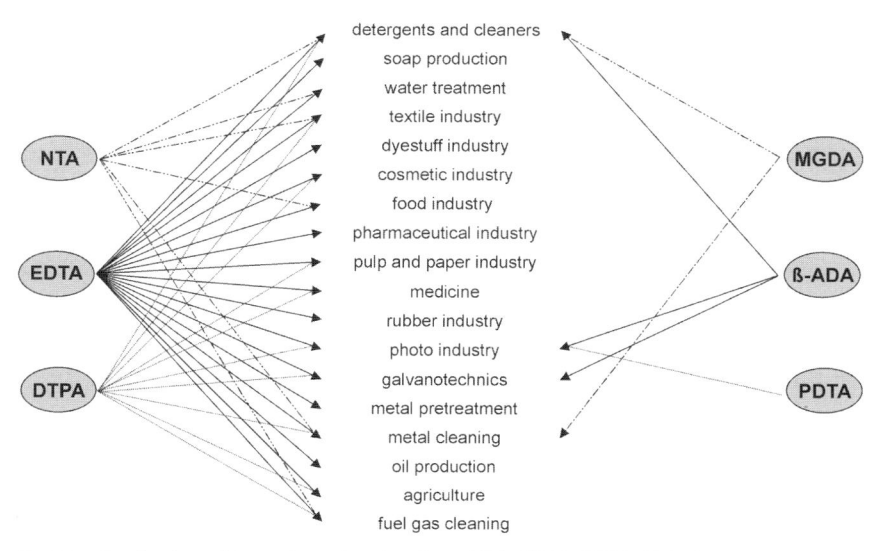

Fig. 7.2 Application of the aminopolycarboxylates NTA, EDTA, DTPA, MGDA, β-ADA, and 1,3-PDTA by industries (from Refs. [1, 14]).

sales), DTPA mainly in the pulp and paper industry (about 75%) [10, 13]. The complexing agents MGDA, β-ADA and 1,3-PDTA, however, are applied in only a small number of branches of industry.

EDTA, NTA, and DTPA are used in huge quantities worldwide. The annual global consumption is estimated to be approximately 200 000 tons (calculated as 100% acid); about half of this is EDTA, and a quarter each are NTA and DTPA [14]. The pattern of consumption in Western Europe and Germany is very similar [13]. Europe consumes approximately 50% of the world's aminopolycarboxylate production volume [10]. Almost half of the amount sold in Europe goes to Germany, Sweden, and the UK [13]. Outside Europe, aminopolycarboxylates are mostly used in North America and to a smaller extent in Japan and South-East Asia [13, 14]. More than 50% of the total aminopolycarboxylic acid consumption is in the washing, cleaning, and pulp and paper industries [13]. In 1998, more than 100 000 t aminopolycarboxylic acids (EDTA, NTA and DTPA) were used in these two fields. On a quantity basis, further important consumer groups are the photographic industries (approx. 15 000 t) and agriculture (approx. 13 000 t) [13]. Although only limited data are available concerning the consumption of MGDA, β-ADA, and 1,3-PDTA, the production of these compounds seems to be relatively low compared to EDTA, NTA, and DTPA. In Germany, approx. 130 t year^{-1} MGDA and 0.15 t year^{-1} (data from 2001) are used in detergents and cleaning agents [15], and approx. 28 t year^{-1} 1,3-PDTA (data from 2000) are used in the German photo industry [13].

7.3
Occurrence and Fate in Wastewaters

The fate of aminopolycarboxylates during wastewater treatment has been investigated intensively. According to this research, the elimination of aminopolycarboxylates in wastewater treatment plants is usually restricted to biodegradation processes. Other elimination steps like sorption to sewage sludge or flocculation turned out to have little or no significance.

7.3.1
Biodegradation

In wastewater treatment plants, NTA is typically easily removable in biologically active steps such as sewage lagoons, the activated sludge process, or trickling filters. A precondition for effective microbial degradation is, however, that the microorganisms have become well adapted, a process that could take a long time: about 1–4 weeks [1, 2, 16]. Wastewater treatment plants with normal throughput times are capable of removing between 70 and >90% of the total NTA load. Neither inhibiting nor competitive effects of other microbially degradable compounds nor a breakthrough of NTA during shock loading were observable [17]. However, the extent of NTA degradation is influenced by its speciation (Fig. 7.3). Hg-NTA and Ni-NTA were described as rather hardly degradable, whereas the Ca-NTA complex

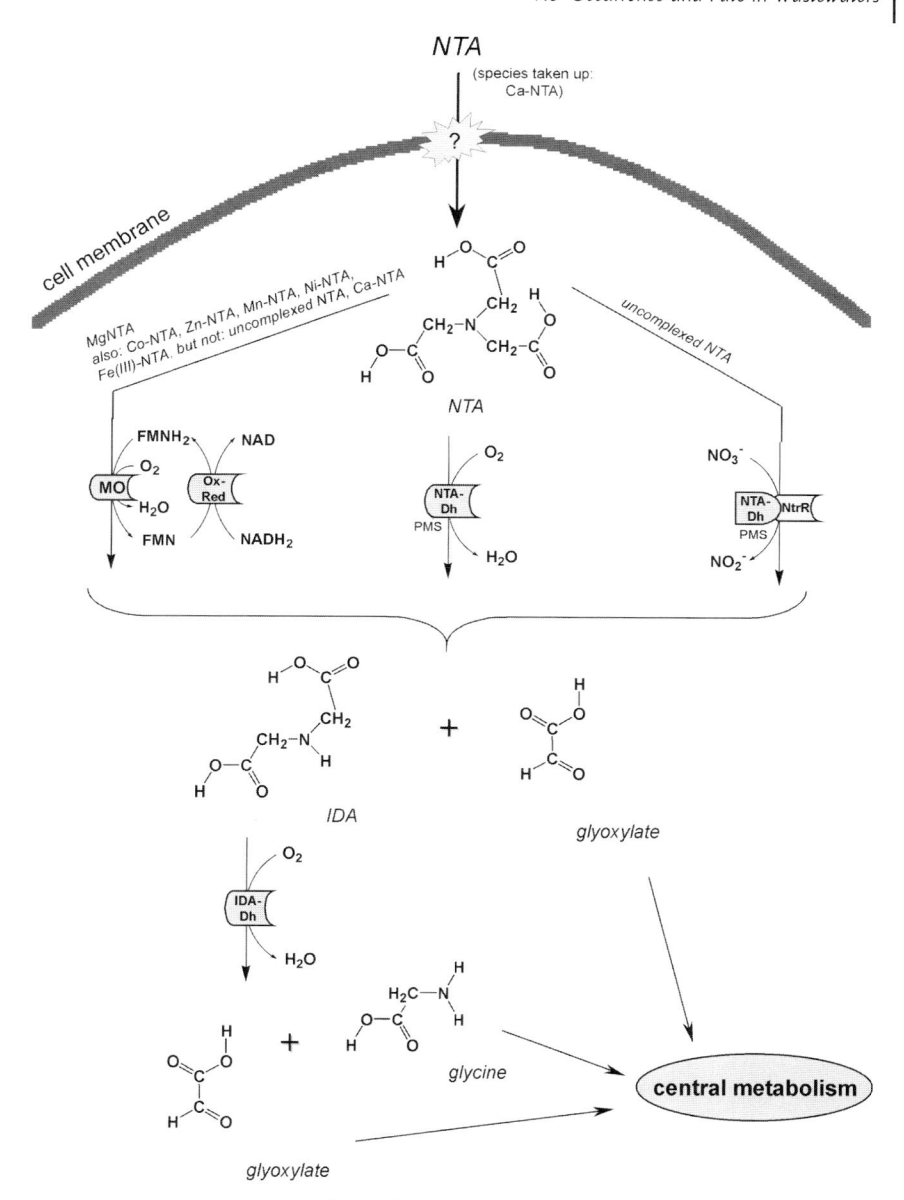

Fig. 7.3 Biochemistry of microbial NTA degradation in Chelatobacter heintzii, Chelatococcus asaccharovus, and the Gram-negative strain TE11 (from Refs, [1, 16, 89]). Catabolism can be performed by three different enzyme systems: a specific monooxygenase/ NADH$_2$: FMN oxido-reductase system (aerobic), a dehydrogenase system with electron transfer to the respiration chain (aerobic), and an oxygen-independent dehydrogenase/nitrate reductase system (denitrifying).

turned out to be more readily utilizable by microorganisms [1, 18]. Under practical conditions, however, hardly degradable metal-NTA complexes experience rapid metal exchange reactions with calcium ions, which are typically available in excess, facilitating their microbial breakdown. The degradation proceeds without any enrichment of intermediates in terms of a complete mineralization (total elimination) [18, 19].

Like all microbial degradation processes, the mineralization of NTA is temperature-dependent, so that at times of lower temperatures during the winter months lower degradation rates were sometimes observed [16, 18, 20]. Since NTA is only slightly adsorbed to sewage sludge, its removal during anaerobic sludge digestion is only of secondary importance. Nevertheless, NTA degradation is also possible under anaerobic conditions, but proceeds comparatively slowly [16, 21]. During sludge digestion, the extent of NTA degradation was reported to be between 8 and 45% [22, 23]. Almost 100% elimination was observed, when NTA-adapted activated sludge was added to the anaerobic sludge process [24]. In small-scale sewage treatment facilities with predominantly anaerobic conditions, NTA was degraded by 70% in summer and by only 20% in winter [17].

According to several investigations at municipal and industrial wastewater treatment plants with biological treatment steps, EDTA demonstrated insufficient degradability or none at all, in contrast to NTA and independently of boundary conditions (summer, winter, different nitrification effectiveness) [1, 4, 5, 16]. Also, laboratory studies with inoculi from industrial and municipal wastewater treatment plants underlined the highly persistent behavior of EDTA [1, 25, 26]. However, recent studies demonstrated that EDTA can be degraded in the biological purification step of some wastewater treatment plants by up to 90% when a number of specific conditions are fulfilled: (a) an alkaline pH value of the wastewater, (b) a relatively high initial EDTA concentration, (c) complexation of EDTA predominantly with alkaline earth metal ions, and (d) a relatively high hydraulic and sludge retention time in the system [1, 27–36].

The conditions mentioned result from a number of effects and influences. Low EDTA concentrations cannot stimulate the adaptation process necessary for microbial degradation. Several investigations indicate that Ca- and Mg-EDTA chelates are microbiologically better degradable than complexes with different heavy metal cations, the latter species being rather unfavorable for the transportation pathway into the microogransims' cells as well as for their subsequent utilization by cellular metabolic enzymes [31, 35–37]. The increased degradation at alkaline pH values can be attributed to the fact that differences in complex stabilities of transition metal ions (in particular Fe(III)-EDTA) and alkaline earth metal ions become smaller with increasing pH values. Alkaline earth metal ions are usually available in sufficient amounts and can successfully compete with transition metal ions at alkaline conditions for their complexation by EDTA. Finally, biologically better degradable EDTA species are formed by cation-exchange reactions within the complex. The relatively high hydraulic retention times in the system (HRT; i.e. the time the liquid remains in the system) required for EDTA degradation could be linked to the slow kinetics of these metal-exchange reactions [37]. EDTA degradation seems to

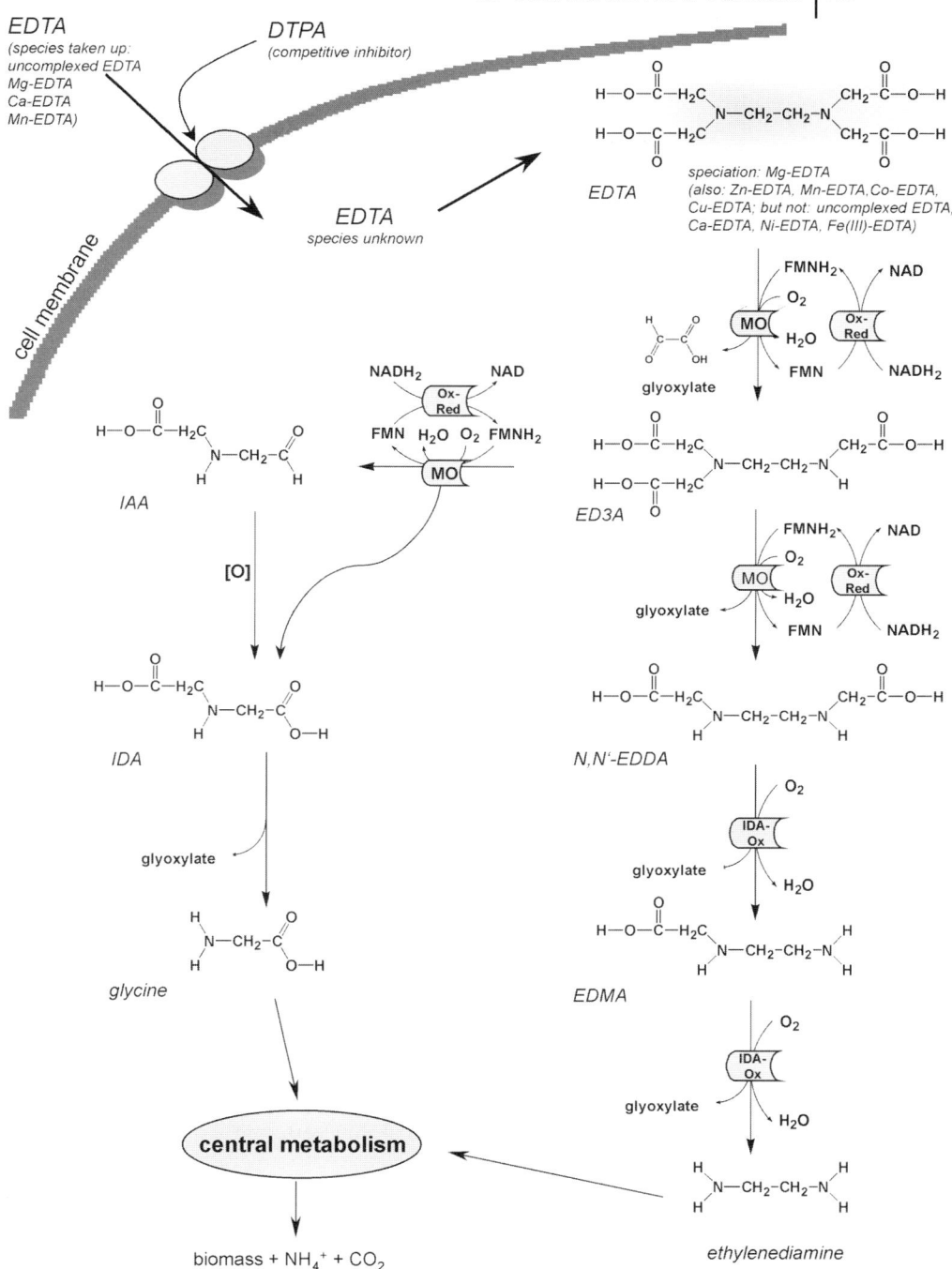

Fig. 7.4 Biochemistry of microbial EDTA degradation in the strains DSM 9103 and BNC1 (from Refs. [1, 16, 89]). Catabolism is initiated by a monooxygenase/$NADH_2$:FMN oxidoreductase system. The metabolites are further degraded by iminodiacetate (IDA) oxidase.

be less in wastewater treatment plants where phosphate is precipitated with Fe(III)-salts than in plants without phosphate precipitation [35, 36, 38]. Because this phosphorus elimination is carried out nowadays mostly during mechanical or biological sewage treatment, microorganisms in the biological purification step in such cases are predominantly faced with the Fe(III)-EDTA complex being rather resistant. Finally, the extent of microbial degradation rises with sludge age or sludge retention time, because the sludge biomass is becoming increasingly multifaceted [39].

In municipal wastewater treatment plants (which receive a major part of EDTA emissions), these conditions are typically not present, and it is assumed that almost no biodegradation of EDTA occurs in these plants. In some industrial plants, however, a partial elimination is expected, since the favored conditions are achieved to a certain degree. This is, in particular, true for industry branches where EDTA is used to mask calcium ions. Considerable elimination of EDTA was indeed directly demonstrated in some paper mills and beverage plants [29, 34–36, 40, 41]. However, industrial wastewater treatment plants possessing the best available techniques and achieving removal rates of about 90% are presently available only at a limited number of sites; at many others less removal is obtained or none at all [29]. In various municipal wastewater treatment plants also, little or no reduction of the total DTPA load was observed [1, 42, 43]. However, microbial DTPA degradation of about 40–70% could be observed in some industrial plants (mostly pulp and paper mills) [40, 44, 45]. Investigations in one industrial treatment plant revealed a higher degree of elimination of DTPA (54–68%) than of EDTA (17–30%) [40]. Investigations with laboratory-scale treatment plants based on activated sludge produced different results. In some studies, 20–50% removal of DTPA was found [46, 47], while in others no removal was noticed at all [25, 48, 49].

The other aminopolycarboxylates covered in this review have been investigated only sporadically [1]. As a consequence, less is known about their potential microbial degradation in the field. According to various laboratory tests, however, MGDA and β-ADA are biodegradable and should thus be eliminable in wastewater treatment plants, whereas 1,3-PDTA does not meet current OECD criteria for ready or inherent biodegradability [1].

7.3.2
Other Treatment Options

Elimination of non-biodegradable aminopolycarboxylates has been attempted (depending on their application and speciation) by a large number of possible techniques (different oxidation processes, electrodialysis, ion exchange techniques, thermal processes, chemical precipitation as acid, membrane filtration, flocculation and precipitation, and degradation with specialized strains of bacteria) [1, 37, 50–52]. As an example, the UV/H_2O_2-approach is described here in more detail, because this process was brought to the market and proved to be a very promising method for the degradation of EDTA, DTPA, NTA, and their respective metal complexes [53–65]. Furthermore, UV/H_2O_2 treatment for degradation of EDTA in partial wastewater streams is meanwhile utilized in a number of industrial branches.

The UV/H_2O_2-process is based on the photolysis of hydrogen peroxide in aqueous solution. By UV treatment, each hydrogen peroxide molecule forms two highly reactive hydroxyl radicals, which can react with organic substrates and mineralize them by radical chain reactions. Degradation processes in the UV/H_2O_2-process proceed considerably faster than in classic oxidation procedures based on chlorine or ozone and typically result in complete destruction of aminopolycarboxylates and their intermediates [56, 58, 60, 62, 66, 67]. The application of hydrogen peroxide is generally carried out over-stoichiometrically to the amount of aminopolycarboxylate. The degradation rate of EDTA, for instance, increases up to a molar H_2O_2/EDTA-ratio of about thirty [54, 55]. In general, the reaction rate increases with increasing light intensity [54, 55]. For treatment of aqueous solutions containing metal-EDTA chelates, the process is operated under acidic pH conditions to avoid precipitation of heavy-metal hydroxides that might cause trouble in the process [53–55]. The process is naturally adversely affected by substances that quench reactive hydroxyl radicals or that show strong self-absorption at the irradiation wavelengths utilized, whereas the presence of iron cations improves the degradation of aminopolycarboxylates considerably (Fenton process) [54, 59, 60, 64, 65]. In studies with spiked drinking water, degradation of aminopolycarboxylates in the UV/H_2O_2-process increased in the order NTA < EDTA << DTPA [62]. The process can be carried out up to complete mineralization. Bearing in mind processing aspects, however, this does not always make sense, because intermediates formed in the process are biodegradable and might be eliminated in other treatment steps at lower cost [59, 67]. The process is suitable for both treatment of wastewaters with concentration levels in the mg L^{-1} range and processing of raw waters with concentration in the µg L^{-1} range [61, 68]. For the application of photochemical methods on an industrial scale, a large number of UV lamps is required in order to obtain the necessary UV dose for high wastewater volumes. The commercial availability of reasonably priced and continuously operating photochemical reactors, meanwhile, allows for an intensive UV-irradiation at angles of 360°, thus making the application of UV/H_2O_2 processes more attractive even when economical aspects are considered [56].

7.3.3
Occurrence in Wastewater

Aminopolycarboxylate concentrations in wastewater have been the subject of many investigations and publications [43]. On the basis of a comprehensive evaluation of these data, Table 7.1 gives information on concentration ranges that can be expected in sewage runoffs of municipal and industrial wastewater treatment plants.

The data demonstrate that especially industrial sewage discharges may contain considerable amounts of aminopolycarboxylates. Consequently, it can be expected that municipal wastewaters that receive industrial wastewater discharges will exhibit higher levels of aminopolycarboxylate pollution. Even within one industrial plant, aminopolycarboxylate discharges can be highly variable with time. This

Tab. 7.1 Aminopolycarboxylates: Concentration ranges in wastewater effluents [43].

	EDTA ($\mu g\ L^{-1}$)	NTA ($\mu g\ L^{-1}$)	DTPA ($\mu g\ L^{-1}$)
Effluents of municipal wastewater treatment plants			
Typical pollution	10–250	1–15	1–30
High pollution	1000	200	300
Effluents of industrial wastewater treatment plants			
Typical pollution	100–20 000	100–2000	50–5000
High pollution	400 000	5000	20 000

variability may be due to different application volumes during usage or the fact that many production processes are discontinuous. The aminopolycarboxylates MGDA, β-ADA, and 1,3-PDTA have rarely been found in municipal and industrial sewage plant discharges [43]. At the Rhine river (Germany) and the Schussen river (Germany) β-ADA has been sporadically detected in industrial (14–480 $\mu g\ L^{-1}$) and municipal (2.8–7.2 $\mu g\ L^{-1}$) wastewaters. For 1,3-PDTA, single positive results were obtained for wastewaters from municipal sewage plants at the Schussen river (1.2–4.1 $\mu g\ L^{-1}$). MGDA has so far not been detected in any of the sewage plant discharges investigated.

In several investigations, aminopolycarboxylate concentrations have also been determined in sewage plant influents. Usually, influent and effluent concentrations of EDTA and DTPA in sewage plants did not differ much. This is particularly true for municipal plants. In some industrial plants with optimized treatment conditions, EDTA and DTPA may be reduced to a certain extent. In the case of NTA, which is more easily degradable, concentrations in effluents were roughly 10 times lower than those in influents of various sewage plants [43].

Tab. 7.2 Possible EDTA concentrations in municipal and industrial wastewater effluents (C_{eff}) and in receiving waters in immediate vicinity to the point of discharge (PEC_{local}) (in accordance with the EU risk assessment [29], worst case scenario).

Kind of discharge	C_{eff} ($\mu g\ L^{-1}$)	$PEC_{local,\ water}$ ($\mu g\ L^{-1}$)
Municipal wastewater	1 000	200
Industrial detergents	5 400	640
Dairy and beverage industry	2 500–25 000	350–2 600
Photo chemicals	4 700–23 000	570–2 400
Textile industry	19 000	2 000
Pulp and paper	4 000–25 000	500–2 600
Oil production	630 000	630
Metal plating	116 000	12 000
Polymer and rubber production	16 000	1 700
Waste disposal	23 000	2 400

In the EU risk assessment, an estimation of EDTA releases was carried out employing model calculations on the basis of monitoring results and application volumes [29]. Starting from the possible input [i.e. the possible concentration in the sewage discharge ($C_{effluent}$)], local EDTA concentrations in receiving waters in the immediate vicinity to sewage discharge points (PEC_{local}, Predicted Environmental Concentration) were assessed on the basis of a reasonable worst-case scenario, taking into account the dilution of the sewage and the background level of the receiving waters. Table 7.2 shows the estimated concentrations classified according to various branches of industry on the basis of a typical wastewater dilution of 1:10 (exception: oil production, 1:1000). However, in some European rivers, the dilution is much lower. In Great Britain for instance, the rivers Thames and Lee show a wastewater portion of 70 and 50%, respectively [6]. It is clear that close to industrial locations relatively high levels of EDTA may occur in the receiving water. Locally, the expected concentration range may even represent a risk for aquatic organisms [29, 69]. In the discharge area of municipal wastewaters, however, this risk is much lower, since pollution levels can be expected to be around two orders of magnitude lower than those of industrial wastewaters [43].

In municipal wastewater treatment plants, phosphorus removal is often performed by precipitation with iron(III) salts. In this process, the high iron concentration, the high stability of the Fe(III)-EDTA complex, and the relatively low pH (locally, after iron salt addition) favor the formation of the Fe(III)-EDTA complex from other metal-EDTA species (e.g. Cu-EDTA or Zn-EDTA) by metal exchange reactions [38, 70, 71]. Following phosphate precipitation, Fe(III)-EDTA and also Fe(III)-DTPA were repeatedly detected in the corresponding effluents of wastewater treatment plants [45, 71, 72]. For example, the Fe(III)-EDTA portion of the total EDTA in effluents of wastewater treatment plants along the Glatt river (Switzerland) was around 60–70% [73]. Other studies demonstrated that, in addition to Fe(III)-EDTA, wastewater treatment plants mainly discharge Ni-EDTA, Mn-EDTA, Al-EDTA, and Ca-EDTA with the effluent stream [38, 74–76].

7.4
Occurrence and Fate in Surface Waters

7.4.1
Occurrence and Speciation

Because of their incomplete elimination in municipal and industrial wastewater treatment plants, aminopolycarboxylates enter surface waters. In industrialized countries in Europe and the world, EDTA has even become the measurable anthropogenic compound at the highest concentration in rivers and lakes, being detected for instance in the Mississippi (USA) at 5–30 µg L^{-1} [77], in the Santa Ana River at 2.4–13 µg L^{-1} [78], in the Zerka River (Jordan) at 900 µg L^{-1} [79], in the Lagoa de Jurtunaiba (Brazil) at 2 µg L^{-1} [80], in various rivers in Great Britain at 6–60 µg L^{-1} [6, 81], in Finnish lakes at 2–390 µg L^{-1} [48, 82], in Swedish lakes at 1–735 µg L^{-1}

[83], in the Rio Odiel (Spain) at 599–2460 µg L^{-1} [84], in the Rhine (Germany) at 1–10 µg L^{-1} [1], in the Theiß (Romania) at 1.2–4.0 µg L^{-1} [85], and in the Ijsselmeer (The Netherlands) at 3–10 µg L^{-1} [86]. As a consequence of political and regulatory action, aminopolycarboxylate concentrations were investigated particularly intensively in Germany [43]. Emission and immission estimates revealed a total release of 580–920 tons EDTA per year to German surface waters currently. Typical concentrations of NTA, EDTA, and DTPA found in German rivers were in the range of 0–20, 0–50, and 0–60 µg L^{-1}, respectively. β-ADA and 1,3-PDTA were found only sporadically at concentrations around 1–5 µg L^{-1}. So far, no positive findings have been reported for MGDA in German surface waters [43].

At the concentrations currently measured in surface waters, aminopolycarboxylates occur most probably solely as metal complexes [87]. The speciation pattern of aminopolycarboxylates originally discharged by wastewater treatment plants, however, changes on mixing with receiving waters according to the local chemical environment. Based on a combination of direct analytical measurements and equilibrium calculations, Nowack evaluated the speciation of EDTA in river water [87]. According to his results, 51% of the total EDTA is present as Zn-EDTA and 32% as Fe(III)-EDTA, minor species being Ca-EDTA (7%), Mn(II)-EDTA (5%), Mg-EDTA (2%), Pb-EDTA (2%), and Ni-EDTA (0.1%). Despite its high stability, the proportion of Cu-EDTA is very low (<1%), since copper is preferentially bound by natural ligands. 2 nmol L^{-1} Cu-EDTA or less were directly measured in river water polluted to different degrees [88]. In another study, Nowack et al. detected 24 nmol L^{-1} Fe(III)-EDTA and 1.5 nmol L^{-1} Ni-EDTA in a river [73]. Considering a number of boundary conditions, model calculations by Sillanpää et al. also established iron and zinc complexes as the dominating EDTA and DTPA species in surface waters [76]. According to Nowack [87], NTA in rivers is predominantly associated with calcium (62% of the total NTA), followed by magnesium (29%) and zinc (3%). The proportion of uncomplexed NTA is about 5%. Only a minor part of the total NTA is complexed with other heavy metal ions.

7.4.2
Fate and Effects

In principle, aminopolycarboxylates may be degraded in surface waters by both biotic and abiotic processes [1, 16, 52, 87, 89]. NTA and MGDA are readily degradable by microbiological processes, whereas EDTA and 1,3-PDTA are only slightly reduced by natural processes [16, 90–92], with β-ADA and DTPA in between [80, 93–96]. Photochemical processes are assumed to have the largest impact on the natural reduction of EDTA [87] (Fig. 7.5). Basically, this pathway is restricted to the decomposition of the Fe(III)-EDTA complex. However, its fraction of the total EDTA speciation in surface waters amounts to roughly 30%, and metal exchange reactions of Fe(III)-EDTA are notably slow [87]. In principle, EDTA also can be degraded by microorganisms, as was shown by studies with certain enriched bacterial cultures [16, 33, 52, 89]. However, the extent of EDTA reduction by this process depends significantly on its metal speciation. In particular, ZnEDTA and Fe(III)-

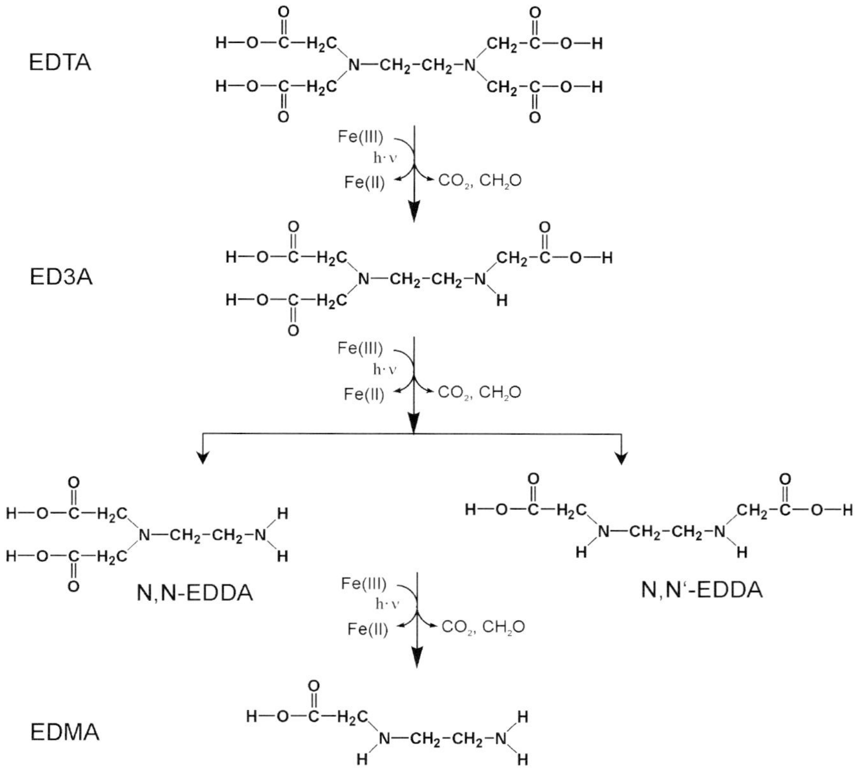

Fig. 7.5 Photochemical degradation of EDTA (from Ref. [1]).

EDTA complexes are relatively resistant to microbial degradation, although they predominate in the environment. This is the basic difference from NTA, of which those species dominating in rivers (e.g. Ca-NTA) are easily attacked by most bacterial strains [97].

Because of their chelating properties, those aminopolycarboxylates that are not (or only slowly) degraded may affect the mobility of metals in the environment [1]. Aminopolycarboxylates that are in contact with sediments can also be adsorbed to them by electrostatic interactions. Furthermore, they can dissolve sediment components [in particular iron and aluminum (hydr)oxides] with formation of the iron and aluminum complexes (in particular in an acidic environment) or remobilize already adsorbed or precipitated metal ions (preferentially zinc) from the sediment (in particular in an alkaline environment) [87]. According to current scientific opinion, effects of aminopolycarboxylic acids on the remobilization of heavy metals are assumed to be rather small in waters with levels in the low to middle µg L^{-1} range. However, local metal mobilization cannot be excluded when shock loadings occur [1].

Some ecotoxicological studies have demonstrated a possible eutrophication potential of aminopolycarboxylates at environmentally relevant concentrations [69].

In principle, this is not a result of any direct impact as a nitrogen source but rather of indirect effects related to the capability of aminopolycarboxylates to complex metal ions. On the one hand, aminopolycarboxylates increase the availability of essential trace elements for algae by the extraction and complexation of trace metals from sediments and humic acids. Thus, aminopolycarboxylates will stimulate algal growth when there is trace element deficiency. On the other hand, aminopolycarboxylates reduce the availability of certain toxic heavy metal ions, resulting in the growth of algae species that normally are inhibited by the toxicity. Furthermore, aminopolycarboxylates could also release phosphates by iron abstraction from insoluble Fe(III) phosphates, with simultaneous phosphate solubilization. Under natural conditions, these eutrophication effects are difficult to quantify, because the concentration of plankton algae is also influenced by natural variations that superimpose on possible aminopolycarboxylate-related effects. However, even though the effects of aminopolycarboxylates on algal growth in surface waters seem small or often even negligible, the possibility that they could become relevant under certain conditions cannot be excluded [69].

The historical and current use of aminopolycarboxylates and their ubiquitous presence in surface waters have prompted many studies on their possibly detrimental impact on aquatic organisms [69]. In general, aminopolycarboxylates do not differ much from each other with respect to their impact on aquatic organisms. Aminopolycarboxylates are unlikely to accumulate in aquatic food chains, and they show relatively low toxicity to aquatic organisms. Typically, acute toxicity from aminopolycarboxylates has been observed only at concentrations above 40 mg L^{-1}, and there is probably no harmful long-term impact at concentrations below 1 mg L^{-1}. Aminopolycarboxylates exert their toxic effects most often by causing metal deficiencies via chelation of trace elements that are essential to an organism. Aminopolycarboxylate toxicity is influenced by water hardness, chemical speciation, and pH. Aquatic organisms living in soft waters may be more adversely affected than those living in hard or marine water. Furthermore, aminopolycarboxylates are less toxic to aquatic organisms when they have been complexed with nontoxic metal ions (as is the case in most natural waters) [69]. Observed aminopolycarboxylate concentrations in surface waters are usually several orders of magnitude below those concentrations that are harmful to aquatic life. However, near certain industrial locations, where highly concentrated wastewater discharge might occur in connection with small surface water recipients, a potential risk to aquatic organisms cannot be excluded because aminopolycarboxylates could reach concentrations locally in the range of a few milligrams per liter [43, 69].

7.5
Occurrence and Fate in Groundwater

Groundwater pollution by aminopolycarboxylates is mainly caused by natural or artificial infiltration of polluted surface water into aquifers. The fate of EDTA,

NTA, and DPTA during riverbank filtration has been studied intensively in Germany on field sites with different redox conditions (aerobic, denitrifying, strictly anaerobic) and residence times [98]. According to these studies, EDTA can be regularly detected in bank filtrates, even though concentrations are in most cases slightly (not more than 50%) lower than those in the corresponding surface water. The impact of travel times and pathway lengths on EDTA elimination seems to be rather low. With regard to the impact of redox conditions, it was postulated that the reduction in EDTA concentrations under aerobic conditions is slightly more pronounced than under denitrifying and anaerobic redox conditions. In cases of simultaneous surface water pollution with DTPA, concentrations of EDTA in bank filtrate samples might be even higher than in the corresponding surface water. This was attributed to the microbial degradation of DTPA generating the aminopolycarboxylates NTA and EDTA as metabolites [80]. Overall, DTPA is slightly more easily degradable than EDTA and is reduced along the aquifer at an intermediate degradation rate. In contrast to EDTA and DTPA, the complexing agent NTA can be easily degraded by microorganisms, and its elimination turned out to be independent of redox conditions and travel times. The entire elimination of NTA takes place already in the first few centimeters to meters of any particular underground passage [98].

During the infiltration of surface water into aquifers, aminopolycarboxylate-metal complexes may change their speciation by interaction with the sediments. In particular, EDTA speciation has been intensively investigated. Interactions of metal-EDTA complexes with sediments decisively depend on pH and the metal ion complexed by the aminopolycarboxylate [87, 99–101]. In principle, the complexed metal ion can be released by exchange with adsorbed or dissolved metal ions (e.g., copper, zinc, lead, and cadmium). Also, the dissolution of iron, aluminum, or manganese oxides with formation of Fe-EDTA, Al-EDTA, or Mn-EDTA may occur [102]. At alkaline pH values, quite common in calcareous aquifers, the remobilization of adsorbed metals by replacement of the initial EDTA central atom [e.g., Fe(III)] predominates. These reactions presumably affect the mobility of zinc, lead, and cadmium in particular. In infiltrated groundwater close to the Glatt river (Switzerland), with a calciferous aquifer, EDTA speciation was predominantly characterized by Zn-EDTA, Ca-EDTA, and Fe(III)-EDTA; in some groundwaters, elevated Mn(II)-EDTA levels also occurred. Furthermore, it could be demonstrated that the Fe(III)-EDTA fraction is reduced during the passage of the water through the aquifer. In all investigated groundwaters, Zn-EDTA was the most relevant EDTA species, representing 40–60% of the total EDTA. The Pb-EDTA fraction was around 3%, while those of Cu-EDTA and Mg-EDTA always were below 1% [100]. During infiltration at low pH, dissolution of iron oxides by various metal-EDTA species takes place. Fe(III)-EDTA is formed in an cation exchange reaction, and the initially complexed metal ion is released and adsorbed by the solid phase [100]. The speciation of aminopolycarboxylates in groundwater is therefore considerably dependent on the individual geological and chemical properties of the infiltration pathway under investigation and is therefore extremely site-specific.

7.6
Behavior during Drinking Water Processing

Passage of water underground is often only the first step of the whole drinking water preparation process and is typically followed by further treatment steps to build up multiple barrier systems. Important treatment steps are in particular ozonation, activated carbon filtration, and chlorine disinfection.

Aminopolycarboxylic acids and their metal complexes can be converted into biodegradable oxidation products by ozone [103]. However, this reduction depends upon the quality of the raw water [1]. Under optimal conditions, ozone doses of 1–2 mg L^{-1} could cause an aminopolycarboxylate elimination of about 70–80% [104, 105]. The elimination improves with increasing reaction time and ozone dose and is catalytically enhanced by the presence of Cu(II) ions [105–108]. However, the extent of degradation is always dependent on the amount of organic compounds in the raw water and their capability of reacting with ozone. It is assumed, in particular at low ozone doses, that ozone reacts at first with humic matter and is only thereafter capable of oxidizing aminopolycarboxylates in competing reactions [1]. Furthermore, the rate of aminopolycarboxylate degradation by ozonation is significantly dependent on chelate speciation. Free, non-complexed EDTA, Ca-EDTA, and Cd-EDTA are, for instance, much more rapidly degraded than Fe(III)-EDTA [108–110].

In principle, aminopolycarboxylates can be adsorbed onto activated carbon by electrostatic interactions [111–115]. In this process, the pH of the aqueous solution is of central importance, because the surface charge of activated carbon, the charge of metal-aminopolycarboxylates, their complex stabilities, and their speciation are all pH dependent. Aminopolycarboxylate-metal complexes are negatively charged over a wide pH range and are preferentially adsorbed by protonated activated carbon. As a consequence, the adsorption of aminopolycarboxylates onto activated carbon is typically favored at acidic pH values [111–114, 116, 117]. Under optimal conditions, adsorption of metal-EDTA or metal-NTA complexes onto fresh or regenerated activated carbon can amount to more than 90% [18, 111, 116, 118, 119]. EDTA turned out to be slightly more readily adsorbable than NTA, but both compounds are relatively weakly retained by activated carbon when compared to other organic substances [104, 120, 121]. The competing adsorption of other water constituents considerably lowers the retaining power of preloaded activated carbon in waterworks in practice. Even relatively small loadings of the activated carbon with raw water (<15 m^3 kg^{-1}) can cause a considerable loss of sorption capacity and result in a breakthrough of aminopolycarboxylates, making activated carbon treatment unsuitable for substantial aminopolycarboxylate removal from an economical standpoint [18, 104]. However, biodegradable aminopolycarboxylates like NTA and MGDA might be eliminated by microbial degradation in biologically active carbon filters.

Low amounts of chlorine, as utilized for drinking water disinfection, do not result in any significant aminopolycarboxylate degradation. Only chlorine concentrations above 10 mg L^{-1} would be capable of causing some significant oxidative destruction [1, 122, 123].

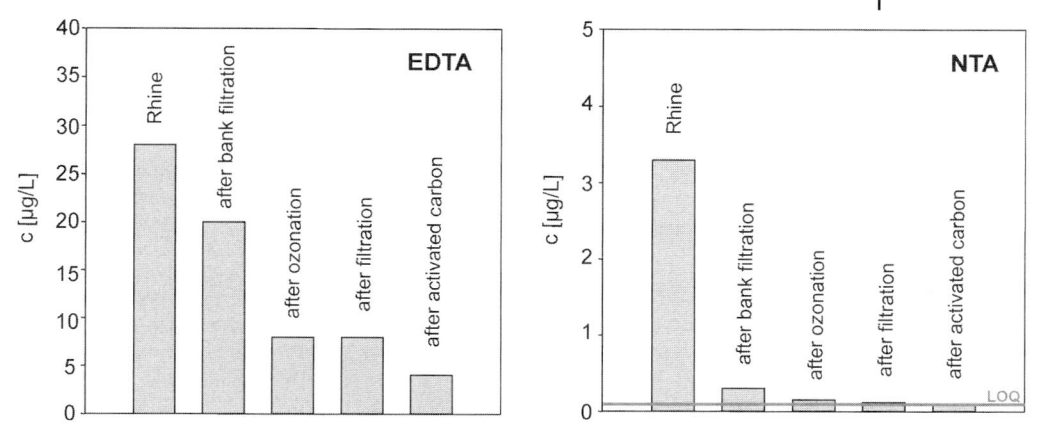

Fig. 7.6 Behavior of EDTA and NTA during drinking water production on the lower Rhine. Presented data are mean values of weekly composite samples (each a mixture of five spot samples) over a time period of two months (data from Ref. [104]).

As an example, Fig. 7.6 shows the behavior of EDTA and NTA in a waterworks on the lower river Rhine [104]. The waterworks obtains its raw water from riverbank filtrate. Further drinking water treatment steps include ozonation and filtration over multilayer filters including activated carbon. The results show that NTA is extensively removed during the bank filtration process, whereas EDTA concentrations are lowered only by about 20%. The concentrations of EDTA and NTA in raw riverbank filtrate were further reduced by using ozonation and activated carbon filtration; however, total removal of EDTA could not be achieved, indicating its significance for drinking water [104].

As a consequence, aminopolycarboxylates have to be considered as relevant to drinking water. According to studies in Germany, levels found in drinking water are decisively dependent on both the water treatment process in operation and the degree of pollution of the raw water [43]. Evaluation of data from 20 locations in Germany revealed that EDTA levels of 1.1–11 µg L^{-1} in the raw water were lowered by between 0% and 70% during drinking water processing. Highest levels of EDTA found over the last few years in German drinking waters were about 7 µg L^{-1}. NTA and DTPA were found only in individual cases, and concentrations were generally lower than 3 µg L^{-1} and 4 µg L^{-1}, respectively. Because of their very low levels in surface waters in Germany, β-ADA, 1,3-PDTA, and MGDA are not found in detectable concentrations in German drinking waters [43]. It is worth mentioning that drinking water levels in the low µg L^{-1}-range do not constitute a risk to human health. NOEL (no observed effect level) values established for mammals are 250 mg EDTA and 10 mg NTA per kg body weight and day [1, 124, 125].

7.7
Conclusions

Aminopolycarboxylates constitute a class of complexing agents that occur in a wide range of domestic products and are used intensively as metal sequestrants in several industrial applications. Because they are highly polar and partially nondegradable, aminopolycarboxylates are released into the aquatic environment in significant quantities, mainly via wastewater. In industrialized countries, aminopolycarboxylates are present in almost all anthropogenically impacted rivers and streams at µg L^{-1} concentrations, and they currently are among the organic pollutants with the highest mass concentrations measured in surface waters. Once released into the environment, the fate of aminopolycarboxylates is determined either by abiotic or biotic processes. Each aminopolycarboxylic acid has to be assessed individually, because no general pattern applicable to all of them has emerged. Both abiotic and biotic degradation processes are strongly influenced by the chelating properties and speciation pattern of each individual aminopolycarboxylic acid. While NTA and MGDA turned out to be easily degradable by microbial processes, EDTA and 1,3-PDTA are hardly degraded at all. β-ADA and DTPA occupy an intermediate position. At concentrations in the low to middle µg L^{-1}-range, aminopolycarboxylates should have no adverse impact on aquatic organisms. However, because of the possible influence of aminopolycarboxylates on eutrophication and their potential hazard for aquatic organisms close to industrial sewage discharge points (i.e. at higher concentrations), efforts should be made to prevent the release of highly concentrated aminopolycarboxylate effluents into the aquatic environment. This should be feasible, because several efficient treatment techniques have been tested and established in the past few years, and a number of easily degradable substitution products have been brought onto the market. Even though aminopolycarboxylates do not present any major risk to mammals, persistent aminopolycarboxylates (in particular EDTA and DTPA) cause an undesirable anthropogenic pollution of drinking water and, according to the precaution principle, their levels should be kept as low as possible. In principle, it is technically easier and economically cheaper to treat highly concentrated partial wastewater streams, thereby avoiding their emission to surface waters, instead of trying to achieve their complete elimination at trace levels during drinking water processing.

References

1 C. K. Schmidt, H.-J. Brauch, *Aminopolycarbonsäuren in der aquatischen Umwelt*, Schriftenreihe Technologiezentrum Wasser (TZW) Vol. 20, Karlsruhe, **2003**.

2 R. L. Anderson, W. E. Bishop, R. L. Campbell, *Crit. Rev. Toxicol.* **1985**, 15, 1–102.

3 BUA, *Nitrilotriessigsäure*, BUA-Stoffbericht 5, VCH, Weinheim, **1987**.

4 BUA, *Ethylendiamintetraessigsäure/Tetranatriumethylendiamintetraacetat*, BUA-Stoffbericht 168, S. Hirzel Wissenschaftliche Verlagsgesellschaft, Stuttgart, **1996**.

5 T. P. Knepper, *Trends Anal. Chem.* **2003**, 22, 708–724.

6 K. Wolf, P. Gilbert in *The Handbook of Environmental Chemistry*, Vol. 3 Part F (Ed.: O. Hutzinger), Springer, Berlin, **1992**, pp. 243–259.

7 G. Anderegg in *Comprehensive coordination chemistry, Vol. 2: ligands* (Eds.: G. Wilkinson, R. D. Gillard, J. A. McCleverty), Pergamon Press, New York, **1987**, pp. 777–792.

8 T. S. West, *Complexometry with EDTA and related reagents*, 3rd ed., BDH Chemicals Ltd Poole, **1969**.

9 J. R. Hart, *Ethylenediaminetetraacetic acid and related chelating agents*, Ullmann's Encyclopedia of Industrial Chemistry (Electronic Release), Wiley-VCH, Weinheim **2001**.

10 R. E. Davenport, F. Dubois, A. DeBoo, A. Kishi, *Chelating agents – CEH Product Review*, Chemical Economics Handbook, SRI International, **2000**.

11 A. E. Martell, *Fed. Proc.* **1961**, 20, 35–39.

12 A. E. Martell, R. M. Smith, *Critical stability constants. Vol. 1: amino acids*, Plenum Press, New York, **1974**.

13 UBA, *Unterlagen zum 15. EDTA-Fachgespräch "Verringerung der Gewässerbelastung durch EDTA" am 8.11.2001*, Umweltbundesamt (UBA), Berlin, **2001**.

14 H.-U. Jäger, W. Schul, *Muench. Beitr. Abwasser Fisch Flussbiol.* **2001**, 54, 207–226.

15 T. P. Knepper, H. Weil, *Vom Wasser* **2001**, 97, 193–232.

16 M. BucheliWitschel, T. Egli, *FEMS Microbiol. Rev.* **2001**, 25, 69–106.

17 G. Metzner in *Aquatische Umweltverträglichkeit von Nitrilotriessigsäure (NTA)* (Ed. Kernforschungszentrum Karlsruhe), Karlsruhe, **1991**, pp. 33–62.

18 H. Bernhardt, *NTA-Studie über die Umweltverträglichkeit von Nitrilotriacetat*, Richarz, St. Augustin, **1984**.

19 N. M. Brouwer, P. M. J. Terpstra, *Tenside Surf. Det.* **1995**, 32, 225–228.

20 M. Dwyer, S. Yeoman, J. Lester, R. Perry, *Environ. Technol.* **1990**, 11, 263–294.

21 R. Perry, P. W. W. Kirk, T. Stephenson, J. N. Lester, *Water Res.* **1984**, 18, 255–276.

22 P. W. W. Kirk, J. N. Lester, R. Perry, *Water Res.* **1982**, 16, 973–980.

23 S. A. Klein, *J. Water Pollut. Control Fed.* **1974**, 46, 78–88.

24 L. Moore, E. F. Barth, *J. Water Pollut. Control Fed.* **1976**, 48, 2406–2414.

25 M. L. Hinck, J. Ferguson, J. Puhaakka, *Water Sci. Technol.* **1997**, 35, 25–31.

26 R. Saunamäki, *Tappi J.* **1995**, 78, 185–192.

27 M. Ek, M. Remberger, A.-S. Allard, *Nordic Pulp Paper Res. J.* **1999**, 14 (2–3), 310–314.

28 M. Ek, M. Remberger, A.-S. Allard, *Biological degradation of EDTA in pulping effluents at higher pH – a laboratory study, report B 1322*, IVL Swedish Environmetal Research Institute, **1999**.

29 EU, *Tetrasodium ethylenediaminetetraacetate (Na_4EDTA)*, EU Risk Assessment, **2001**.

30 N. Gschwind, *gwf Wasser Abwasser* **1992**, 133, 546–549.

31 L. Henneken, B. Nörtemann, D. C. Hempel, *Appl. Microbiol. Biotechnol.* **1995**, 44, 190–197.

32 L. Henneken, B. Nörtemann, D. C. Hempel, *J. Chem. Technol. Biotechnol.* **1998**, 73, 144–152.

33 T. Klüner, *Chemie und Biochemie des mikrobiellen EDTA-Abbaus*, Cuvillier Verlag, Göttingen, **1996**.

34 C. G. van Ginkel, K. L. Vandenbroucke, C. A. Stroo, *Bioresour. Technol.* **1997**, 59, 151–155.

35 C. G. van Ginkel, H. Kester, C. A. Stroo, A. van Haperen, *Water Sci. Technol.* **1999**, 40 (11–12), 259–265.

36 C. G. van Ginkel, J. Virtaphoja, J. Steyaert, R. Alén, *Tappi J.* **1999**, 82, 138–142.

37 M. Sillanpää, K. Pirkanniemi, *Environ. Technol.* **2001**, 22, 791–801.

38 P. M. Nirel, P.-E. Pardo, J.-C. Landry, R. Revaclier, *Water Res.* **1998**, 32, 3615–3620.

39 A. C. Alder, H. Siegrist, K. Fent, T. Egli, E. Molnar, T. Poiger, C. Schaffner, W. Giger, *Chimia* **1997**, 51, 922–928.

40 M. Sillanpää, *Chemosphere* **1996**, 33, 293–302.

41 U. Kaluza, P. Klingelhöfer, K. Taeger, *Water Res.***1998**, 32, 2843–2845.

42 H.-B. Lee, T. E. Peart, K. L. E. Kaiser, *J. Chromatogr. A* **1996**, 738, 91–99.

43 C. K. Schmidt, M. Fleig, F. Sacher, H.-J. Brauch, *Environ. Pollut.* **2004**, 131, 107–124.

44 R. Järvinen, *Water Sci. Technol.* **1997**, 35 (2–3), 139–145.
45 D. E. Richardson, G. H. Ash, P. E. Harden, *J. Chromatogr. A* **1994**, 688, 47–53.
46 Bayerisches Landesamt für Wasserwirtschaft, Institut für Wasserforschung, *Untersuchungen zu ökologischen Verhalten von ausgewählten organischen Komplexbildnern*, final report, **1996**.
47 F. Nispel, W. Baumann, G. Hardes, *Korrespondenz Abwasser* **1990**, 37, 707–709.
48 A. Langi, M. Priha, T. Tapanila, E. Talka, *Das Papier* **1998**, 10A, V28–V34.
49 E. Magnus, G. E. Carlberg, H. Hoel, *Nordic Pulp Paper Res. J.* **2000**, 15, 29–36.
50 F. H. Frimmel in *Detergents in the aquatic environment* (Ed.: M. J. Schwuger), Marcel Dekker, New York, **1997**, pp. 289–312.
51 C. Oviedo, D. Rodríguez, *Quim. Nova* **2003**, 26, 901–905.
52 B. Nörtemann, *Appl. Microbiol. Biotechnol.* **1999**, 51, 751–759.
53 A. J. Chaudhary, J. D. Donaldson, S. M. Grimes, M. Hassan, R. J. Spencer, *J. Chem. Technol. Biotechnol.* **2000**, 75, 353–358.
54 H. Kroder, K.-H. Bauer, D. Saur, *Vom Wasser* **1994**, 82, 269–279.
55 Y. Ku, L.-S. Wang, Y.-S. Shen, *J. Hazard. Mat.* **1998**, 60, 41–55.
56 A. Kunz, P. Peralta-Zamora, N. Durán, *Adv. Environ. Res.* **2002**, 7, 197–202.
57 J. Pröter, K. Pflugbeil, *KA-Wasserwirtschaft Abwasser Abfall* **2000**, 47, 1820–1824.
58 M. Sörensen, F. H. Frimmel, *Z. Naturforsch. B* **1995**, 50, 1845–1853.
59 M. Sörensen, *PhD thesis*, Universität Karlsruhe (Germany), **1996**.
60 M. Sörensen, S. Zurell, F. H. Frimmel, *Acta Hydrochim. Hydrobiol.* **1998**, 26, 109–115.
61 M. Sörensen, J. Weckenmann, R. Hofmann, J. Pagel, A. Weber, F. Harder, *Galvanotechnik* **2002**, 8, 2127–2133.
62 T. A. Ternes, K. Stolte, K. Haberer, *Vom Wasser* **1997**, 88, 243–256.
63 M. Sörensen, F. H. Frimmel, *Acta Hydrochim. Hydrobiol.* **1996**, 24, 185–188.
64 M. Sörensen, U. Tanner, G. Sagawe, F. H. Frimmel, *Acta Hydrochim. Hydrobiol.* **1996**, 24, 132–136.
65 M. Sörensen, F. H. Frimmel, *Water Res.* **1997**, 31, 2885–2891.
66 R. Fandrich, R. Kümmel, K. Hagen, H. Tonn, A. Allendorf, *gwf Wasser Abwasser* **1998**, 139, 783–791.
67 D. Rodríguez, A. Mutis, M. C. Yeber, J. Freer, J. Baeza, H. D. Mansilla, *Water Sci. Technol.* **1999**, 40 (11–12), 267–272.
68 M. Gülbas, *Metalloberfläche* **1994**, 48, 310–318.
69 C. K. Schmidt, H.-J. Brauch, *Environ. Toxicol.* **2004**, 19, 620–637.
70 H. B. Xue, S. Jansen, A. Prasch, L. Sigg, *Environ. Sci. Technol.* **2001**, 35, 539–546.
71 F. G. Kari, *PhD Thesis*, ETH Zürich (Switzerland), **1994**.
72 W. Buchberger, S. Mülleder, *Mikrochim. Acta* **1995**, 119, 103–111.
73 B. Nowack, F. G. Kari, S. U. Hilger, L. Sigg, *Anal. Chem.* **1996**, 68, 561–566.
74 W. W. Bedsworth, D. L. Sedlak, *Environ. Sci. Technol.* **1999**, 33, 926–931.
75 F. G. Kari, W. Giger, *Water Res.* **1996**, 30, 122–134.
76 M. Sillanpää, M. Orama, J. Rämö, A. Oikari, *Sci. Total Environ.* **2001**, 267, 23–31.
77 L. B. Barber, J. A. Leenheer, W. E. Pereira, T. I. Noyes, G. K. Brown, C. F. Tabor, J. H. Writer. *Organic contamination of the Mississippi River from municipal and industrial wastewater. Contaminants in the Mississippi River*, U.S. Geological Survey Circular 1133, Reston, Virginia, **1995**.
78 W.-H. Ding, J. Wu, M. Semadeni, M. Reinhard, *Chemosphere* **1999**, 39, 1781–1794.
79 A. M. van Dijk-Looyaard, A. C. de Groot, P. J. C. M. Janssen, E. A. Wondergem, *EDTA in drink – en oppervlaktewater*, Rapport Nr. 718629006, Rijksinstituut voor Volksgezondheid en Milieuhygiene, Bilthoven, **1990**.
80 M. Stumpf, T. A. Ternes, B. Schuppert, K. Haberer, P. Hoffmann, H. M. Ortner, *Vom Wasser* **1996**, 86, 157–171.
81 J. H. N. Garland, J. Capon, J. Cox, A. Collier, M. G. Kibblewhite, *Survey of concentrations of NTA and EDTA in UK*

rivers and at sewage treatment works. Laboratory Report 912 (CA), Warren Spring Laboratory, Stevenage, **1992**.
82 M. Sillanpää, J. Sorvari, M.-L. Sihvonen, *Chromatographia* **1996**, 42, 578–582.
83 M. Remberger, A. Svenson, *Fate of EDTA and DTPA in aquatic environments receiving waste water from two pulp and paper mills, B-1256*, IVL Swedish Environmental Research Institute, Stockholm, **1997**.
84 C. Kowalik, J. W. Einax, *Vom Wasser* **2000**, 94, 229–243.
85 IAWD, *Annual Report 1999/2000*. International Association of Water Supply Companies in the Danube River Catchment Area (IAWD), Vienna, **2001**.
86 RIWA, Jaarverslag 2001–2002 – De Rijn. Vereniging van Rivierwaterbedrijven Sectie Rijn (RIWA), Nieuwegein, **2003**.
87 B. Nowack, *Environ. Sci. Technol.* **2002**, 36, 4009–4016.
88 A. A. Ammann, *Anal. Bioanal. Chem.* **2002**, 372, 448–452.
89 T. Egli, *J. Biosci. Bioeng.* **2001**, 92, 89–97.
90 P. Pitter, V. Sukora, *Chemosphere* **2001**, 44, 823–826.
91 B. Potthoff-Karl, T. Greindl, A. Oftring, *SÖFW J.* **1996**, 122, 392–397.
92 V. Sukora, P. Pitter, I. Bittnerová, T. Lederer, *Water Res.* **2001**, 35, 2010–2016.
93 L. Nitschke, A. Wilk, C. Cammerer, G. Lind, G. Metzner, *Chemosphere* **1997**, 34, 807–815.
94 H. Bolton, S. W. Li, D. J. Workman, D. C. Girvin, *J. Environ. Qual.* **1993**, 22, 125–132.
95 UBA, *Unterlagen zum 5. EDTA-Fachgespräch "Verringerung der Gewässerbelastung durch EDTA" am 25.09.1991*, Umweltbundesamt (UBA), Bonn-Bad Godesberg, **1991**.
96 S. Metsärinne, P. Rantanen, R. Aksela, T. Tuhkanen, *Chemosphere* **2004**, 55, 379–388.
97 J. M. Vanbriesen, B. Rittmann, L. Xun, D. Girvin, H. Bolton, *Environ. Sci. Technol.* **2000**, 34, 3346–3353.
98 C. K. Schmidt, F. T. Lange, H.-J. Brauch, Proc. 4th *International Conference on Pharmaceuticals and Endocrine Disrupting Chemicals in Water*, October 13–15, **2004**, Minneapolis, Minnesota, pp. 195–205.
99 J. C. Friendly, D. B. Kent, J. A. Davis, *Environ. Sci. Technol.* **2002**, 36, 355–363.
100 B. Nowack, H. Xue, L. Sigg, *Environ. Sci. Technol.* **1997**, 31, 866–872.
101 J. M. Zachara, S. C. Smith, L. S. Kuzel, *Geochim. Cosmochim. Acta* **1995**, 59, 4825–4844.
102 B. Nowack, F. G. Kari, H. G. Krüger, *Water Air Soil Pollut.* **2001**, 125, 243–257.
103 U. von Gunten, *Water Res.* **2003**, 37, 1443–1467.
104 H.-J. Brauch, S. Schullerer, *Vom Wasser* **1987**, 69, 155–164.
105 S. Schullerer, H.-J. Brauch, *Vom Wasser* **1989**, 72, 21–29.
106 H.-J. Brauch, F. Sacher, E. Denecke, T. Tacke, *gwf Wasser Abwasser* **2000**, 141, 226–234.
107 D. W. Sundstrom, J. S. Allen, S. S. Fenton, F. E. Salimi, K. J. Walsh, *J. Environ. Sci. Health* **1996**, A31, 1215–1235.
108 C. Fabjahn, R. Davies, K. Marschall, *Galvanotechnik* **1976**, 67, 643–645.
109 M. S. Korhonen, S. E. Metsärinne, T. A. Tuhkanen, *Ozone Sci. Eng.* **2000**, 22, 279–286.
110 E. Gilbert, S. Hoffmann-Glewe, *Water Res.* **1990**, 24, 39–44.
111 D. Bhattacharyya, C. Y. R. Cheng, *Environ. Progress* **1987**, 6, 110–118.
112 C. Chang, Y. Ku, *J. Hazard. Mater.* **1994**, 38, 439–451.
113 C. Chang, Y. Ku, *Sep. Sci. Technol.* **1998**, 33, 483–501.
114 A. J. Rubin, D. L. Mercer, *Sep. Sci. Technol.* **1987**, 22, 1359–1381.
115 K. H. Chu, M. A. Hashim, *J. Chem. Technol. Biotechnol.* **2000**, 75, 1054–1060.
116 J. P. Chen, S. Wu, *J. Chem. Technol. Biotechnol.* **2000**, 75, 791–797.
117 C. Chang, Y. Ku, *Sep. Sci. Technol.* **1995**, 30, 899–915.
118 A. H. M. Linckens, J. K. Reichert, *Vom Wasser* **1982**, 58, 27–37.
119 T. W. Marrero, S. E. Manahan, J. S. Morris, *J. Radioanal. Nucl. Chem.* **2002**, 254, 193–200.
120 Kernforschungszentrum Karlsruhe GmbH, *Aquatische Umweltverträglichkeit von Nitrilotriessigsäure (NTA)*, Karlsruhe, **1991**.
121 R. Schick, *Proc. EDTA im Bodenseeraum*, November 29–30, **1994**, Friedrichshafen, Germany.

122 M. Grawehr, B. Sener, T. Waltimo, M. Zehnder, *Int. Endod. J.* **2003**, 36, 411–415.
123 W. Erlmann, M. Foery, H. Jantschke, *Galvanotechnik* **1990**, 81, 1249–1258.
124 WHO in *Guidelines for drinking-water quality. Vol 2. Health criteria and other supporting information*, (Ed.: World Health Organization), Geneva **1996**, pp. 565–573.
125 WHO in *Guidelines for drinking-water quality. Addendum to Vol. 2. Health criteria and other supporting information.* (Ed.: World Health Organization), Geneva **1998**, pp. 111–122.

8
Amines

Hilmar Börnick and Torsten C. Schmidt

8.1
Introduction

8.1.1
Characterization of Amines and Choice of Compounds

Nitrogen-organic compounds, and in particular amines, are highly relevant anthropogenic pollutants because of their broad application, their high production volume, and the many possible entry paths into the water cycle.

Structurally, amines are derived from ammonia/ammonium. As a function of the number of hydrogen atoms that are substituted by hydrocarbon groups, the amines are classified as primary, secondary, tertiary, or quaternary amines. Depending on the molecular structure and the character of the substituents, there are other possibilities for classifying the amines, e.g., into the groups of aliphatic, aromatic, aromatic heterocyclic, and alicyclic compounds. Further compounds such as amino acids, amides, nitrosamines, and aminoacetic acids also contain amino groups and can be formally considered as amines. However, such compounds are not covered here.

Amines comprise a wide variety of compounds, each with a wide variety of properties. Therefore, it is necessary to limit this monograph to a selected group of amines, and these will be characterized by relatively high polarity (log K_{ow} up to about 3) and high recalcitrance. Recalcitrance is difficult to evaluate because of the dependence of degradation on highly variable boundary conditions, e.g., redox milieu or pH. Aniline, for example, is known to be easily degradable, but under specific anoxic conditions it has been proven to be stable [1, 2], and it is therefore included here. Furthermore, the amines considered in this chapter are relevant in terms of eco- and human toxicity, industrially produced in high quantities [High Production Volume (HPV) chemicals as defined by OECD], and known to occur in the hydrosphere.

Organic Pollutants in the Water Cycle. T. Reemtsma and M. Jekel (Eds.).
Copyright © 2006 WILEY-VCH Verlag GmbH & Co. KGaA, Weinheim
ISBN: 3-527-31297-8

Of interest in this chapter are

1. aromatic amines such as toluidines, chloroanilines, nitroanilines, and benzidines,

2. aliphatic and alicyclic amines and diamines such as dimethylamine, isobutylamine, ethylenediamine, pyrrolidine, morpholine, piperazine, and thiourea,

3. nitrogen-containing aromatic heterocycles such as pyridine, quinoline, and benzotriazole,

4. amphoteric aromatic amines such as aminobenzoic acids, aminosulfonic acids, and aminophenols.

Some of the compounds belonging formally to groups 1 to 4 above, e.g., longer-chain and heavily branched aliphatic amines or multiply chlorinated (n > 3) aromatic amines, are not discussed here because of their low polarity (log $K_{ow} \gg 3$). Amines that are complexing agents, pesticides, or pharmaceuticals are dealt with in other chapters of this monograph.

8.1.2
Properties

Except for quaternary ammonium compounds, amines are bases. This is because of the free electron pair of the nitrogen atom that can accept protons. The basicity of the amines increases with increasing electron density at the nitrogen atom.

Aromatic amines show lower basicity than aliphatic compounds because of the interaction of the free electron pair of the nitrogen with the delocalized π-orbital system. This additional stabilization is impeded by protonation.

Acceptor substituents at the aromatic ring such as -Cl or -NO$_2$ cause an additional decrease in the basic character. Donor groups such as -CH$_3$ or -OR that occur in the meta or para position lead to an increase in the basicity of aromatic amines. However, donor substituents in the ortho position can sterically impede the protonation and consequently decrease the basicity. This behavior is shown by the acidity constant, pK_a, of protonated aniline and toluidines (Fig. 8.1).

Amines are characterized by amphoteric properties when the molecules contain additional acidic groups such as hydroxyl, carboxyl, or sulfoxyl substituents. De-

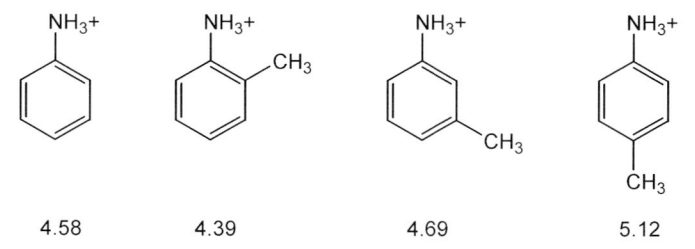

| 4.58 | 4.39 | 4.69 | 5.12 |

Fig. 8.1 Dependence of the acidity constant on the presence and position of the methyl group (pK_a-values at 25 °C [3]).

pending on the pH value, the amphoteric compounds can be anionic (deprotonated at the acidic group), cationic (protonated at the amino group), non-ionic, or even zwitterions. Accordingly, knowledge of the acidity constant(s) is indispensable for the characterization of the behavior of amines.

Owing to their rather strong ability to interact via hydrogen bonds, amines are relatively polar compounds. Environmental partitioning is influenced by substituents as well as by the number of carbon atoms and the cyclic, aliphatic, branched, or aromatic structure of the carbon skeleton. The presence of an amino group causes a higher boiling point, a higher water solubility, a lower Henry's law constant, K_H, and a higher mobility in the water cycle in comparison with hydrocarbons. Note, however, that, depending on the milieu parameters, the amino group can also reduce the mobility of a molecule by specific interactions with solids via covalent bonding to carbonyl moieties or cation exchange.

The volatility of amines in aqueous solution is in most cases relatively low. Hence, the atmosphere is hardly considered here. Only a few, low-molecular-weight aliphatic amines possess high Henry's law constants, e.g., dimethylamine, with 1.79 Pa m^3 mol^{-1}.

Selected physico-chemical properties of the amines of interest are summarized in Table 8.1.

Various amines, mainly aromatic compounds, show both acute and chronic toxicity. High potential health risk is caused by absorption of amines through the gastrointestinal tract, the skin, and the lungs. Numerous studies, in particular with regards to tobacco smoke, have shown the formation of hemoglobin adducts by many aromatic amines [5] and the disturbance of blood formation [6]. LD$_{50}$ values reported for aliphatic, alicyclic, and aromatic amines range between 100 and 2 000 mg kg^{-1} body weight.

Several aromatic amines cause damage of DNA that can lead to the genesis of malignant tumors. Electron-donating substituents in the ortho or para position can increase the carcinogenic potential [7]. Indeed, it has been reported that the epidemiology of aromatic amine carcinogenesis is essentially connected with the epidemiology of human cancer caused by industrial activities [8].

Some of the best known amines inducing cancer in humans and animals are benzidine and its derivatives [9]. A cancer-causing effect has also been proved for a large number of other anilines, e.g., 2-nitroaniline [10], 1- and 2-naphthylamine, 2-methylaniline, 2,4-diaminotoluene [11], 4-chloroaniline, *ortho*-toluidine, and 4,4'-methylendianiline [12]. For smokers, correlations between blood or urine levels of aromatic amines and cancer incidents have been shown numerous times. Recently, Skipper et al. even suggested that many incidents of bladder cancer in non-smokers might be caused by environmental exposure to aromatic amines [13]. Similarly, many heterocyclic amines are known mutagens and carcinogens [14].

In contrast, aliphatic and alicyclic amines are generally more problematic because of odor and taste impairment even at very low concentrations [15]. For example, the odor threshold of trimethylamine in water is about 0.2 μg L^{-1} [16]. Greim et al. demonstrated the dependence of different toxic effects on the structure of 37 aliphatic amines [17]. They proved that the dermal uptake of primary

Tab. 8.1 Representative amines and some of their physical properties [4].

Compound	b.p.	K_H	C_{sat}^w	pK_a	log K_{OW}
Aniline	184	2.05E−01	3.6E+04	4.6	0.90
2-Chloroaniline	209	5.46E−01	8 160	2.66	1.90
2-Nitroaniline	284	5.98E−03	1 470[b]	−0.28	1.85
3-Nitroaniline	306	8.01E−04	1 200[c]	2.47	1.37
2,5-Dichloroaniline	251	1.62E−01	230	2.05	2.75
3,4-Dichloroaniline	272	1.48E+00	92[a]	2.97	2.69
2-Chloro-4-nitroaniline	n.s.	9.67E−04	933	−0.94	2.12
3-Chloro-4-methylaniline	243	1.58E−01	832	4.05	2.27
N-Ethylaniline	203	9.91E−01	2 410	5.12[c]	2.16
Diphenylamine	302	3.43E−01	53[a]	0.78[c]	3.50
4,4′-Methylendianiline	398	5.67E−06	1 000	n.s.	1.59
4,4′-Diaminobiphenyl	401	5.24E−06	322	4.66[a]	1.34
1,2-Diaminobenzene	257	7.30E−04	4.04E+04[d]	4.47	0.15
4-Aminodiphenylamine	354	3.76E−05	1 450	5.20	1.82
4-Aminoazobenzene	366	8.82E−06	32	2.82	3.41
1,8-Naphthylenediamine	205	6.66E−06	850	4.44	1.78
1,3-Diamino-4-methylbenzene	292	8.02E−05	7.48E+04	n.s.	1.40
Morpholine	128	1.18E−01	1.0E+06	8.49	−0.86
Ethylenediamine	117	1.75E−04	1.0E+06	9.92	−2.04
Diethanolamine	269	3.92E−06	1.0E+06[a]	8.96	−1.43
Piperazine	146	2.23E−04	1.0E+06	9.73	−1.50
Triethanolamine	335	7.14E−08	1.0E+06	7.76	−1.00
Triisopropanolamine	305	9.90E−07	8.3E+05[a]	8.06	−1.22

K_H Henry's law constant at 1013 hPa and 25 °C (Pa m³ mol⁻¹)
C_{sat}^w water solubility (mg L⁻¹) at 25 °C, 20 °C[a], 30 °C[b], 24 °C[c], or 35 °C[d]
pK_a dissociation constant of the protonated amine at 25 °C, 20 °C[a], or 24 °C[c]
log K_{OW} n-octanol/water partition coefficient
n.s. not specified

amines decreases with longer chains and that secondary amines and in particular tributylamine are easily and rapidly resorbed. The toxicity diminished essentially with the protonation of the amines.

Secondary amines have an additional potential health risk because of the formation of nitrosamines from reaction with nitrite. Nitrosamines are proven carcinogens. The nitrosation reaction takes place at very low concentrations and depends on boundary conditions such as pH, temperature, steric influences of the alkyl groups, and the presence of catalysts or inhibitors [17]. A well-known precursor for nitrosamine formation is dimethylamine [18–20].

8.2
Production

The typical properties of amines such as high reactivity and basicity are responsible for their manifold and broad applications in industry.

The production and use of amines is strongly connected to the development of artificial dye production on an industrial scale, which started in the mid-19th century. At present, more than 3 000 azo dyes, including dyestuffs and pigments, are used [12]. More than 500 000 tons of azo dyes were produced in 1994 [21]. Important aromatic amines of toxicological relevance that are starting material or intermediates for dye production are aniline, chloroanilines, naphthylamines, methylanilines, benzidines, phenylenediamines, and others.

Tab. 8.2 Average production amounts and uses of different amines [26–29].

Compound	Area	Year	Amount per year [tons]	Major use/ starting material
3,4-Dichloroaniline	Worldwide	1988	26 000	Azo dyes, pesticides
	Western Europe	1991	12 000	
4-Chloroaniline	Western Europe	1989	6 500	Pesticides, dyes, pharmaceuticals
1,2-Diaminobenzene	Germany	1990	4 000	Corrosion inhibitor, pigments, dyes, pesticides
N,N-Dimethylaniline	Wordwide	1990	15 000	Dyes
	Germany	1991	2 000	
4-Aminodiphenylamine	Worldwide	1991	80 000	Protectives against rubber aging
	Western Europe	1991	25 000	
	Germany	1991	12 500	
4,4′-Methylenedianiline	USA	1988	424 000	Production of polyurethanes
	Japan	1988	146 000	
Benzotriazole	USA	Annually	9 000	Corrosion inhibitor
Diphenylamine	Worldwide	1980s	40 000	Explosives and propellants
Diethanolamine	USA	1984	75 400	Raw material of detergents, drilling oil
	Germany	1990	20 500	
Ethylenediamine[a]	USA	1992	>150 000	Fungicides, complexing and bleaching agents, textile softener
	Japan	1992	15 000	
	Netherlands	1992	>50 000	
N,N-Dimethyl-propanediamine	Germany	1994	15 000	Hardener for epoxides, binder, curing agent
Triethylenetetramine	Worldwide	1990	23 000	Hardener for epoxides, paper and textile auxiliary
N,N'-Diphenyl-guanidine	Worldwide	End of 1980s	7 500	Vulcanizing accelerator in rubber industry
Methylamines[a]	North America	Annually	170 000	Pharmaceutics, ion exchange resins, pesticides, solvents
	Western Europe	Annually	210 000	

[a] mixture of different derivatives

Many aromatic amines are used in large amounts as intermediates in the production of pesticides, pharmaceuticals, and plastics. With respect to production volume of amines, the production of isocyanates as raw material for the polyurethane production is dominant.

Aromatic amines such as aniline, mono- and dichloroanilines, nitroanilines and 2-phenylendiamine are important starting materials for the manufacturing of pesticides.

The production amount of lower aliphatic amines (not including ethanolamines and methylamines) worldwide is more than 400 000 tons per year [7]. Amines with short (number of C-atoms < 5) as well as long (n > 12) chain length are of particular economic importance. Because of their surface-active behavior, longer-chain amines are used for the manufacture of flotation agents, cationic or non-ionic surfactants, invert soaps, or antifoaming agents [22].

Aliphatic amines like methylamine and dimethylamine are of importance for the production of pharmaceutical products, pesticides, dyes, and rubber. Some further applications are given in Table 8.2.

Mixtures of amines (e.g., diethanolamine, methyldiethanolamine, piperazine) are commonly used for the removal of CO_2 from gases or for the extraction of CO_2 or H_2S in gas sweetening processes [23]. The alicyclic amine morpholine is used as a versatile solvent, anticorrosive agent, rubber additive, and raw material in drug and pesticides production [24]. Quaternary aliphatic amines are produced for the manufacture of fabric softening agents, hair rinses, and disinfectants [25].

8.3
Emission

When discussing the emission of amines into the environment, we must distinguish between *local* or *point sources* on the one hand, and *diffuse sources* on the other. As discussed above, amines play an important role as intermediates or parts of formulations. Typical point sources are therefore discharges of industrial wastewaters into surface water. In many countries worldwide, this source was substantially reduced by the construction of wastewater treatment plants over the last few decades. However, during handling of HPV chemicals, some losses due to unintended releases and subsequent exposure to the environment are inevitable. Further point sources include industrial waste sites (e.g., production of explosives, gasworks) and seepage water from landfills that contain a multitude of amines. In these cases it is usually not possible to distinguish between a release of amines or that of a precursor with subsequent formation of amines. Heterocyclic aromatic amines such as quinoline and acridine are important constituents of coal tar and creosote that are ubiquitous pollutant sources in groundwater [30].

However, with regard to the environmental mass flux, in particular of aromatic amines, it might be more important that they can be formed in the environment via transformation of precursors, e.g., reduction or hydrolysis of nitroaromatics, azo dyes, isocyanates, and pesticides. If these precursors are themselves emitted, the result will be a more diffuse input of amines into the water cycle. In general, amines that have been utilized as production intermediates can be formed in the environment once the product is degraded. Illustrative examples of typical precursors and their corresponding amine transformation products that have been identified in the environment are listed in Table 8.3.

Figure 8.2 summarizes important pathways of amines entering the environment. Both direct input via industrial waste and formation from precursors have to be considered.

Tab. 8.3 Typical precursors and their corresponding amines.

Precursor	Amine(s)	Reference
Alachlor, metolachlor (herbicides)	2,6-Diethylaniline, 2-ethyl-6-methylaniline	31
Naptalam (pesticide)	1-Naphthylamine	32
Diuron, propanil (herbicides)	3,4-Dichloroaniline	26
Azodisalicylate (azo compound)	5-Aminosalicylic acid, 4-phenylenediamine	12
CI Direct Black 38 (azo dye)	Benzidine	33
Toluenediisocyanate (intermediate for polyurethan production)	2,4-Diaminotoluene	12
Trinitrotoluene (explosive)	Aminodinitrotoluenes, diaminonitrotoluenes	34
Dimethylformamide (e.g. solvent)	Dimethylamine	35

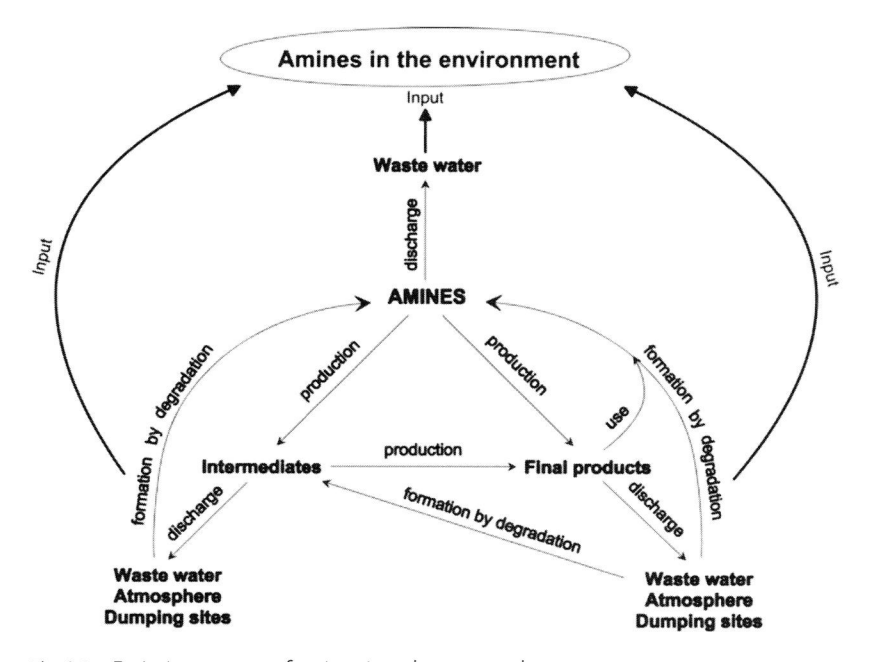

Fig. 8.2 Emission sources of amines into the water cycle.

One of the main application fields of aromatic amines is their use as starting materials for the production of azo dyes. Also, aromatic amines are formed in the degradation process of azo dyes, especially during biological wastewater treatment [9, 12, 36–41]. The use of azo dyes for textile coloring and for paper, plastics, pharmaceuticals, and food production is accompanied by the emission of azo dyes into the environment. A high proportion of the azo dyes processed in the textile industry are assumed to be discharged into the wastewater [42]. Another pathway of amines into the hydrosphere is emission from dumping or washing of colored textiles [12].

A general trend in azo dye production is to substitute the most toxic starting compounds such as benzidine. The substitutes, however, are usually more polar (e.g., aminosulfonates). Owing to their higher mobility, polar compounds of lower toxicity could be more difficult to remove during the wastewater treatment [12].

Benzidine, a typical intermediate in the manufacturing process and a product of the degradation of azo dyes, is one of the most toxic amines. The only significant introduction of benzidine into the environment is due to industrial production of dyes via wastewater, sludge, or solid deposits [9].

At facilities for the production, storage, or deposition of nitroaromatic explosives, aromatic amines, especially nitroanilines, are abundant contaminants because of the stepwise reduction of nitro substituents that occurs in the environment [34]. The TNT metabolites 4-amino-2,6-dinitrotoluene and 2,4-diamino-6-nitrotoluene are poorly degradable. More recently, Schmidt et al. proved the occurrence of several amino-substituted toluenesulfonic acids and aminobenzoic acids at a former military site in Germany [43, 44]. They demonstrated possible degradation schemes and rated the amino products to be quite stable. In soil sample eluates from another former military site, Bruns-Nagel et al. found the polar 2,4,6-trinitrotoluene metabolites 2-amino-4,6-dinitrobenzoic acid and 4- and 2-aminodinitrotoluenes [45].

An example of the formation of amines from intermediates is the reduction of isocyanate in water to 2,4-diaminotoluene. Isocyanate is a starting material for the production of polyurethane. The formation of aromatic amines has been described in foodstuffs due to hydrolysis of isocyanates from packaging materials [46]. 2,4-Diaminotoluene was also found in urine of women with breast implants made of polyurethane [47].

Finally, dimethyl- and methylamines are produced by biological degradation of quaternary amines in wastewater treatment plants [25].

Besides the direct and indirect industrial sources of amines, there is a potential for the formation of carcinogenic amines by the combustion of organic material with high protein content (e.g., forest fires and cigarette smoke), by car exhaust emissions [48], and by the consumption of meat [49]. Furthermore, primary amines can originate from photochemical metabolism of dissolved organic nitrogen compounds [50]. Volatile aliphatic amines have also been detected in precipitation [51].

8.4
Wastewater and Wastewater Treatment

The occurrence and removal of aliphatic, alicyclic, and aromatic amines within wastewater treatment processes have been investigated intensively. Some selected results are highlighted in this chapter, including biological treatment, oxidative techniques, and other methods such as membrane processes or adsorption.

8.4.1
Occurrence in Wastewater

On account of high production quantities and manifold applications in very different industrial branches, amines are often identified in industrial wastewater and also in municipal sewage. Some examples listed in Table 8.4 illustrate the appearance of amines in very different concentration ranges in different types of wastewater.

Among other examples, the content of benzidines and different anilines in industrial wastewaters of an azo dye manufacturer and the behavior of these compounds along the treatment path is given in Table 8.4 [52]. The treated effluent was discharged into Lake Ontario, Canada.

8.4.2
Biological Treatment

The most common methods for the removal of amines from effluents are biological wastewater treatment technologies, both aerobic and combined anaerobic/aerobic. The latter approach has also been proposed for the treatment of soil contaminated with nitroaromatic compounds [61]. The processing, the quality, the adaptation of bacteria, and the reactor type are decisive parameters for the removal efficiency.

Treatment plants that receive process water effluents from the chemical industry have a high potential for the formation of amines. This effect has often been documented for effluents containing azo dyes (see also Section 8.3) [38–41]. To avoid that risk, anaerobic/aerobic sequential reactor systems seem to be an efficient method. In such a reactor system, a successful removal of color took place during the anaerobic treatment. The content of aromatic amines and other organic compounds was reduced to a large degree during the subsequent aerobic treatment step [41].

Similarly, Ekici et al. observed a reductive cleavage of the azo group under anaerobic conditions [38]. The reaction products that also contained aromatic amines (e.g., 2-naphthylamine) were only slowly or not further degraded. Under aerobic conditions, most of the azo dye metabolites considered were quickly degraded by oxidation of the substituents or of the side branches. However, 2,4-diaminotoluene and 2,4-diaminoanisole were still rather recalcitrant.

Tab. 8.4 Maximum concentrations of different amines in various wastewaters.

Sample	Compound	Concentration [µg L^{-1} or µg kg^{-1}]	Reference
Wastewater treatment plant,			
– Industrial water	Benzidine	13.4	52
	3,3'-Dichlorobenzidine	654	
– Raw effluent	2,4-Dimethylaniline	2760	
	2,4,5-Trichloroaniline	388	
– Primary influent	2,6-Dimethylaniline	811	
– Final effluent	4-Chloro-2-nitroaniline	0.64	
	Aniline	1,270	
– Primary sludge	3,4-Dichloroaniline	483	
Pharmaceutical wastewater	5-Chloro-2-methylaniline	200	53
Wastewater of an industrial plant	3-Aminobenzenesulfonate	190 (influent)	54
		67 (effluent)	
Wastewater samples discharged into rivers	Morpholine	20	55
	Piperidine	14	
	Piperazine	0.2	
Wastewater of an industrial plant	Methylamine	30	56
	Dimethylamine	70	
	Diethylamine	30	
Wastewater of an industrial plant	Methylamine	3100	57
	Dimethylamine	2700	
Wastewater from rubber additives factory	Aniline	6200	58
Printworks wastewater	3,4-Dichloroaniline	88.2	59
Wastewater from industrial wastewater treatment plant	2-Amino-1-naphthalen-sulfonate	0.18	60
	3,3'-Dichlorobenzidine	0.12	

Successful removal of poorly degradable amines was often achieved by adaptation of microorganisms. For example, nitroanilines are assumed to be hardly degradable under aerobic conditions. Nevertheless, by allowing acclimatization of biological sludge, a fast degradation of 4-nitroaniline was observed [10]. After gradual adaptation, the microorganisms were also able to eliminate 3- and 4-nitroaniline simultaneously.

In contrast to the results of Ekici et al. [38], Razo-Flores found anaerobic mineralization to be efficient for aromatic amines, in particular with carboxyl, hydroxyl, or methoxy groups [2]. Experiments both with and without adaptation of granular sludge resulted in good degradability of 2-aminophenol and 2- and 3-aminobenzoate. Aniline and in part also 4-aminophenol were stable under methanogenic conditions. However, nitrate- or sulfate-reducing conditions allow the degradation of aniline [62, 63].

Aromatic amines containing sulfo substituents in the meta position are known to be difficultly biologically degradable. An example is the compound 3-aminobenzenesulfonate, an important component of textile wastewaters. Kölbener et al. showed a biological elimination of this compound in laboratory trickling filters [64]. The efficiency was again dependent on the constitution and the acclimatization of the applied activated sludge.

For methyldiethanolamine, aerobic degradation did not succeed within standardized batch experiments, whereas experiments with flow-through reactors resulted in a removal of about 96% within 28 days. This observation was also explained by the adaptation of the microorganisms within the continuous flow-through reactor [23].

The ingredients of petroleum refinery wastewater, monoethylamine and isopropylamine in the presence of phenol, were studied using laboratory scale biological fluidized bed reactors [65]. These amines were completely degraded in aerobic and anoxic systems, though microbial activity was higher under oxic conditions.

In conclusion, the redox conditions are of main importance for the extent of biological degradation of amines.

8.4.3
Oxidative Technologies

Chemical oxidation was also considered for the removal of amines in wastewater treatment. Wastewater containing aniline und *para*-chloroaniline was treated with ozone at different pH values to compare direct ozonation with indirect ozonation by the formation of hydroxyl radicals. Both methods were efficient. Treatment of aniline in wastewater at low pH values creates primarily nitrobenzene, which is itself a regulated contaminant. Therefore, treatment with ozone at alkaline pH, resulting in a higher number of oxidation products such as azoxybenzene and 2-pyridine-carboxylic acid, is recommended for the removal of aniline from wastewater [66]. Ozonation is also efficient for the removal of 4-nitroaniline. Compared to other aromatic nitro-hydrocarbons such as nitrobenzene, 4-methylnitrobenzene, and 4-dinitrobenzene, 4-nitroaniline is more susceptible to ozonation because of the electron-donating amino-group in the para position, which increases the electron density of the ring and therefore improves the electrophilic attack of ozone. Intermediates of the degradation of 4-nitroaniline are easily oxidizable [67].

For the oxidative degradation of pyridine and other amino-containing compounds, ozonation or photocatalysis using both TiO_2 and ZnO turned out to be suitable [68–70].

By the application of Fenton's reagent to aniline-containing wastewater, aniline was polymerized and precipitated as polyaniline, which can easily be separated from the water phase [71].

8.4.4
Other Treatment Technologies

Acid-activated montmorillonite has been used as an adsorber for 4-aminobenzoic acid in wastewater. The best removal was found at lower temperatures and under acid conditions [72].

Wastewater from ore flotation containing primary aliphatic amines and hydrofluoric acid was effectively treated by neutralization, coke filtration, and/or adsorption onto zeolite or activated alumina. Coke filtration was the most effective treatment, about 70% of amines being removed [73].

Several membrane technologies are suitable to recover and reuse aromatic amines from the wastewater cycle cost-effectively. An example is the so-called membrane aromatic recovery system (MARS). High recovery efficiencies with relative pure streams of recovered aniline and phenol was documented for MARS by Ferreira et al. [74].

8.5
Surface Water

8.5.1
Occurrence

The analysis of amines at low levels in aqueous matrices is difficult. Therefore, the first occurrence of data on amines in surface waters dates back only to the 1970s. These investigations were closely related to the determination of pesticide metabolites [75, 76, 87]. In these studies, the determination of aromatic amines was performed using a spectrophotometric method. Such a procedure of determining total amounts of primary aromatic amines based on the diazotization and subsequent coupling of analytes was also used later as a screening method [77, 78]. However, this method does not allow a specific analysis of individual anilines, and even the total concentration of anilines could be strongly biased [79]. Therefore, beginning in the mid-1970s, more sensitive and selective gas and liquid chromatographic methods together with efficient and selective concentration processes have been applied for the determination of amines in surface waters.

Relatively high concentrations of aromatic amines were detected in the river Rhine and its tributary streams (see Table 8.5). At the end of the 1970s, the concentrations of numerous contaminants, including anilines, in the river Rhine noticeably decreased because of the reduction of inputs via wastewater discharges. Nevertheless, Scholz and Palauschek (1988) still determined various substituted anilines in the $\mu g\ L^{-1}$ range in the river Main and found concentrations ranging from 1 to 5 $\mu g\ L^{-1}$ in most cases, including 3,3′-dimethylbenzidine [80]. From the 1990s on, no or very low contents of aromatic amines could be detected in the river Rhine and later also in its tributary streams, as, for example, the river Wupper and the river Emscher [81].

In the river Elbe and its tributaries, investigations of anilines were initiated in the 1990s. Franke et al., Müller et al., and Eppinger identified numerous aromatic amines, including nitroanilines, chloroanilines, alkylanilines, and naphthylamines [82–84]. By means of a non-target GC/MS screening of pentane extracts of river Elbe water, 2,6-dichloroaniline, 2,4,6-trichloroaniline, and the whitening agent 7-diethylamino-4-methyl-coumarin were identified by Theobald et al. [85]. The last substance was characterized as a specific pollution indicator of river Elbe water, and its concentration clearly decreased at the beginning of 1990. Wastewater discharges from dye production in Czechoslovakia were suspected to be an important source of different aromatic amines in the river Elbe. Supporting this hypothesis, high contents of anilines were determined in a small tributary close to a chemical manufacture. For example, in 1995 17.6 μg L^{-1} of 3,4-dichloroaniline and 85 μg L^{-1} of 4-chloroaniline were measured in this tributary. Because of the later commissioning of a wastewater treatment plant, the amount of anilines decreased significantly [84, 86].

A similar decrease in the amine concentration to that observed in the river Rhine two decades earlier was observed in the river Elbe and its tributary streams in the nineties. This was due on the one hand to the close-down of numerous factories as a result of the political and economic changes in the former socialist countries and on the other hand to the installation and modernization of effluent treatment, which resulted in a general reduction of contaminant concentrations in the receiving rivers.

As well as aromatic amines, amphoteric, aliphatic, and alicyclic amines have been detected in surface waters, although data on occurrence are limited.

Yang et al. determined several aliphatic and alicyclic amines in overlying waters from a coastal salt marsh (Flax pond, New York), including methylamine (1.9 μg L^{-1}), dimethylamine (8.1 μg L^{-1}), ethylamine (1.3 μg L^{-1}), and pyrrolidine (2.6 μg L^{-1}) [98]. In pore water, the amine concentrations were nearly the same or higher, with the exception of pyrrolidine. These amines represent metabolic intermediates in decomposition processes and are widely distributed in marine sediments.

Gerecke and Sedlak found that dimethylamine and comparable compounds predominantly occur in contaminated surface waters and groundwaters [20]. These compounds are regarded as N-nitroso-dimethylamine-(NDMA-)precursors (see Chapter 10). Their increased concentrations in municipal wastewater effluents originate from human excretions and from the broad industrial use of dithiocarbamates, which hydrolyze to dimethylamine (DMA). So far, there is no information on background concentrations of DMA and other NDMA-precursors in pristine waters.

8.5.2
Transformation and Degradation Processes

Figure 8.3 schematically shows transformation pathways of amines in surface waters. The potential environmental sinks for amines are further discussed in Sections 8.5.2.1 and 8.5.2.2. The majority of the processes given below occur similarly in other environmental compartments.

Tab. 8.5 Selected examples of aromatic amines in surface waters.

Surface water	Year	Compound	Concentration [µg L^{-1}]	Reference
Rhine (river)	1972	2-Chloroaniline	3.1–3.8	75
		3,4-Dichloroaniline	2.4–2.5	
Rhine (river)	1976–1978	4-Nitroaniline	0.5	88
		3,4-Dichloroaniline	0.7	
Ijssel (river)	1979	4-Chloroaniline	0.11–0.29	89
		3,4-Dichloroaniline	0.19–0.69	
Meuse (river)	1979	3,4-Dichloroaniline	0.07–2.1	89
		Aniline	0.55–2.4	
Rhine (river)	1985	N,N-Dimethylaniline	1.4[a]	81
		2-Chloroaniline	3.7[a]	
Llobregat (river)	[b]	N,N-Dimethylaniline	1–100	90
		3-Nitroaniline	0.5–10	
Rhine (river)	1990	Aniline	3.0[a]	81
		5-Chloro-2-methylaniline	2.5[a]	
Wupper (river)	1992	3-Chloroaniline	1.2[a]	81
		3-Trifluoromethylaniline	5.8[a]	
Seine (river)	1991	4-Chloroaniline	0.065–0.072	91
Emscher (river)	1992	2-Chloro-4-nitroaniline	2.1[a]	81
		3-Chloro-4-methylaniline	2.6[a]	
Elbe (river)	1994	2-Aminonaphalene-4,8-disulfonate	9.8	92
		2-Amino-4-bromoanthraquinone-2-sulfonate	2.0	
Elbe (river)	[b]	7-Aminonaphthalene-1,3-disulfonate	1.1–6.4	93
Elbe (river)	1996	3,4-Dichloroaniline	0.79	86
		ortho-Toluidine	1.2	
Elbe (river)	1998	N-Ethylaniline	0.33[a]	84
		2-Chloro-5-nitroaniline	0.35[a]	
Sea water	[b]	4-chloro-2-nitroaniline	0.098[a]	94
Seveso (river)	2002	2-Amino-1-naphthalene-sulfonate	0.046–0.11	60
Suwa (lake)	[b]	Triethylamine	0.25	95
		Tributylamine	0.23	
Rhine (river)	[b]	Dimethylamine	3.0	56
		Morpholine	2.0	
Mulde (river)	1996–1998	Diethanolamine	4.6	96
Elbe (river)		Morpholine	2.5	
Ganga (river)	[b]	Diethylamine	70	97

[a] maximum values in the respective year
[b] not specified

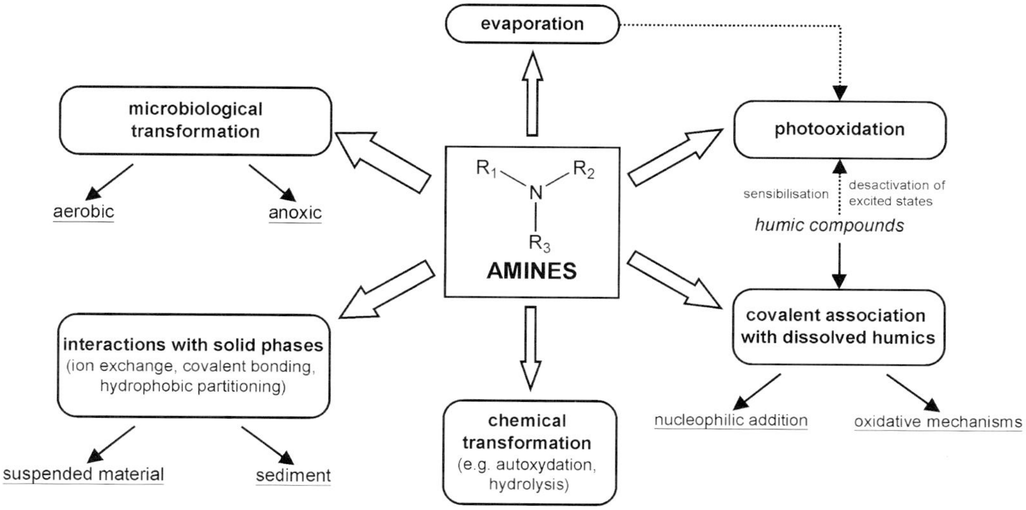

Fig. 8.3 Elimination pathways of amines in surface waters. A semiquantitative assessment of the processes was made and is represented by different arrow strengths, with stronger arrows representing more relevant processes.

Especially biological degradation, covalent interactions, and in case of near-surface areas photochemical reactions are of relevance for the elimination of amines under natural environmental conditions. Evaporation is of less importance because of the limitations of this process for compounds with a relatively high volatility.

8.5.2.1 Abiotic Processes

Amines can be oxidized by dissolved oxygen, for example, to hydroxylamines, quinones, or nitro compounds [7]. However, in comparison to biological degradation, such so-called autoxidation processes can often be neglected for the amines considered because of slow reaction rates. For aniline, Lyons et al. estimated a degradation rate of autoxidation in pond water of less than 0.005 d^{-1} [99]. Furthermore, most amines are stable to hydrolysis in water.

In lakes and rivers, the direct and/or indirect photochemical degradation of amines by sunlight can be significant. Depth of light penetration, temperature, and presence of other water constituents are important boundary conditions influencing photo-transformation.

Hwang et al. estimated photo-transformation half-lives in surface estuarine water for aniline and chloroanilines to be in the range 2 to 125 h [100]. Photolysis rates in estuarine water containing humic compounds were observed to be lower than those in pure water. The authors explained this by quenching effects of humics. The investigations also revealed higher rates in warm and light-intensive seasons and lower rates of aniline transformation than those of chloroanilines.

In contrast to the results of Hwang et al. [100], Zepp et al. proved a photosensitive effect of humic substances on the photoreaction of aniline [101]. A possible cause of these contradictory results could be the use of different humic compounds. The photosensitized reaction of anilines involved hydrogen transfer from the nitrogen of amines to the sensitizer. The resulting anilino free radicals can lead to the formation of different reaction products, including azobenzenes.

As an example of indirect photolysis, Larson and Zepp have shown the reaction of carbonate radicals, generated by photolysis, with different aniline derivatives [102]. In contrast to OH radicals, the electrophilic carbonate radicals react selectively with electron-rich aromatic compounds, and the oxidation of aniline is 10^5 times mae effective than that of benzene. The results show that the carbonate radical reaction could be a significant pathway of removal of even non-biodegradable anilines in the aquatic environment.

During sunlight-induced photodegradation of amines, partially stable metabolites can be formed; as an example, during irradiation of neutral air-saturated solutions of 4-chloroaniline or N-(4-chlorophenyl)-benzenesulfonamide at wavelengths of more than 290 nm, the formation of 4-chloronitrosobenzene and 4-chloronitrobenzene was found (Fig. 8.4) [103]. These authors postulated the formation of an N-peroxy radical as an intermediate.

Aromatic amines are known to associate with humic substances. Besides ionic interactions and hydrophobic partitioning, there are two types of covalent binding: a nucleophilic addition of the amino group to electrophilic moieties of humic molecules (e.g., quinone or other carbonyl groups) or oxidative mechanisms resulting in interactions between formed anilino radical intermediates and radical species of NOM. Biphasic kinetics is considered typical for the covalent binding between anilines and humic compounds: a rapid, often reversible sorption process and a slower irreversible binding process. Weber et al. examined the rate of binding of aniline and dissolved organic matter (DOM) and found dependence on pH, oxygen content in water, DOM, and number of highly reactive covalent binding sites of DOM [104]. Because the latter exist in very small quantities, covalent binding of aromatic amines to DOM is not a relevant elimination path of anilines in surface waters, but has been proposed to be in bottom sediment (see Section 8.6.2). However, if aromatic amines are present at very low concentrations in water, binding to dissolved humic associates could be more relevant, as has been shown by Di Corcia et al. [105]. By using ^{15}N NMR-spectrometry, Thorn et al. have succeeded for the first time in producing direct analytical evidence for the condensation of aniline with quinone structures of humic compounds [106].

8.5.2.2 Biological Degradation

The following statements with respect to the biological degradation are relevant not only for surface waters but also for wastewater treatment processes and in part for the mineralization/transformation of amines in soil and groundwater systems.

In the literature can be found numerous investigations into biological degradation, especially of simple aromatic amines such as aniline, monochloroanilines,

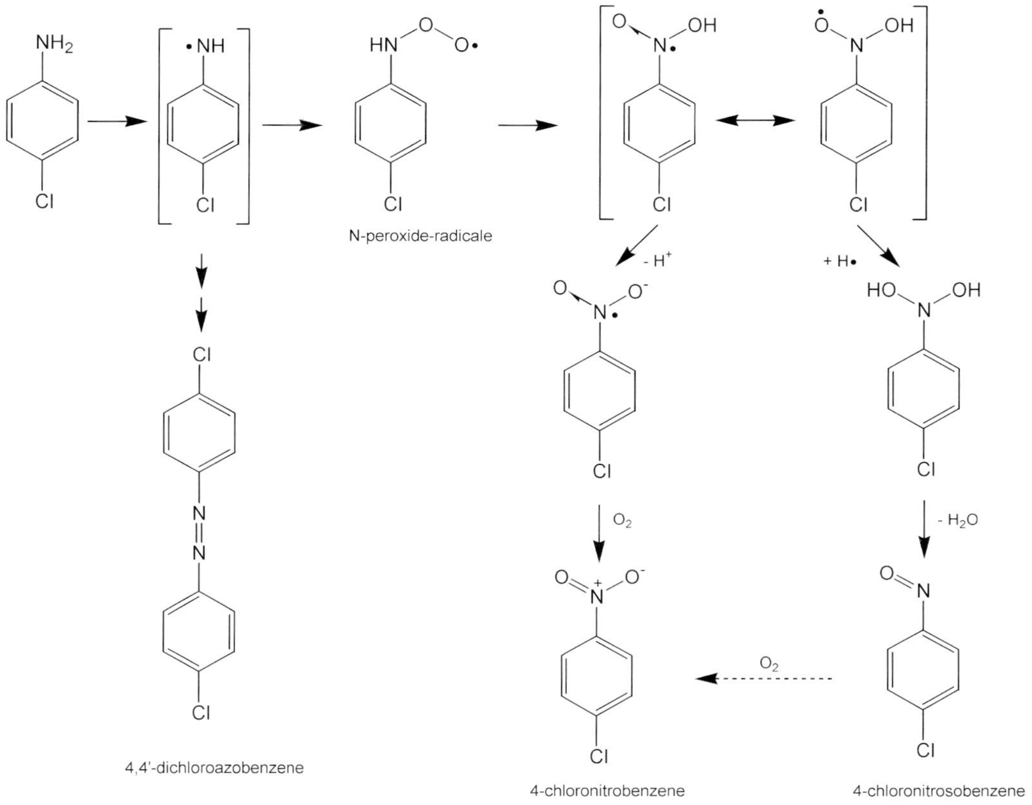

Fig. 8.4 Proposed mechanism of the phototransformation of 4-chloroaniline (adapted from Miller and Crossby [103]).

and dichloroanilines in surface waters, using different laboratory test methods or even field scale studies [107–109]. In contrast to those compounds, only very few data are available for other amines.

Extensive investigations concerning the paths of degradation were, for example, described by Lyons et al., Pitter and Chudoba, Fetzner, Delort and Combourieu, and Díaz et al. [99, 110–113]. Experiments were generally performed using relatively high initial concentrations of amines and pure cultures without the presence of co-substrates.

Anilines seem to be generally biodegradable because, in contrast to halogens, the amino substituent can be readily biotransformed into a desirable hydroxy group. In detail, the oxidative biodegradation of aromatic amines usually results from a formation of pyrocatechols by dioxygenases and dehydrogenases and a subsequent ortho cleavage of the rings, as described by Lyons et al. and Stockinger et al. for aniline, 4-chloroaniline, and 3-chloro-4-methylaniline [99, 114]. According to Zeyer et al., a meta cleavage of aromatic rings is also possible [115].

8 Amines

The degradation of primary and secondary aliphatic amines is generally considered to start with dehydrogenation and the formation of imines. Such a biochemical oxidation of aliphatic amines is catalyzed by flavo- and metalloproteins. The imines are further hydrolyzed to aldehydes and ammonia (see Fig. 8.5). Compounds with tertiary and quaternary nitrogen are biologically more stable. The readily biodegradable trimethylamine represents an exception. With this compound, oxidative demethylation causes a cleavage of C-N bonds, and this is followed by an oxidative deamination affected by amine oxidases [110].

Under aerobic conditions, polymerization of aromatic amines represents another pathway of biological reaction in the environment (Fig. 8.6). After reaction to free radicals, especially certain aromatic amines are oxidatively coupled to humic-like polymers by peroxidases, laccases, and tyrosinases from plants or various fungi [116]. Such polymers can be very recalcitrant in the environment.

Boundary conditions like concentration and composition of the biocenosis (pure or mixed cultures), the concentration of the compounds considered, the presence of other substrates, temperature, pH-value or redox conditions, and the structure of the amine are decisive parameters for the type and rate of degradation processes. Thus, results of degradation tests are not generally transferable to other conditions. For example, Gerike and Fischer investigated the degradation of 2-chloroaniline and proved non-biodegradability using standard degradation tests like the closed-bottle test, the MITI test, or the OECD test, whereas the Zahn-Wellens test resulted in nearly 100% degradability [117].

Under aerobic conditions, the presence of an electron-donating amino group improves the electrophilic attack of aromatic rings (Fig. 8.7). Therefore, different simple aromatic amines such as aniline or methylanilines would be expected to be easily degradable in the presence of oxygen. By enlarging the ring systems, e.g., in the

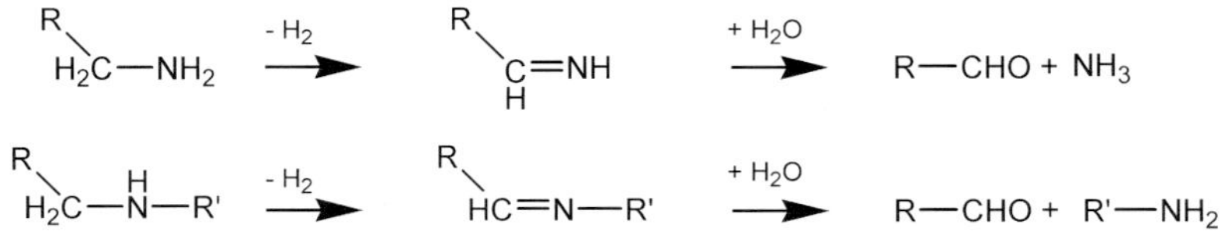

Fig. 8.5 Biochemical dehydrogenation and subsequent hydrolysis of primary and secondary amines.

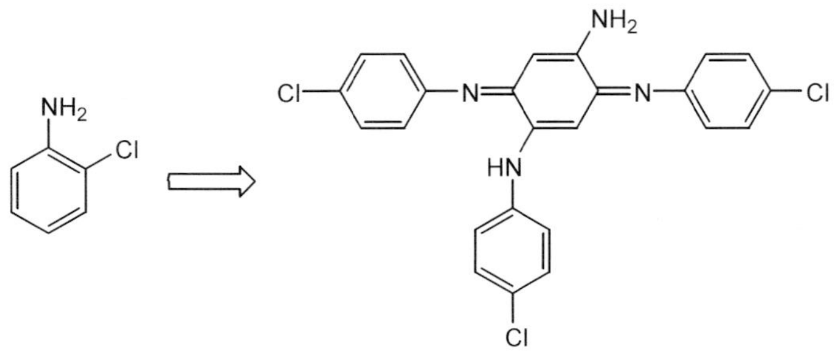

Fig. 8.6 Polymerization of 4-chloraniline (adapted from Field et al. [116]).

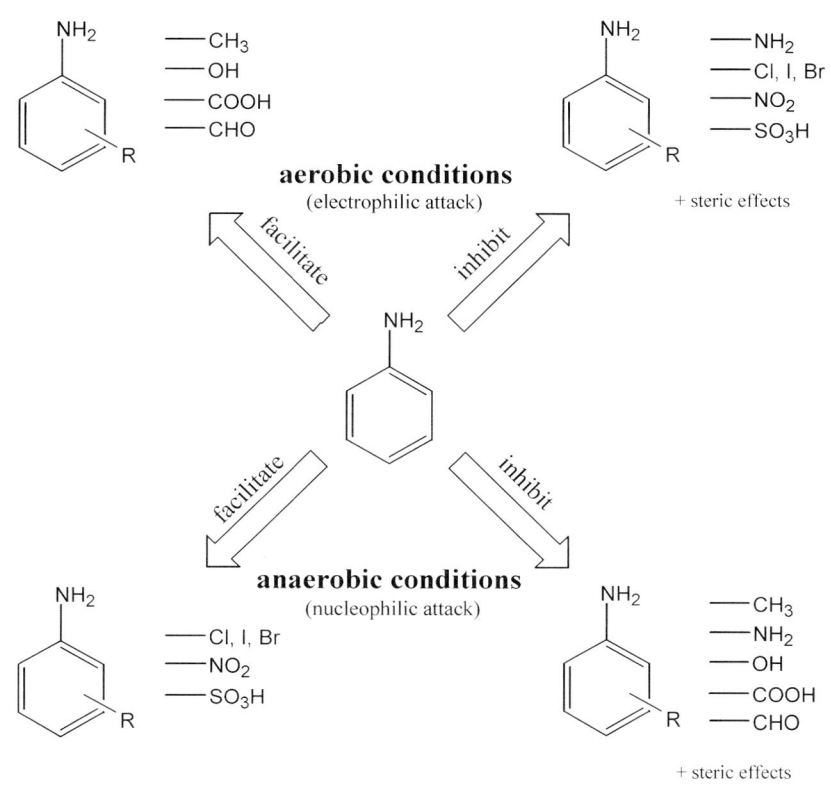

Fig. 8.7 Influence of substituent type on the biodegradability of aromatic amines under aerobic and anaerobic conditions [110, 116].

case of naphthylamines, or in the presence of electron-withdrawing substituents on the aromatic ring (type, number, and position are of importance), the rate of amine degradation can decrease significantly. Alkyl substituents on the N atom of the amino group or the occurrence of more than one aromatic ring (benzidine, azo dyes) often lead to a noticeable decrease in biodegradability [110]. It is generally accepted that chloro, sulfo or nitro groups on the ring decrease the degradation rate of arylbenzenes, while carboxy or hydroxy substituents increase it. Amines with degradation-inhibiting substituents situated in the meta and particularly the para position tend to be more persistent [118].

In contrast to highly electrophilic compounds such as nitro- or azo-aromatics, aromatic amines are often characterized as poorly degradable under anaerobic conditions. For example, aniline was found to be stable under methanogenic conditions [2, 62, 119]. However, one additional group may facilitate an anaerobic transformation. This also applies to N-substituted anilines. More complex compounds with a greater number of different substituents, e.g., sulfo, amino, hydroxo, and ethoxy groups (often reductive degradation products of azo dyes), were rather poorly degraded by anaerobic microorganisms, probably because of steric hindrance [116].

Simple aliphatic amines are metabolized very well and very fast. Degradation decreases with increase in chain length and branching [120]. Primary diamines are well degradable, while the biodegradability decreases in the presence of ethanediamine groups in the molecule. Tertiary amines are rather recalcitrant, because of the absence of hydrogen at the nitrogen atom [110].

In the case of alicyclic amines, the presence of a further heteroatom in the ring is of special importance for biodegradability. For example, the degradation rate of cyclohexylamine is substantially higher than the degradation rates of morpholine or piperazine, which have two heteroatoms (O or N) in their rings. Besides steric effects, this difference is caused by a lower electron density on the reactive places of morpholine and piperazine compared with amines with one heteroatom in the alicyclic ring [121]. The ring cleavage of alicyclic compounds is more complicated than that of aromatic amines, and thus their degradation rate is often lower [110].

Zoeteman et al. investigated the persistence of amines and other compounds directly in rivers and lakes [88]. The authors pointed out the differences between bench-scale and field-scale studies and also between the diverse systems investigated. One conclusion of their investigation was that bench-scale experiments often overestimate the degradation potential. Furthermore, the studies resulted in a classification of the compounds investigated based on their persistence, methoxyaniline, aniline, and 4-chloroaniline being classified as easily to slightly degradable in rivers.

The degradability of aniline in pond water with and without the addition of sewage sludge was investigated by Lyons et al. [108]. After 2 weeks, 90% of aniline was eliminated by the added sewage sludge inoculum, whereas no significant loss of aniline was observed in the pure pond water system.

For 3-chloroaniline added to river Oconee water, a half-life of 4.8 months was estimated using enriched microorganisms from pond and river water without adaptation. After 72 h adaptation of bacteria, a very considerable reduction in the half-life (to 3 h) for this biodegradation was found [122].

Laboratory batch tests with shaking flasks for the investigation of degradation behavior of aniline proved to be comparable to field experiments on the river Rhine [109]. Taking into account dilution as an additional process, a field half-life of 0.38 days (temperature range from 15 to 22 °C) was estimated. In these investigations, laboratory and field experiments resulted in comparable half-lives (laboratory: 0.46 days at 15 °C, 0.35 days at 20 °C).

8.6
Groundwater

8.6.1
Occurrence

There is little information on the occurrence of amines in groundwater except for point source-related emissions (landfills, ammunition wastes). Therefore, the reported spectrum of compounds may be biased, in particular towards degradation intermediates of nitroaromatic explosives, which are well known to be transformed in the environment, primarily via reduction of the nitro group.

In groundwater influenced by ammunition wastes, many authors reported on the presence of aromatic amines (see, e.g., Refs. [123–125]). Often, the dominant anilines still bear at least one nitro group. More recently, the occurrence of amphoteric anilines either from manufacturing wastes or via formation in the environment has been documented [126, 44]. In some samples, compounds such as 2-amino-4,6-dinitrobenzoic acid were shown to be the major metabolites of 2,4,6-trinitrotoluene.

In contrast to the compound spectrum found in ammunition wastes, groundwater influenced by landfills shows rather methyl-substituted and halogenated anilines [126]. N-heterocyclic compounds have been found primarily in the vicinity of former gasworks sites. Johansen et al. reported the occurrence of a number of compounds at various sites, including quinoline, indole, pyrrole, methylpyridines, methylquinolines, acridine, and carbazole in concentrations up to 150 µg L^{-1} [127]. At another site, Zamfirescu and Grathwohl found up to a few µg L^{-1} of methylquinolines and acridine [30]. Furthermore, oxidized metabolites of these compounds were identified.

For aliphatic amines, even less information on groundwater occurrence is available. Mishra et al. determined up to 0.46 µg L^{-1} diethylamine in groundwater samples (Jabalpur, India) but without giving further information on the type of sample [97]. Schmalz et al. reported very high concentrations (up to 10 mg L^{-1}) dimethylamine in groundwater at a contaminated industrial site [35]. However, because of the analytical method used, interference by ammonium could have led to an overestimation of the concentration [35].

8.6.2
Sorption and Transformation Processes

In this section, only those properties and reactions of amines specific to groundwater systems will be addressed that have not already been covered in Section 8.5.2. Schwarzenbach et al. have conceptualized potential interactions of organic compounds in soil-water systems using the example of 3,4-dimethylaniline, as shown in Fig. 8.8 [128].

Speciation in the dissolved phase (depending on the pK_a of the corresponding acid) is an important parameter, since at environmental pH some anilines, but especially the more basic aliphatic amines, will be present as cations to a substantial extent. The ions could then undergo cation exchange as an important process leading to interaction with solid phases. As discussed in Section 8.5.2.1, amines could form covalent bonds with reactive moieties in organic matter. However, in groundwater this process will be much more important than in surface water because of the higher solid/water ratios in subsurface systems. As for all organic molecules, non-specific partitioning into such organic matter will also be of importance for the sorption and thus the transport of amines.

Cowen et al. investigated sorption of the most important polyurethane precursors and degradation products 2,4-TDA, 2,6-TDA and 4,4'-MDA. K_{OC} values for

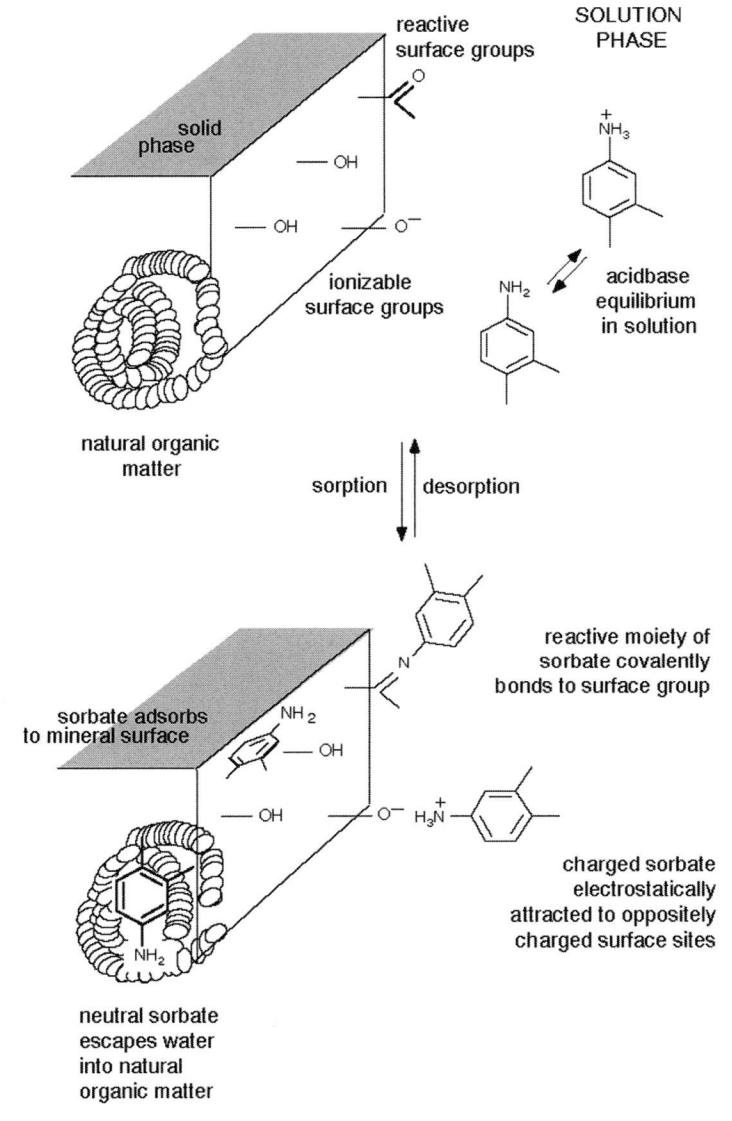

Fig. 8.8 Some sorbent-sorbate interactions possibly controlling the association of a chemical (3,4-dimethylaniline) with natural solids. Reproduced with permission from Schwarzenbach et al., 2003 [128].

these compounds ranged between 500–1 300 for the TDA isomers and 3 800–5 700 for 4,4′-MDA [129]. Whereas Cowen et al. found no difference in sorption under anaerobic and aerobic conditions, Elovitz and Weber reported irreversible sorption of diaminonitrotoluenes only under aerobic conditions, and hardly any sorption under anaerobic conditions [130]. In the last decade, sorption of aniline has been intensively studied by the research groups of Jafvert and Lee (see, e.g., Refs. [131–134]). They found in general a biphasic sorption behavior. Short-term sorption seems to take place primarily by ion exchange of the charged species and nonspecific partitioning into organic matter, whereas long-term sorption seems to involve primarily nucleophilic addition to carbonyl moieties in the soil organic matter and oxidation-induced polymerization. Recently, Donaldson and Nyman introduced a multiparameter model for the sorption of benzidines that allows to estimate the contributions from partitioning, covalent bonding, and ion exchange [135]. Depending on the sediment type and concentration used, the relevance of the three processes differed substantially. Because of the various sorption mechanisms possible for amines, their association with soil and sediment is frequently stronger than that for other polar compounds with comparable log K_{OW}.

Interestingly, there are hardly any studies on sorption of aliphatic amines. For methylamines, Wang and Lee determined very low sorption coefficients (unitless K ranged from 2 to 5) in marine sediments assuming a linear partition model [136].

Degradability of amines has already been discussed in Sections 8.4 and 8.5, and therefore we focus here on few additional studies in groundwater environments. Kromann and Christensen investigated the degradability of 19 organic chemicals, including the amines 3- and 4-nitroaniline, aniline, 2-methylaniline, and 4-methylaniline, in a test field downstream of a landfill. Contrary to other studies on anaerobic degradation of anilines, it was shown that only nitroanilines were easily degradable under anaerobic conditions with low methane production [1]. A possible explanation is that degradation rates differ, depending on the dominant terminal electron acceptor. Biodegradation experiments with ^{14}C-labeled compounds showed initial fast degradation under aerobic conditions. However, this slowed down rapidly, resulting in less than 40% degradation based on ^{14}C loss after 1 year. Under methanogenic conditions, no degradation was observed in the time frame of the experiments [129]. In addition to the ring oxidation and cleavage mechanism usually found for anilines, Travkin et al. (2002) recently found evidence for the deamination of 3,4-dichloroaniline in bacteria enrichment cultures under nitrate-reducing conditions [137], adding another possible pathway for aniline removal in the environment.

8.7
Drinking Water and Drinking Water Treatment

8.7.1
Occurrence

There is little information about the occurrence of amines in samples from the water treatment process or from drinking water. Several aromatic amines such as aniline, 2-methylaniline, 4-chloroaniline, 2,5-dichloroaniline, and 3,4-dichloroaniline were determined in river bank filtrates of the Rhine and the Elbe as well as in drinking water in the concentration range ng L^{-1} to lower µg L^{-1} [76, 86, 88].

8.7.2
Treatment

River bank filtration can be considered as a first step in the treatment of polluted river water. The processes during subsurface transport have already been discussed in Section 8.6. According to Sontheimer, organic compounds, which are not eliminated during bank filtration need to be classified as relevant for waterworks [138]. Laboratory-scale test filter experiments are suitable for simulating the degradation behavior of sum parameters such as dissolved organic carbon (DOC) and individual organic micro-pollutants under conditions similar to those in the field [139]. As an example, test filters were used by Knepper et al. to determine the relevance of polar compounds for waterworks and drinking water [140]. In this study, aromatic sulfonates containing amino substituents were found to be relevant for drinking water because of their poor removal.

Investigations of the behavior of aromatic amines in the lower $\mu g\ L^{-1}$-range using river water-adapted test filters resulted in a classification of degradability ranging from easy (aniline, 2-methylaniline and monochloroanilines) to very difficult (nitroanilines, higher chlorinated amines). Furthermore, the results pointed to a strong enhancement of biodegradation of 2,4,6-trichloroaniline because of adaptation of microorganisms and a two-stage degradation kinetics of 2-nitroaniline indicating co-metabolic effects. Moreover, a correlation between degradation rate and acidity constant of protonated amines was found [141].

The biodegradability of several aliphatic and alicyclic amines was determined using river Elbe water-adapted test filter systems [96, 142]. Morpholine and piperazine (with two heteroatoms in the ring), *tert*-butylamine, and cyclohexylamine were characterized to be relevant for waterworks.

Sediment column experiments (organic carbon content of the sediment: 0.02%) to simulate the transport behavior of different amines during river bank filtration showed a high mobility of aromatic amines with retardation factors (R) ranking from 1.0 (aniline) to 1.7 (2,4,6-trichloroaniline) and a stronger retention for aliphatic amines (R between 1.7 and 2.2) [143, 144]. The latter is due to ionic interactions between protonated amines and the negatively charged mineral surface.

Aromatic amines such as aminonaphthalene sulfonates or naphthylamines may be removed by activated carbon adsorption [145, 146]. However, in laboratory tests the adsorption capacity of activated carbon for aliphatic amines such as methylamine or morpholine was found to be very low [147]. Freundlich coefficients, determined on the basis of isotherms, confirm these statements [148]. Furthermore, it was shown that flocculation is an ineffective treatment process to remove neutral arylamines, cationic aliphatic, and alicyclic amines from water [86, 96].

In contrast, ozonation seems to be an efficient method to remove amines from raw waters. The direct reaction of molecular ozone proceeds quickly with non-protonated amines [149]. The reaction of secondary oxidants, as for example OH radicals, depends on the molecular weight, with small compounds showing higher stability.

Investigations into degradation rates of polar alicyclic amines in neutral solution using ozone showed the order: piperazine > morpholine >>> piperidine > pyrrolidine [150]. In addition to the molecular structure, the fraction of amines in neutral form and therefore the pH value of the treated water and the pK_a values of the compounds have a strong influence on the oxidation rate. By the addition of radical scavengers Pietsch et al. demonstrated that direct oxidation with ozone dominated at a pH value of 7 [150]. Other authors also characterized ozonation as a suitable treatment step to remove aromatic and aliphatic amines [151, 152].

Chlorine reacts with olefinic systems via electrophilic addition and with aromatic systems via electrophilic substitution. Substituents increasing the electron density in the ring, e.g., amino groups, promote such reactions. Therefore, polar persistent aromatic amines pose a relatively high risk of formation of disinfection by-products [153].

The oxidative degradation of aliphatic and alicyclic amines using chlorine-containing agents proceeds slowly. With the exception of morpholine, treatment with chlorine dioxide was ineffective to remove 9 investigated amines under drinking water treatment conditions. During disinfection with chlorine, only the concentration of dimethylamine decreased significantly [96].

8.8 Conclusions

Amines are produced and utilized industrially in large quantities. Therefore, a path into the environment by sewage discharge is generally suggested. However, the high relevance for the environment results primarily from the fact that amines can also originate from a number of mostly anthropogenic precursors such as pesticides, azo dyes, or explosives. The formation of aliphatic amines may also result from various natural sources.

Biodegradation is an important process for the elimination of amines both during water treatment and in the environment, where the boundary conditions, in particular the redox milieu and the type of microorganisms, affect the extent and rate of biotransformation. Depending on the system investigated, abiotic transformations, including cation exchange, phototransformation, and formation of covalent bonds, can considerably influence the fate of amines.

On account of numerous investigations in Europe (river Elbe and river Rhine and tributaries), much is known about the occurrence of aromatic amines and their behavior during water treatment. For other regions, much less information is available. However, this is not an indication of an absence of anthropogenic amines in these regions, since at all recent or former industrial sites, in particular those with chemical industries, amines are likely contaminants in surface and/or groundwaters. Although the substance class of amines is currently not as much in the focus of environmental research as for example pesticides or pharmaceuticals, their relevance for the water cycle unquestionably exists.

References

1. A. Kromann, T. H. Christensen, *Waste Manag. Res.* **1998**, 16, 437–445.
2. E. Razo-Flores, B. Donlon, G. Lettinga, J. A. Field, *FEMS Microbiol. Rev.* **1997**, 20, 525–538.
3. S. Hauptmann, *Organische Chemie, Verlag für Grundstoffindustrie*, Leipzig, **1985**.
4. Syracuse Research Corporation, Interactive PhysProp Database Demo, can be found under http://www.syrres.com/esc/
5. M. S. Bryant, P. Vineis, P. L. Skipper, S. R. Tannenbaum, *Proc. Natl. Acad. Sci. USA.* **1988**, 85, 9788–9791.
6. U. Stephan, P. Elstner, K. Müller, Toxikologie, Bibliographisches Institut, Leipzig, **1987**.
7. Ullmann's Enzyclopedia of Industrial Chemistry, Vol. A2 (Ed.: W. Gerhartz), Wiley-VCH, New York, Weinheim, **2000**.
8. J. H. Weisburger, *Mutat. Res.* **1997**, 376, 261–266.
9. G. Choudhary, *Chemosphere* **1996**, 32, 267–291.
10. A. Saupe, *Chemosphere* **1999**, 39, 2325–2346.
11. L. Fishbein in *The Handbook of Environmental Chemistry*, Vol. 3C (Ed.: O. Hutzinger), Springer, Berlin, **1984**, pp. 1–40.
12. H. M. Pinheiro, E. Touraud, O. Thomas, *Dyes Pigm.* **2004**, 61, 121–139.
13. P. L. Skipper, S. R. Tannenbaum, R. K. Ross, M. C. Yu, *Cancer Epidem. Biomar. Prevent.* **2003**, 12, 503–507.
14. H. Kataoka, *J. Chromatogr. A* **1997**, 774, 121–142.
15. W.-L. Gong, K. J. Sears, J. E. Alleman, E. R. Blatchley III, *Environ. Toxicol. Chem.* **2004**, 23, 239–244.
16. J. E. Amoore, E. Hautala, *J. Appl. Toxicol.* **1983**, 36, 272–290.
17. H. Greim, D. Bury, H.-J. Klimisch, M. Oeben-Negele, K. Ziegler-Skylakakis, *Chemosphere* **1998**, 36 , 271–295.
18. W. A. Mitch, D. L. Sedlak, *Environ. Sci. Technol.* **2004**, 38, 1445–1454.
19. J. Choi, S. E. Duirk, R. L. Valentine, *J. Environ. Monit.* **2002**, 4, 249–252.
20. A. C. Gerecke, D. L. Sedlak, *Environ. Sci. Technol.* **2003**, 37, 1331–1336.
21. A. Stolz, *Appl. Microbiol. Biotechnol.* **2001**, 56, 69–80.
22. *Römpp Chemielexikon Vol. 1* (Ed.: J. Falbe), Thieme, New York, **1989**.
23. M. Fürhacker, A. Pressl, R. Allabashi, *Chemosphere* **2003**, 52, 1743–1748.
24. T. Schräder, G. Schuffenhauer, B. Sielaff, J. R. Andreesen, *Microbiology* **2000**, 146, 1091–1098.
25. N. Nishiyama, Y. Toshima, Y. Ikeda, *Chemosphere* **1995**, 30, 593–603.
26. BUA Reports Vol. 89, 91, 96, 131, 132, 140, 153, 158, 184 (Ed.: GDCh-Advisory Committee on Existing Chemicals of Environmental Relevance), Hirzel, Stuttgart, *1992–1995*.
27. D. S. Hart, L. C. Davis, L. E. Erickson, T. M. Callender, *Microchem. J.* **2004**, 77, 9–17.
28. O. Drzyzga, *Chemosphere* **2003**, 53, 809–818.
29. *Ullmann's Encyclopedia of Industrial Chemistry*, Vol. A16 (Eds.: B. Elvers, S. Hawkins, G. Schulz), Wiley-VCH, New York, Weinheim, **2000**.
30. D. Zamfirescu, P. Grathwohl, *J. Contam. Hydrol.* **2001**, 53, 407–427.
31. O. Osano, W. Admiraal, D. Otieno, *Environ. Toxicol. Chem.* **2002**, 21, 375–379.
32. T. G. Díaz, M. I. Acedo, A. M. de la Pena, M. S. Pena, F. Salinas, *Analyst* **1994**, 119, 1151–1155.
33. D. T. Sponza, M. Isik, *Process Biochem.* **2005**, 40, 35–44.
34. J. C. Spain, *Annu. Rev. Microbiol.* **1995**, 49, 523–555.
35. V. Schmalz, H. Börnick, V. Neumann, E. Worch, *Vom Wasser* **2001**, 97, 63–74.
36. R. F. Straub, R. D. Voyksner, J. T. Keever, *Anal. Chem.* **1993**, 65, 2131–2136.
37. E. Idaka, T. Ogawa, H. Horitsu, *Bull. Environ. Contam. Toxicol.* **1987**, 39, 100–107.
38. P. Ekici, G. Leopold, H. Parlar, *Chemosphere* **2001**, 44, 721–728.
39. R. M. Melgoza, A. Cruz, G. Buitron, *Wat. Sci. Technol.* **2004**, 50 (2), 149–155.
40. C. B. Shaw, C. M. Carliell, A. D. Wheatley, *Water Res.* **2002**, 36, 1993–2001.

41 M. Isik, D. T. Sponza, *J. Hazard. Mater.* **2004**, 114, 29–39.
42 J. Riu, I. Schönsee, D. Barcelo, C. Ràfols, *Trends Anal. Chem.* **1997**, 16, 405–419.
43 T. C. Schmidt, K. Steinbach, E. von Löw, G. Stork, *Chemosphere* **1998**, 37, 1079–1090.
44 T. C. Schmidt, U. Buetehorn, K. Steinbach, *Anal. Bioanal. Chem.* **2004**, 378, 926–931.
45 D. Bruns-Nagel, T. C. Schmidt, O. Dryzyga, E. von Löw, K. Steinbach, *Environ. Sci. Pollut. Res.* **1999**, 6, 7–10.
46 C. Brede, I. Skjevrak, H. Herikstad, *J. Chromatogr. A* **2003**, 983, 35–42.
47 K. Shanmugam, S. Subrahmanyam, S. V. Tarakad, N. Kodandapani, D. F. Stanly, *Anal. Sci.* **2001**, 17, 1369–1374.
48 Z. Moldovan, J. M. Bayona, *Rapid Commun. Mass Spectrom.* **2000**, 14, 379–389.
49 E. G. Snyderwine, R. Sinha, J. S. Felton, L. R. Ferguson, *Mutat. Res.* **2002**, 506–507, 1–8;
50 D. J. Koopmans, D. A. Bronk, *Aquat. Microb. Ecol.* **2002**, 26, 295–304.
51 K. Gorzelska, J. N. Galloway, K. Watterson, W. C. Keene, *Atmos. Environ., Part A: General Topics* **1992**, 26A, 1005–1018.
52 F. I. Onuska, K. A. Terry, R. J. Maguire, *Water Qual. Res. J. Canada* **2000**, 35, 245–261.
53 Chlorotoluidine, BUA Report Vol. 55 (Ed.: GDCh-Advisory Committee on Existing Chemicals of Environmental Relevance), Hirzel, Stuttgart, **1990**.
54 B. Altenbach, W. Giger, *Anal. Chem.* **1995**, 67, 2325–2333.
55 J. Pietsch, S. Hampel, W. Schmidt, H.-J. Brauch, E. Worch, *Fresenius J. Anal. Chem.* **1996**, 355, 164–173.
56 F. Sacher, S. Lenz, H.-J. Brauch, *J. Chromatogr. A* **1997**, 764, 85–93.
57 J. Verdú-Andrés, P. Campíns-Falcó, R. Herráez-Hernández, *Chromatographia* **2002**, 55, 129–134.
58 C. T. Yan, J. F. Jen, *Chromatographia* **2004**, 59, 517–520.
59 J. F. Peng, J. F. Liu, G. B. Jiang, C. Tai, M. J. Huang, *J. Chromatogr. A* **1072**, 3–6.
60 R. Loos, G. Hanke, S. J. Eisenreich, *J. Environ. Monit.* **2003**, 5, 384–394.
61 D. Bruns-Nagel, O. Drzyzga, K. Steinbach, T. C. Schmidt, E. von Löw, T. Gorontzy, K.-H. Blotevogel, D. Gemsa, *Environ. Sci. Technol.* **1998**, 32, 1676–1679.
62 M. A. De, O. A. O'Connor, D. S. Kosson, *Environ. Toxicol. Chem.* **1994**, 13, 233–239.
63 S. Schnell, F. Bak, N. Pfennig, *Arch. Microbiol.* **1989**, 152, 556–563.
64 P. Kölbener, U. Baumann, A. M. Coor, T. Leisinger, *Water Res.* **1994**, 28, 1855–1860.
65 Van T. Nguyen, W. K. Shieh, *Wat. Sci. Tech.* **1995**, 31 (1), 185–193.
66 J. Sarasa, S. Córtes, P. Ormad, R. Gracia, J. L. Ovelleiro, *Water Res.* **2002**, 36, 3035–3044.
67 W. Baozhen, Y. Jun, *Aqua* **1986**, 2, 86–93.
68 R. Andreozzi, A. Insola, V. Caprio, M. G. Damore, *Water Res.* **1991**, 25, 655–659.
69 P. Pichat, *Wat. Sci. Technol.* **1997**, 35 (4), 73–78.
70 J. Prousek, A. Klcova, *Chem. Listy* **1997**, 91, 575–579.
71 B. Funke, M. Kolb, P. Jaser, R. Braun, *Acta Hydrochim. Hydrobiol.* **1994**, 22, 6–9.
72 O. M. Hocine, M. Boufatit, A. Khouider, *Desalination* **2004**, 167, 141–145.
73 J. Naumcyzk, L. Szpyrkowicz, F. Zilio-Grandi, *Toxicol. Environ. Chem.* **1992**, 34, 113–121.
74 F. C. Ferreira, S. Han, A. Boam, S. Zhang, A. G. Livingston, *Desalination* **2002**, 148, 267–273.
75 P. A. Greve, R. C. C. Wegman, *Schriftenreihe Verein Wasser-, Boden- u. Lufthyg.* **1975**, 46, 59–80.
76 H. Kußmaul, M. Hegazi, *Vom Wasser* **1975**, 44, 31–47.
77 G. Norwitz, P. Keliher, *Talanta* **1984**, 31, 295–297.
78 H. Börnick, V. Hultsch, T. Grischek, D. Lienig, E. Worch, *Vom Wasser* **1996**, 87, 305–326;
79 T. C. Schmidt, U. Bütehorn, K. Steinbach, R. Kotke, D. Bruns-Nagel, E. von Löw, *Acta Hydrochim. Hydrobiol.* **2000**, 28, 117–122.
80 B. Scholz, N. Palauschek, *Fresenius Z. Anal. Chem.* **1988**, 331, 282–289.

81 LWA (Landesamt für Wasser und Abfall) Nordrhein-Westfalen: *Water Quality Reports* (in German: Gewässergüteberichte) from 1984 to 1992. Düsseldorf, **1985–1993**.
82 S. Franke, S. Hildebrandt, J. Schwarzbauer, M. Link, W. Francke, *Fresenius J. Anal. Chem.* **1995**, 353, 39–49.
83 S. Müller, J. Efer, W. Engewald, *Fresenius J. Anal. Chem.* **1997**, 357, 558–560.
84 P. Eppinger, *PhD thesis*, Dresden University of Technology (Germany), **2000**.
85 N. Theobald, W. Lange, W. Gählert, F. Renner, Fresenius *J. Anal. Chem.* **1995**, 353, 50–56.
86 H. Börnick, *PhD thesis*, Dresden University of Technology (Germany), **1998**.
87 R. C. C. Wegman, P. A. Greve, *Pestic. Monit. J.* **1978**, 12, 149–162.
88 B. C. J. Zoeteman, K. Harmsen, J. B. H. J. Linders, C. F. H. Morra, W. Slooff, *Chemosphere* **1980**, 9, 231–249.
89 R. C. C. Wegman, G. A. L. de Korte, *Water Res.* **1981**, 15, 391–394.
90 J. Rivera, F. Ventura, J. Caixach, M. de Torres, A. Figueras, J. Guardiola, *Int. J. Environ. Anal. Chem.* **1987**, 29, 15–35.
91 V. Coquart, M.-C. Hennion, *Chromatographia* **1993**, 37, 392–398.
92 S. Fichtner, F. T. Lange, W. Schmidt, H.-J. Brauch, *Fresenius J. Anal. Chem.* **1995**, 353, 57–63.
93 S. J. Kok, E. M. Kristenson, C. Gooijer, N. H. Velthorst, U. A. T. Brinkman, *J. Chromatogr. A* **1997**, 771, 331–341.
94 K. Kadokami, D. Jinya, T. Iwamura, T. Taizaki, *J. Environ. Chem.* **1998**, 8, 435–453.
95 T. Tsukioka, H. Ozawa, T. Murakami, *J. Chromatogr.* **1993**, 642, 395–400.
96 J. Pietsch, F. Sacher, W. Schmidt, H.-J. Brauch, *Water Res.* **2001**, 35, 3537–3544.
97 S. Mishra, V. Singh, A. Jain, K. K. Verma, *Analyst* **2001**, 126, 1663–1668.
98 X. H. Yang, C. Lee, M. L. Scranton, *Anal. Chem.* **1983**, 65, 572–576.
99 C. D. Lyons, S. Katz, R. Bartha, *Appl. Environ. Microbiol.* **1984**, 48, 491–496.
100 H.-M. Hwang, R. E. Hodson, R. F. Lee, *Water Res.* **1987**, 21, 309–316.
101 R. G. Zepp, G. L. Baughman, P. F. Schlotzhauer, *Chemosphere* **1981**, 10, 109–117.
102 R. A. Larson, R. G. Zepp, *Environ. Toxicol. Chem.* **1988**, 7, 265–274.
103 G. C. Miller, D. G. Crosby, *Chemosphere* **1983**, 12, 1217–1227.
104 E. J. Weber, D. L. Spidle, K. A. Thorn, *Environ. Sci. Technol.* **1996**, 30, 2755–2763.
105 A. Di Corcia, A. Costantino, C. Crescenzi, R. Samperi, *J. Chromatogr. A* **1999**, 852, 465–474.
106 K. A. Thorn, P. J. Pettigrew, W. S. Goldenberg, E. J. Weber, *Environ. Sci. Technol.* **1996**, 30, 2764–2775.
107 R. V. Subba-Rao, H. E. Rubin, M. Alexander, *Appl. Environ. Microbiol.* **1982**, 43, 1139–1150.
108 C. D. Lyons, S. E. Katz, R. Bartha, *Bull. Environ. Contam. Toxicol.* **1985**, 35, 696–703.
109 L. Toräng, P. Reuschenbach, B. Müller, N. Nyholm, *Chemosphere* **2002**, 49, 1257–1265.
110 P. Pitter, J. Chudoba, *Biodegradability of Organic substances in the Aquatic Environment*, CRC Press, Boca Raton, **1990**.
111 S. Fetzner, *Appl. Microbiol. Biotechnol.* **1998**, 49, 237–250.
112 A. M. Delort, B. Combourieu, *J. Ind. Microbiol. Biotechnol.* **2001**, 26, 2–6.
113 E. Díaz, A. Ferrández, M. A. Prieto, J. L. García, *Microbiol. Mol. Biol. Rev.* **2001**, 65, 523–569.
114 J. Stockinger, C. Hinteregger, M. Loidl, A. Fersch, F. Streichsbier, *Appl. Microbiol. Biotechn.* **1992**, 38, 421–428.
115 J. Zeyer, A. Wasserfallen, K. N. Trimmis, *Appl. Environ. Microbiol.* **1985**, 50, 447–453.
116 J. A. Field, A. J. M. Stams, M. Kato, G. Schraa, *Anton. Leeuw. Int. J. G.* **1995**, 67, 47–77.
117 P. Gerike, W. K. Fischer, *Ecotoxicol. Environ. Saf.* **1981**, 5, 45–55.
118 M. Alexander, B. K. Lustigman, *J. Agric. Food Chem.* **1966**, 14, 410–413.
119 N. S. Battersby, V. Wilson, *Appl. Environ. Microbiol.* **1989**, 55, 433–439.

120 S. Paul, *PhD thesis*, Dresden University of Technology (Germany), **2001**.
121 J. Pietsch, *PhD thesis*, Dresden University of Technology (Germany), **1997**.
122 D. F. Paris, N. L. Wolfe, *Appl. Environ. Microbiol.* **1987**, 53, 911–916.
123 M. Godejohann, A. Preiss, K. Levsen, G. Wunsch, *Chromatographia* **1996**, 43, 612–618.
124 T. Zimmermann, W. J. Ensinger, T. C. Schmidt, *Anal. Chem.* **2004**, 76, 1028–1038.
125 U. Lewin, L. Wennrich, J. Efer, W. Engewald, *Chromatographia* **1997**, 45, 91–98.
126 T. C. Schmidt, M. Less, R. Haas, E. von Löw, K. Steinbach, *Int. J. Environ. Anal. Chem.* **1999**, 74, 25–41.
127 S. S. Johansen, A. B. Hansen, H. Mosbaek, E. Arvin, *Ground Water Monit. Remediat.* **1997**, 17, 106–115.
128 R. P. Schwarzenbach, P. M. Gschwend, D. M. Imboden, *Environmental Organic Chemistry*, Wiley, Hoboken, **2003**, p. 279.
129 W. F. Cowen, A. M. Gastinger, C. E. Spanier, J. R. Buckel, R. E. Bailey, *Environ. Sci. Technol.* **1998**, 32, 598–603.
130 M. S. Elovitz, E. J. Weber, *Environ. Sci. Technol.* **1999**, 33, 2617–2625.
131 J. R. Fabrega, C. T. Jafvert, H. Li, L. S. Lee, *Environ. Sci. Technol.* **1998**, 32, 2788–2794.
132 J. R. Fábrega-Duque, C. T. Jafvert, H. Li, L. S. Lee, *Environ. Sci. Technol.* **2000**, 34, 1687–1693.
133 L. S. Lee, A. K. Nyman, H. Li, M. C. Nyman, C. T. Jafvert, *Environ. Toxicol. Chem.* **1997**, 16, 1575–1582.
134 H. Li, L. S. Lee, *Environ. Sci. Technol.* **1999**, 33, 1864–1870.
135 F. P. Donaldson, M. C. Nyman, *Environ. Toxicol. Chem.* **2005**, 24, 1022–1028;
136 X. C. Wang, C. Lee, *Geochim. Cosmochim. Acta* **1990**, 54, 2759–2774.
137 V. Travkin, B. P. Baskunov, E. L. Golovlev, M. G. Boersma, S. Boeren, J. Vervoort, W. J. H. van Berkel, I. Rietjens, L. A. Golovleva, *FEMS Microbiol. Lett.* **2002**, 209, 307–312.
138 H. Sontheimer, *DVGW-Schriftenreihe Wasser* **1988**, 60, 27–50.
139 E. Worch, T. Grischek, H. Börnick, P. Eppinger, *J. Hydrol.* **2002**, 266, 259–268.
140 T. P. Knepper, F. Sacher, F. T. Lange, H.-J. Brauch, F. Karrenbrock, O. Roerden, K. Lindner, *Waste Manag.* **1999**, 19, 77–99.
141 H. Börnick, P. Eppinger, T. Grischek, E. Worch, *Water Res.* **2001**, 35, 619–624.
142 S. Paul, H. Börnick, E. Worch, *Vom Wasser* **2001**, 96, 29–42.
143 H. Börnick, T. Grischek, E. Worch, *Fresenius J. Anal. Chem.* **2001**, 371, 607–613.
144 S. Paul, H. Börnick, T. Grischek, E. Worch, *Vom Wasser* **2001**, 97, 125–134.
145 F. H. Frimmel, M. Assenmacher, M. Sörensen, G. Abbt-Braun, G. Grabe, *Chem. Eng. Process.* **1999**, 38, 601–610.
146 R. A. K. Rao, M. Ajmal, R. Ahmad, B. A. Siddiqui, *Environ. Monit. Assess.* **2001**, 68, 235–247.
147 F. Sacher, F. Karrenbrock, T. P. Knepper, K. Lindner, *Vom Wasser* **2001**, 96, 173–192.
148 H. Sontheimer, J. C. Crittenden, R. S. Summers, *Activated Carbon for Water Treatment*, Braun, Karlsruhe, **1988**, pp. 649–660.
149 J. Hoigné in *Handbook of Environmental Chemistry*, Vol. 5 C (Ed.: J. Hrubec), Springer, Berlin, Heidelberg, **1988**, pp. 83–141.
150 J. Pietsch, W. Schmidt, H.-J. Brauch, E. Worch, *Ozone: Sci. Eng.* **1999**, 21, 23–37.
151 J.-W. Kang, H.-S. Park, R.-Y. Wang, M. Koga, K. Kadokami, H.-Y. Kim, E.-Z. Lee, S.-M. Oh, *Wat. Sci. Technol.* **1997** (12), 36, 299–307.
152 W.-J. Huang, G.-C. Fang, C.-C. Wang, *Sci. Tot. Environ.* **2005**, 345, 261–272.
153 M. Sörensen, H. J. Reichert, F. H. Frimmel, *Acta Hydrochim. Hydrobiol.* **2001**, 29, 301–308.

9
Surfactant Metabolites

Thomas P. Knepper and Peter Eichhorn

9.1
Introduction

Surfactants constitute a broad group of chemicals which play important and vital roles in a great variety of fields. Applications as diverse as cleaning, food, metallurgy, pharmacy, medicine, paints and varnishes, mining, and many others utilize the characteristic properties provided by surfactants. Soap was the first man-made surface active agent (surfactant) known. The replacement of soap by synthetic alkylbenzene sulfonates (ABS) in most laundry products was essentially driven by two factors: the undesired formation of insoluble calcium and magnesium salts of soap, which precipitated in the washing liquor and deposited on the clean surface of the laundry, and the lower production costs of ABS. The widespread use of ABS, mainly in laundry detergents, and their subsequent discharge into the sewer, however, led to the unexpected effect of strong foam formation in sewage water, treated sewage, and even in river water [1, 2]. This observation was directly related to the physical properties of the surfactant which had originally been responsible for its great success.

Shortly after the appearance of the described environmental problems, it was recognized that tetrapropylene-derived ABS was fairly resistant to biodegradation because of the presence of a branched alkyl chain in the hydrophobic moiety. Legal restrictions introduced in the mid-1960s, in Germany for example, prescribing a certain degree of primary degradation for anionic surfactants [3], or voluntary industrial bans as in the United States, led to the development and introduction of an alternative anionic surfactant into the detergent market: the linear alkylbenzene sulfonates (LAS). A modification in the chemical structure by substituting the branched alkyl chain by a linear one substantially improved the biodegradability. Quickly after the changeover from ABS to LAS in the mid-1960s, foam-related environmental problems disappeared. Along with this, the levels of surfactants in aquatic environments dropped significantly [4].

Since that time an enormous number of surfactants covering a wide range of chemical and physico-chemical properties have been developed for quite universal

Organic Pollutants in the Water Cycle. T. Reemtsma and M. Jekel (Eds.).
Copyright © 2006 WILEY-VCH Verlag GmbH & Co. KGaA, Weinheim
ISBN: 3-527-31297-8

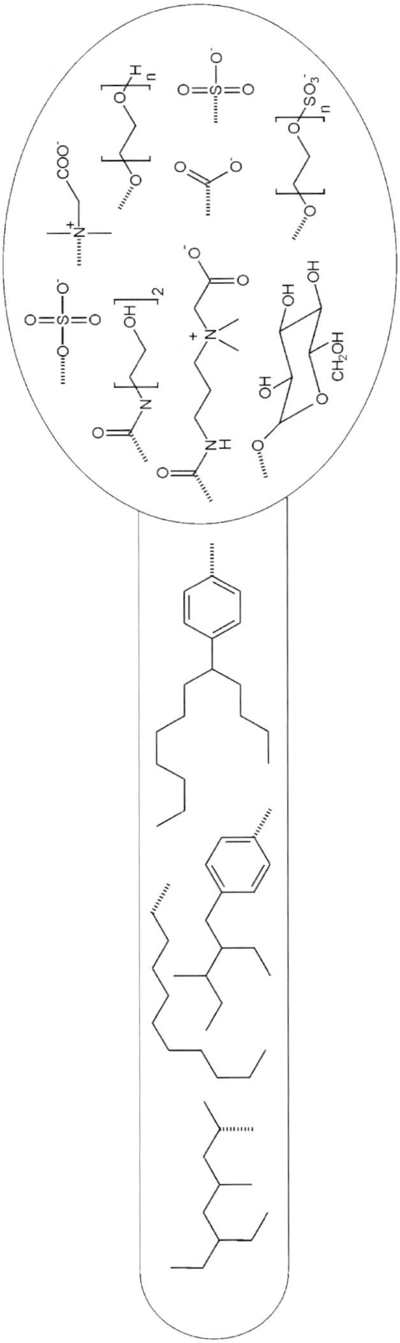

Fig. 9.1 Schematic tail-head model of surfactant molecule showing examples of some of the important hydrophobic and hydrophilic groups used.

as well as specific tasks in domestic and industrial applications. The criteria for selection of a surfactant for industrial production are directly connected with the feasibility of large-scale production. This is determined by several factors including availability and costs of raw materials, cost of manufacture, and performance of the finished products. In addition to these aspects, environmental considerations also play an increasingly important role.

All surfactants have in common an asymmetric skeleton with a hydrophobic and a hydrophilic moiety (Fig. 9.1). The hydrophobic part generally consists of a linear or branched alkyl chain, which is then linked to a hydrophilic group. Because of this bifunctionality, both parts of the surfactant molecule interact differently with water, the most commonly used solvent. On one hand, the hydrophilic group is surrounded by water molecules, which results in good solubility. On the other hand, the hydrophobic moieties are repulsed because they interfere with the strong interactions between the water molecules. Hence the molecules are repelled out of the aqueous phase and accumulate at the interfaces. According to the charge of their hydrophilic moiety, surfactants can be classified into four categories: anionic, nonionic, cationic, and amphoteric [5].

9.2
Aerobic Biodegradation of Surfactants

9.2.1
Introduction

Regulations requiring a certain degree of biodegradability of surfactants were first released in Germany shortly after the low degradability of ABS (Fig. 9.2b) was recognized, and have since also been established in all Member States of the European Union (EU) [6, 7, 8]. The general fate of surfactants after use is disposal into the sewage system and discharge to wastewater treatment plants (WWTP), where they are subsequently submitted to biodegradation. In the initial step of microbial action, known as the primary degradation, minor alterations in the chemical structure of a surfactant molecule may occur, often resulting in the loss of the surface-active properties. This is usually paralleled by a decrease in the aquatic toxicity, since, for the majority of the substances, the surface activity itself is often responsible for the adverse interactions with biological surfaces, e.g., cell membranes.

To reduce the risk of potentially environmentally harmful surfactants, a requirement for a minimum primary degradation amounting to 80% for anionic and nonionic surfactants was stipulated as far back as 1977 [9]. However, within this early regulation no restrictions were included regarding cationic or amphoteric surfactants, as these did not occupy a significant market share when the laws came into force.

Methods to monitor the primary degradation of anionic and non-ionic surfactants traditionally used a group-specific analysis based on the formation of a surfactant complex which was subsequently quantified by photometric or titrimetric

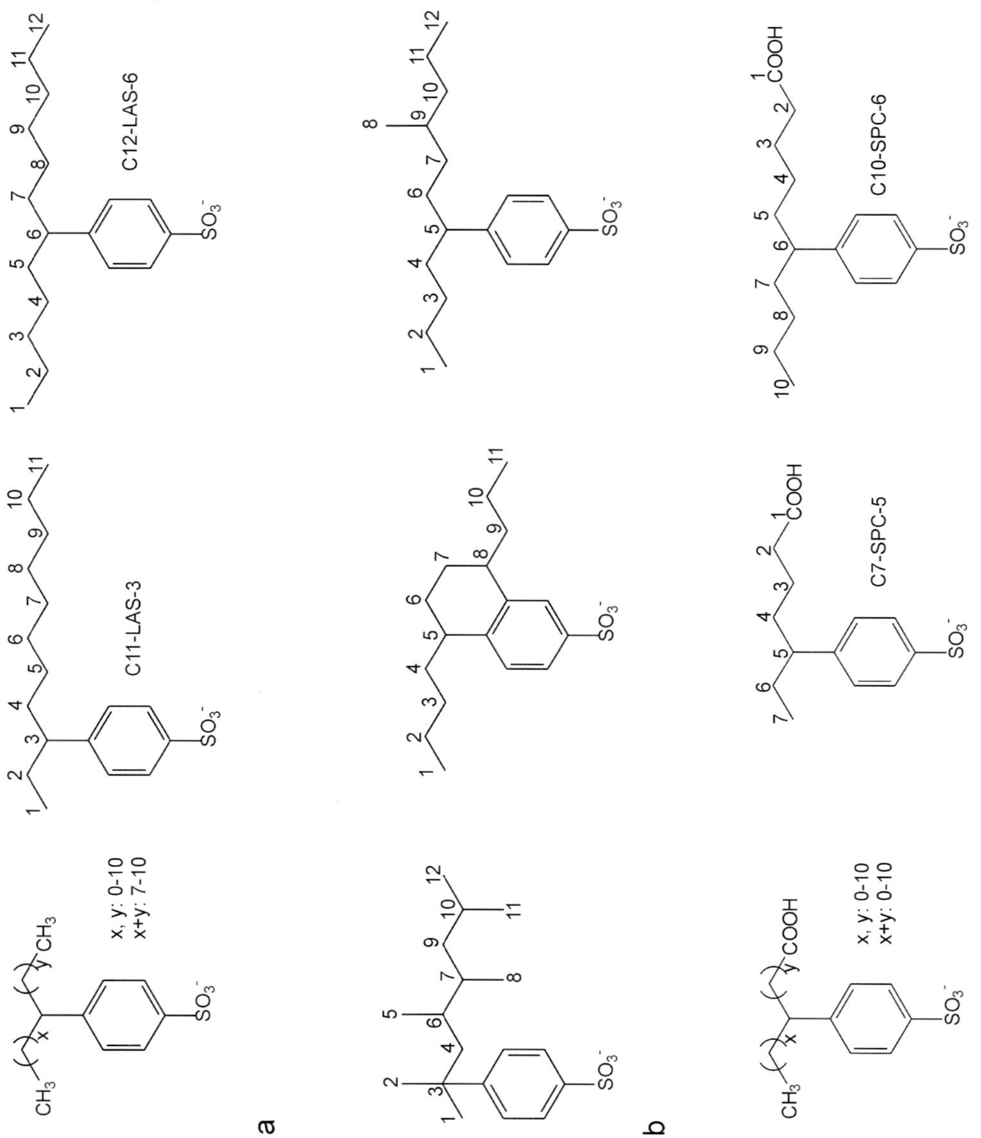

Fig. 9.2 (a) General structure of LAS (left) and structures of two components (middle and right) illustrating the nomenclature used to identify individual species; (b) possible structures of branched C_{12}-alkylbenzene sulfonate (ABS) (left), C_{11}-dialkyltetralin sulfonate (DATS) (middle), and C_{12}-single methyl-branched linear alkylbenzene sulfonate (iso-LAS) (right); (c) general structure of sulfophenyl carboxylates (SPC) (left) and the structure of two possible components deriving from LAS biodegradation (middle and right) illustrating the nomenclature used to identify individual species.

methods [10, 11]. Anionic surfactants were complexed with the cationic dye methylene blue and extracted with chloroform prior to photometric analysis. The major anionic surfactants such as LAS, alkyl sulfates (AS) and alkylether sulfates (AES) give a positive response as methylene blue-active substances (MBAS), but detection can be adversely affected by many kinds of interference [12, 13]. Similar problems are encountered in the analysis of non-ionic surfactants, which are determined as bismuth iodide-active substances (BiAS). Furthermore, the MBAS and BiAS complexes are usually only formed from intact surfactant molecules, i.e. degradation intermediates do not contribute to the overall value. The fate of surfactants, however, is not sufficiently described if the primary degradation is the only parameter taken into account. Knowledge about the ultimate transformation is also of considerable importance, as has become particularly evident in the case of the nonylphenol ethoxylates (NPEO). For this purpose, the total degradation of a compound, which can occur either by mineralization or assimilation, is measured by a substance-independent sum parameter such as the evolution of carbon dioxide, the oxygen consumption, or the removal of the dissolved organic carbon (DOC).

A series of international standard methodologies have been established by the Organization of Economic Co-operation and Development (OECD) to assess the biodegradability of surfactants [14]. These tiered tests distinguish between "ready biodegradability", "inherent biodegradability", and "simulation biodegradability". They differ mainly in the comparability with real environmental conditions, the concentration of the test compound and the test duration, the inoculum used, the measured analytical parameter, and the pass level. In the first stage the surfactant is submitted to the ready biodegradability tests (OECD 301A to F). If it meets the pass level, i.e. 60% CO_2 formation or 70% DOC removal of the theoretically possible amounts within 28 days, it is concluded that with high likelihood the examined substance is (ultimately) degraded under environmental conditions. In case of failure to reach this designated pass level, a test of inherent biodegradability under ideal conditions is then required (OECD 302A to C). A positive result in this test defines the compound as inherently biodegradable, i.e. it has the potential to be degraded. The third tier has to be initiated if the pass level in the former two tests has not been attained. In this instance the biodegradability under WWTP conditions is tested (OECD 303A and 304A).

For the study of the aerobic biodegradability of surfactants in aqueous media, in addition to the standardized OECD methods [14], a large variety of assays have been developed. These can differ in instrumental and technical set-up, measured parameters, type and origin of test medium and inoculum used, test duration, and substrate concentration [15–24]. More sophisticated test systems for simulating the elimination in surface waters were developed by Schöberl et al. [25] and Boeije et al. [26] based on a staircase model. The use of trickling filters allowed for determination of the degree of bioelimination and mineralization [27–29]. A biologically active fixed-bed bioreactor (FBBR) operated under aerobic conditions was employed for investigating both the primary degradation of the targeted surfactants and their breakdown pathway [30].

9.2.2
Anionic Surfactants

Linear Alkylbenzene Sulfonates

The linear alkylbenzene sulfonates (LAS) generally consist of a mixture of homologs and isomers. Individual components are classified by the length of the alkyl substituent and by the position of attachment of the sulfophenyl ring (Fig. 9.2a). In formulations, the alkyl chain length may range from C10 to C13, with an average distribution of 11.4 to 11.7, and the position of the phenyl ring can vary from the 2- to the 7-carbon depending on the alkyl chain length. Commercial blends usually contain some sulfonated impurities (<1%), namely ABS, dialkyltetralin sulfonates (DATS), and singly methyl-branched LAS (iso-LAS) (Fig. 9.2b). The levels of these impurities depend on the synthetic route used.

As a consequence of the vast amounts of LAS entering the aquatic environment via wastewater discharges, their environmental fate and distribution has been the subject of thorough investigation for more than three decades. Hence, an enormous body of literature exists regarding the analytical methods used for their determination in various environmental compartments, investigations of their biodegradability under both laboratory and field conditions, and assessment of their toxic potential toward aquatic organisms [31, 32, 33]. In essence, LAS are one of the best-studied classes of organic substances produced on an industrial scale with respect to their environmental behaviour.

A detailed description of the aerobic degradation of LAS was provided by Swisher [15] and Schöberl [34]. The majority of the laboratory studies published indicated that metabolism starts with oxygenation of one of the terminal methyl groups of the alkyl chain and the conversion of the alcohol to a carboxylic group (ω-oxidation), releasing a sulfophenyl carboxylate (SPC; Fig. 9.2c) with the same number of carbon atoms in the side chain as in the original LAS isomer (Fig. 9.3) [35, 36]. The conversion rates of individual components in a commercial LAS mixture are dependent on the molecular structure. For example, the length of the alkyl chain is positively correlated with the primary degradation rate, and phenyl isomers substituted at central positions of the alkyl chain are degraded more slowly than other isomers [37, 38]. Both effects are a direct consequence of the enzymatic attack on the hydrophobic moiety. The relationship between surfactant structure and the biodegradation has been termed "Swisher´s distance principle", which, in summary, states that increased distance between the xenobiotic arylsulfonate moiety and the far end of the alkyl chain increases the speed of primary degradation [15].

It is generally accepted that ω-oxidation of LAS is followed by successive stepwise oxidative shortening of the alkyl chain by two carbon units, termed β-oxidation (Fig. 9.3) [35, 39]. The resulting very short-chain SPCs are further broken down by ring cleavage, which is considered the rate-determining step of the whole process, and by desulfonation, which completes the degradation. Although β-oxidation is certainly the most important mechanism for the destruction of the alkyl chain in LAS, some indication has been given that removal of a single carbon, i.e. α-oxidation, may also occur to a minor extent. This alternative route was proposed in or-

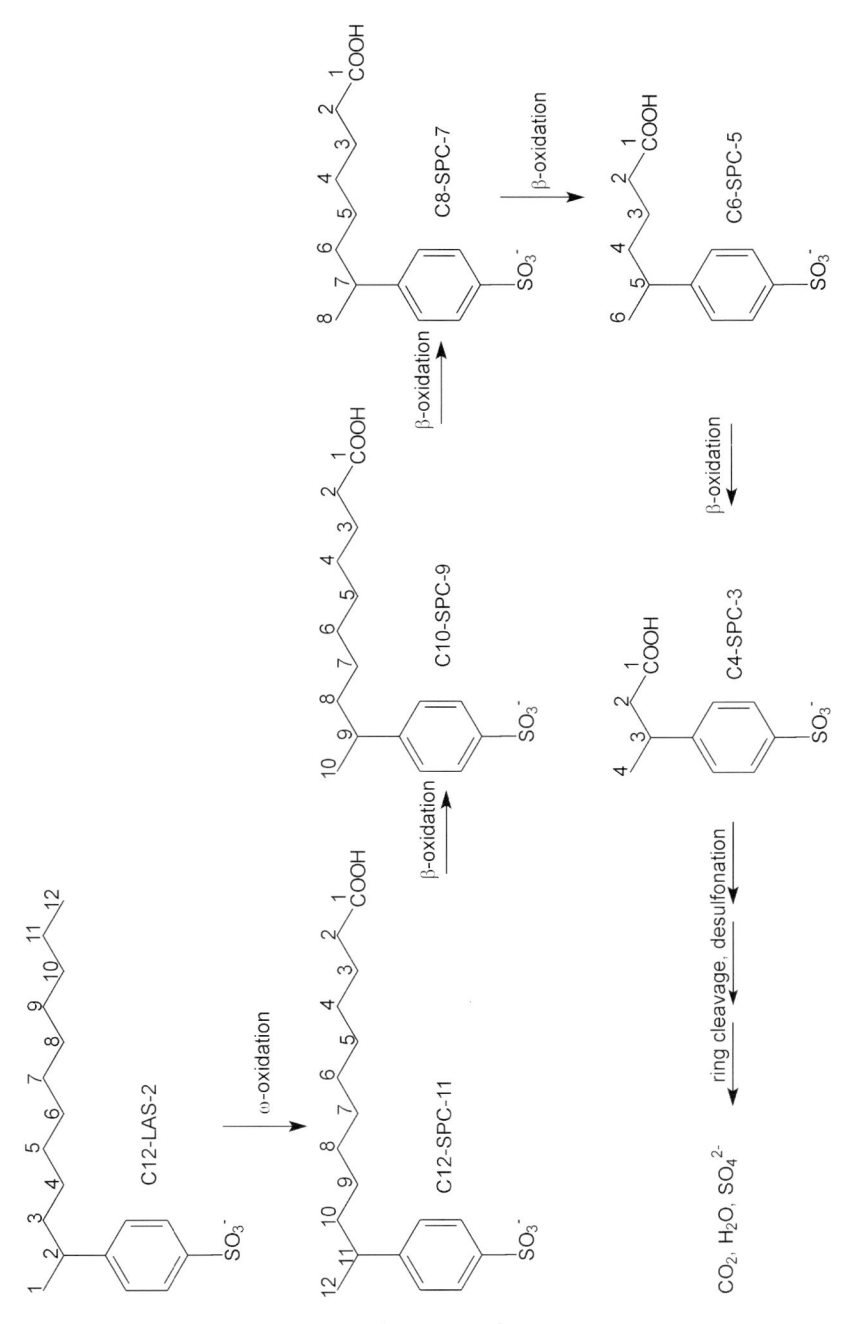

Fig. 9.3 Aerobic degradation pathway of LAS shown for C_{12}-LAS-2.

der to explain the isolation of C_5-SPC as a by-product in the degradation of C_{12}-LAS [40] and the detection of a multitude of both C-even and C-odd SPC intermediates observed during the degradation of C_{11}-LAS [41].

According to the scheme shown in Fig. 9.3, for the pure C_{12} homolog as the starting compound, exclusively C-even SPCs are expected to be formed. Direct support for the scheme was provided by Hrsak and Begonja [42] and Dong et al. [43], who used pure cultures of bacteria, which attacked C_{12}-LAS-2 and C_{12}-LAS-3 (see Fig. 9.2a for nomenclature), respectively. The SPCs formed were subject to chain shortening in pure or mixed cultures. The ring cleavage mechanism of an SPC, via 4-sulfocatechol, was established [44], and this was suspected to be common to the general degradation of LAS, because mixed cultures which degrade LAS contain high levels of 4-sulfocatechol 1,2-dioxygenase [43]. LAS congeners are usually racemic mixtures of optically active compounds [15], and their transient degradative intermediates are also usually optically active [45]. Most of these intermediates are transients, but some can have long half lives [46]. Some LAS congeners are evidently subject to attack at both methyl groups on the alkyl chain as sulfophenyl dicarboxylates (SPdC) are also detected during the degradation of LAS [46].

The observation of SPC and SPdC during the degradation of LAS, however, still did not prove either ω-oxygenation or β-oxidation of SPC as the sole significant mechanism, as Schleheck et al. [47] realized when they isolated the first heterotrophic bacterium able to attack commercial LAS. They could not distinguish

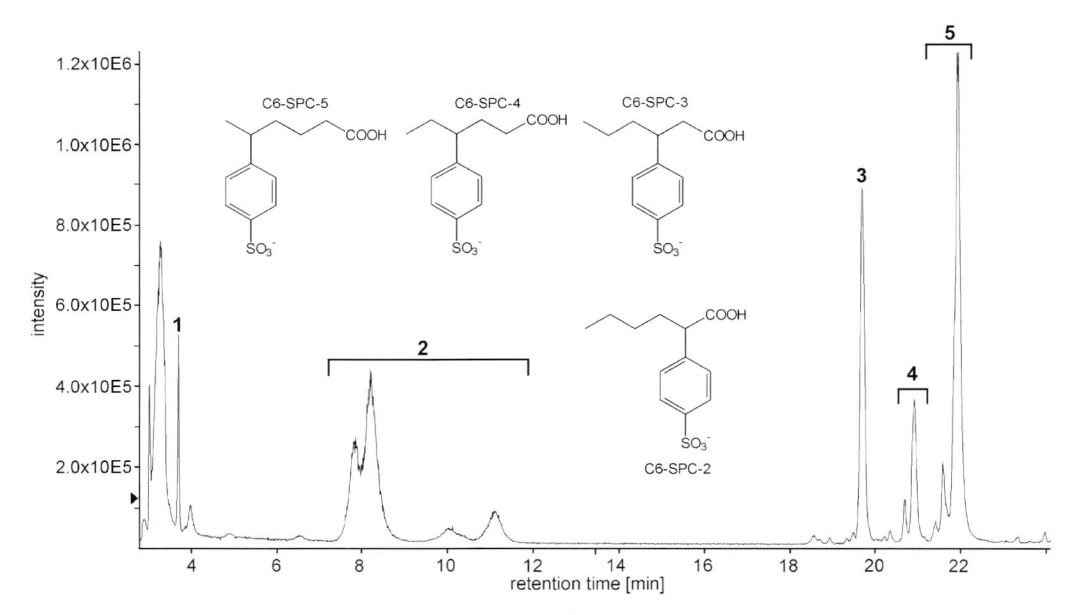

Fig. 9.4 (–)-LC-ESI-MS chromatogram of an FBBR degradation experiment on LAS. Peak numbering: (1) C_4-SPC, (2) C_6-SPC, (3) C_8-SPC, (4) C_{10}-SPC, (5) C_{12}-SPC. Only the time window where the SPCs are eluting is shown. C_{12}-LAS eluted at 27.5 min under the selected conditions.

between ω-oxygenation and subterminal oxygenation with subsequent Baeyer-Villiger oxygenation and hydrolysis to yield an SPC. Furthermore, they were working with a pure culture and not with an environmentally realistic microbial mixture.

In order to resolve the oxidative mechanisms of LAS, Eichhorn and Knepper [48] used the following experimental approach: (a) the material subjected to the biodegradation study used was a pure LAS homolog with very low amounts of structurally related synthetic by-products in the test medium, which could potentially disturb the chemical analysis of LAS, and for which their corresponding intermediates could also complicate the determination of SPC species [49]; (b) the experimental set-up was based on a mixed microbial community from a natural setting other than isolated cultures, as this guaranteed a higher environmental relevance of the outcomes; and (c) the applied analytical protocol included a sample preparation method allowing for accurate quantification of all individual degradation products. The samples collected from the FBBR were analyzed by liquid chromatography-electrospray ionization-mass spectrometry (LC-ESI-MS) for the occurrence of SPC [48]. The C-even SPCs were by far the most prominent intermediates (Fig. 9.5), whereas the C-odd SPCs were found at much lower amounts. The concentrations of the two long-chain homologs, C_{12}-SPC and C_{10}-SPC, reached a maximum and were then rapidly converted into the shorter-chain homologs. In all cases, C_6-SPC was the most abundant homolog, while its direct precursor, C_8-SPC, and its suc-

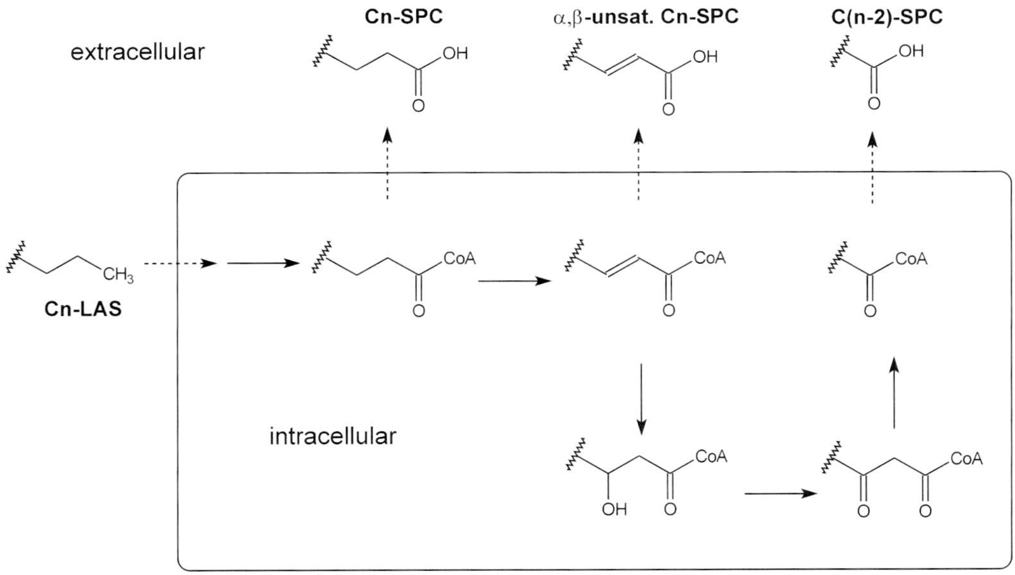

Fig. 9.5 Presumed major pathway for intracellular aerobic biodegradation of LAS (as acetyl-CoA-derivatives) to SPC via ω-oxidation followed by successive oxidative shortening of the alkyl chain by two carbon units (β-oxidation). Intermediates of β-oxidation, such as SPC-2H formed by enzymatic dehydrogenation, are also transported out of the cell after cleavage of the CoA-ester.

cessor, C_4-SPC, were detected at substantially lower levels. It was observed that (a) the kinetics of the degradation of individual SPC homologs decreased with shortening of the oxidized side chain, i.e. higher concentrations of metastable SPCs with shorter alkyl chains were detected, and (b) within a set of phenyl isomers for each SPC homolog, some species show substantial resistance to further degradation. Hence, an accumulation in the test liquor occurred, which was dominated by the shorter chain SPCs.

Besides the evidence from the repetitive cycles of β-oxidation in C_{12}-LAS biotransformation, further support for β-oxidation could be gathered if other intermediates occurring via this pathway were detectable. Beta-oxidation, though known for decades, is still a subject of active research [50, 51]. The class of compound that has been identified to date, albeit in mammalian systems, is the α,β-unsaturated carboxylate. Indications of the formation of analogous species in microbial metabolism of LAS were reported by Knepper and Kruse [30] during biotransformation of commercial LAS surfactant on an FBBR. A further successful attempt to detect α,β-unsaturated SPCs (SPC-2H) during the degradation of LAS was undertaken by Eichhorn and Knepper [48] (Fig. 9.5). These intermediates corresponded to the β-oxidation products of the C-even SPC thereby directly supporting β-oxidation as a mechanism for all relevant SPC. The route of metabolism of LAS observed in the FBBR was thus ω-oxidation followed by β-oxidations.

Secondary Alkane Sulfonates
Secondary alkane sulfonates (SAS; $CH_3–CH_2–CH(SO_3^-Na^+)–(CH_2)_n–CH_3$) were shown to be readily biodegradable under aerobic conditions [15, 52], with ω-oxidation and oxidative desulfonation being the significant steps in the conversion [53]. As such, in most aquatic environmental compartments ultimate degradation is considered likely. However, sorption to particulate matter prior to microbial degradation may be a relevant removal process, especially in WWTP [53, 54]. The fate of SAS in sludge removed from WWTP will thus be dependent on the extent of aerobic digestion in the plant and the aerobic nature of the subsequent compartment to which it is transferred, since degradation under anaerobic conditions is limited.

Alkyl Sulfates
The aerobic degradation of alkyl sulfates (AS; $CH_3–(CH_2)_n–O–SO_3^-Na^+$) is rapid and extensive [15, 52, 55, 56], and it has been speculated that a broad range of microorganisms must be capable of catalyzing the process [15]. The degradation is thought to proceed by enzymatic cleavage of the sulfate ester bond, followed by subsequent oxidation steps [15, 53], although none of the steps require molecular oxygen [53]. The products obtained were all amenable to ultimate biodegradation, and no recalcitrant metabolites were observed. AS adsorption to sludge does not lead to preservation, as anaerobic digestion is also extensive for this surfactant class [15, 55]. Furthermore, biodegradability of linear and branched derivatives was shown to be equally rapid and extensive [57], and available data indicate that all AS

homologs are readily degradable irrespective of the length of their alkyl chain. Greater than 99% removal of AS isomers has been described in several studies using a variety of detection methods including MBAS [58], ^{14}C radiolabeling [55], and LC-MS [59].

Alkyl Ether Sulfates

Ready aerobic biodegradation of alkyl ether sulfates (AES; $CH_3-(CH_2)_n-O-(CH_2CH_2O)_m-SO_3^-Na^+$) has been described [52], with ω/β-oxidation and cleavage of the sulfate and ether bonds attributed to the process [15]. However, molecular oxygen is not necessary for the latter two steps, and primary and ultimate degradation has been described under both aerobic and anaerobic conditions [53].

Fatty Acid Esters

Fatty acid esters (FES) are readily degraded in aerobic environments [60], by ω- and β-oxidation steps, followed by desulfonation, such that extensive mineralization is achieved. Persistence from aquatic sources can only be envisaged in cases where adsorption leads to removal of the surfactant from the aerobic conditions, as anaerobic degradation has not been observed in any study to date [60]. Aerobic degradation in sludge-amended soils, however, was found to be rapid [60].

9.2.3
Non-ionic Surfactants

Nonylphenol Ethoxylates

The biodegradation of NPEO was generally believed to start with a shortening of the ethoxylate chain, leading to short-chain NPEO containing one or two ethoxylate units ($NPEO_1$ and $NPEO_2$). Further transformation proceeds via oxidation of the ethoxylate chain, producing mainly nonylphenoxy ethoxy acetic acid (NPE_2C) [61] and nonylphenoxy acetic acid (NPE_1C) (Fig. 9.6a) [62]. Little has been reported about the further degradation of NPEC. In a semi-quantitative biodegradation study by Di Corcia et al. [63], the formation of doubly carboxylated metabolites was described, with both the alkyl and ethoxylate chains oxidized (CAPEC). Ding et al. detected these intermediates in river and wastewater [64, 65].

It was generally assumed that the endocrine disruptor nonylphenol (NP) is the most persistent metabolite of NPEO. However, experimental data on the formation of the metabolite NP from NPEO is surprisingly scarce, and NP has been reported to be formed mostly under anaerobic conditions [66, 67].

The purpose of a study performed by Jonkers et al. [68] was to elucidate the aerobic biodegradation route of NPEO. For NPEO a relatively fast primary degradation of >99% was observed after 4 days. Contrary to the generally proposed degradation pathway of EO-chain shortening, it could be shown that the initiating step of the degradation is ω-carboxylation of the individual ethoxylate chains: NPEC metabolites with long carboxylated EO-chains were identified (Fig. 9.6b). Further degra-

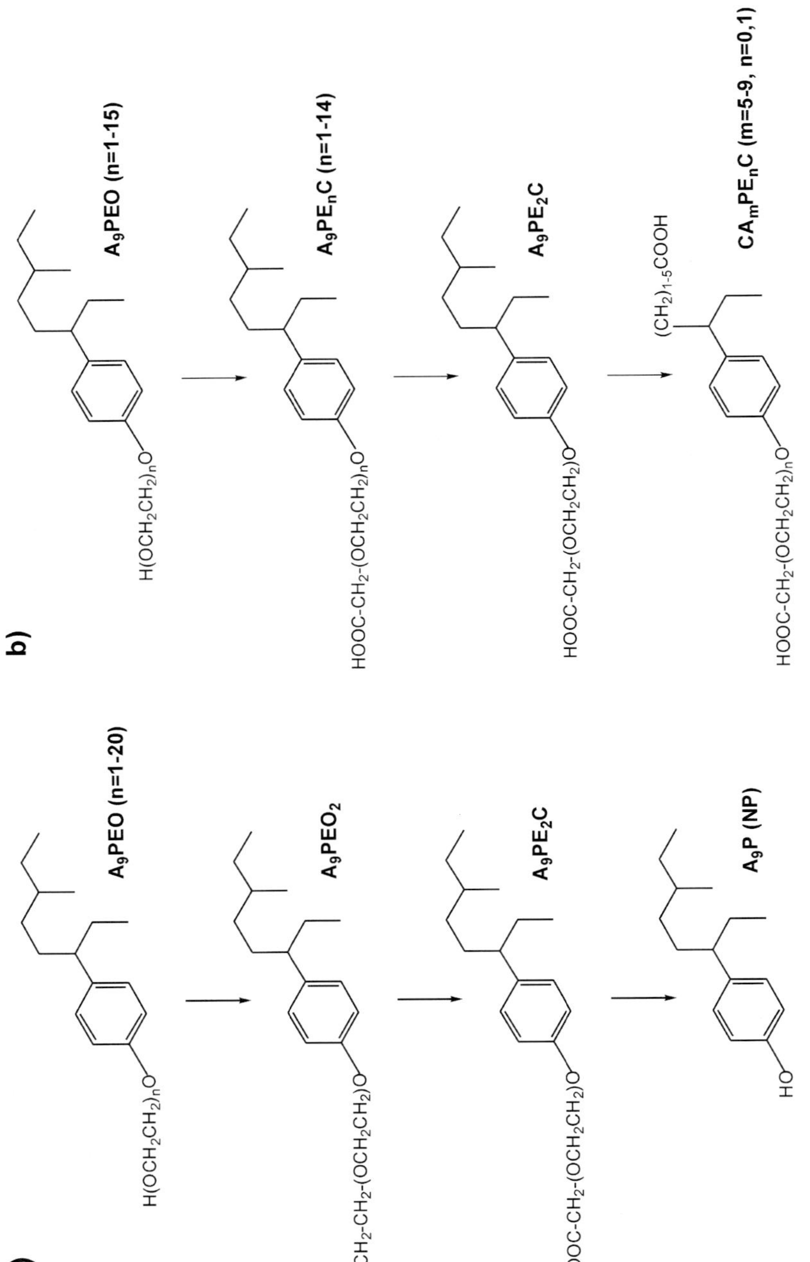

Fig. 9.6 (a) Generally proposed aerobic biodegradation pathway of NPEO (adapted from [62]); (b) newly proposed aerobic biodegradation pathway (adapted from [68]).

dation proceeded gradually, forming short-chain carboxylated EO. It could be shown that the oxidation of the nonyl chain proceeded concomitantly with this degradation, leading to metabolites having both a carboxylated ethoxylate and an alkyl chain of variable length (CAPEC). The identity of the CAPEC metabolites was confirmed by the fragmentation pattern obtained with LC-ESI-tandem mass spectrometry [68].

Alkyl Glucamides

Stalmans et al. [69] applied standardized degradation tests using a batch-activated sludge system and the modified Sturm test, and thereby obtained a high mineralization rate for a commercial alkyl glucamide (AG) mixture. In both tests the DOC removal exceeded 98% and primary degradation was greater than 99%. Eichhorn and Knepper [70] investigated the biodegradability and the metabolic pathway of C_{10}-AG on a fixed-bed bioreactor. This exclusively proceeded via ω-oxidation of the alkyl chain, resulting in the formation of carboxylic acids, which were then further broken down through β-oxidations. The corresponding pathway relevant to AG is shown in Fig. 9.7. Of the putative carboxylic acids theoretically possible, only the C_4-glucamide acid could be detected [70]. The lack of detection of the C_6-, C_8-, and C_{10}-carboxylic acids was attributed to their short lifetime in solution or to the possibility that they were not released from the cells into the bulk solution. It was also likely that they were rapidly further degraded to shorter homologs, with similar observations made for long-chain SPC, which undergoes rapid conversion into shorter-chain species during LAS metabolism. The relative longevity of the C_4-glucamide acid was explained by the fact that, in contrast to the preceding steps, a subsequent β-oxidation would not lead to a regular carboxylic acid, but to an α-keto acid. It was suggested that the final conversion of C_4-glucamide acid occurred as an intracellular process. This assumption was supported by the findings of Stalmans et al. [69], in which high DOC removal and CO_2 formation in standardized biodegradation tests of AG was reported.

Alkyl Polyglucosides

Alkyl polyglucosides (APG) consist of a complex mixture of a variety of homologs and isomers, including stereoisomers, binding isomers, and ring isomers within the glucose moiety. In order to assess the aerobic and anaerobic biodegradability of APG, a series of standard laboratory tests with activated sludge were carried out by Steber et al. [71] using both discontinuous screening tests and continuous test systems. Evaluation of the results of sum parameters analysis [DOC removal, chemical oxygen demand (COD), and biological oxygen demand (BOD)] allowed them to classify APG as "readily biodegradable" according to OECD definitions. Ultimate biodegradation was proved by the coupled-units test (OECD 303A), giving a DOC removal of 89%, and even the anaerobic biodegradability testing (ECETOC screening test) indicated complete transformation into CO_2 and CH_4, amounting to 84% theoretical gas production. The findings of Steber et al. [71] were confirmed in two

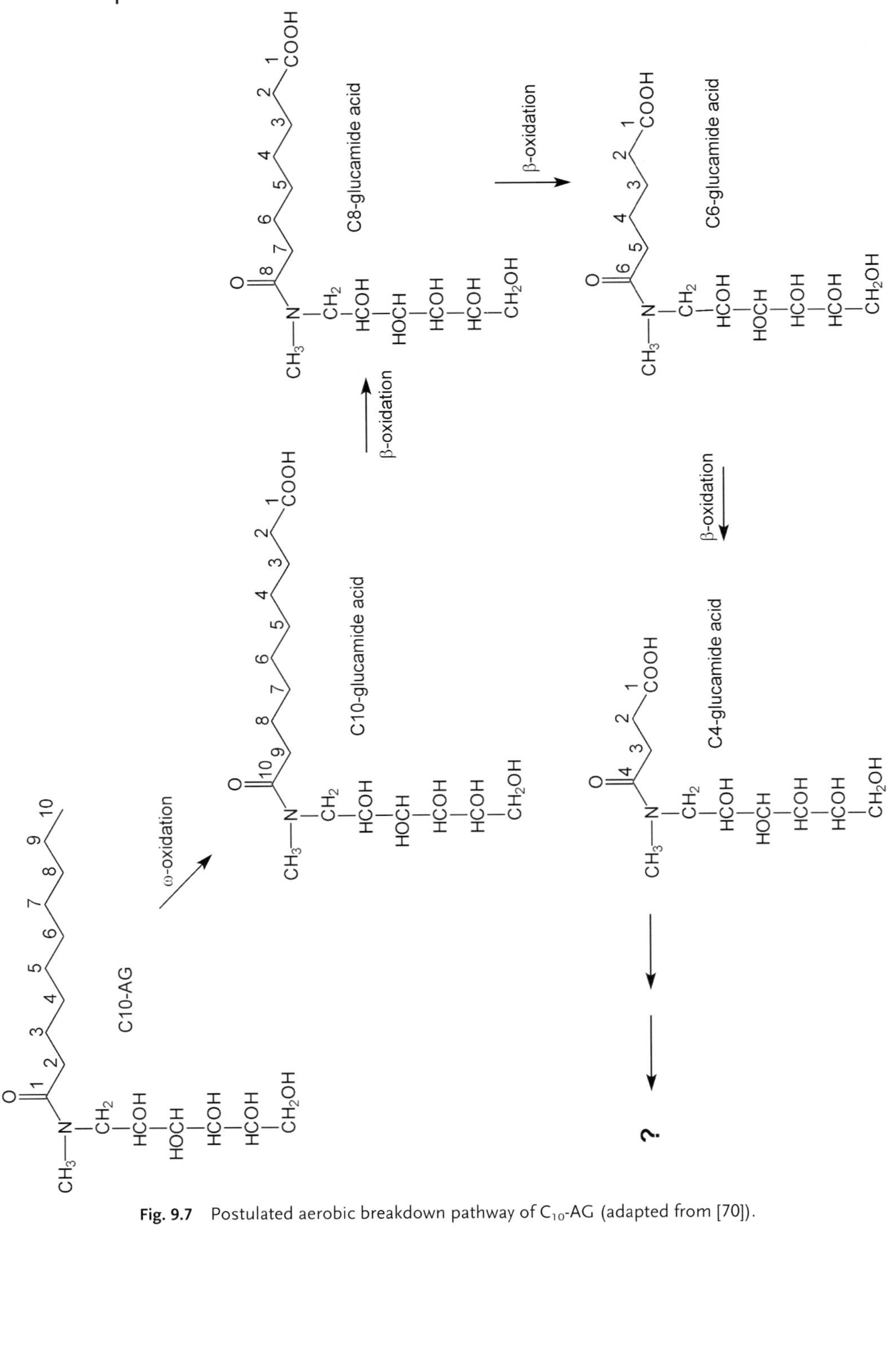

Fig. 9.7 Postulated aerobic breakdown pathway of C_{10}-AG (adapted from [70]).

other studies likewise working with OECD standard tests [72, 73]. However, all these test methods were exclusively based on sum parameters, and hence they did not yield any information regarding the metabolic pathway of APG. Eichhorn and Knepper [74] applied more specific methods to investigate the aerobic biodegradation of C_8-, C_{10}-, and C_{12}-β-monoglucoside in river water, utilizing a fixed-bed bioreactor set-up and mass spectrometric analysis. A breakdown pathway of the APG was postulated on the basis of the metabolism of LAS and AG, which are degraded by _-oxidation of the alkyl chain resulting in carboxylic acids (see Fig. 9.3). These acids in turn are further degraded by β-oxidation releasing C_2-units. The corresponding mechanism for APG is shown in Fig. 9.8 (pathway I). The search for the putative "polyglucoside alcanoic acids" was not successful. An alternative degradation mechanism was thus proposed (Fig. 9.8, pathway II), involving the cleavage of the glucosidal bond, which led to glucose and the fatty alcohol intermediates. The glucose so generated was believed to be rapidly further metabolized via pyruvate,

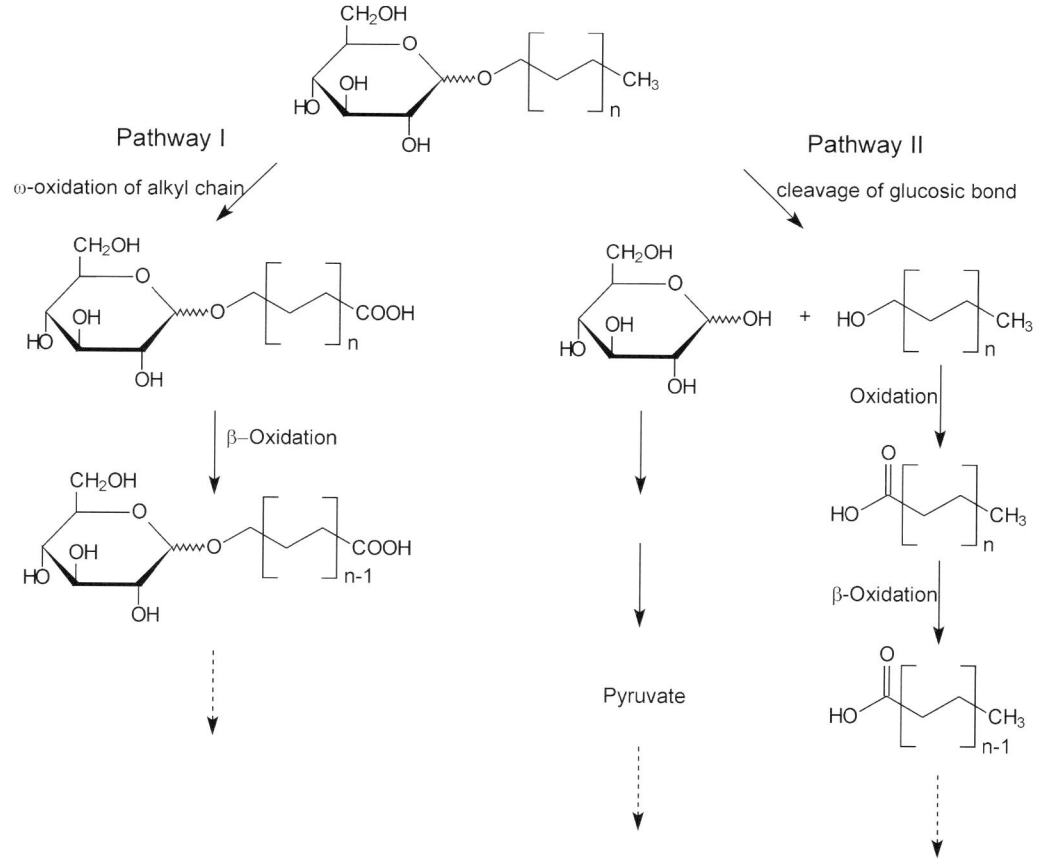

Fig. 9.8 Possible degradation pathways of alkyl glucopyranosides; n = 8, 10, 12, 14, 16 (adapted from [74]).

and the fatty alcohol was oxidized to the corresponding acid, which could subsequently undergo classical fatty acid degradation. The lack of detection of fatty acids, which were also investigated by LC-(–)-ESI-MS, was attributed to their rapid intracellular metabolism.

Alcohol Ethoxylates
Alcohol ethoxylates (AE) have been developed as an environmentally friendly replacement of alkylphenol polyethoxylates (APEO), and as such the data available regarding their biodegradability is relatively extensive. The linear derivatives exhibit good degradation [75], but the branched analogs are converted to a much lesser extent [76]. The process is thought to proceed by cleavage of the hydrophobe-hydrophile bond and subsequent ω/β-oxidation of the alcohol [77]. The proposed mechanisms for the aerobic biodegradation of AE are shown in Fig. 9.9 [78] including (a) the central cleavage of the molecule leading to the formation of polyethylene glycols (PEG) and aliphatic alcohols, (b) the ω/β oxidation of the terminal carbon of the alkyl chain, and (c) the hydrolytic shortening of the terminal carbon of the polyethoxylated chain. PEG and aliphatic alcohols are degraded independently. PEG biodegradation proceeds by successive depolymerization of the ethoxy chain via non-oxidative and oxidative cleavage of C_2-units. The process leads to the formation of shorter-chain neutral PEG, mono- and dicarboxylated metabolites of PEG (-MCPEG and DCPEG), alcohol ethoxylates carboxylated on the polyethoxy chain (AEC) or in the alkyl chain (CAE), or alcohol ethoxylates carboxylated on both ends (CAEC). Depending on the aquatic system in which PEGs are found, their half-lives can vary from weeks to months.

9.2.4
Amphoteric Surfactants

Cocamidopropyl Betaines
Few data have been published on the degradability of betaines, such as cocamidopropyl betaines (CAPB, Fig. 9.10), and the limited information available is somewhat contradictory. Swisher reported that total degradability was rather poor (45–58%) [15], whilst Brunner et al. [79] obtained mixed results with a series of the amphoteric surfactants in an extended OECD 302B test and a laboratory trickling test filter. While the two compounds cocoamphodiacetate and cocoamphodipropionate were mineralized to only a minor extent, CAPB and cocoamphoacetate proved to be nearly totally degradable. Eichhorn and Knepper [80] studied the aerobic biodegradation of four C-even CAPB homologs (C_8 through C_{14}) on an FBBR, with detection of the CAPB and potential metabolites performed by LC-ESI-MS. The degradation rates of the four homologs could be correlated to the length of the hydrocarbon chain, with higher alkyl homologs, i.e. the more lipophilic ones, being degraded faster. As such a relationship had been described for alkyl homologs of LAS (Swisher´s distance principle), FBBR samples from CAPB degradation experiments were screened for the presence of intermediates analogous to those

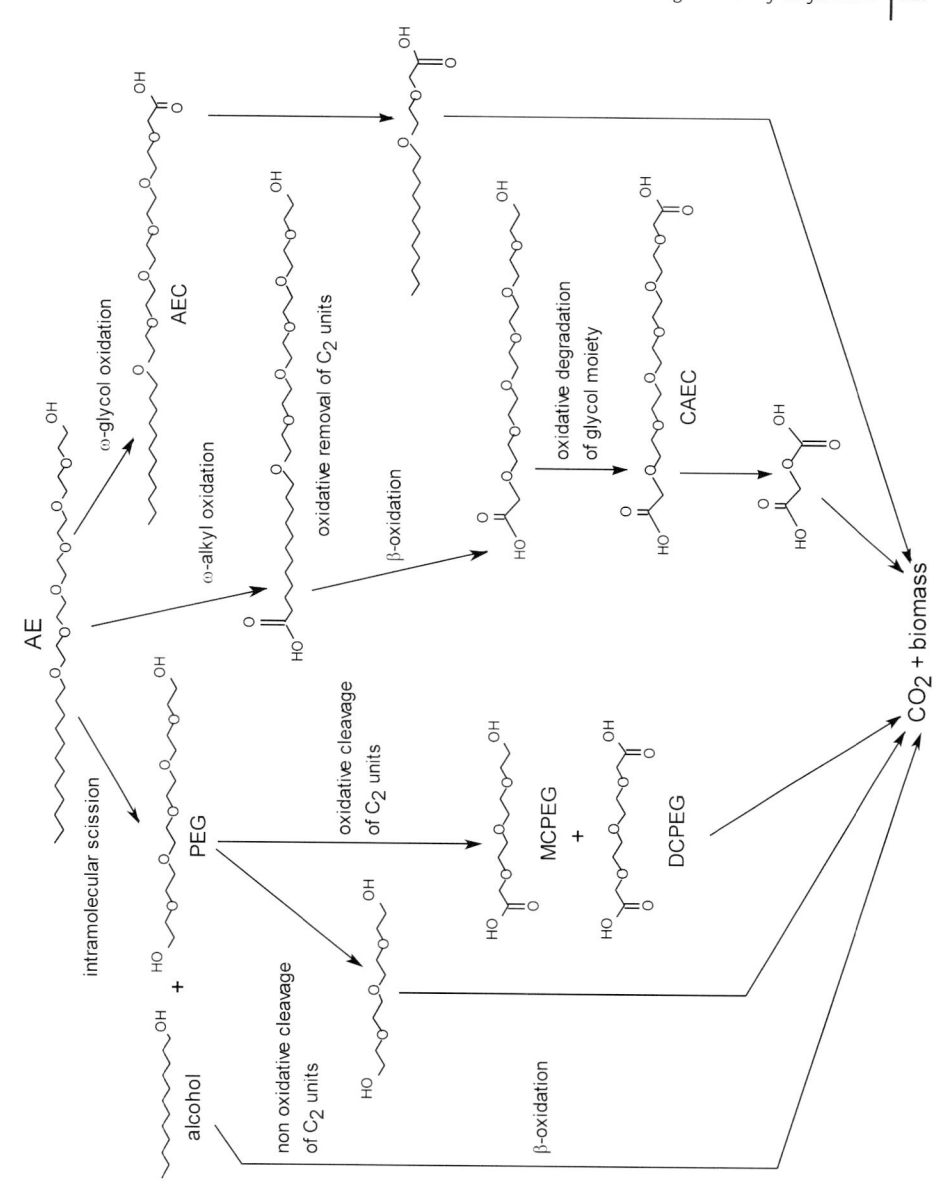

Fig. 9.9 Proposed biodegradation pathways for aliphatic alcohol ethoxylates under aerobic conditions (adapted from [78]).

formed by LAS breakdown. However, no carboxylated intermediates like the hypothetical C-even acids C_{14} through C_4 were observed, nor could any other biotransformation product be detected in the FBBR samples. It was hypothesized that the intact surfactant molecules were taken up by the cells and rapidly degraded intra-

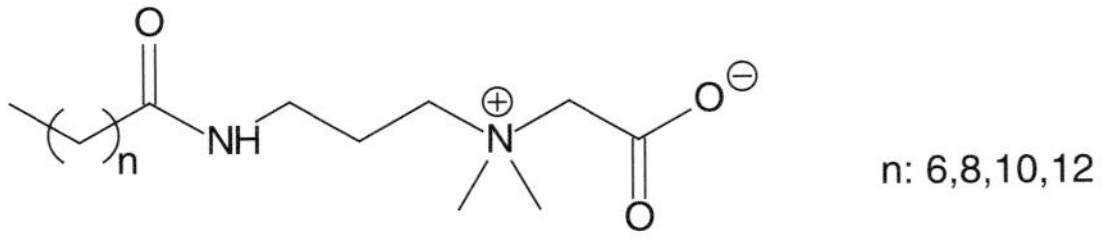

Fig. 9.10 General chemical structure of cocamidopropyl betaine.

cellularly without release of any metabolite. This assumption was supported by the results of Brunner et al. [79], who demonstrated high ultimate degradation of CAPB.

9.3
Surfactants and Metabolites in Wastewater Treatment Plants

Currently, under optimized conditions, more than 98% of surfactants can be eliminated by conventional biological wastewater treatment. Even if such high elimination rates are achieved, the principal problem is the formation of recalcitrant metabolites from the parent surfactants. Some compounds, e.g. APEO, are eliminated only incompletely and form metabolites that are more resistant to further degradation and more toxic than their parent compounds. Discharges of the sewage effluents into rivers and lakes and the disposal of sludge onto agricultural fields are likely to cause diffuse contamination in the aquatic environment. In order to reduce the concentration of surfactants and metabolites in the wastewater effluents and to increase and complete their elimination in the treatment process, different new biochemical and physical procedures are being investigated today. However, the majority of the WWTPs in Europe still employ physico-chemical and biological treatment. Very few data about the occurrence of surfactants in wastewaters and their elimination in WWTP are available with the exception of a few compounds such as LAS and APEO, which have been frequently and systematically analyzed over the past two decades.

9.3.1
Linear Alkylbenzene Sulfonates and their Degradation Products

The degradation of LAS, as it occurs in the environment and in WWTPs, yields the SPCs, mostly with six to ten carbon atoms in the side chain (Table 9.1). The biodegradation of LAS occurs already in the sewer system, as confirmed by the detection of SPCs in the influent samples of WWTPs. Three WWTPs in Spain were monitored for the occurrence of LAS and their degradation products [81]. The C_6- to C_{12}-SPCs were detected in raw influents in total concentration of 6–19 µg L^{-1}. Higher levels between 49 and 158 µg L^{-1} were found in the corresponding effluents after microbial formation from LAS. The most abundant species were short- to mid-chain intermediates (C_7- to C_9-SPCs). Di Corcia and Samperi [82] determined sig-

Tab. 9.1 Reported concentrations of LAS and SPC (values in parentheses) in influents and effluents of wastewater treatment plants (adapted from [141]).

Country	Influent (µg L^{-1})	Effluent (µg L^{-1})	Reference
Germany	1900–8000	65–115	142, 143
The Netherlands	3400–8900	19–71	144, 145
UK	15100	10	148
Spain	2020–2170 (6–19)	10–91 (49–158)	81
	988–1309	136–197	146
	1423	17.8	147
Italy	4600	68	148
	3140–8400 (35–304)	13–115 (80–260)	83
	3400–10700	21–290	106
		(220–1150)	82
	1850–5580	40–1090	149
United States	1800–7700	<1–1500	92, 150, 151

nificantly higher levels of SPCs (calculated as a sum of the various isomers and homologs of SPCs originating from both LAS and DATS) in effluent samples of an Italian WWTP. Concentrations ranged from 220 to 1,150 µg L^{-1}. In another study, monocarboxylated and dicarboxylated metabolites of LAS and DATS were separately monitored in three activated sludge WWTPs from the area of Rome [83]. The concentration of mono- and dicarboxylated SPCs in influents ranged from 35 to 304 µg L^{-1} and from 7 to 60 µg L^{-1}, respectively. The effluent concentrations ranged from 80 to 260 µg L^{-1} for SPCs and from 20 to 106 µg L^{-1} for SPdCs. The total concentrations of carboxylated metabolite of DATS ranged from 9 to 115 µg L^{-1} in influents and from 152 to 390 µg L^{-1} in treated wastewater.

9.3.2
Alkylphenol Ethoxylates and their Degradation Products

The APEOs are among the most widely used surfactants, although their environmental acceptability is strongly disputed. The widespread occurrence of APEO-derived compounds in treated wastewaters raises concerns about their impact on the environment following disposal of effluents into the aquatic system. Studies have found that their neutral and acidic metabolites are more toxic than the parent compounds and possess the ability to mimic natural hormones by interacting with the estrogen receptor [84]. These findings have raised public concern, and a voluntary ban on APEO use in household cleaning products began in 1995, with restrictions on industrial cleaning applications in 2000 [85]. In Western Europe and the United States, the APEOs in household detergents have been completely replaced by AEs. As a result, a general trend of declining APEO concentrations has been observed, especially in the Scandinavian countries, The Netherlands, Switzerland

and the UK. For example, concentrations measured in Swiss effluents prior to the voluntary ban on NPEO surfactants in laundry detergent formulations ranged from <10 to 200 µg L^{-1} (sum of NP, NPEO$_1$ and NPEO$_2$), while present-day levels (data from 1998) show a tenfold decrease [86]. However, mainly because of their lower price, APEOs are still being used in substantial amounts in institutional and industrial applications. De Voogt et al. [87] and Belfroid et al. [88] compared the levels of APEO in domestic and industrial wastewaters in The Netherlands. The average concentration of APEO in twelve samples of domestic wastewater collected in 1999 was 32.3 µg L^{-1}, while in three samples of industrial wastewater average concentration reached 7600 µg L^{-1}.

Wastewater-derived alkylphenolic compounds have been extensively studied. The concentrations of NPEO, as the strongly prevalent sub-group of APEO, determined in the influents of WWTPs (Table 9.2), varied widely among various WWTPs from less than 30 to 1035 µg L^{-1}. However, values went up to 22 500 µg L^{-1} in wastewaters stemming from industry, such as tannery, textile, pulp, and paper industries. Levels of octylphenol polyethoxylates (OPEO) are significantly lower, comprising approximately 5–15% of total APEO in WWTP influents, which is consistent with their lower commercial use.

NPEO metabolites, NP and NPEC, could be detected already in WWTP influents at levels of up to 40 µg L^{-1} (high values up to 250 µg L^{-1} were detected in industrial wastewaters) (Table 9.2). The presence of NP and NPEC in the influents was not only attributed to the biodegradation of NPEO in the sewer system but also to the discharge of NP and NPEC itself, from application in other fields. For example NP was also used as an ingredient of pesticide formulations and NPE$_1$C as a corrosion-inhibiting agent. A comprehensive monitoring of NP in effluents from eight WWTPs with activated sludge secondary treatment in the United States detected concentrations from 1.2 to 23 µg L^{-1} [89], while maximum concentration found in 40 WWTP in Japan did not exceed 1.7 µg L^{-1} [90].

The more polar NPECs are mainly discharged via secondary effluents, and concentrations reported were approximately 2–10 times higher than in primary effluents. Di Corcia et al. [91] determined the concentration of CAPEC in treated effluents of five major activated sludge WWTPs in Rome. Unexpectedly, and contrary to the general belief that NPECs are the refractory metabolites, the results showed that CAPECs were the dominant products of the NPEO biotransformation. The concentrations were as follows: 2.5–24 µg L^{-1} CA$_8$PEC, 0.3–2.0 µg L^{-1} CA$_7$PEC, 2.3–16 µg L^{-1} CA$_6$PEC, 0.3–2.1 µg L^{-1} CA$_5$PEC, 0.1–1.4 µg L^{-1} CA$_4$PEC and 0.1–0.6 µg L^{-1} CA$_3$PEC. By averaging data relative to five WWTPs over 4 months, relative abundances of NPEO (n_{EO}=1 and 2), NPEC, and CAPEC were found to be respectively 10±2%, 24±5% and 66±7%.

9.3 Surfactants and Metabolites in Wastewater Treatment Plants

Tab. 9.2 Concentration ranges (in μg L^{-1}) of NPEO, NPEC, and NP detected in influents and effluents of wastewater treatment plants (adapted from Ref. [141]).

Country	NPEO		NPEC		NP		Reference
	Influent	Effluent	Influent	Effluent	Influent	Effluent	
Sweden	–	0.5–27[h]	–	–		0.5–3	152
Denmark	30	<0.5	–	–	–	<0.5	153
	25	2.4	–	–	–	–	154
Germany	120–270	24–48	–	1.7–5.4	–	0.3–1	142
The Netherlands	<LOD–125	<LOD–2.2[d]	–	–	<LOD–19[dd]		
	50–22.500[e]				<LOD–40[e]	<LOD–1.5[d]	87, 88
Belgium	–	–	–	–	6 (122)[g]	<1.0	155
United Kingdom	–	1–60	–	–	–	<0.2–5.4	156
	–	45±16[f]	–	–	–	3±0.85	157
Switzerland	96–430	<0.1–35	–	4–16	–	<0.1–3.8	158, 159, 160, 161
Spain	33–820	<0.2–49	<0.2–14[a]	13–113	<0.5–22	<0.5–21	81, 162, 163
	27–2,120[b]	10–24	<0.4–219	13–33	17–251[b]	15–225[b]	147, 164
	140	2	–	–	58	0.65	165
Italy	–	–	–	–	2–40	0.7–4	82
	29–145	1.7–6.6	–	0.6–15	–	–	91
	127–221	2.2–4.1	–	–	–	–	166
US	–	<1.0–330	–		–		167
						<1.0–33	168
				142–272[i]			169
	–	160–460	–	–	–	11–23	170
Canada						0.8–15	171
Japan	–	–	<0.1–1119		<0.1–1.7		90, 172
	<0.1–60 (n_{EO}=1–3)						
	<0.1–245 (n_{EO}=4–18)						

[a] NPE$_1$C
[b] WWTP receiving >60% of industrial wastewater (mainly textile)
[c] WWTP receiving mainly industrial WW
[d] WWTP receiving domestic WW, data 1999
[e] Industrial WW(TP), data 1999
[f] NPEO$_1$
[g] effluent of a textile plant
[h] NPEO$_1$ and NPEO$_2$
[i] NPE$_{1-4}$C

9.4
Surfactants and Metabolites in Surface Waters

9.4.1
Linear Alkylbenzene Sulfonates and their Degradation Products

In a study investigating the impact of WWTP effluents on the concentrations of LAS-related compounds in surface waters [83], the spectrum of target compounds included (besides LAS and SPC) DATS (see Fig. 9.2b) and their carboxylated biotransformation products (DATS(d)C). Analytes were extracted from water samples by solid-phase extraction on graphitized carbon black cartridges and quantitatively determined by LC-(−)-ESI-MS. In samples collected 100 m upstream the treatment plant concentrations of LAS, SP(d)C, DATS and DATS(d)C were found to be 1.5, 0.84, 0,16 and 1.9 µg L^{-1}, respectively. Discharge of the secondary effluent containing 43, 184, 40 and 260 µg L^{-1} of each sulfonated compound resulted in elevated concentrations in the samples taken 100 m downstream the plant: 36 (LAS), 154 (SP(d)C), 33 (DATS) and 216 µg L^{-1} (DATS(d)C). For a more detailed description of the fate of SPC in the receiving waters, the authors distinguished between SPC with linear alkyl side chains and those carrying one methyl branch in the carbon chain (termed iso-SPC; refer to Fig. 9.2 for structure of iso-LAS). The latter species, eluting under RP conditions before their linear counterparts, largely predominated over SPC in the effluent samples, showing a mean number of carbon atoms of 7.6.

The studies conducted by Trehy et al. [92] on river waters likewise comprised the four surfactant-derived compounds LAS, SPC, DATS, and DATSC. Concentrations were determined in the two compartments upstream and downstream of ten United States WWTPs using either trickling-filter or activated-sludge treatment. On average, the samples collected upstream of the plants contained 16 µg L^{-1} LAS and 9.3 µg L^{-1} SPC, determined as their trifluoroethyl esters by gas chromatography-mass spectrometry. The values below the plant outlet were 35 and 31 µg L^{-1}, respectively. In the river water the mean level of DATS increased from 3.2 to 19 µg L^{-1}, while the concentration of their carboxylated degradation products rose from 19 to 55 µg L^{-1} after emission of treated sewage into the stream. Higher levels of DATSC in comparison to SPC were attributed to a slower biodegradation of the first, possibly due to the steric hindrance of the two-ringed structure.

9.4.2
Alkylphenol Ethoxylates and their Degradation Products

The monitoring studies conducted to assess the removal of APEO during wastewater treatment illustrated the capacity of treatment plants to efficiently reduce the surfactant loads in sewage, thereby reducing significantly the proportion of these compounds that reach surface waters. In view of the large-scale usage of APEO and its subsequent discharge with domestic and industrial sewage water, this is of utmost importance for protecting aquatic life forms in receiving water bodies. Residues of APEO and their degradation products, however, are detectable in the receiving rivers [93, 94, 95, 96, 97, 98]. In samples from the river Rhine taken in July 2001, the concentrations of NPEO were invariably below 0.1 µg L^{-1}, whereas NP was present at levels between 0.17 and 0.30 µg L^{-1}. NPEC concentrations were significantly higher than both NPEO and NP, with values up to 1.1 µg L^{-1}. In the river Meuse, the NPEO and NP concentrations varied between 0.07 and 0.34 µg L^{-1} and 0.13 and 0.38 µg L^{-1}, respectively. Thus, they fell within the same range as those observed in the river Rhine. Detected levels of NPE$_{1-6}$C ranged from less than the LOD up to 3.4 µg L^{-1}, with NP$_3$EC – NP$_6$EC being significantly higher than NP$_1$EC and NP$_2$EC [99, 100].

9.5
Surfactants and Metabolites in Subsoil/Underground Passage

Direct infiltration of (treated) wastewater through sand beds and infiltration of sewage water from laundry ponds or septic systems are among the major contamination sources of surfactants into subsurface and groundwaters. Most studies published on this issue were released by research groups from the United States, where these treatment techniques have found some use.

9.5.1
Alkylbenzene Sulfonates and their Degradation Products

Thurman and colleagues [101] determined LAS and ABS in groundwater samples collected 500 and 3000 m downfield of sewage infiltration ponds. Quantitative analysis of the groundwater samples showed that ABS was present in the 3000 m well (downgradient) at 2300 µg L^{-1}, thereby accounting almost quantitatively for the MBAS found (2500 µg L^{-1}). LAS was not detectable in the sample from this well, but was determined at 300 µg L^{-1} in the 500 m well (MBAS: 400 µg L^{-1}), where ABS in turn was absent. The detection of the alkyl-branched isomers in the 3000 m well samples gave indications that the groundwater contamination was older than 1965 (the last year of domestic ABS use), i.e. the poorly biodegradable compound had survived more than 20 years in the aquifer.

A field study was conducted by Larson et al. [102] to characterize the impact of effluent discharges to a sandy soil about 0.5 m below the surface. A 2.5 m thick unsaturated zone and a 3–4 m thick unconfined sand/gravel aquifer underlay this field. LAS concentrations in the effluent plume decreased over a distance of 10 m from 10 000 to 30 µg L^{-1}. A further object of study was a laundromat pond exposed to LAS-containing sewage for more than 25 years. A clay layer separated the natural pond from the vadose zone made up of porous sand. Measurements of LAS levels as function of soil depth beneath the pond showed a rapid decrease from about 220 mg kg^{-1} at 30 cm to approximately 20 mg kg^{-1} at 2.0 m. At a depth of 3.6 m, no LAS surfactant was detectable (LOD not specified). In the near groundwater, LAS was likewise absent. Overall, the authors concluded from their outcomes that biodegradation was the principal factor limiting the accumulation of LAS in soil and groundwater compartments.

Field and co-workers [103] identified persistent anionic surfactant-derived chemicals in groundwater contaminated by rapid infiltration of secondary effluents through sand beds. The infiltration had caused a 3500 m effluent plume in a shallow water table aquifer. The target analytes comprising LAS, SPC, DATS, and their carboxylated metabolites (DATSC) were isolated from groundwater samples by ion exchange, and the extracts obtained were subsequently purified by desalting and liquid-liquid extraction, yielding one fraction containing LAS and DATS, the other SPC and DATS. The groundwater concentrations of LAS (C_{10} to C_{13}) and SPC (C_3 to C_{13}) were estimated to be 3±1 and 8±8 µg L^{-1}, respectively. The levels of DATS and DATSC, for which authentic standards were not available, were determined on the basis of C_4-SPC and C_9-LAS (assuming equivalent response factors) giving 5±1 and 27±3 µg L^{-1}, respectively. In comparison with the LAS concentration in the sewage effluent, the value decreased by 97% until the groundwater well was reached, located at about 500 m downgradient from the infiltration beds. The LAS removal was attributed to both sorption and biodegradation occurring during infiltration through the vadose zone, but not during transport in the groundwater. During the infiltration/groundwater transport, a preferential removal of higher alkyl chain homologs as well as of external phenyl isomers was detected. The LAS metabolites in turn showed a similar homolog pattern in the sewage effluent and the groundwater, dominated by the mid-chain length C_5 to C_8. The higher ratio of DATS to LAS in groundwater (1.7) compared with the ratio determined in sewage effluent (0.2) indicated the preferential breakdown of LAS in the subsurface soil. The survival of LAS and its derivatives in the groundwater was assumed to be due to unfavorable biogeochemical conditions (such as nutrient limitations, below-threshold concentrations, or decreased microbial populations). On the assumption that no retardation through adsorptive effects occurred during infiltration/groundwater transport, a residence time for LAS in the groundwater well was calculated to be between 2.7 and 4.6 years.

In a following study [104] conducted on the same contamination site, the fate and transport of alkylbenzene sulfonates and DATS were investigated. Groundwater samples collected from a longitudinal transect of wells along the effluent plume were analyzed after concentration on SPE-C_2 and clean-up on SAX cartridges by HPLC-FL and FAB-MS. In the groundwater samples, LAS occurred at levels below the LOD established by HPLC-FL (10–20 µg L^{-1}), but traces could be identified in selected samples applying FAB-MS. Determination of DATS residues in the groundwater samples gave levels likewise in the range 10–20 µg L^{-1}. Comparison of the DATS homolog distribution in sewage effluent and groundwater indicated a shift to the shorter alkyl chain length likely attributable to increased partitioning or biodegradation of the more hydrophobic components during infiltration/groundwater transport. In the sample taken from the well located 3000 m downgradient from the infiltration point, 2100 µg L^{-1} ABS was measured by HPLC-FL, representing 90% of MBAS (2330 µg L^{-1}). Its presence corresponded to a groundwater contamination that had occurred prior to 1965.

The goal of a further study [105] was to determine the transport and biodegradation behavior of LAS under *in-situ* conditions in a sewage-contaminated aquifer by conducting natural gradient tracer tests. A technical mixture of the anionic surfactant was dosed into two geochemically distinct zones of the sand and gravel aquifer: into the oxic zone characterized by high dissolved oxygen (about 8 mg L^{-1}) and into the transition zone with only 1 mg L^{-1} dissolved oxygen. At two monitoring wells situated at 4.6 and 9.4 m downgradient from the injection well, samples were withdrawn for chemical analysis of LAS and its biotransformation products. Breakthrough curves were recorded and compared with that of the unretained tracer bromide, showing that LAS almost co-migrated with bromide in the transition zone, but was retarded relative to inorganic anion in the oxic zone. Moreover, the composition of the breakthrough front was enriched by the shorter alkyl chain homologs, which was attributed to sorptive processes in this zone, since hydrophobic interactions of LAS with organic soil matter preferentially slowed down the transport of the higher homologs. The nearly unretained migration of the LAS homologs in the transition zone in turn correlated well with the low organic carbon content in the aquifer sediment (<0.01%). As for the occurrence of biodegradation intermediates, namely SPC, no such compounds were detectable in the oxic zone (proved by GC-MS analysis of trifluoroethylated groundwater extracts). The absence of metabolites in conjunction with an unchanged LAS homolog/isomer pattern suggested that no biodegradation took place in the oxic zone. In contrast to the observations in the oxic zone, SPC could be identified in the transition zone. Their appearance went hand in hand with a decline of dissolved oxygen (down to 0.1 mg L^{-1}) associated with the oxidative conversion of LAS into the carboxylated metabolites. The total mass of metabolites determined was almost equal to that of the mass of total LAS lost. During the 40-day experiment, the formed SPCs appeared to be stable and were not completely mineralized. The differences in biodegradability of LAS in the two aquifer zones were assumed to be related to the exposure of indigenous microbial communities to the surfactant. Apparently, microorganisms in the oxic zone had not been previously exposed to sewage-contaminated groundwater, whereas the occurrence of LAS residues in the transition zone had led to a bacterial acclimatization. However, once the dissolved oxygen in the water was depleted, both LAS and its biotransformation products seemed to persist.

In an area where domestic wastewater was discharged through cesspools, Crescenzi et al. [106] were able to identify LAS and DATS in a groundwater sample. In the groundwater samples taken from a 20-m deep well, the LAS concentration amounted to 0.3 µg L^{-1}, while the DATS level totaled 0.9 µg L^{-1} (LOD 0.1 µg L^{-1}). Of the four DATS homologs determined (C_{10}–C_{13}), the lightest one was identified as the most abundant species. The change in the homolog pattern as compared to the one typical of wastewater was assumed to be due to a preferential removal of the longer-chain homologs as a consequence of stronger sorption onto soil particles and/or faster biodegradation.

The sorption and transport of LAS and the cationic DTDMAC were determined in the upper soil horizons (78–97% sand, 0.03–0.5% C_{org}) and in the aquifer (95–99% sand, 0.04–0.50% C_{org}) below a septic tank tile field [107]. Monitoring results showed that both LAS and DTDMAC were substantially removed during transport in the unsaturated zone. The LAS level dropped from 13 850 to 1640 µg L^{-1} within 0.5 m from the surface and was further reduced to <10 µg L^{-1} before entering the groundwater at a depth of 2.5 m. The concentrations of dissolved DTDMAC decreased 25-fold from initially 4570 to 195 µg L^{-1} at 0.5 m and were below the LOD of 5 µg L^{-1} upon entering the groundwater. In the tile field gravel an LAS concentration of 21 mg kg^{-1} was determined, which decreased to nondetectable levels 5 cm beneath the gravel bed. Somewhat higher values were found for DTDMAC, reflecting its higher sorption tendency: about 60 mg kg^{-1} in the field gravel, 10 mg kg^{-1} at 7.5 cm below the surface, and less than 1 mg kg^{-1} at 10 cm depth. Measurements of sorption coefficients gave K_d values for DTDMAC one order of magnitude higher than those for LAS. Experiments carried out on a series of core samples collected throughout the leachate plume indicated that sorption of LAS was controlled by organic carbon and clay content in the upper soil layers. No clear correlation between soil composition and sorption of DTDMAC could be established. With respect to the elimination of the cationic surfactant in the upper soil horizons, the stronger retention was believed to allow adequate time for biodegradation, as DTDMAC had been shown to degrade more slowly than LAS.

The behavior of SPC during drinking water production was investigated by Eichhorn et al. [108] in two waterworks, one of which was performing artificial groundwater recharge of surface water previously submitted to flocculation and granular active carbon filtration. The analytical method for the determination of C_5- to C_{12}-alkyl homologs comprised isolation of the sulfonates on RP-C18 material at pH 3.0 and subsequent ion-pair HPLC separation with mass spectrometric detection. Whereas the total SPC concentration amounted to 0.25 µg L^{-1} in the water samples collected before infiltration into the subsoil, a level of 0.12 µg L^{-1} was determined in the water from the recovery well located at a distance of about 170 m from the infiltration well, which corresponds to a travel time of 4–6 weeks. Although the samples, taken in triplicate over a period of 6 weeks, did not correspond to each other, the results indicated that the very polar degradation products of LAS were quite resistant during the subsoil passage.

9.5.2
Alkylphenol Ethoxylates and their Degradation Products

Rudel and co-workers [109] identified alkylphenols and other estrogenic phenolic compounds in groundwater impacted by infiltrations of secondary effluents and septage from septic tanks. The area was characterized by sand, gravel, and boulders deposited by glaciers. The monitoring wells with depths of between 12–24 m were located about 150–275 m downgradient of the infiltration beds. No wastewater had been discharged to these beds during the six months before sample collection. The analytical procedure comprised liquid-liquid extraction with dichloromethane and either NP-HPLC-UV determination (OP/NPEO$_{1-6}$) or GC-MS analysis after silylation (NPEO$_1$, NPEO$_2$, and NPE$_1$C). The sum concentration of OP/NPEO in the untreated wastewater and the septage samples ranged from 1350 to 11 000 µg L^{-1}. After treatment, the water contained 5.5 µg L^{-1} of NPEO$_1$, 0.8 µg L^{-1} of NPEO$_2$ and 42 µg L^{-1} of the carboxylated metabolite NPE$_1$C. In one of two samples collected, the tetra- to hexaethoxylates were found at levels between 12 and 48 µg L^{-1}. In four groundwater samples analyzed, OP/NPEO$_2$ was detected at levels between 14 and 38 µg L^{-1}, while traces of OP/NPEO$_3$ and OP/NPEO$_4$ occurred occasionally in the low µg L^{-1} range (4–5 µg L^{-1}, below the limit of quantification of the HPLC method). Because of the limited number of samples examined, no firm conclusions could be drawn in view of the source of contamination (treated effluent or septic systems), but the findings suggested that further studies were required to study the impact of septic tanks on groundwater serving as drinking water supplies.

The fate of NPEO was examined in two household septic systems of different design and hydro-geological settings [110]. The first site (A) consisted of clay-dominated glacial till of low permeability, while the second (B) was characterized by sand and gravel of high permeability. At both locations, the groundwater table was 10 m below the land surface. After a one-year period of discharging NPEO-containing sewage to the septic tanks, the tank effluent from site A contained about 15 000 µg L^{-1} NPEO of which some 5000 µg L^{-1} corresponded to the two short-chain ethoxymers NPEO$_1$ and NPEO$_2$. In tank B effluent, the sum concentration of NPEO$_{1-2}$ was about 900 µg L^{-1}. The higher oligomers NPEO$_{3-16}$ totaled 4000 µg L^{-1}. Analysis of the septage leachate of site A showed an NPEO reduction between 94.3 and 99.6%. NPEC levels were found in the range from 148 to 1340 µg L^{-1}. In water samples taken from lysimeters beneath the leach field laterals, only traces of NPEO were detectable (<LOD to 2.7 µg L^{-1}). Analysis of selected samples for NPEC gave values of <LOD to 0.5 µg L^{-1}. In all analyzed groundwater samples, NPEO-derived compounds were absent. The same held true for the occurrence of NPEC in groundwater from site B, which was *a priori* more permeable for infiltrating effluents. The non-ionic surfactant was occasionally detected at sub-µg L^{-1} levels in waters from the observation wells, but no clear evidence was obtained that these originated from subsoil leaching (instead laboratory contamination was suspected).

Zoller [111] discussed the influence of activated sludge/aquifer treatments on the quality of reused groundwater. After secondary sewage treatment and intermittent flooding of the effluent in spreading basins, the water was infiltrated through the saturated zone to recharge groundwater supplies. In the surroundings of the recharge area, recovery wells pumped water from the aquifer, which was used for agricultural irrigation. Despite an efficient removal of non-ionic surfactants during the wastewater treatment (including polishing ponds), reducing the level on average from 7500 to 300 µg L^{-1}, between 22 and 25 µg L^{-1} were still found in the reclaimed water. In view of the presence of non-ionic surfactants in reused water, which represented an issue of environmental and possibly health risk concern, the author stressed that the use of reclaimed effluents for irrigation and for aquifer recharge should be carried out with caution.

Zoller and co-workers [112] conducted a study aimed at determining the extent to which Israel's groundwater in the vicinity of sewage-polluted surface waters was affected by infiltration of NPEO-contaminated water. The wells with depths of

50–156 m were located at distances of between 40 and 300 m from the riverbank of two investigated rivers. Colorimetric analyses of non-ionic surfactants in surface and adjacent groundwaters indicated their insufficient removal in the subsurface environment. In comparison with the levels determined in the streams (1600 to 2600 µg L^{-1}), the groundwaters contained between 120 and 780 µg L^{-1} non-ionic surfactant, representing elimination rates of 92 to 54%. The data gathered revealed an inverse relationship between the distance of a given well from the stream and the surfactant concentration measured. The authors concluded that neither naturally occurring biodegradation nor adsorption processes onto soil were efficient enough to prevent contamination of groundwaters.

In subsequent work, Zoller [113] aimed at mapping the country's surface, sea, and groundwater according to their non-ionic surfactant content. A further objective was to establish the connection between surface input and output into receiving surface water, seawater, and groundwater as a function of the local physicochemical and hydro-geological parameters. In water wells sampled during two campaigns, the non-ionic surfactant concentrations ranged from 22 to 1800 µg L^{-1} (n=8), confirming the findings of the previous study. It was suggested that surface and subsurface pollution by poorly biodegradable surfactants was not a specific point problem, but a general contamination problem in Israel.

The occurrence of alkylphenol carboxylates (APEC) as an indicator of anthropogenic impact on wastewater was determined by Fujita and Reinhard [114] in groundwater artificially recharged with tertiary treated effluents. Treated wastewater from two treatment trains applying different technologies contained sum concentrations of APEC, including their brominated analogs, of 5.1 and 0.5 µg L^{-1}. After blending of the effluents with local groundwater, taken from a depth of more than 260 m, at a ratio of 1:1, the water was directly injected into the aquifer. In a sample from a monitoring well located about 15 m downgradient of the infiltration site, the APEC level amounted to 7.0 µg L^{-1}. Preliminary analytical data from the nearest production well (about 500 m downgradient) indicated the presence of APEC at an estimated level of 1 to 10 µg L^{-1}. The findings suggested that APEC persisted in groundwater during the 1–2 months travel from the injection location to the monitoring well and was apparently not eliminated during transport to the production well, corresponding to a residence time of 2–3 years.

Schaffner and colleagues [115] presented preliminary results on the behavior of short-chain NPEO during bank filtration of river water into groundwater, being a first step in the treatment of river water for public water supplies. The water from the small heavily polluted stream (Glatt River), which upstream received discharges from several municipal wastewater treatment plants, infiltrated into a quaternary fluvioglacial valley fill aquifer. The concentrations of NPEO$_1$ and NPEO$_2$ were monitored during one year in samples (n=17) from observation wells located at various distances from the riverbank (2.5 to 14 m). The levels of NPEO$_1$ and NPEO$_2$ in the river water amounted on average to 7.5 and 8.2 µg L^{-1}, respectively. With increasing distance of the observation well from the bank, the levels of both ethoxymers decreased: the mean value for NPEO$_1$ dropped from 1.0 (2.5 m) to 0.1 µg L^{-1} (14 m), while NPEO$_2$ exhibited a slightly better removal during the first

2.5 m (0.4 µg L^{-1}) of the subsoil passage. In the sample taken from the 14-m well, NPEO$_2$ was detected at 0.1 µg L^{-1}.

In subsequent work [116] the same research group presented a more detailed description of the fate of persistent NPEO-derived metabolites during infiltration of contaminated river water into groundwater. The concentrations found in the river and the adjacent groundwater showed a temporal and spatial variation. Mean levels in the groundwater were considerably lower, indicating efficient removal of nonylphenolic compounds during most of the year. On average, the concentrations of NPEO$_1$ and NPEO$_2$ in the river were 7.8 and 8.4 µg L^{-1}, respectively. After reaching the first well at 2.5 m from the bank, the levels were 0.91 µg L^{-1} for NPEO$_1$ and 0.33 µg L^{-1} for NPEO$_2$. During the further underground passage, the values dropped to 0.04 and <0.01 µg L^{-1}, respectively. In contrast to the substantial removal of the ethoxylates in the section between the river and the first observation well, the carboxylates NPE$_1$C and NPE$_2$C showed a much slower elimination. Upon reaching the first well, the concentrations had been reduced on average from 14.7 to 10.9 µg L^{-1} for NPE$_1$C and from 24.7 to 22.4 µg L^{-1} for NPE$_2$C. In the samples collected from the 14-m well, mean values of 4.5 and 5.1 µg L^{-1} were obtained for the carboxylates. The higher mobility compared to those of NPEO was attributed to a higher water solubility and their resistance to biodegradation under aerobic conditions. As for the occurrence of NPEO in the water from the pumping station situated at 130 m from the river, both NPEO$_1$ (<0.1 to 3.4 µg L^{-1}) and NPEO$_2$ (<0.1 to 1.7 µg L^{-1}) were detectable at trace levels. Examinations of seasonal variations in the groundwater sampled from the 2.5-m well showed a trend of highest levels in winter and lowest in summer. These findings correlated well with the general trend of concentration changes in the river.

In a field study, Ahel [117] investigated the infiltration of organic compounds from a heavily polluted river and a wastewater canal into the alluvial aquifer. Sampling wells were situated at distances of between 3 and 120 m from the surface waters. NPEO-derived metabolites were among the most abundant contaminants in the analyzed groundwaters. Measurements of NPEO$_1$, NPEO$_2$, and NPEC indicated relatively efficient removal during infiltration, reducing the concentrations of 0.4, 0.2, and 5 µg L^{-1}, respectively, found in the river to <0.1 µg L^{-1} for NPEO and to 0.05 µg L^{-1} for NPEC in the well located 120 m from the bank. On the other field site situated adjacent to the wastewater canal, the elimination efficiencies during a short underground passage (well 3 m from the river) ranged from 98.3 and 99.8%. Residual concentrations of NPEO$_1$, NPEO$_2$, and NPEC in the groundwater amounted 0.078, 0.01, and 0.005 µg L^{-1}.

9.5.3
Perfluorinated Surfactants

It was only recently that the use of aqueous film-forming foams (AFFF) containing perfluorinated surfactants was recognized as a source of groundwater pollution.

One of the first groundwater contaminations with perfluorinated surfactants was reported by Levine et al. [118] at a fire training site which had served for approximately eleven years for fire-fighting exercises using water and AFFF. The site was characterized by a very shallow groundwater table at depths ranging from 0.6 to 3 m below the land surface. The aquifer consisted of clean, fine-grained quartz and clayed sandy soils. The groundwater samples were extracted with diethyl ether employing liquid-liquid extraction and derivatized with diazomethane to yield the methyl esters. Analysis was performed by GC-MS. At a distance of about 45 m downgradient of the fire training pit, the surfactant concentration reached a maximum level of around 11500 µg L^{-1}, indicating surfactant transport in the aquifer.

A more in-depth description of the extent of groundwater pollution with perfluoroalkanoic acids (PFAA) was given by Moody and Field [119] in a study conducted at the same site [118]. For chemical analysis, the samples were concentrated on SAX disks and the extracts methylated with methyl iodide. GC-MS operated in electron impact mode was employed for quantitative measurements of the methyl esters (LOD: 18 µg L^{-1}), while confirmation of the identity of the target compounds was achieved by electron capture negative ionization GC-MS. The water samples collected from three wells contained total PFAA concentrations varying between 124 and 298 µg L^{-1}, with the highest levels in wells located in the close vicinity of the fire training pit. The two major components were C_6-PFAA (46–52%) and C_8-PFAA (34–40%); the odd homolog C_7-PFAA accounted for the rest. The authors had performed studies on a second field site, likewise used for fire-training activities. Monitoring wells were situated within a radius of 120 m from the fire pit, where the groundwater table was between 2 and 3 m below ground surface. The highest PFAA concentration (7090 µg L^{-1}) was measured in the groundwater near the burn pit, whereas in a well located downgradient of this a concentration of 540 µg L^{-1} were determined. On average, 89% of the total PFAA was C8-PFAA. As information on the exact chemical composition of surfactants used in AFFF formulations was not available, the authors discussed, as the origin of the detected PFAAs, their presence as active ingredients or by-products in AFFF mixtures or their formation during combustion or through (bio)chemical degradation. In any case, the PFAA concentrations found at both field sites confirmed the poor biodegradability of perfluorinated surfactants deriving from AFFF, which was attributed to the strength of the carbon-fluorine bond and the rigidity of the perfluorocarbon chain.

9.6
Surfactants and Metabolites in Drinking Waters

9.6.1
Alkylphenol Ethoxylates (APEO) and their Degradation Products

In view of the fact that in The Netherlands the drinking water supply is largely dependent on water from the rivers Meuse and Rhine, the quality assessment of these rivers is a long existing goal of RIWA (Association of River Waterworks, The Netherlands) and the associated water companies, with permanent monitoring for a variety of organic micropollutants in place [120, 121]. During a sampling campaign carried out in 1999, APEO and degradation products such as NP were determined in raw, process, and drinking waters for the first time [122]. Of the compound groups analyzed, only the NPEO were found above the limit of quantification (LOQ). NPEO were found above LOD in 5 out of 23 samples of process water; in these five, levels ranged from 1.0 to 3.8 µg L^{-1}. In 4 out of 22 drinking water samples NPEO were found above the LOD, ranging from 1.2 to 2.1 µg L^{-1}. OP and OPEO were invariably below the LOD, and NP was also generally below the LOD with the exception of two samples of drinking water, where the levels found were between the LOD and LOQ (0.2–0.5 µg L^{-1}). The values observed were consistent with the generally low levels of these compounds found in the rivers Rhine and Meuse. A clear distinction between the values found in the river Meuse and the Rhine river basin was not observed. Analysis of samples from the Rhine and Meuse basins in the same period showed that levels in general were below the LOD, with occasionally detectable values in the range 1.0–3.5 µg L^{-1} observed [122, 123]. These findings were confirmed in a study by Jonkers et al. [124], in which total dissolved APEO concentrations in the downstream area of the Rhine river of between 0.2 and 0.9 µg L^{-1}, and concentrations of NP averaging 0.1 µg L^{-1} with little variation, were reported.

In United States drinking water, Sheldon and Hites [125] reported in the 1970s that levels of NP were below the LOD, whereas OP was found at a level of 0.01 µg L^{-1}. Clark et al. [126] in 1992 reported levels of 0.08 (NPEO), 0.15 ($NPEO_2$), and 0.5 ($NPEO_{3-7}$) µg L^{-1}, respectively. $OPEO_2$ was found at a level of 0.002 µg L^{-1}. Clark et al. were also the first to describe, in addition to APEO, determinations of NPEC in drinking water samples. They reported values of 0.16 (NPE_2C), 0.06 ($NPE_{3-7}C$), 0.04 (OPE_2C), and 0.012 ($OPE_{3-4}C$) µg L^{-1}, respectively. At Cape Cod, U.S.A., a study was carried out by Rudel et al. [109] to investigate the impact of septic systems as a source of APEOs and their degradation products in groundwater. In this study, NP was detected in all sewage samples at concentrations above 1000 µg L^{-1}. In groundwater downgradient of an infiltration bed for secondary treated effluent, NP, OP, and ethoxylates were present at about 30 µg L^{-1}. NPE_1C and $NP/OP(EO)_4$ were detected in some drinking water wells at concentrations ranging up to 33 µg L^{-1}.

Tap water samples from the City of Barcelona, supplied by a waterworks withdrawing its raw water from the highly polluted Llobregat river, were analyzed by Guardiola et al. [127]. They reported maximum levels of 0.14 µg L^{-1} for NP, 1.1 µg L^{-1} for NPEO and 0.25 µg L^{-1} for $NPEO_2$. The same research group showed that during breakpoint chlorination of the raw water, formation of brominated NP was observed [128, 129]. This was ascribed to high levels of bromide ions together with the presence of NP in the surface water. In the tap water of Barcelona, monobrominated NP (BrNP) was found in later investigations. In more recent in-depth investigations on this issue [130], a series of chlorinated and brominated by-products of alkylphenolic surfactants and their degradation intermediates formed during chlorine disinfection of raw water were identified using mass spectrometry

9.6.2
Behavior of Sulfophenyl Carboxylates during Drinking Water Production

In a study conducted by Eichhorn et al. [108], the Barcelona waterworks located at the Llobregat river was compared with a drinking water facility on the River Rhine in Germany in terms of their capacity to eliminate SPC employing different treatment technologies. The measured concentrations in the raw waters averaged 5.0 and 1.8 µg L^{-1} in the Llobregat river and the river Rhine, respectively. In the Spanish waterworks, neither prechlorination nor flocculation followed by rapid sand filtration had an impact on the elimination of SPC. After ozonation and blending with groundwater, an SPC reduction of about 50% was achieved, which was in part attributed to the mixing. No significant removal of SPC was observed during the subsequent stage of granular activated carbon (GAC) filtration. In the final step of the water treatment process at the Llobregat waterworks, disinfection of the water with chlorine was performed. This, however, did not lead to any measurable removal, resulting in SPC levels in the finished drinking water of about 2 µg L^{-1}.

Different behavior was observed at the Rhine waterworks, with much better reduction of the SPC burden. While flocculation with iron(III) chloride did not reduce the SPC concentration, in accordance with the results obtained for the flocculation process at the Llobregat waterworks, a marked reduction was observed during rapid sand filtration, during which the SPC level dropped from 1.8 to 0.25 µg L^{-1}. It was suggested that the specific elimination of SPC was brought about by microorganisms residing on the sand grains. Such an impact was not observed during the corresponding step in the Llobregat waterworks, which was hypothesized to be because of differences in the contact times (about 10-fold less in the latter) and also because prechlorination of the raw water carried out in the Spanish waterworks most likely slowed down (if not completely inhibited) biological processes in the subsequent treatment stages. The next stage in the Rhine water treatment, the GAC filtration, did not result in any measurable reduction of SPC. After the subsequent subsoil passage of about 180 m length, the concentration of SPC amounted to 0.12 µg L^{-1}. A nearly complete elimination of the SPC homologs was then accomplished by slow sand filtration. The residues detected in the water after this step and after the final disinfection with chlorine dioxide were found to range well below the quantification limit of the applied analytical method (0.03 µg L^{-1}). Comparing the capacities of both waterworks with respect to SPC elimination, it was evident that the polar pollutants exhibit a high potential to escape the purification processes. Only the sand filtration steps, with their supposed biological activity, contributed significantly to reducing the SPC concentrations.

9.7
Risk Assessment

9.7.1
Linear Alkylbenzene Sulfonates and their Metabolites

Concentration ranges reported for the occurrence of LAS in fresh and marine waters are frequently found in the low to medium $\mu g\ L^{-1}$ range [131]. Values in the high $\mu g\ L^{-1}$ range are determined at locations where sewage treatment is either inadequate or lacking. In comparison to the PNEC of 270 $\mu g\ L^{-1}$, calculated by HERA (Human & Environmental Risk Assessment on ingredients of European household cleaning products) [132], most environmental concentrations are below this threshold value, indicating that LAS residues routinely detected in rivers or coastal waters do not pose a risk to aquatic organisms. With regard to the PNEC of sediment-bound LAS (8.1 $mg\ kg^{-1}$ according to HERA [132]), the concentrations are generally below this value [17, 20–22, 32–34] provided that there is efficient removal of LAS during wastewater treatment prior to discharge into the receiving bodies of water. Average concentrations well above the PNEC were reported for sampling sites in the proximity of wastewater treatment plant (WWTP) effluents using trickling-filter (TF) technology [131]. Elevated LAS levels were also determined in sediment samples in various areas of Spain, where coastal regions were strongly impacted by industrial and urban activities [131]. At such hot spots, there must be concern about possible adverse effects on benthic organisms.

The ecotoxicity of SPC to aquatic organisms (fish and *Daphnia*) was explored, showing that the LC_{50} was increased by a factor of 200 to 300 through carboxylation of one terminal methyl group of the alkyl chain and by another 10 to 20 times upon shortening the alkanoate chain from 11 to 7 and then to 5 [133]. This was attributed to the fact that the toxicity of LAS is directly associated with its surface-active properties, since oxidation of the alkyl chain leads to the loss of surface activity. This is in line with the findings of Moreno and Ferrer [134] monitoring the toxicity towards *Daphnia magna* during biodegradation of various LAS blends. The metabolites formed during this process were far less toxic than the parent molecule irrespective of the mean molecular weight of the type of LAS assayed. The residual material from trickling-filters studies of Kölbener et al. [135] showed no detectable impact on the growth of the algae *Selenastrum capricornutum* and the mobility of *Daphnia magna*. As for the ability of SPC to disturb the endocrine system of aquatic organisms or humans (the latter exposed to SPC upon consumption of drinking water containing traces of the carboxylates), no evidence of such effects were found. In the recombinant estrogen yeast screen, Routledge and Sumpter [136] proved that the range of SPC homologs assayed were not estrogenic. These findings were corroborated by Navas et al. [137], who used the vitellogenin assay. The two SPC homologs C5 and C11 did not induce any positive response in the tests.

9.7.2
Alkylphenol Ethoxylates and their Metabolites

Several countries and agencies have performed risk assessments for alkylphenols (AP) and APEO based on different approaches. Perhaps the most elaborate has been the assessment carried out by Environment Canada [138]. In this document, three different approaches were used after an extensive literature study of effects documented and environmental concentrations measured in the Canadian aquatic environment. Thus, based on the most sensitive endpoint found in the literature (LC_{50} of nonylphenol (NP) for winter flounder: 17 µg L^{-1} [139]), and applying an uncertainty factor of 100, a predicted no-effect concentration (PNEC) of 0.17 µg L^{-1} was derived for NP. Analogously, no-effect concentrations (NEC) were derived for NPEO and NPEC, these being 1.1 µg L^{-1} for both $NPEO_1$ and $NPEO_2$ and 9.9 µg L^{-1} for $NPEO_1$. The risks of NP have been evaluated in a recent risk assessment report from the EC states [140]. According to this document, currently there exists no limit value for NP in the European Union. The document provides a PEC from model calculations at the regional scale of 0.6 µg L^{-1}, as well as a PNEC of 0.33 µg L^{-1}. Hence, for NP the risk quotient (PEC/PNEC) amounts to 1.8.

9.8
Conclusions

Following their use in aqueous systems, today's surfactants often reach wastewater treatment plants (WWTPs) in extreme amounts, where they are almost completely eliminated. But still, particularly the more persistent compounds and their metabolites can be detected in raw water that is treated for use as for drinking water. Today, this does not happen as often as it did in the past. Nevertheless, some surfactants can still be found in relatively remote sediments across Western Europe and in effluents from WWTPs in relatively high concentrations. The broad variety of their chemical structures combined with their excellent water solubility, their surface-active nature, and the persistence of some of their known metabolites make them a group of environmental pollutants that need to be addressed with high priority.

In freshwater and marine surface waters in Europe, mainly in the Mediterranean countries, concentrations of nonylphenol (NP) still exceed the PNEC of 0.33 µg L^{-1}. As proven for Germany and The Netherlands stronger regulation combined with state-of-the-art wastewater treatment could well solve these problems. At present, alkylphenol polyethoxylates (APEO) are no longer used in household detergents in the Western world and represent only a minor fraction of the whole nonionic surfactants group. Even though APEO is a group of surfactants which is no longer of any significant commercial importance, there is still a need for risk assessment, since these compounds are still present in the aquatic environment owing to the recalcitrant nature of some of their metabolites. Also, more attention should be given in the future to other nonionic surfactants such as alcohol ethoxylates (AE).

As far as linear alkylbenzenesulfonates (LAS) are concerned, most environmental concentrations are below the threshold value for aquatic organisms and thus do not pose a risk, while in sludge-amended soils, harmful effects of LAS on soil organisms cannot totally be ruled out directly after application of sewage sludge. Regarding the risk assessment of sulfophenyl carboxylates (SPC), it can be concluded that these polar metabolites pose little or no risk to the aquatic environment since the measured concentrations in river water are lower than the PNEC values.

References

1 H. L. Webster and J. Halliday, *Analyst* 84 (1959) 552.
2 W. K. Fischer and K. Winkler, *Vom Wasser* 47 (1976) 81.
3 Anonymous. *German Detergent Act, Verordnung über die Abbaubarkeit von Detergentien in Wasch- und Reinigungsmitteln*. BGBl I, 12.12.1962: 698–701.
4 W. T. Sullivan and R. D. Swisher, *Environ. Sci. Technol.* 2 (1968) 194.
5 T. P. Knepper and J. L. Berna, In: *Analysis and fate of surfactants in the aquatic environment*. T. P. Knepper, D. Barceló, P. de Voogt (Eds.), Elsevier, Amsterdam 2003.
6 EEC, *Council Directive of 22 November 1973 on the approximation of the laws of the Member States relating to detergents*. 73/405/EEC, Official Journal of the European Communities, No. L 347/53.
7 EEC, *Council Directive of 31 March 1982 amending Directive 73/405/EEC on the approximation of the laws of the Member States relating to methods of testing the biodegradability of nonionic surfactants (82/242/EEC)*. Official Journal of the European Communities, No. L 109/1.
8 EEC, *Council Directive of 31 March 1982 amending Directive 73/405/EEC on the approximation of the laws of the Member States relating to methods of testing the biodegradability of anionic surfactants (82/243/EEC)*, Official Journal of the European Communities, No. L 109/18.
9 *Verordnung über die Abbaubarkeit anionischer und nichtionischer grenzflächenaktiver Stoffe in Wasch- und Reinigungsmitteln* vom 30.01.1977, BGBl. 1, 244.
10 *DIN German Standard Method for the Examination of Water, Wastewater and Sludge; General Measures of Effects and Substances (Group H). (A) Determination of Methylene Blue Active Substances (H23-1), (B) Determination of Bismuth Active Substances (H23-2), (C) Determination of the Disulfine Blue Active Substances (H20)*, VCH Weinheim, Germany, 1992.
11 Anionic surfactants as MBAS, *Standards Methods for the Examination of Water and Wastewater*. American Public Health Association, 512B, 1985.
12 Nederlandse Vereniging van Zeepfabrikanten NVZ, *Environmental data review of soap*, Bongerts, Kuyer and Huiswaard, consulting engineers, Delft, The Netherlands, 1994.
13 L. H. M. Vollebregt and J. Westra, *Environmental effects of surfactants*, Onderzoeks- en Adviescentrum Chemie Arbeid Milieu, University of Amsterdam, 1998.
14 OECD, *Guidelines for testing of chemicals*. Organization for Economic Co-operation and Development (OECD), Paris, 1993.
15 R. D. Swisher, *Surfactants and Biodegradation*. Marcel Dekker (1987), New York, NY, USA.
16 L. Cavalli, G. Cassani, M. Lazzarin, C. Maraschin, G. Nucci and L. Valtorta, *Tenside Surf. Deterg.* 33 (1996) 393.

17 A. Di Corcia, F. Casassa, C. Crescenzi, A. Marcomini and R. Samperi, *Environ. Sci. Technol.* 33 (1999) 4112.

18 A. Marcomini, G. Pojana, C. Carrer, L. Cavalli, G. Cassani and M. Lazzarin, *Environ. Toxicol. Chem.* 19 (2000) 555.

19 A. Di Corcia, C. Crescenzi, A. Marcomini and R. Samperi, *Environ. Sci. Technol.* 32 (1998) 711.

20 A. Moreno, J. Ferrer, J. Bravo, J.L. Berna and L. Cavalli, *Tenside Surf. Det.* 35 (1998) 375.

21 A. Remde and R. Debus, *Chemosphere* 32 (1996) 1563.

22 C. A. Staples, J. B. Williams, R. L. Blessing and P. T. Varineau, *Chemosphere* 38 (1999) 2029.

23 C. Crescenzi, A. DiCorcia, A. Marcomini and R. Samperi, *Environ. Sci. Technol.* 31 (1997) 2679.

24 M. Huber, U. Meyer and P. Rys, *Environ. Sci. Technol.* 34 (2000) 1737.

25 P. Schöberl, W. Guhl, N. Scholz and K. Taeger, *Tenside Surf. Det.* 35 (1998) 279.

26 G. M. Boeije, D. R. Schowanek and P. A. Vanrolleghem, *Water Res.* 34 (2000) 1479.

27 U. Baumann, M. Benz, E. Pletscher, K. Breuker and R. Zenobi, *Tenside Surf. Det.* 36 (1999) 288.

28 P. Kölbener, U. Baumann, T. Leisinger and A. M. Cook, *Environ. Toxicol. Chem.* 14 (1995) 571.

29 J. Mampel, T. Hitzler, A. Ritter and A. M. Cook, *Environ. Toxicol. Chem.* 17 (1998) 1960.

30 T. P. Knepper, P. Eichhorn, L. S. Bonnington, In: *Analysis and fate of surfactants in the aquatic environment.* T. P. Knepper, D. Barceló, P. de Voogt (Eds.), Elsevier, Amsterdam 2003, p. 525.

31 J. Blasco, M. Hampel, I. Morreno-Garrido, In: *Analysis and fate of surfactants in the aquatic environment.* T. P. Knepper, D. Barceló, P. de Voogt (Eds.), Elsevier, Amsterdam 2003, p. 827.

32 N. Olea, M. F. Fernandez, A. Rivas, F. Olea-Serranoctants, In: *Analysis and fate of surfactants in the aquatic environment.* T. P. Knepper, D. Barceló, P. de Voogt (Eds.), Elsevier, Amsterdam 2003, p. 887.

33 P. de Voogt, P. Eichhorn, T. P. Knepper, In: *Analysis and fate of surfactants in the aquatic environment.* T. P. Knepper, D. Barceló, P. de Voogt (Eds.), Elsevier, Amsterdam 2003, p. 913.

34 P. Schöberl, *Tenside Surf. Det.* 26 (1989) 86.

35 R. L. Huddleston and R. C. Allred, *Dev. Ind. Microbiol.* 4 (1963) 24.

36 W. J. Payne and V. E. Feisal, *Appl. Microbiol.* 11 (1963) 339.

37 E. A. Setzkorn, R. L. Huddleston and R. C. Allred, *J. Am. Oil Chem. Soc.* 41 (1964) 826.

38 R. D. Swisher, *J. Water Poll. Control Fed.* 35 (1963) 877.

39 R. D. Swisher, *J. Water Poll. Control Fed.* 35 (1963) 1557.

40 G. Baggi, D. Catelani, A. Colombi, E. Galli and V. Treccani, *Ann. Microbiol Enzymol.* 24 (1974) 317.

41 C. R. Eggert, R. G. Kaley and W. E. Gledhill (1979) *Proc. of the Workshop: Microbial Degradation of Pollutants in Marine Environments*, Bourquin A. W., Pritchard P. H. (Eds.) U.S. EPA Reprot No. 600/9-79-012 451.

42 D. Hrsak and A. Begonja, *J. Appl. Microbiol.* 85 (1998) 448.

43 W. Dong, S. Radajewski, P. Eichhorn, K. Denger, T. P. Knepper, J. C. Murrell and A. M. Cook, *J. Appl. Microbiol.* 96 (2004) 630.

44 S. Schulz, W. Dong, U. Groth and A. M. Cook, *Appl. Environ. Microbiol.* 66 (2000) 1905.

45 C. Kanz, M. Nölke, H.-P. Fleischmann, E. Kohler, W. Giger, *Anal. Chem.* 70 (1998) 913.

46 A. Di Corcia, F. Casassa, C. Crescenzi, A. Marcomini and R. Samperi, *Environ. Sci. Technol.* 33 (1999) 4112.

47 D. Schleheck, W. Dong, K. Denger, E. Heinzle and A. M. Cook, *Appl. Environ. Microbiol.* 66 (2000) 1911.

48 P. Eichhorn and T. P. Knepper, *Environ. Toxicol. Chem.* 21 (2002) 1.

49 P. Kölbener, A. Ritter, F. Corradini, U. Baumann and A. M. Cook, *Tenside Surf. Det.* 33 (1996) 149.

50 S. Eaton, K. Bartlett and M. Pourfarzam, *Biochem. J.* 320 (1996) 345.

51 R. J. A. Wanders, R. Vreken, M. E. J. de Boer, F. A. Wijburg, A. H. van Gennip and L. Ijlst, *J. Inherit. Metab. Dis.* 22 (1999) 442.
52 P. Schölberl, K. J. Bock and L. Huber, *Tenside Surf. Det.* 25 (1988) 86.
53 M. J. Scott and M. N. Jones, *Biochim. Biophys. Acta* 1508 (2000) 235.
54 A. M. Bruce, J. D. Swanwick and R.A. Owensworth, *J. Proc. Inst. Sew. Purif. Pt.* 5 (1966) 661.
55 J. Steber, P. Gode and W. Guhl, *Soap Cosmet. Chem. Spec.* 64 (1988) 44.
56 O. R. T. Thomas and G. F. White, *Biotechnol. Appl. Biochem.* 11 (1989) 318.
57 N. S. Battersby, L. Kravetz, J. P. Salanitro, Proc. 5th CESIO World Surfactant Congress, Firenze, May 2000, pp. 1397–1407.
58 W. Fischer and P. Gericke, *Water Res.* 9 (1975) 1137.
59 M. L. Cano, J. P. Salanitro, K. A. Evans, A. Sherren and L. Kravetz, 92nd AOCS Annual Meeting and Expo, Minneapolis, May 13, 2001.
60 J. Steber and P. Weirich, *Tenside Surf. Det.* 2 (1989) 406.
61 S. S. Talmage, *Environmental and human safety of major surfactants: alcohol ethoxylates and alkylphenol ethoxylates. A report to the Soap and Detergent Association*; Lewis Publishers: Boca Raton, FL, 1994.
62 B. Thiele, K. Günther and M. Schwuger, *Chem. Rev.* 97 (1997) 3247.
63 A. Di Corcia, A. Costantino, C. Crescenzi, E. Marinoni and R. Samperi, *Environ. Sci. Technol.* 32 (1998) 2401.
64 W. H Ding, Y. Fujita, R. Aeschimann and M. Reinhard, *Fresenius J. Anal. Chem.* 354 (1996) 48.
65 W. H. Ding and C. T. Chen, *J. Chromatogr. A* 824 (1998) 79.
66 W. Giger and P. H. Brunner, *Science* 225 (1984) 623.
67 J. Ejlertsson, M. Nilsson, H. Kylin, A. Bergman, L. Karlson, M. Öquist and B. Svensson, *Environ. Sci. Technol.* 33 (1999) 301.
68 N. Jonkers, T. P. Knepper and P. de Voogt, *Environ. Sci. Technol.* 35 (2001) 335.
69 M. Stalmans, E. Matthijs, E. Weeg and S. Morris, *SÖFW Journal* 13 (1993) 10.
70 P. Eichhorn and T. P. Knepper, *J. Mass Spectrom.* 35 (2000), 468.
71 J. Steber, W. Guhl, N. Stelter and F. R. Schröder, *Tenside Surf. Det.* 32 (1995) 515.
72 M. T. Garcia, L. Ribosa, E. Campos and J. Sanchez Leal, *Chemosphere* 35 (1997) 545.
73 T. Madsen, G. Petersen, C. Seiero and J. Torslov, *J. Am. Oil Chem. Soc.* 73 (1996) 929.
74 P. Eichhorn and T..P. Knepper, *J. Chromatogr. A* 854 (1999) 221.
75 D. B. Knaebel, T. W. Federle and J. R. Vestal, *Environ. Toxicol. Chem.* 9 (1990) 981.
76 L. Kravetz, J. P. Salanitro, P. B. Dorn and K. F. Guin, *J. Am. Oil Chem. Soc.* 68 (1991) 610.
77 T. Balson and M. S. B. Felix, in D. R. Karsa and M. R. Porter (Eds.), *Biodegradability of surfactants*, Blackie Academic & Professional, Glasgow, United Kingdom, 1995, p. 204.
78 A. Marcomini and G. Pojana, *Analusis* 25 (1997) 35.
79 C. Brunner, U. Baumann, E. Pletscher and M. Eugster, *Tenside Surf. Det.* 37 (2000) 276.
80 P. Eichhorn and T. P. Knepper, *J. Mass Spectrom.* 36 (2001) 677.
81 P. Eichhorn, M. Petrović, D. Barceló and T. P. Knepper, *Vom Wasser* 95 (2000) 245.
82 A. Di Corcia and R. Samperi, *Environ. Sci. Technol.* 28 (1994) 850.
83 A. Di Corcia, L. Capuani, F. Casassa, A. Marcomini and R. Samperi, *Environ. Sci. Technol.* 33 (1999) 4119.
84 A. M. Soto, H. Justicia, J. W. Wray and C. Sonnenschein, *Environ. Health Perspect.* 102 (1991) 380.
85 R. Renner, *Environ. Sci. Technol.* 31 (1997) 316 A.
86 M. Ahel and W. Giger, *Am. Chem. Soc. Nat. Meeting Extended Abst.* 38 (1998) 276.
87 P. De Voogt, O. Kwast, R. Hendriks and C. C. A. Jonkers, *Analusis* 28 (2000) 776.
88 A. C. Belfroid, P. de Voogt, E. G. van der Velde, G. B. J. Rijs, G. J. Schafer and A.

D. Vethaak, in A. D. Vethaak, B. van der Burg, A. Brouwer (Eds) *Endocrine-disrupting compounds: wildlife and human health risks*, RIZK, The Hague, 2000, pp 31-37.
89 L. B. Barber, G. K. Brown and S. D. Zaugg, *ACS Symposium Serie* 747 (2000) 97.
90 M. Fujita, M. Ike, K. Mori, H. Kaku, Y. Sakaguchi, M. Asano, H. Maki and T. Nishihara, *Water Sci. Technol.* 42/7-8 (2000) 23.
91 A. Di Corcia, R. Cavallo, C. Crescenzi and M. Nazzari, *Environ. Sci. Technol.* 34 (2000) 3914.
92 M. L. Trehy, W. E. Gledhill, J. P. Mieure, J. A. Adamove, A. M. Nielsen, H. O. Perkins and W. S. Eckhoff, *Environ. Toxicol. Chem.* 15 (1996) 233.
93 M. Ahel and W. Giger, *Anal. Chem.* 57 (1985) 1577.
94 C. G. Naylor, J. P. Mieure, W. J. Adams, J. A. Weeks, F. J. Castaldi, L. D. Ogle and R. R. Romano, *J. Am. Oil Chem. Soc.* 69 (1992) 695.
95 D. T. Bennie, C. A. Sullivan, H. B. Lee, T. E. Peart and R. J. Maguire, *Sci. Total Environ.* 193 (1997) 263.
96 M. A. Blackburn, S. J. Kirby and M. J. Waldock, *Mar. Poll. Bull.* 38 (1999) 109.
97 P. L. Ferguson, C. R. Iden and B. J. Brownawell, *Environ. Sci. Technol.* 35 (2001) 2428.
98 A. Tabata, S. Kashiwa, Y. Ohnishi, H. Ishikawa, N. Miyamoto, M. Itoh, and Y. Magara, *Water Sci. Technol.* 43/2 (2001) 109.
99 N. Jonkers, P. Serné and P. de Voogt (2001) *Report to RIZA*, University of Amsterdam, Amsterdam.
100 Hessisches Landesamt fuer Umwelt und Geologie, Wiesbaden, Germany, unpublished results 2000 and 2001.
101 E. M. Thurman, T. Willoughby, L. B. Barber and K. A. Thorn, *Anal. Chem.* 59 (1987) 1798.
102 R. J. Larson, T. W. Federle, R. J. Shimp and R. M. Ventullo, *Tenside Surf. Det.* 26 (1989) 116.
103 J. A. Field, J. A. Leenheer, K. A. Thorn, L. B. Barber, C. Rostad, D. L. Macalady and S. R. Daniel, *J. Contam. Hydrol.* 9 (1992) 55.
104 J. A. Field, L. B. Barber, E. M. Thurman, B. L. Moore, D. L. Lawrence and D. A. Peake, *Environ. Sci. Technol.* 26 (1992) 1140.
105 C. J. Krueger, L. B. Barber, D. W. Metge and J. A. Field, *Environ. Sci. Technol.* 32 (1998) 1134.
106 C. Crescenzi, A. di Corcia, E. Marchiori, R. Samperi and A. Marcomini, *Water Res.* 30 (1996) 722.
107 D. C. McAvoy, C. E. White, B. L. Moore and R. A. Rapaport, *Environ. Toxicol. Chem.* 13 (1994) 213.
108 P. Eichhorn, T. P. Knepper, F. Ventura and A. Diaz, *Water Res.* 36 (2002) 2179.
109 R. A. Rudel, S. J. Melly, P. W. Geno, G. Sun and J. G. Brody, *Environ. Sci. Technol.* 32 (1998) 861.
110 C. G. Naylor, B. E. Huntsman, J. G. Solch, C. A. Staples and J. B. Williams, *Proceedings 5th World Surfactant Congress*, Firenze, 2000.
111 U. Zoller, Water Res., 28 (1994) 1625.
112 U. Zoller, E. Ashash, G. Ayali, S. Shafir and B. Azmon, *Environ. Intern.* 16 (1990) 301.
113 U. Zoller, *J. Environ. Sci. Health* A27 (1992) 1521.
114 Y. Fujita, W.-H. Ding and M. Reinhard, *Water Environ. Res.* 68 (1996) 867.
115 C. Schaffner, M. Ahel and W. Giger, *Water Sci. Technol.* 19 (1987) 1195.
116 M. Ahel, C. Schaffner and W. Giger, *Water Res.* 30 (1996) 37.
117 M. Ahel, *Bull. Environ. Contam. Toxicol.* 47 (1991) 586.
118 A. D. Levine, E. L. Libelo, G. Bugna, T. Shelley, H. Mayfield and T. B. Stauffer, *Sci. Total Environ.* 208 (1997) 195.
119 C. A. Moody and J. A. Field, *Environ. Sci. Technol.* 33 (1999) 2800.
120 RIWA (2000) *Organic micropollutants in Rhine and Meuse – Monitoring with HPLC/UV-fingerprint*, Amsterdam.
121 RIWA (2000) *Inventory and toxicological evaluation of organic micropollutants – Revision 1999*, Amsterdam.
122 R. T. Ghijsen and W. Hoogenboezem (2000) *Endocrine disrupting compounds in the Rhine and Meuse basin: occurrence in surface, process and drinking water*, RIWA, Amsterdam.

123 P. de Voogt, R. Hendriks and O. Kwast (2002), In: D. Vethaak et al., *Estrogens and xeno-estrogens in the aquatic environment of The Netherlands*. RIZA/RIKZ report 2002.001, Lelystad/The Hague, ISBN 9036954010.
124 N. Jonkers, R. W. P. M. Laane and P. de Voogt, *Environ. Sci. Technol.* 37 (2003) 321.
125 L. S. Sheldon, R. A. Hites, *Environ. Sci. Technol.* 13 (1979) 574.
126 L. B. Clark, R. T. Rosen, T. G. Hartman, J. B. Louis, I. H. Suffet, R. L. Lippincott and J. D. Rosen, *Intern. J. Environ. Anal. Chem.* 47 (1992) 167.
127 A. Guardiola, F. Ventura, L. Matia, J. Caixach and J. Rivera, *J. Chromatogr.* 562 (1991) 481.
128 F. Ventura, A. Figueras, J. Caixach, I. Espalder, J. Romero, A. Guardiola and J. Rivera, *Water Res.* 22 (1988) 1211
129 F. Ventura, J. Caixach, A. Figueras, I. Espalder, D. Fraisse and J. Rivera, *Water Res.* 23 (1989) 1191.
130 M. Petrovic, A. Diaz, F. Ventura and D. Barceló, *Anal. Chem.* 73 (2001) 5886.
131 P. Eichhorn In: *Analysis and fate of surfactants in the aquatic environment*. T. P. Knepper, D. Barceló, P. de Voogt (Eds.), Elsevier, Amsterdam 2003, p. 695.
132 HERA (2001) *Human and Environmental Risk Assessment*, CEFIC, Brussels: http://www.heraproject.com/ExecutiveSummaryPrint.cfm?ID=22
133 R. D. Swisher, W. E. Gledhill, R. A. Kimerle and T. A. Taulli, *Proceedings Surfactant Congress*, Vol. 7, 1976.
134 A. Moreno and J. Ferrer, *Tenside Surf. Det.* 28 (1991) 129.
135 P. Kölbener, U. Baumann, T. Leisinger and A. M. Cook, *Environ. Toxicol. Chem.* 14 (1995) 561.
136 E. J. Routledge and J. P. Sumpter, *Environ. Toxicol. Chem.* 15 (1996) 241.
137 J. M. Navas, E. González-Mazo, A. Wenzel, A. Gómez-Parra and H. Segner, *Mar. Poll. Bull.* 38 (1999) 880.
138 M. R. Servos, R. J. Maguire, D. T. Bennie, H. B. Lee, P. M. Cureton, N. Davidson, R. Sutcliffe and D. F. K. Rawn (2000) *Supporting document for nonylphenol and its ethoxylates*. National Water Research Institute 00-029, Environment Canada, Burlington, p. 1–217.
139 S. Lussier, D. Champlin, J. LiVolsi, S. Poucer, R. Pruell and G. Thursby (1996), in [138].
140 JRC (2002), *4-Nonylphenol (branched) and nonylphenol: summary risk assessment report*. Special publication I.02.69, Joint Research Centre – EC, Ispra, available at: http://www.europa.eu.int/comm/enterprise/chemicals/markrestr/studies/nonylphenol.pdf
141 M. Petrovic and D. Barceló, In: *Analysis and fate of surfactants in the aquatic environment*. T. P. Knepper, D. Barceló, P. de Voogt (Eds.), Elsevier, Amsterdam 2003, p. 655.
142 H.-Q. Li, F. Jiku and H. Fr. Schroeder, *J. Chromatogr. A* 889 (2000) 155.
143 P. Spengler, W. Körner and J. W. Metzger, *Vom Wasser* 93 (1999) 141.
144 T. C. Feijtel, J. Struijs and E. Matthijs, *Environ. Toxicol. Chem.* 18 (1999) 2645.
145 T. C. Feijtel, E. Matthijs, A. Rottiers, G. B. J. Rijs, A. Kiewiet and A. de Nijs, *Chemosphere* 30 (1995) 1053.
146 J. Riu, P. Eichhorn, J. A. Guerrero, T. P. Knepper and D. Barceló, *J. Chromatogr. A* 889 (2000) 221.
147 M. Castillo, M. C. Alonso, J. Riu, M. Reinke, G. Klöter, H. Dizer, B. Fisher, P. D. Hansen and D. Barceló, *Anal. Chim. Acta* 426 (2001) 265.
148 J. Waters and T. C. J. Feijtel, *Chemosphere* 30 (1995) 1939.
149 M. S. Holt, K. K. Fox, M. Burford, M. Daniel and H. Bickland, *Sci. Total Environ.* 210/211 (1998) 255.
150 D. C. McAvoy, S. D. Dyer, N. J. Fendinger, W. S. Eckhoff, D. L. Lawrence and W. M. Begely, *Environ. Toxicol. Chem.* 17 (1998) 1705.
151 D. C. McAvoy, W. S. Eckhoff and R. A. Rapaport, *Environ. Toxicol. Chem.* 12 (1993) 977.
152 N. Paxéus, *Water Res.* 30 (1996) 1115.
153 B. N. Jacobsen and T. Guildal, *Water Sci. Technol.* 42/7–8 (2000) 315.
154 A. Cohen, K. Klint, S. Bowadt, P. Persson and J. A. Jönsson, *J. Chromatogr. A* 927 (2001) 103.

155 T. Tanghe, G. Devriese and W. Verstraete, *J. Environ. Qual.* 28 (1999) 702.
156 T. P. Rodgers-Gray, S. Jobling, S. Morris, C. Kelly, S. Kirby, A. Janbakhsh, J. E. Harries, M. J. Waldock, J. P. Sumpter and C. R. Tyler, *Environ. Sci. Technol.* 34 (2000) 1521.
157 C. M. Lye, C. L. J. Frid, M. E. Gill, D. W. Cooper and D. M. Jones, *Environ. Sci. Technol.* 33 (1999) 1009.
158 M. Ahel, W. Giger and M. Koch, *Water Res.* 28 (1994) 1131.
159 M. Ahel, W. Giger, E. Molnar and S. Ibric, *Croat. Chem. Acta* 73 (2000) 209.
160 M. Ahel, W. Giger, E. Molnar, S. Ibrić, C. Ruprecht and C. Schaffner, *Division of Environmental Chemistry Preprints of Extended Abstract* 38 (1998) 267.
161 M. Ahel, E. Molnar, S. Ibric and W. Giger, *Water Sci. Technol.* 42/7-8 (2000) 15.
162 D. Barceló, J. Dachs and S. Alcock (Eds.) *BIOSET: Biosensors for Evaluation of the Performance of Wastewater Treatment Works*, Final report, 2000.
163 M. Petrovic and D. Barceló, *Fresenius J. Anal. Chem.* 368 (2000) 676.
164 M. Castillo, E. Martinez, A. Ginebreda, L. Tirapu and D. Barceló, *Analyst* 125 (2000) 1733.
165 C. Planas, J. M. Guadayol, M. Doguet, A. Escalas, J. Rivera and J. Caixach, *Water Res.* 36 (2002) 982.
166 C. Crescenzi, A. Di Corcia and R. Samperi, *Anal. Chem.* 67 (1995) 1797.
167 S. A. Snyder, T. L. Keith, D. A. Verbrugge, E. M. Snyder, T. S. Gross, K. Kannan and J. P. Giesy, *Environ. Sci. Technol.* 33 (1999) 2814.
168 R. C. Hale, C. L. Smith, P. O. de Fur, E. Harvey, E. O. Bush, M. J. La Guardia and G. G. Vadas, *Environ. Toxicol. Chem.* 19 (2000) 946.
169 J. A. Field and R. L. Reed, *Environ. Sci. Technol.* 30 (1996) 3544.
170 J. R. Todorov, A. A. Elskus, D. Schlenk, P. L. Ferguson, B. J. Brownawell and A. E. McElroy, *Marine Environ. Res.* (2002) 54, 691.
171 H. B. Lee and T. E. Peart, *Anal. Chem.* 67 (1995) 1976.
172 T. Isobe, H. Nishiyama, A. Nakashima and H. Takada, *Environ. Sci. Technol.* 35 (2001) 1041.

10
Trihalomethanes (THMs), Haloacetic Acids (HAAs), and Emerging Disinfection By-products in Drinking Water

Christian Zwiener

10.1
Introduction

10.1.1
Disinfection – Fields of Application

Disinfectants are intended to kill bacteria and to inactivate viruses. But disinfectants also react with many other inorganic and organic materials present in water and thereby form unwanted and partly toxic disinfection by-products (DBPs). Trihalomethanes (THMs) are the most prominent DBPs in chlorinated drinking water found by Rook in 1974 [1]. To date, several hundred DBPs from the reaction of disinfectants with natural organic matter have been identified in laboratory scale studies and from samples obtained after full-scale treatment of drinking water and wastewater [2].

The concentration and the species of DBPs formed by disinfection are dependent on several parameters (Fig. 10.1):

- concentration and kind of disinfectant (e.g., chlorine, chlorine dioxide),
- concentration and kind of organic constituents of water (e.g., natural organic matter (NOM) or humic and fulvic acids (HA, FA), anthropogenic compounds),
- concentration and kind of inorganic constituents of water such as bromide or iodide ions,
- reaction temperature,
- reaction time,
- pH,
- alkalinity.

A general scheme of DBP formation and its key parameters is given in Fig. 10.1. NOM has been recognized as the major, naturally occurring precursor of the formation of organohalogenated DBPs.

Organic Pollutants in the Water Cycle. T. Reemtsma and M. Jekel (Eds.).
Copyright © 2006 WILEY-VCH Verlag GmbH & Co. KGaA, Weinheim
ISBN: 3-527-31297-8

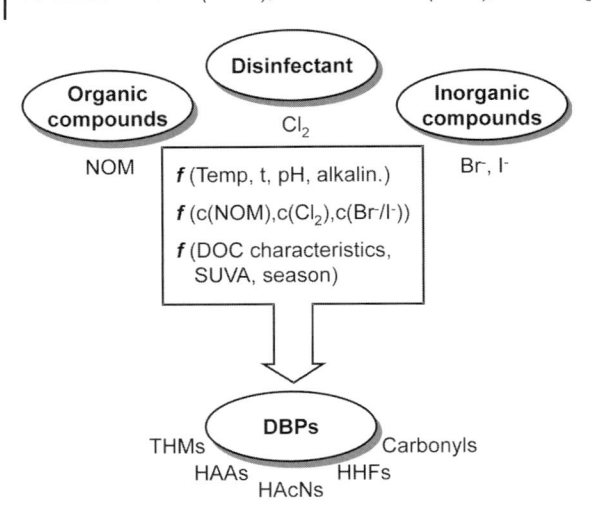

Fig. 10.1 General scheme of DBP formation by reactions of disinfectants with water constituents.

Disinfectant: Cl_2
Organic compounds: NOM
Inorganic compounds: Br^-, I^-

f (Temp, t, pH, alkalin.)
f (c(NOM), c(Cl_2), c(Br^-/I^-))
f (DOC characteristics, SUVA, season)

DBPs: THMs, HAAs, HAcNs, HHFs, Carbonyls

Disinfection and hence DBP formation is an important issue in drinking water and swimming pool water treatment. By far the most research work and publications deal with DBPs in drinking water treatment.

Mainly in the past, DBPs had also to be considered as an environmental issue. For example, a chlorine bleach was widely used in kraft pulp mills. The chlorination stage bleaching liquors therefore contained considerable amounts of chlorination by-products like dioxins and other toxic compounds. Several of these have been identified as mutagens. The extremely strong mutagen MX, 3-chloro-4-(dichloromethyl)-5-hydroxy-2(5H)-furanone, is the most important one and can account for up to 60% of the total mutagenicity of chlorination bleaching liquors and also of chlorinated drinking water [3, 4]. Now, most pulp processes have switched to less problematical bleaching processes using oxygen-based chemicals (e.g., oxygen, peroxides, and ozone).

Disinfection of wastewater treatment effluents with chlorine and other chlorine-based chemicals can be another source of DBPs entering the aqueous environment. Since wastewater effluents contain considerable amounts of dissolved organic matter and dissolved organic nitrogen, chlorination results in the formation of undesirably large amounts of chlorinated and non-chlorinated toxic DBPs. A comparison of the DBP formation potentials (DBPFP) of wastewater and surface water revealed higher DBPFPs in wastewaters due to higher precursor concentrations of dissolved organic carbon (DOC), but normalized on a DOC basis less reactivity of the effluent-derived matter [5]. For example, around 100 000 t year^{-1} of chlorine was used in the 1970s in the United States for the disinfection of wastewater, resulting in the discharge of more than 5000 t year^{-1} of chlorine-containing DBPs into the environment [6]. Furthermore the chlorination of wastewater effluents can lead to the formation of over 100 ng L^{-1} of the potent carcinogen N-nitrosodimethylamine (NDMA) by reaction of monochloramine with organic nitrogen-containing precursors [7]. Therefore, wastewater disinfection with chlorine is not considered to be a sustainable procedure, and – if such disinfection is necessary at all – has mostly been replaced by other processes (e.g., application of UV irradiation [8]). The surface water quality is generally an important issue of drinking water with respect to indirect potable reuse. For example, contaminants in rivers and lakes can enter the drinking water supply via bank filtration.

Artificial groundwater recharge or aquifer storage and recovery (ASR) can be a further source of DBPs if chlorination pretreatment of the source water used for ASR is applied. ASR is primarily used in arid zones and areas of high population growth with water scarcity. ASR of chlorinated source water can therefore cause on one hand a direct discharge of DBPs into the groundwater. On the other hand substantial increase in DBPs can occur by reaction of residual chlorine with natural organic matter. However, the DBP increase can be masked by concomitantly occurring natural attenuation of DBPs in the aquifer. On a field site in South Australia, THM half-lives were found to vary between 1 and 65 days, whereas attenuation of haloacetic acids (HAA) was rapid (<1 day) [9]. THM attenuation was shown to be highly dependent on the type of DBP compound and on geochemical parameters, in particular on the prevailing redox potential. Chloroform was found to be the most persistent compound, bromoform the least. A 2–5 fold decrease in THM half-lives was observed under methanogenic conditions compared to nitrate-reducing conditions. However, against the background of DBP discharge into groundwater and formation of DBPs in the groundwater, the question arises whether it is wise to inject chlorinated water into groundwater, since the latter is particularly vulnerable to contamination. Groundwater recharge by soil aquifer treatment (SAT) takes advantage of further processes, such as adsorption, oxidation/reduction, filtration, and biodegradation. However, the removal of organic trace compounds and microorganisms is of course also dependent on the particular prevailing conditions and therefore may not be completely achieved for all water constituents by SAT [10].

10.1.2
Disinfection in Drinking Water Treatment

A good drinking water quality and the organization of wastewater disposal were the major preconditions of an improved quality of life in the 20th century. Cholera and typhoid were the major diseases and caused illness and death in the 19th century in Europe. Since 1900, the average life expectancy has been increased by more than 30 years, mainly as a result of improved hygiene and public health. Drinking water disinfection was one of the major hygienic measures. From the very beginning, chlorine and ozone proved to be useful disinfectants, and chlorine is still one of the major disinfectants used worldwide.

THMs are the most abundant class of chlorination DBPs (Fig. 10.2). Research has shown that haloacetic acids (HAAs) can be present at concentrations equal to or greater than the concentration of THMs [11]. Thus, it was understood that THMs were but only one class of halogenated DBPs produced by water chlorination. If bromide ions are present in chlorinated water, they can be oxidized to hypobromous acid (HOBr), which itself is an even more potent halogenation reagent

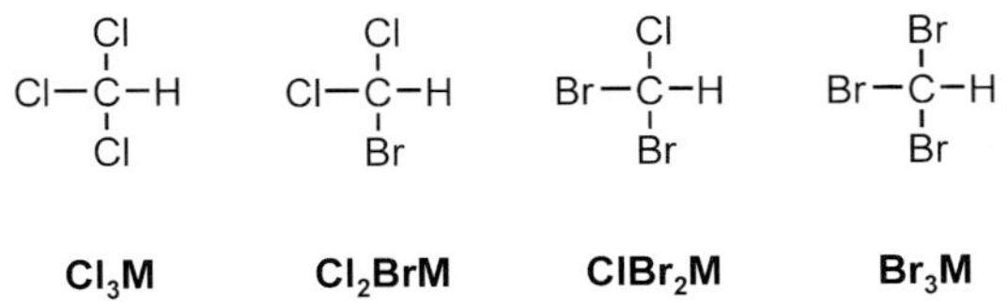

Cl_3M: Cl-C(Cl)(Cl)-H
Cl_2BrM: Cl-C(Cl)(Br)-H
$ClBr_2M$: Br-C(Cl)(Br)-H
Br_3M: Br-C(Br)(Br)-H

Fig. 10.2 Structures of the four chlorinated and brominated THMs (Cl_3M: trichloromethane; Cl_2BrM: dichlorobromomethane; $ClBr_2M$: chlorodibromomethane; Br_3M: tribromomethane).

than hypochlorous acid (HOCl), the active disinfectant of chlorine in water. The same principle applies for iodide ions with weak oxidants like chloramine [12]. With stronger oxidants like ozone or chlorine, hypoiodous acid is rapidly further oxidized to iodate (IO_3^-), which does not form iodoorganic byproducts.

THMs comprise on the order of 20–25% of the total organic halide (TOX) concentration in chlorinated drinking water (Table 10.1) [13]. Still around 50% of the TOX after chlorination cannot be accounted for. With ozonation and chlorine dioxide application, more than 90% of the TOX formed are unknown [14]. This means that DBP formation and occurrence are not yet fully understood, and it is currently not known what chemical DBPs may be responsible for biological effects recently observed [15].

After chloroform was found to cause cancer in laboratory animals, THMs were regulated in several countries. Furthermore, DBPs have been linked to cancers of the digestive and urinary tract, most notably bladder cancer [20]. More recently DBPs have been linked to adverse reproductive and developmental outcomes, including increased risk of spontaneous abortion [21]. This is a very important result, since adverse reproductive and developmental outcomes are a result of acute (short-term) exposure, in contrast to the cancer endpoint, which has to be considered as a result of chronic exposure over a lifetime.

Therefore, a major task of sustainable water treatment is to find a compromise between minimization of the health risk of pathogenic microorganisms and minimization of the formation of DBPs. This means application of the minimum disinfectant dose to obtain a maximum of hygienic safety. Regulatory measures have to take into consideration both decreasing risk with increasing disinfectant levels

Tab. 10.1 Proportions of DBP formation potentials of THMs and HAAs for different types of water and isolated humic and fulvic acid fractions [16–19].

	THM (% of TOXFP)	CX_3COOH (% of TOXFP)	CHX_2COOH (% of TOXFP)	Ref.
Lake Constance	26	n.a.	n.a.	16
Lake Kinneret, winter	55	13	15	17
River Ruhr	39	n.a.	n.a.	16
Fulvic acid	20	16	5.3	18
Humic acid	20	21	6.1	18
Commercial humic acid	n.a.	50 (all HAAs)	38	19

n.a. data not available

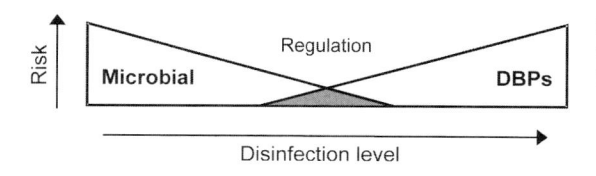

Fig. 10.3 Balancing the risk of microbial contamination and DBP formation (from Ref. [22]).

(disinfectant efficiency) and at the same time increasing risk due to increasing levels of DBPs formed by disinfection (Fig. 10.3).

10.2 Regulations

The concentrations of DBPs are in general regulated in drinking water by surrogate parameters like THMs or HAAs, the most prominent chlorination by-products (Fig. 10.4). In Germany and the rest of Europe, THMs are the only surrogate parameter used. In the United States, THMs and HAAs are taken into consideration.

Fig. 10.4 Structures of selected HAAs (Cl_3AA: trichloroacetic acid; Cl_2AA: dichloroacetic acid; Br_3AA: tribromoacetic acid).

The World Health Organization (WHO) has set quality target values based on an additional cancer risk of 1/100 000 for several compounds, including species of THMs, HAAs, haloacetonitriles, and 2,4,6-trichlorophenol [23]. A comparison of the current regulations on DBPs in the United States, Europe, and Germany with the WHO quality target values is shown in Table 10.2. THM levels of the regulations generally are below the WHO target values. Remarkably, very low WHO target values are suggested for dichloroacetic acid, chloral hydrate, and trichloroacetonitrile. These values may not be met directly by reducing target THMs to the proposed levels, since the key parameters and disinfectants responsible for their formation can be different from those of THM formation.

In other countries, THM guidelines are set at far higher concentrations. e.g., 350 µg L^{-1} in Canada, or 250 µg L^{-1} in Australia and New Zealand [24, 25].

10.2.1 European Union

According to the 98/83/EEC Directive (1998) on drinking water quality, parametric values for THMs of 100 µg L^{-1} are set on the basis of an additional lifetime cancer risk of 1/1 000 000 (according to COM (94) 612 Council Directive, adopted 1995) [26]. According to the previous Directive 80/778/EEC, the member states of the EU

Tab. 10.2 Regulations concerning DBPs compared to WHO guidelines [23–28].

DBP	US EPA MCL ($\mu g\ L^{-1}$)	EU Standards Standard value ($\mu g\ L^{-1}$)	German DWO Parameter value ($\mu g\ L^{-1}$)	WHO Guidelines ($\mu g\ L^{-1}$)
Total THMs	80	100	50 (10)[a]	–
HAAs(5)	60	–	–	–
Bromate	10	10	10	25
Chlorite	1000	–	200	200
Chloroform	–	–	–	200
Bromodichloromethane	–	–	–	60
Dibromochloromethane	–	–	–	100
Bromoform	–	–	–	100
Dichloroacetic acid	–	–	–	50
Trichloroacetic acid	–	–	–	100
Chloral hydrate	–	–	–	10
Dichloroacetonitrile	–	–	–	90
Dibromoacetonitrile	–	–	–	100
Trichloroacetonitrile	–	–	–	1
Cyanogen chloride	–	–	–	70
2,4,6–Trichlorophenol	–	–	–	200
Formaldehyde	–	–	–	900

[a] No further THM monitoring in the distribution system is required if TTHM is lower than 0.010 mg L^{-1} in the finished water at the waterworks.

Tab. 10.3 National regulations for DBPs in European countries [29].

Country	Total THMs ($\mu g\ L^{-1}$)	$CHCl_3$ ($\mu g\ L^{-1}$)	$CHCl_2Br$ ($\mu g\ L^{-1}$)
Germany	10[a] (50)		
Switzerland[b]	25		
Austria	30		
France		30	
Italy	30		
Luxemburg	50		
Sweden	50		
Belgium	100		
England	100		
Ireland	100		
Norway	100		
Finland		200	60

[a] If 10 $\mu g\ L^{-1}$ can be met in the finished water at the water works, it is assumed that the THM levels at the tap of the consumer are below the parametric value of 50 $\mu g\ L^{-1}$.
[b] Switzerland is not a Member State of the EU.

were allowed to adopt their own regulations, which should be at least as stringent as those in the Directive [24].

Therefore the national regulations for DBPs in European countries show concentration limits for THMs in the region of 10 µg L^{-1} in Germany and 100 µg L^{-1} (e.g., in Belgium, Ireland, England, Norway; see Table 10.3).

10.2.2
Germany

The German Drinking Water Regulation (TrinkwV 2001) imposes an obligation to treat all raw waters where microbial contamination cannot be excluded (TrinkwV 2001, §5 chapter 4). The treatment has to rely on a multi-barrier system comprising the protection of raw water sources, steps of particle removal by flocculation/filtration, and a disinfection step. There is no general requirement to disinfect drinking water with chemical disinfectants [30].

Groundwater, therefore, is in most cases not disinfected if it already meets the bacteriological requirements. Surface water treatment generally has a disinfection step at the end of the treatment process to take advantage of a low chlorine demand and an already reduced DOC concentration due to preceding treatment steps.

Since it is not allowed to use chlorine or chlorine dioxide for oxidation purposes in the treatment process, the point of application of chlorine disinfection chemicals is generally at the end of the treatment process. However, ozone is being used for both disinfection and oxidation purposes. Ozonation is therefore typically applied at different points of the treatment process. Commonly, ozonation is followed by biofiltration to reduce the bioavailable carbon (assimilable organic carbon AOC) and by safety disinfection.

In Germany, groundwater is by far the most important raw water source (64%), followed by surface water (27%, including bank filtrate) and spring water (9%) [31].

For disinfection, the use of chlorine, chlorine dioxide, UV irradiation, and ozone is allowed, but not chloramine, because of the low disinfection efficiency. The use pattern of disinfectants reveals the dominant role of chlorine (chlorine gas and hypochlorite solution; Table 10.4). Alternative disinfectants are being used much less but with an increasing trend. More than 50% of the drinking water in Germany is

Tab. 10.4 Disinfectants used for drinking water in Germany [32, 30].

Disinfectant	Use (%)
Hypochlorite	26.4
Chlorine gas	14.0
Chlorine dioxide	4.8
UV radiation	2.9
Ozone	0.6
Without disinfection	51.3

not being disinfected, according to data from a study of 984 German waterworks with 3094 facilities in 1992 [32, 33].

The allowed concentration ranges for the different disinfectants are fairly low (Table 10.5). To guarantee sufficient disinfection capacity, minimum residue levels of the disinfectants are also stipulated. The average dosages of chlorine-based disinfectants applied are commonly below 0.3 mg L^{-1} [32].

Tab. 10.5 Concentrations of disinfectants and DBPs according to the German Drinking Water Regulation.

Disinfectant	Maximum addition (mg L^{-1})	Minimum value after treatment (mg L^{-1})	Maximum residual concentration (mg L^{-1})
Chlorine	1.2	0.1	0.3 free chlorine 0.05 or 0.01 THMs[a]
Chlorine dioxide	0.4	0.05	0.2 chlorine dioxide 0.2 chlorite
Ozone	10	–	0.05 ozone 0.05 THMs 0.01 bromate

[a] 0.05 mg L^{-1} total THMs at the consumer tap or 0.01 mg L^{-1} in the finished water leaving the the waterworks.

10.2.3
United States

The current regulation in the United States is the Stage 1 DBP Rule of 1998. This sets maximum concentration levels (MCLs) of THMs at 80 µg L^{-1} and five HAAs at 60 µg L^{-1} [28]. The five HAAs are monochloro-, dichloro-, and trichloroacetic acid, and monobromo-, and dibromoacetic acid. The maximum residue levels of chlorine, chlorine dioxide, and chloramine are 4 mg L^{-1}, 0.8 mg L^{-1} and 4 mg L^{-1}, respectively, and are therefore much higher than the residual concentrations allowed in Germany. Typical application doses of chlorine range between 2 mg L^{-1} and 6 mg L^{-1}, but can also reach peak values at 17 mg L^{-1} [34]. Chlorine dioxide is applied between 0.5 mg L^{-1} and 2 mg L^{-1}.

In a Stage 2 Rule to be promulgated in 2005 the MCLs remain at 80 µg L^{-1} for THMs and 60 µg L^{-1} for five HAAs. But compliance will be based on a locational running annual average, with the aim of reducing peaks in DBP concentrations in the distribution system and therefore reducing acute exposure to high DBP levels.

10.3
Reactants Leading to DBP Formation

The typical reactants leading to DBP formation are both the disinfectants and so-called DBP precursors. Natural organic matter (NOM) – commonly measured as total organic carbon (TOC) or dissolved organic carbon (DOC) – is a major organic precursor, whereas bromide (Br$^-$) and iodide (I$^-$) serve as further inorganic precursors of halogens.

The reactant concentrations are typically in the milligrams per liter range. Disinfectant doses of sub-mg L^{-1} up to several mg L^{-1} are commonly used to inactivate microorganisms. DOC concentrations of natural water resources range from the low mg L^{-1} level of groundwaters to several mg L^{-1} in lake and river waters, and can reach several tens of mg L^{-1} in the brown waters of swamps and bog lakes.

10.3.1
Disinfectants

Chlorine was introduced for disinfecting drinking water in the early 1900s and is still the most common disinfectant used. Despite its high disinfection efficiency and ease of use, alternatives to chlorine that produce lower THM levels are of interest. The most widely used alternative disinfectants are chlorine dioxide, chloramine, and ozone. Their disinfection efficiencies decrease in the order: ozone > chlorine dioxide > chlorine >> chloramine.

10.3.1.1 Chlorine

Chlorine is used in form of the gas, in granular or powdered form as calcium hypochlorite [Ca(OCl)$_2$], or in liquid form as sodium hypochlorite (NaOCl: bleach). Dissolved in water, all chlorine chemicals react to hypochlorous acid [HOCl: Eq. (1)], which acts as a potent disinfecting and oxidizing agent (E^0 = 1.49 V in water at 25 °C).

$$Cl_2 + H_2O \rightarrow HOCl + H^+ + Cl \quad \text{(Eq. 1)}$$
$$HOCl \rightleftharpoons H^+ + OCl^- \quad \text{(Eq. 2)}$$

Therefore, HOCl reacts with a wide variety of compounds such as NOM, but also with other reducing agents like bromide (Br$^-$), iodide (I$^-$), sulfite (SO$_3^-$), nitrite (NO$_2^-$), hydrogen sulfide (H$_2$S), manganese (II), or iron (II). These reactions are important from the point of view of DBP formation as well as of fast consumption of chlorine, which reduces the disinfection capacity. Hypochlorous acid is a weak acid with a pK$_a$ of approximately 7.5 at 25 °C, which can dissociate into a hypochlorite anion (OCl$^-$) and a proton [H$^+$: Eq. (2)]. Since non-dissociated HOCl is the major disinfectant species, the disinfection efficiency is pH dependent and dominant at pH values below 7.5. Decomposition reactions – particularly in more alkaline solutions – produce chlorite (ClO$_2^-$) and chlorate (ClO$_3^-$), which are therefore found in aged hypochlorite solutions. In certain cases, inorganic DBPs in the form of chlorite and chlorate may be dosed together with the disinfectant.

A further important reaction of HOCl is the oxidation of bromide to give hypobromous acid, HOBr [pK_a = 8.7, Eq. (3)], which itself is a more potent halogenating agent than HOCl.

$$HOCl + Br^- \rightarrow Cl^- + HOBr \qquad (Eq.\ 3)$$

10.3.1.2 Chlorine Dioxide

Chlorine dioxide is a potentially explosive gas, which cannot be compressed or stored. It has to be produced directly at the point of use from solutions of sodium chlorite with HCl or chlorine, respectively [Eqs. (4) and (5)]. However, solutions with concentrations of more than 30 g L^{-1} also tend to explode. Only diluted aqueous solutions are quite stable at neutral or acidic pH if kept cool, well sealed, and protected from sunlight. They can be therefore handled without hazard. Chlorine dioxide does not hydrolyze in water.

$$5\ NaClO_2 + 4\ HCl \rightarrow 4\ ClO_2 + 5\ NaCl + 2\ H_2O \qquad (Eq.\ 4)$$
$$2\ NaClO_2 + Cl_2 \rightarrow 2\ ClO_2 + 2\ NaCl \qquad (Eq.\ 5)$$

Chlorine dioxide is a much more specific and selective reactant than chlorine and has a lower redox potential (E^0 = 0.95 V). Less chlorine dioxide is thus required for disinfection, and it produces fewer by-products than chlorine. However, as it has an odd number of electrons, ClO$_2$ most commonly reacts by an electron transfer mechanism, producing chlorite ions. Therefore, chlorine dioxide is also a powerful oxidizing agent and reacts with iodide, sulfide, iron (II), and manganese (II). Most importantly, bromide is generally not oxidized under conditions of water treatment by chlorine dioxide. Therefore, bromide ions will not give rise to brominated DBPs with chlorine dioxide [35]. However, the redox potentials of bromide and ClO$_2$ can become similar at suitable concentration ratios of the active redox pairs, and therefore bromide oxidation cannot be completely excluded.

In drinking water treatment, chlorine dioxide disinfection is very often accompanied by (a) the formation of odor, and (b) the formation of by-products [36].

Because of the commonly applied chlorite/chlorine process for ClO$_2$ production, the use of chlorine dioxide is to a certain extent accompanied by the addition of chlorine. This means that chlorine is the second main component applied and is not merely a residue in the ClO$_2$ stock solution. An advantage of this two-component disinfectant is its robustness and flexibility, since in cases of changing water quality or if a more efficient disinfection is needed, the chlorine concentration can easily be increased. On the other hand, the formation of chlorine dioxide is accompanied by a complex range of secondary reactions and an undefined formation of odor, which does not make the process of chlorine dioxide disinfection easily controllable.

10.3.1.3 Ozone

Ozone is a prickly smelling, colorless, and reactive gas. The diamagnetic O_3 molecule has a bent symmetrical conformation with a mean O-O bond length of 127.8 pm and a bond angle of 116.8°. Ozone has to be produced directly at the point of use from oxygen or air by an endothermic reaction initiated by an electrical discharge. Ozone is three times more soluble in water than oxygen. Dissolved in water, ozone has a high redox potential (E^0 = 2.07 V in acidic solution) and can react as a powerful oxidizing agent in two ways (a) directly, and (b) in the form of highly reactive hydroxyl radicals (OH-radicals) formed, for example, by hydroxide ion-catalyzed decomposition reactions of ozone in water [37, 38]. Therefore, both reaction pathways always have to be considered. Molecular ozone is a selective oxidant and predominantly attacks electron-rich sites of organic molecules by 1,3-dipolar cycloaddition (Criegee mechanism) or electrophilic substitution. Furthermore, radical-type chain reactions which consume ozone can occur concurrently with direct ozone reactions. For example, the half-life of a 5-mg L^{-1} ozone solution in pure water at pH 8 is about 20 to 30 min because of a hydroxide ion-catalyzed decomposition. The half-life can be doubled in the presence of 5 mmol L^{-1} carbonate ions, acting as typical OH-radical scavengers. The ratio of the concentrations of OH-radicals and ozone are in the range of 10^{-9} to 10^{-7}, typical for secondary-phase ozonation, whereas values higher than 10^{-7} can be observed during the initial phase of ozonation or during advanced oxidation processes (AOPs) [37]. In AOPs, OH-radicals are intentionally produced, for example, by the combination ozone/ hydrogen peroxide.

The primary reasons to use ozone in drinking water treatment are for taste and odor control, to remove color, iron, and manganese, for micro flocculation and oxidative flocculation, and to remove organic compounds. The kinetics of ozone reactions is favorable for many organic and inorganic compounds [39], and ozone is a very efficient disinfection agent. Because of the ozone demand of natural organic matter (NOM, [40]) a combination of preozonation, filtration, and prior main ozonation is favorable for the disinfection of surface waters [41]. The major organic ozonation by-products are acids, aldehydes, and ketoacids [42]. Bromate is the major inorganic by-product with carcinogenic potential, and is formed by reactions of ozone and OH-radicals with bromide ions [43, 44].

Microorganisms can utilize refractory organic compounds after oxidation with ozone. In drinking water treatment, therefore, a biologically active filtration is required after oxidation to reduce easily assimilable carbon and to prevent the growth of biofilms and an increased biological activity in the distribution system.

10.3.1.4 Chloramine

Mixing ammonia and free chlorine produces chloramine. Ammonia can be already present in the raw water or can be added in form of ammonium chloride, ammonium sulfate, or aqueous ammonia. The hypochlorous acid reacts with ammonia to form monochloramine, dichloramine, and trichloramine. These reactions are governed by pH and the chlorine-to-nitrogen weight ratio and predominantly yield monochloramine at pH levels commonly present at drinking water plants (7.0–9.5).

Chloramine has the lowest oxidizing potential but also the lowest efficiency for disinfection. It is therefore a poor primary disinfectant and not approved for application in drinking water disinfection in Germany. However, because of its lower reactivity, chloramine remains longer in the distribution system and is therefore used as a secondary disinfectant. Iodide is rapidly oxidized to hypoiodous acid in the presence of chloramine. Further reactions form organoiodine compounds like iodo-THMs and iodo-acids [12, 45]. In the United States, approximately 300 drinking water plants currently use chloramine. Chloramine produces considerably lower levels of THM (less than 3%) and TOX (9–48%) than chlorine alone [46]. This applies if ammonia is already in the source water or applied before chlorine in water treatment. If the free chlorine is applied first and then the ammonia, much higher THM and TOX levels are found because of the formation potential of free chlorine. Therefore chloramine as a secondary disinfectant in combination with ozone or chlorine dioxide appears to be attractive for minimizing DBP formation [47].

10.3.2
Organic DBP Precursors

10.3.2.1 Natural Organic Matter (NOM)

Naturally occurring organic material or natural organic matter (NOM) is suggested to be a major precursor of organic DBP formation in drinking water treatment. NOM consists of a mixture of humic substances (humic and fulvic acids) and hydrophilic material (non-humic matter, comprising proteins, aminosugars, sugars, and polysaccharides). The role of NOM in DBP formation is affected by both its concentration and its characteristic properties [48]. Several approaches have been made to characterize NOM [49]:

- fractionation (hydrophilic/hydrophobic; high/low molecular weight),
- UV absorbance, fluorescence, ^{13}C NMR spectroscopy,
- pyrolysis GC-MS, electrospray ionization-MS [50–53]
- elemental analysis and statistical calculation of properties.

Several studies have focused also on the DBP formation potential – the yield of DBPs after the reaction of chlorine with NOM [18, 54–57]. The degree of aromaticity of NOM was found to greatly affect the yield of chlorination DBPs. UV absorbance at 254 nm (UVA_{254}) was found to be a good and in the first instance easily measurable surrogate for aromaticity [58]. In more detail, the activated aromatic structures – for example 1,3-dihydroxyphenyl moieties – are suggested to be the main points of electrophilic chlorine attack.

Numerous models have been developed to predict total THM (TTHM) formation. They very often comprise TOC concentration and UVA_{254} as important parameters [47, 59–61], like the EPA Water Treatment Plant Simulation Program [Eq. (6); Malcolm Pirnie 1992]

$$TTHM = 0.00309 \, [(TOC)(UVA_{254})]^{0.440} \times (Cl_2)^{0.409} \, (t)^{0.265} \, (T)^{1.06} \, (pH - 2.6)^{0.715} \, (Br^- + 1)^{0.036} \quad \text{(Eq. 6)}$$

The DBP formation potentials of different organic model compounds – typical ingredients of the dissolved organic matter of raw waters – were investigated by Knepper et al. [62]. The compounds comprise humic acids (a sodium salt of humic acid from Aldrich, a potassium salt of humic acid from Roth, an IHSS humic acid from the Suwannee River), proteins from bovine serum albumin, and lysocyme (hen's egg protein), amino acids glycine, glutamic acid, and tryptophan [63], an amino sugar D(+)-glucosamine, and the sugars D(+)-glucose and D(+)-saccharose. Model solutions with concentrations of 5 mg L^{-1} were chlorinated with chlorine solutions at a Cl_2/C ratio of 1:0.6 for the humic acids, proteins and sugars, and, because of the formation of chloramines, at a higher Cl_2/C ratio of 1:0.28 for the amino acids and amino sugars [62]. The results revealed that sugars and amino sugars make no appreciable contribution to the formation of adsorbable organic halogens (AOX), while humic acids, the amino acid tryptophan, and the proteins are the main contributors. Only 22%, 23%, and 3% of the measured AOX could be linked to humic acids, amino acids, and proteins, respectively (Table 10.6). The major non-volatile fraction of DBPs can be attributed to HAAs for humic acids and proteins and to chloral hydrate and dichloroacetonitrile for amino acids.

Tab. 10.6 DBP formation of different model fractions after 4 h reaction time (data from Ref. [62]).

Model DOC (5 mg L^{-1})	Humic acids	Amino acids	Proteins
Chlorine dose (mg L^{-1})	2.8	8.4	2.8
Residual chlorine (mg L^{-1})	0.7	0.1	0.5
AOX (µg L^{-1})	205	181	119
TTHM (µg L^{-1})	66	37	23
Neutral DBPs (µg L^{-1})	8	23	3
THAA (µg L^{-1})	38	23	8

Relatively little is known about the kinetics of NOM reactions with ozone. The decrease in ozone concentration is influenced by the type and concentration of NOM and positively correlated with the UV absorbance (a surrogate of non-saturated bonds in NOM). Ozone produces a variety of organic by-products, such as aldehydes and ketoacids. These are highly biodegradable and potentially hazardous and may produce increased amounts of chlorinated by-products upon secondary chlorination [64]. A model to estimate the potential for total aldehyde formation in source waters upon ozonation was developed by Siddiqui et al. [65].

10.3.2.2 **Micropollutants**
Single contaminants present in raw water or drinking water can also serve as reaction partners of disinfectants and therefore can form halogenated and oxidized DBPs including THMs and HAAs. In particular, compounds having activated sites

for halogen attack, like activated aromatic rings with hydroxyl groups, are suitable candidates to produce halogenated by-products. Disinfection by-products of micropollutants reported in a recent review comprise substances of the compound classes of pesticides, surfactants, estrogens, pharmaceuticals, and cyanobacterial toxins [66].

10.4
Occurrence of DBPs

The formation and occurrence of chlorination DBPs has been well studied because of the widespread application of chlorination disinfection for many years. Ozonation by-products have recently received much attention. DBPs of chlorine dioxide and chloramines have been studied relatively less. Table 10.7 gives an overview of the major classes of DBPs commonly found during disinfection with chlorine, ozone, chlorine dioxide, and chloramine.

Tab. 10.7 Different types of disinfectants and their important groups of DBPs produced (data from Refs. [14, 22, 42, 45, 67]).

Class of DBPs	Example	Cl_2	O_3	ClO_2	NH_2Cl
Trihalomethanes (Cl,Br-THMs)	Chloroform	+[a]	+[b]		+
Iodo-THMs	Iodoform				+
Haloacetic acids (HAA)	Chloroacetic acid	+			+
Haloacetonitriles (HAcN)	Chloroactetonitrile	+	+		
Inorganic compounds	Bromate, Hypobromite, Chlorite, Chlorate		+	+	
Hal. hydroxyfuranones	MX	+		+	+
Haloketones	Chloropropanone	+	+	+	
Halonitromethane	Chloropicrin	+			
Carbonyls	Formaldehyde, Acetone	+	+	+	
Carboxylic acids	Acetic acid	+	+	+	
Aldo and Ketoacids	Glyoxylic acid		+	+	
Iodo-acids	Iodoacetic acid				+
Nitrosamines	NDMA	+			+

[a] 4 regulated chloro/bromo THMs, but there will be nine compounds if iodomethanes are included.
[b] Bromoform is produced in the presence of bromide.

10.4.1
Trihalomethanes (THMs) and Halogenated Acetic Acids (HAAs)

10.4.1.1 German Drinking Water

The last comprehensive survey of the application of disinfectants and the occurrence of DBPs in German drinking water was conducted in 1992 [32]. The data represent the situation in 1991. The study of the occurrence of THMs in 994 German waterworks reports median concentrations of THMs to be below 5 µg L^{-1}. In particular, higher median values of THMs were found with the use of chlorine gas (about 3 µg L^{-1} THM), compared to around 1.2 µg L^{-1} with hypochlorite and 0.6 µg L^{-1} with chlorine dioxide [32]. Drinking water from ground and spring water resources revealed lower THM mean values than those of bank filtration or artificial recharge (1.5 times higher) and those of rivers (2 times higher), reservoirs, and lakes (4 times higher). Generally, THM concentrations exceeded a level of 10 µg L^{-1} in only about 10 to 15% of all facilities. Occasional maximum values of THMs, however, are 90 µg L^{-1} for hypochlorite, 33 µg L^{-1} for chlorine, and 50 µg L^{-1} and 20 µg L^{-1} for the combinations of chlorine with chlorine dioxide and ozone, respectively. Unfortunately no data on the concentrations of DBP precursors were available in that study.

A comparison of the frequency distributions of THM concentrations in German and United States water works reveal generally ten times lower concentration levels in German water works (Table 10.8). Median values of THMs of the German study [32] were compared with four United States studies of the years 1975 (NORS: National Organics Reconnaissance Survey), 1976/77 (NOMS: National Organic Monitoring Survey), 1987 (AWWARF: American Water Works Association Research Foundation), and 1988 (USEPA/CDHS: United States Environmental Protection Agency/State of California Department of Health Services).

Tab. 10.8 Median values of THMs in German and United States drinking waters [32].

	THM (µg L^{-1})	Number of samples	Reference
German survey			Haberer 1994 [32]
Chlorine	3.0	239	
Hypochlorite	1.0	351	
Chlorine dioxide	0.5	75	
U.S. studies			
	41	80	NORS 1975
	45	113	NOMS 1976/77
	55	727	AWWARF 1987
	40	35	USEPA/CDHS 1988
EPA 600	range 30–50	n.a.	USEPA 2002 [34]

n.a. data not available

10.4.1.2 European Drinking Water

DBPs in drinking water from major waterworks in several European Countries reveal THMs as the most abundant DBPs, with considerably higher concentration levels in treated surface water than those in treated groundwater (Table 10.9). The data were compiled from a number of papers of two decades (1980–2000) and reflect the situation in France (10 papers), Italy (6), The Netherlands (5), Belgium (3), Spain (4), and Germany (1) [24]. The high dispersion of the concentration values is due to varying water quality of the resources and different treatment processes used.

The occurrence of halogenated DBPs in drinking waters exceeding the concentration levels set by regulation has been observed in European countries only for drinking water from surface water, not from groundwater [24, 68]. The percentages of non-compliance were 25% for dichlorobromomethane, 18% for dibromochloromethane, and 33% for bromoform.

A further example is from Spain. THMs and HAAs were measured in 88 drinking water samples from four Spanish regions, Asturias, Alacante, Barcelona, and Tenerife (Canary Islands). Except in Tenerife, where groundwater is used, surface water is the main water resource [69]. DBP concentrations in drinking waters clearly increase with decreasing raw water quality (revealed by the surrogate parameters chloride and COD values). DBP speciation with respect to chlorinated and brominated THMs revealed the highest proportion of chloroform (60% of TTHM) in As-

Tab. 10.9 DBPs in drinking waters of European Countries (data from Ref. [24]).

	Median	Minimum	Maximum	N
Surface waters				
TOX	22	0.03	34	5
THM	47	n.d.	390	19
Residual Cl_2	0.2	n.d.	2.3	7
Groundwater				
TOX	–	22.7	22.7	1
THM	9	0.3	14	3
Residual Cl_2	0.3	n.d.	0.5	4

N Number of observations; n.d. not detected

Tab. 10.10 Average concentrations of water quality parameters and DBPs in Spanish drinking waters (data from Ref. [69]).

	Tenerife	Asturias	Barcelona	Alicante
Chloride (mg L^{-1})	72	14	280	360
COD (mg O_2/L)	0.60	4.2	5.1	17
TTHMs (µg L^{-1})	8	22	64	86
THAAs (µg L^{-1})	3	15	36	50

turias, whereas in the Mediterranean regions (Barcelona and Alacante) and in Tenerife higher proportions of brominated and bromo-chlorinated species occurred, 70–80% and >90%, respectively (Table 10.10). Salt-water intrusion into the groundwater of Tenerife is suggested; however, no bromide concentrations were reported. Data are consistent with those of a European survey (Table 10.9).

10.4.1.3 Canadian Drinking Water

In a study of small-community drinking water supplies in Canada, a clear correlation of source water quality, measured by TOC, with representative concentrations of THMs and HAAs was found [70]. Source waters with the highest TOC concentrations (15 mg L^{-1}) had the highest average THM concentrations (200 µg L^{-1}, measured as THM3, which represents the sum of trichloro-, bromodichloro-, and chlorodibromomethane) and HAA concentrations (180 µg L^{-1}, measured as HAA3, which represents the sum of monochloro-, dichloro-, and trichloroacetic acid). Therefore DBP concentrations cannot comply with the average benchmarks of 100 µg L^{-1} or 80 µg L^{-1} THMs and 60 µg L^{-1} HAAs, which was suggested to be because of the poor water quality and the limited availability of technical and financial resources in smaller communities.

In the Quebec City region, seasonal variations of river water quality and temperature and their effect on DBP formation in finished drinking water and in the distribution system were studied [71]. Results reveal lowest THM and HAA levels in winter (THM3 5.1 µg L^{-1}), whereas THM levels were five times higher in summer and fall (THM3 106 µg L^{-1}), and HAA levels were four times higher in spring. The average concentrations for THM3 and HAA2 in the distribution system were 43.9 µg L^{-1} and 37.7 µg L^{-1}, respectively. The seasonal differences found in Quebec were considerably higher than those in more temperate environments reported from recent studies in the United States and Europe, where the differences between cold and warm water were about two-fold [72, 73, 74].

Increasing levels of THMs with increasing residence time in the distribution system were found, whereas HAA levels decreased after 5 h residence time [71]. HAA degradation is associated with microbial activity and a decarboxylation pathway [75, 76, 77]. Kinetic studies revealed that the decomposition of $BrCl_2AA$, Br_2ClAA, and Br_3AA in water at neutral pH follows a first-order reaction, with rate constants of 0.0011, 0.0062, and 0.040 day^{-1} at 23 °C, respectively; and 0.000028, 0.00014, and 0.0016 day^{-1} at 4 °C, respectively [78]. Drinking water matrix and pH values between 6 and 9 were found to be insignificant for HAA decomposition.

10.4.1.4 United States Drinking Waters

The most recent comprehensive DBP occurrence study in United States drinking waters was conducted by a collaboration of the USEPA's National Exposure Research Laboratory (NERL), the University of North Carolina (UNC), and the Metropolitan Water District of Southern California (MWDSC) in the years from 2000 to 2002 [34].

Together with regulated DBPs, Information Collection Rule DBPs and a selection of 50 high priority DBPs were quantified.

The results of the nationwide occurrence study show a wide range of THM and HAA concentrations in twelve treatment plants across the United States using different disinfectant combinations. The plants use source waters with low to high TOC (3 to 13 mg L^{-1}) and low to high bromide (0.02 to 0.4 mg L^{-1}) levels. Depending on source water quality, used treatment processes, and disinfectant dosages and combinations, TOX values from 20 to 280 µg L^{-1}, THM values from 2 to 164 µg L^{-1}, and HAA values from 8 to 130 µg L^{-1} were found, taking into consideration all samples from different seasons of the year and different plants (Table 10.11) [34]. Further results of the study report on the stability and fate of DBPs in the real distribution system or in a simulated distribution system (SDS). THMs, HAAs, and haloacetonitriles were found to be stable in the distribution system and increased in concentration in the presence of free chlorine.

The results on the occurrence of DBPs in drinking waters reveal that the four regulated THMs are not necessarily an indicator of the formation and control of other halogenated DBPs and that the use of alternative disinfectants can decrease THM concentrations, but may even increase the concentration of some other DBPs.

Tab. 10.11 Examples of DBP concentrations in finished drinking water in the United States (data from [34]).

Plant	1	2	9	10
Disinfection	O_3–Cl_2–NH_2Cl	Cl_2–NH_2Cl	Cl_2–NH_2Cl	Cl_2–H_2Cl
TOX (µg L^{-1})	21–145	91–200	64–66	175–237
THM4 (µg L^{-1})	9–16	32–99	6–8	15–164
HAA9 (µg L^{-1})	5–12	41–93	6–21	21–130
Ozone dose (mg L^{-1})	1.8–2.5	–	–	–
Chlorine dose (mg L^{-1})	2.2–4.7 (FE)	1.1–1.5 (PI) 0.6–0.8 (FI) 1.9–3.7 (FE)	2–2.9 1.9–2.9 0.24 (PE)	10–17 (PI) 0.8–3.4 (FI) 5–14 (PE)
Ammonia (mg L^{-1})	0.5–0.7	0.4–0.6	0–1.9	0–2
TOC (mg L^{-1})	3–4.5	3–4.5	3.4–5	4–5.4
Br$^-$ (mg L^{-1})	0.12–0.4	0.12–0.4	0.06–0.19	0.05–0.08

FE Filter effluent; FI Filter influent; PE Plant effluent; PI Plant influent

10.4.2
Emerging Organic DBPs

DBPs other than the currently regulated THMs and HAAs are covered in this section. These comprise the compound classes of halonitriles, carbonyls, halogenated hydroxyfuranones, halonitromethanes, nitrosamines, iodinated THMs, and acids.

Furthermore, the transformation products of trace contaminants (e.g., pesticides and pharmaceuticals) and of high-molecular-weight NOM are considered.

Toxicologically important compounds have been found among the brominated DBPs, which are much more carcinogenic than their chlorinated analogs. Recently, iodinated DBPs (iodo-THMs and iodo acids) were also identified in disinfected drinking water [34, 12, 45]. Preliminary studies indicate that iodinated compounds may be more toxic than their brominated analogs [79].

10.4.2.1 Halonitriles

Haloacetonitriles (HANs) are predominantly formed during chlorination and chloramination in the low µg L^{-1} range (0.01–5 µg L^{-1}) [2, 80, 81, 82]. HANs have to be considered as one of the major DBP classes, even though HAN concentrations represent, for example, only 3% of the total surveyed DBPs for Utah water treatment plants [82], or 5% of the average THM concentration in different drinking waters of the Netherlands [80].

Typically, dichloroacetonitrile, a weak bacterial mutagen, is the most predominant HAN species at low bromide levels (20 µg L^{-1} Br$^-$ or less). At higher bromide levels of source water, bromochloroacetonitrile is the second most prevalent compound (Fig. 10.5). Furthermore dibromo-, trichloro- and bromodichloroacetonitrile were found in chlorinated and chloraminated waters. In chloraminated waters, low levels of residual chlorine are believed to cause formation of HAN, but not of chloramine itself. The most important source of nitrogen in natural waters is suggested to be proteins and their hydrolysis products. Dibromoacetonitrile is a bacterial mutagen (Ames assay) and dichloroacetonitrile acts as teratogen and is therefore a developmental toxicant [83].

Cyanogen halides are another important compound class of nitriles formed by chlorination, ozonation, and chloramination [2]. Cyanogen chloride (CNCl) is a respiratory irritant. Although health risks due to chronic exposure to CNCl are not yet well defined, this compound has been listed in the USEPA Drinking Water Priority List [84], and utilities are required to monitor CNCl concentrations in their water supplies under the Information Collection Rule [85]. CNCl is produced after the chlorination and chloramination of amino acids and upon chloramination of formaldehyde, but it is not stable in the presence of free chlorine [86, 87]. Cyanogen bromide (CNBr) can be produced upon ozonation of bromide-containing waters. During ozonation, increasing CNBr concentrations (0.5, 3.6, and 6.1 µg L^{-1})

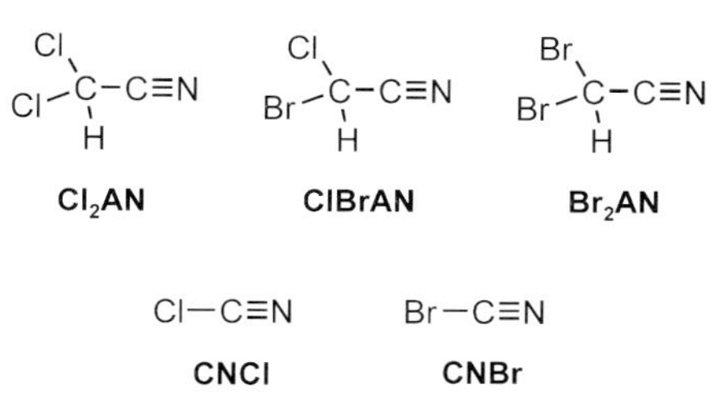

Fig. 10.5 Structures of halonitriles (Cl$_2$AN: dichloroacetonitrile; ClBrAN: chlorobromoacetonitrile; Br$_2$AN: dibromoacetonitrile; CNCl: cyanogen chloride; CNBr: cyanogen bromide).

were found with increasing Br⁻ concentrations (<0.01, 0.11, and 0.76 mg L^{-1}), respectively [88].

Furthermore, dichloro- and trichloropropenenitrile were found upon chlorination, and benzenenitrile and cyanopyridine upon ozonation [2].

10.4.2.2 Carbonyls

Aldehydes, ketones, and ketoacids are non-halogenated DBPs not unique to ozonation, but also associated with chlorination and treatment with other oxidants. A one-year pilot study revealed that aldehyde concentrations increased by 144% upon ozonation, and by 56% upon chlorination [89]. Formaldehyde, acetaldehyde, glyoxal, and methylglyoxal were found to be the predominant four aldehydes found upon ozonation of different source waters (Fig. 10.6) [90, 91]. Ozonation with ozone doses between 1 and 9.2 mg L^{-1} produced aldehyde concentrations between 3 and 30 μg L^{-1} of formaldehyde, between 2 and 65 μg L^{-1} of acetaldehyde, between 3 and 15 μg L^{-1} of glyoxal, and between 3 and 35 μg L^{-1} of methylglyoxal in waters with TOC concentrations between 1 and 9.2 mg L^{-1} [92]. The formation of carbonyl compounds (34 μg L^{-1}) such as formaldehyde (3.4–9 μg L^{-1}), acetaldehyde (4.5 μg L^{-1}), acetone (3.2 μg L^{-1}), and n-valeraldehyde (7–15 μg L^{-1}) was also shown after the application of chlorine dioxide [93]. Commonly, the occurrence of non-halogenated DBPs is quite similar for the different disinfectants, indicating a similar mechanism of oxidation [94].

Both formaldehyde and acetaldehyde are mutagenic and can induce tumors in rats in inhalation studies. However, there is insufficient evidence to establish risk in humans. Acetaldehyde is not expected to be a health concern at the low μg L^{-1} levels typically found in drinking waters [83].

Further carbonyls were identified in drinking water samples and in laboratory-scale ozonation of fulvic acid, humic acid, and natural waters. Among them are aliphatic aldehydes up to tetradecanal (C14) as the longest chain aldehyde, butanedial, and ethylglyoxal, aliphatic ketones (up to heptadecadienone), and diketones [2, 95]. Six non-halogenated aldehydes and ketones, namely 2-hexenal, 5-keto-1-hexanal, cyanoformaldehyde, methylethyl ketone, 6-hydroxy-2-hexanone, and dimethylglyoxal are among the high-priority DBPs included in the United States Nationwide DBP occurrence study [67, 34]. LC-MS analysis after derivatization with 2,4-

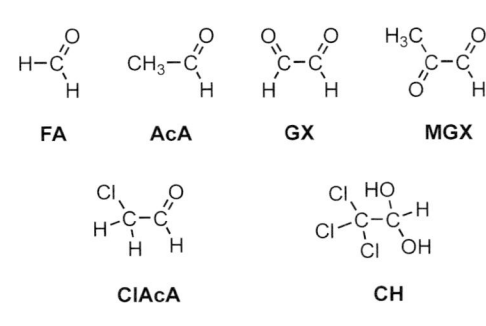

Fig. 10.6 Structures of carbonyl DBPs (FA: formaldehyde; AcA: acetaldehyde; GX: glyoxal; MGX: methylglyoxal; ClAcA: chloroacetaldehyde; CH: chloral hydrate).

dinitrophenylhydrazine (DNPH) revealed highly polar DBPs like 1,3-dihydroxyacetone, pyruvic acid, glyoxylic acid, and ketomalonic acid [96]. LC-MS-MS compound class-specific screening for carbonyls revealed further oxoacids, hydroxycarbonyls, and dicarbonyls that have yet to be identified [97, 98]. Notably the formation of nonhalogenated DBPs like aldehydes and acids can explain, in part, the increase in biodegradable DOC (BDOC) in ozonated water. Commonly, the BDOC can be removed by biologically active filtration (e.g., GAC filtration [99], sand filtration, and anthracite filtration [2]).

Upon chlorination, whether chlorination or chloramination alone or as a secondary disinfectant after ozonation, numerous halogenated carbonyls are also produced. These comprise more than twenty haloketones (e.g., di-, tri-, and pentachloropropanones, which are bacterial mutagens [83]) and more than ten haloaldehydes (e.g., di- and trichloroacetaldehyde, chlorinated propanals, and propenals) [2]. Trichloroacetaldehyde, which occurs in water in the form of chloral hydrate, a hydrolysis product of trichloroacetaldehyde, is a predominant DBP of chlorination and is typically found at µg L^{-1} concentrations (median 3.0 µg L^{-1} [81]). Chloroacetaldehydes are bacterial mutagens (Ames assay): trichloroacetaldehyde is a weak bacterial mutagen [83] and 2,3,3-trichloropropenal is a strong bacterial mutagen.

10.4.2.3 Halogenated Hydroxyfuranones

The high mutagenic activity in the Ames assay brought halogenated hydroxyfuranones (HHF) into the focus of toxicological interest. The compound MX (see Fig. 10.7) – the most popular representative of HHFs – is revealed as the most potent genotoxic compound in the Ames assay (*Salmonella* strain TA 100), comparable only to that of aflatoxins. The contribution of MX to the observed mutagenicity has been determined to be in the range between 3 and 60% in Finnish, United States, UK, and Japanese drinking water extracts [100, 101]. The hydroxyl group and the dichloromethylgroup have been identified to be responsible for the mutagenic activity of MX and other HHFs [102, 103]. This reveals MX as an outstanding example of the overlap of information received from chemical analysis and biological effect testing (mutagenicity).

MX is also present in the open-ring forms (oxobutenoic acid) at pH values typical for drinking water (Fig. 10.7: Z-MX and its isomeric form E-MX). Further oxidized (ox-MX) and reduced forms (red-MX) and other chlorinated and brominated (BMX) HHFs have been identified (Fig. 10.7).

Naturally occurring aquatic humic substances are assumed to be the main precursors for mutagens formed by disinfection. A dose-related increase in mutagenicity was observed upon treatment with either chlorine or chlorine dioxide of both commercial humic acid and natural humic substances isolated from water. Generally, less mutagenicity was found after treatment with chlorine dioxide than after chlorination [104, 105, 106].

MX concentrations and mutagenic activity have been investigated in tap water samples from 36 surface water systems in Massachusetts [107]. Higher MX levels than previously reported (up to 80 ng L^{-1}) were found. Even higher levels of MX,

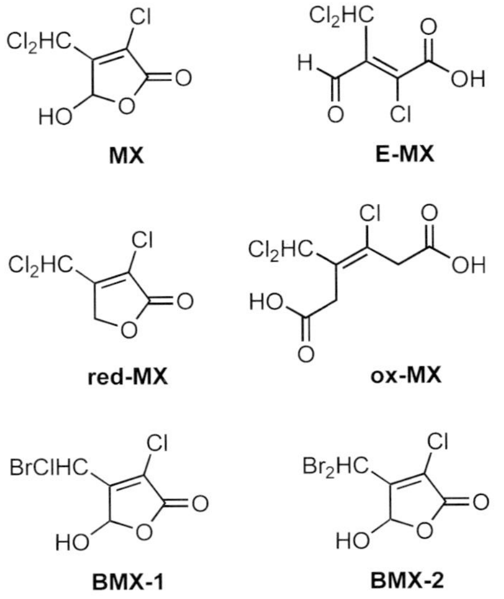

Fig. 10.7 Structures of halogenated hydroxyfuranones (HHFs).

frequently above 100 ng L^{-1} and as high as 310 ng L^{-1}, were found in finished drinking waters across the United States [34]. In high-bromide source water disinfected with chlorine dioxide, chlorine, and chloramines, the brominated forms BMX-1 and BMX-2 have also been found at levels of 170 ng L^{-1} and 200 ng L^{-1}, respectively. In a survey of 11 German waterworks using groundwater, surface water and bank filtrated surface water, MX has been found in two thirds of all samples at concentrations between 1.4 and 27.3 ng L^{-1} [108].

10.4.2.4 Halonitromethanes

Halonitromethanes were found to be more cytotoxic and genotoxic to mammalian cells than most currently regulated DBPs including bromate [109]. In particular, bromonitromethanes have become an important issue. For example dibromonitromethane is at least one order of magnitude more genotoxic to mammalian cells than MX.

Tribromonitromethane (bromopicrin) was found at levels < 2 µg L^{-1} in ozonated water containing high bromide concentrations even over ten years ago [110]. Ozone and bromide ions were suggested to play an important role in halonitromethane formation although they were also found in water treated with chlorine or chloramine only, but at much lower levels [111]. In low-bromide waters, chlorinated halonitromethanes dominated (chloro-, dichloro-, and trichloronitromethane). Increasing bromide levels considerably increased the occurrence of brominated species (bromo-, dibromo-, bromochloro-, bromodichloro-, dibromochloro-, and tribromonitromethane) [2]. In finished drinking waters, concentration ranges for individual halonitromethanes were found to be between 0.1 and 3 µg L^{-1} [34]. The most prevalent species observed were dichloro-, bromochloro-, bromodichloro-, and dibromochloronitromethane (Fig. 10.8).

Fig. 10.8 Structures of halonitromethanes (ClNM: chloronitromethane; Cl$_2$NM: dichloronitromethane; Cl$_3$NM: trichloronitromethane or chloropicrin; Br$_2$NM: dibromonitromethane; Br$_3$NM: tribromonitromethane or bromopicrin).

10.4.2.5 *N*-Nitrosamines

N-Nitrosodimethylamine (NDMA) has been recognized as a probable human carcinogen and is generally present at 10 ng L^{-1} or less in chlorinated drinking water. NDMA was initially found at levels up to 0.3 µg L^{-1} in chlorinated Canadian drinking waters [112], but it now reaches levels of 100 ng L^{-1} or higher in chlorinated wastewater [113]. Recently, highest NDMA concentrations were detected between 2 and 180 ng L^{-1} in chloraminated finished drinking waters in Canada. Furthermore, *N*-nitrosopyrrolidine (2–4 ng L^{-1}) and *N*-nitrosomorpholine could be identified (Fig. 10.9) [114]. NDMA commonly occurs as a DBP from disinfection by chloramines or chlorine. Experiments to reveal the formation mechanism used dimethylamine (DMA) as model precursor. NDMA formation is suggested to involve (a) the formation of monochloramine, (b) the formation of 1,1-dimethylhydrazine intermediate from the reaction of DMA with monochloramine, and (c) the oxidation of 1,1-dimethylhydrazine by monochloramine to form NDMA [115, 116]. Experiments with ^{15}N-labeled monochloramine revealed it as a source of nitrogen in the nitroso group of NDMA. Furthermore, natural organic matter accounted for a significant fraction of NDMA precursors in waters from lakes and reservoirs, in groundwaters, and in isolated natural organic matter [117].

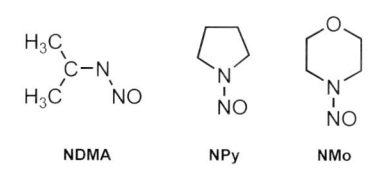

Fig. 10.9 Structures of N-nitrosamines (NDMA: N-nitrosodimethylamine; NPy: N-nitrosopyrrolidine; NMo: N-nitrosomorpholine).

10.4.2.6 **Iodinated THMs and Acids**

Iodinated DBPs were recently found in drinking waters which contained high bromide/iodide concentrations and which were disinfected with chloramines [34]. Iodo-DBPs are of concern, since they have been shown to possess even higher toxicity than brominated DBPs. For example, iodoacetic acid (IAA) has about 3× the cytotoxicity in *S. typhimurium* of bromoacetic acid (BrAA) and more than 50× that of chloroacetic acid (ClAA), and is the most toxic and genotoxic DBP in mammalian cells [45].

Preliminary data on the occurrence of iodo-acids in five drinking water plants in the United States working with chloramine disinfection reveal iodoacetic acid in two out of five plants at levels up to 1.7 µg L^{-1}, and bromoiodoacetic acid in four out of five plants at sub-µg L^{-1} levels [118]. These results suggest that iodoacids could occur in other drinking waters disinfected with chloramines.

Iodo-THMs are responsible for taste and odor problems. The organoleptic threshold concentrations of CHI$_3$ are between 0.03 µg L^{-1} and 1 µg L^{-1}. I-THMs are generally formed by reactions of hypoiodous acid with natural organic matter. HOI is produced in analogy to HOBr from oxidation of iodide by disinfectants (ozone, chlorine, chloramine). The highest formation potential of CHI$_3$ was found for chloramine. Since chloramine is not able to oxidize HOI to IO$_3^-$ like ozone or chlorine, it therefore provides higher steady-state concentrations and contact times of HOI [12]. The results of an United States-wide study show the occurrence of individual iodo-THMs at concentration levels between 0.2 and 15 µg L^{-1} [34].

10.4.2.7 Missing DBPs

Previous DBP research was mostly focused on volatile and semi-volatile low-molecular-weight compounds using analytical methods based on gas chromatography (GC) [119]. Therefore, more than 50% of the total organic halide (TOX) in chlorinated drinking water has not been accounted for [120]. In chlorine dioxide-treated water, more than 70% of the TOX are unknown, in chloraminated water more than 80%, and in ozonated water more than 90% [14]. Furthermore, about half of the assimilable organic carbon (AOC) in ozonated drinking water is unknown [121]. These cicumstances suggest that high-molecular-weight DBPs and polar DBPs are likely to be found in the "missing" DBP fraction.

In fact, fractionation of TOX from fulvic acid chlorination with ultrafiltration (UF) membranes demonstrates the occurrence of high-molecular-weight DBPs in chlorinated waters (14% in the range between 500 and 5000 Da, 3% between 5000 and 50 000 Da, and 9% > 50 000 Da [122]). The fractions showed about 47% of TOX to be < 500 Da. Similar results were found for actual chlorinated and chloraminated drinking water (Fig. 10.10 [123]).

In a fascinating experiment, radiolabeled chlorine (^{36}Cl) was used as a probe for chlorine incorporation into natural organic matter [124]. In this study, ^{36}Cl-labeled

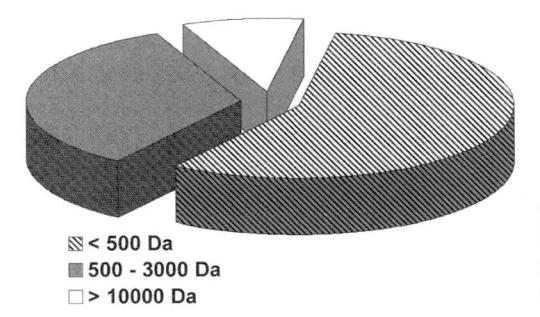

Fig. 10.10 Molecular size fractions of total organic halogen (TOX) after chlorination and chloramination of drinking water [123].

▧ < 500 Da
■ 500 - 3000 Da
☐ > 10000 Da

HOCl was reacted with Suwannee River fulvic acid. Size exclusion chromatography (SEC) of the high-molecular-weight fraction revealed a highly dispersed molecular-weight distribution of ^{36}Cl in TOX that was similar to that of dissolved organic carbon (DOC) profiles, with an average molecular weight around 2000 Da. Follow-up experiments using electrospray ionization (ESI)-MS/MS revealed the chlorine content of high-molecular-weight fractions [125, 126].

10.4.2.8 Transformation Products of Micropollutants

Several new studies have investigated by-product formation from the chlorination or ozonation of micropollutants, including pesticides, alkylphenol ethoxylates (APEOs) and their metabolites, bisphenol A, estrogens (e.g., ethinylestradiol, estradiol), an antibacterial agent, and cyanobacterial toxins.

Halogenated and oxidized by-products were found from reactions with pesticides. The herbicide isoproturon reacts by substitution with chlorine and by oxidation with chlorine dioxide [127]. Reaction products of chlortoluron were formed by the chlorination and hydroxylation of the aromatic ring [128]. For isoxaflutole, a major by-product was a benzoic acid metabolite which is the same non-biologically active degradation product that isoxaflutole forms under natural, environmental conditions [129]. S-triazine herbicides (prometryne, terbutryne, ametryne, and desmetryne) produced three sulfoxide and sulfone by-products by chlorination, with no halogenated products observed. Chlorine dioxide was less reactive, yielding only the sulfoxide by-product [130]. The herbicide atrazine formed ammeline (the major product), hydroxydeethylatrazine, hydroxydeisopropylatrazine, and hydroxyatrazine upon advanced oxidation (ozone/hydrogen peroxide) [131].

In particular, bisphenol A, estrogens (ethinylestradiol, estradiol), and alkylphenoxyethoxylates (APEOs) and their metabolites (alkylphenols and alkylphenolcarboxylates) have the common property of suitable structural moieties for chlorine attack. They have been reported to produce halogenated by-products. Upon chlorination, bisphenol A produced monochloro-, dichloro-, trichloro-, and tetrachloro-derivatives, with chlorine attacking at the ortho positions to the OH groups on both aromatic rings (Fig. 10.11) [132]. 4-Chloroethinylestradiol (4-ClEE2) and 2,4-dichloroethinylestradiol (2,4-diClEE2) were identified as the two major reaction products of the chlorination of 17β-ethinylestradiol (EE2), an oral contraceptive [133]. 17β-Estradiol (E2), a natural hormone, formed 2,4-dichloro-17β-estradiol, monochloroestrone, 2,4-dichloroestrone, and four by-products formed by ring cleavage [134]. Upon ozonation, EE2 formed eleven ozonation reaction products, including adipic acid, cyclohexanone, 1-hydroxycyclohexane-1-carboxylic acid, 1-hydroxycyclopentane-1-carboxylic acid, and other carboxylic acids containing cyclohexane or cyclo-

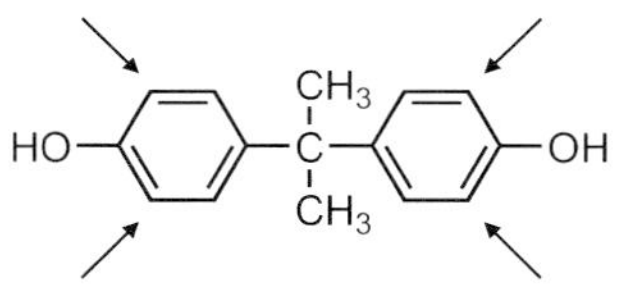

Fig. 10.11 Activated sites for chlorine attack in Bisphenol A.

pentane rings [135]. Significantly diminished estrogenic activity was observed in the water following ozonation, which indicates that ozonation may be promising for controlling estrogens (EE2, E2) in drinking water and wastewater.

Halogenated products of alkylphenol ethoxylates (APEOs) and their metabolites were identified in sludge from drinking water treatment [136]. DBPs include brominated and chlorinated nonylphenol ethoxylates (XNPEOs), octylphenol ethoxylates (XOPEOs), nonylphenols (XNP), and nonylphenoxycarboxylates (XNPECs). Brominated acidic metabolites made up 97% of the halogenated by-products.

Reaction products of the antibacterial agent sulfamethoxazole with chlorine are formed by ring-chlorination, or rupture of the sulfonamide group to form 3-amino-5-methylisoxazole, sulfate, and N-chloro-p-benzoquinoneimine [137].

Microcystin-LR (MC-LR), a cyanobacterial toxin, is oxidized by chlorine dioxide attack at the two conjugated double bonds in the Adda residue, forming products with two hydroxyl groups and with one hydroxyl group and one hydrogen [138]. However, the slow reaction rate of MC-LR with ClO_2 would preclude ClO_2 as a suitable oxidant for the degradation of microcystins in drinking water treatment.

The role of activated sites in aromatic compounds in the formation of halogenated products is revealed by a comparison of the AOX formation potentials of naphthalene-2-sulfonate (NS), 2-aminonaphthalene-1-sulfonate (ANS), and 4,4´-diaminostilbene-2,2´-disulfonate (DSDS) (Fig. 10.12). Whereas no AOX formation has been found upon the chlorination of NS, 76 µg AOX per mg carbon were found for ANS, and 1105 µg AOX per mg carbon for DSDS [139]. Ozonation prior to chlorination could further increase the AOX formation potential for NS and ANS, presumably by hydroxylation and therefore further activation of the aromatic ring.

Finally, the active ingredients of sunscreens typically contain activated aromatic moieties in their structures like benzophenone or dibenzoylmethane, very suitable for incorporating chlorine. Those compounds show also high THM formation potentials: up to 200 µg chloroform per mg carbon [140].

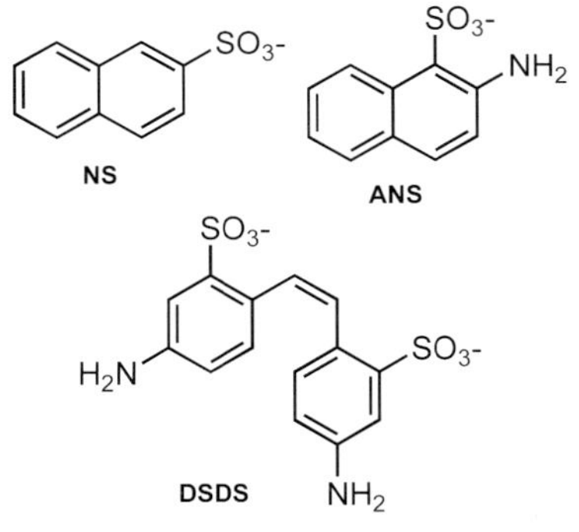

Fig. 10.12 Structures of aromatic sulfonates (NS: naphthalene-2-sulfonate; ANS: 2-aminonaphthalene-1-sulfonate; DSDS: 4,4´-diaminostilbene-2,2´-disufonate).

10.4.3
Inorganic Disinfection By-products

10.4.3.1 Chlorite and Chlorate

Chlorite (ClO_2^-) and chlorate (ClO_3^-) have been identified as the major inorganic DBPs of chlorine dioxide. Chlorite is the predominant by-product formed from between 40% and 70% of applied or consumed chlorine dioxide under conditions typical for drinking water treatment and distribution [141, 36]. Chlorate was produced from up to 0.3 mg per mg of chlorine dioxide consumed or applied [142].

Drinking water treatment with chlorine dioxide produced chlorite concentrations of between 650 and 1300 µg L^{-1} (ClO_2 doses between 1.5 and 195 mg L^{-1} [34]).

Chlorite formation gives cause for concern, since chlorite causes anemia in rats, though at high exposure levels. In the German Drinking Water Regulation, therefore, chlorite is limited to 0.2 mg L^{-1} in drinking water. Chlorine dioxide application is limited to a maximum concentration of 0.4 mg L^{-1}. This is in order to observe the chlorite limit, since a typical 50% transformation of the applied chlorine dioxide to chlorite is assumed.

Chlorite formation is considerably dependent on the residual chlorine dioxide level in the water. It is caused not only by the oxidation of organic matter, but also by a complex mechanism of reactions between chlorine dioxide and other chlorine species of the stock solution like chlorine, chlorite, and chlorate. A so-called minimum chlorine dioxide dosage (MCDD), limiting the residual chlorine dioxide concentration to 0.05 mg L^{-1} after a 30-min contact time, should guarantee minimizing DBP formation without compromising disinfection efficiency [143]. Using chlorine dioxide at MCDD for groundwater and bank filtrate after different pretreatment with ozone and activated carbon, the molar ratio of chlorite formation was observed to be in the range between 40% and 80% of initial chlorine dioxide concentration. Pretreatment by ozonation in combination with activated carbon reveals higher ratios of chlorite formation.

Hypochlorite solutions can be another source of chlorite and chlorate in drinking water disinfection. Decomposition reactions of hypochlorite, in particular in more alkaline solutions, produce chlorite (ClO_2^-) and chlorate (ClO_3^-), which are therefore found in considerable amounts in aged hypochlorite solutions. In these cases in particular, chlorate has been found to be dosed together with the disinfectant and is therefore not a product of reactions of the disinfectant with water constituents [34].

10.4.3.2 Bromate

Bromate is a potential human carcinogen, and therefore a maximum concentration of 10 µg L^{-1} in water destined for human consumption was established by a European guideline proposition in 1995 and in the European Drinking Water Directive 98/83/EC (1998). United States regulation also has an annual average bromate standard of 10 µg L^{-1}.

Bromate generally forms by reactions between oxidants or oxidative disinfectants and bromide ions in water. Thus, bromate is a major by-product of ozonation.

The elucidation of the first mechanisms of bromate formation by ozone oxidation of bromide goes back to the 1980s [43]. In a direct mechanism, molecular ozone oxidizes bromide (Br^-) to hypobromite (BrO^-), and subsequently to bromite (BrO_2^-), which itself is quickly oxidized to bromate (BrO_3^-). Furthermore, in a direct-indirect mechanism, BrO^- (HOBr) can be attacked by an OH-radical to form a BrO radical, which reacts with itself to give bromite (BrO_2^-). Finally, in an indirect-direct mechanism, OH-radicals attack bromide (Br^-) to give Br and BrO radicals. Bromate formation through direct reactions with molecular ozone contributes in the range of 30–80% [44] or up to 100% [144] to the overall bromate formation in NOM-containing waters. Up to 65% was reported for NOM-free waters [144].

Commonly, the formation of bromate increases with increasing ozone dose [91]. However, taking the ozone demand of different waters into account, $c \cdot t$ values (ozone concentration times reaction time) are being recognized to be a more accurate parameter of bromate formation. Laboratory scale-derived relationships of $c \cdot t$ versus bromate are shown to be linear [145]. At typical $c \cdot t$ values of *Cryptosporidia* inactivation, less than 10% (typically between 2 and 6%) of bromate is formed on a molar basis of bromide, taking ozone dosages of 1–4 mg L^{-1} and $c \cdot t$ values between 2.5 and 10 mg·min L^{-1} into account [34].

A survey in Switzerland revealed bromate levels higher than 10 µg L^{-1} in only 2 water samples from 84 waterworks, with bromide levels mostly below 0.5 µg L^{-1}. Only in four raw waters out of 21 did bromide levels exceed 25 µg L^{-1} [38]. Two inventories in France showed, however, that almost 50% of the distributed water (13 facilities out of 47) contained bromate levels equal to or higher than 10 µg L^{-1} in summer, whereas in winter the figure was only 6–7% (4 facilities out of 43) [42].

A further but minor source of bromate can be sodium hypochlorite with bromate contamination in the sub-µg L^{-1} level [146].

10.5
Measures to Control DBPs

The formation of disinfection by-products follows the rules of any chemical reaction. The yield therefore depends on the reaction partners (type and concentrations) and the reaction conditions (such as temperature, pH, and competing parameters) [147]. To control the formation of reaction by-products means to control the reaction partners and the reaction conditions. In the practice of drinking water treatment, the reaction conditions can be varied typically only within a very limited range (pH, temperature). Thus, the type and concentration of disinfectants and of precursors (NOM, bromide, and others) remain as the major control parameters. Strategies for DBP control therefore comprise

- source control,
- precursor removal,
- minimization of disinfectant concentrations,
- use of alternative disinfectants.

10.5.1
Source Control

Source control may comprise control of nutrient inputs (e.g., algae growth control), watershed management (e.g., stormwater management), saltwater intrusion control (to control TOC and bromide), or the application of the concept of aquifer storage and recovery (ASTR, to compensate for seasonal variations of the source water quality) [64].

The major technologies of precursor removal are commonly enhanced coagulation, granular activated carbon (GAC) adsorption, or membrane filtration [47]. Enhanced coagulation with aluminum and ferric salts at optimal pH has been widely used to remove the more hydrophobic fraction of NOM [148]. The technical bulletin W 296 "Reducing the THM formation of the German Association of Gas and Water (DVGW Deutsche Vereinigung des Gas- und Wasserfachs)" recommends DOC removal down to concentrations below 2 mg L^{-1}. The remaining DOC should be highly refractory or not biodegradable, in order to prevent microbiological growth even with low residual disinfectant concentrations [147]. This is especially important with oxidative treatment using ozone and advanced oxidation [149]. Ozone decreased the THM and AOX formation potentials more efficiently than the peroxone process (OH-radical-mediated process) [150].

Since most techniques to remove precursors do not remove bromide ions, a shift to higher ratios of bromide to TOC occurs upon NOM removal, and therefore a shift to more brominated DBPs can occur [148]. For example, while reducing the formation of the chlorinated THMs an increase of brominated THMs was observed [151].

10.5.2
Disinfection Control

To minimize disinfectant concentrations applied in water treatment without compromising disinfection efficiency, the disinfection step is moved to the end of the treatment. This is practiced in most German waterworks to take advantage of DOC and microorganism removal before disinfection is used. This results in lower disinfectant demand and therefore lower disinfectant dosages required to keep the minimum residual disinfectant concentration in the finished water and the distribution system. In the case of very high microbial quality of the source water, constant over the years, and a well-maintained distribution system, disinfection of drinking water is not necessary at all. In Germany, about half of the waterworks surveyed in 1992 applied drinking water treatment without disinfection [32]. An investigation into water treatment of bank filtrate from the rivers Elbe and Saale in Central Germany revealed that the chlorine dose is the major parameter for THM formation. However, because of a high bromide concentration, neither reduction of the DOC below the 2 mg L^{-1} level nor ozonation could reduce THM formation very much [152].

Another means of minimizing DBPs is the use of alternative disinfectants to chlorine. These comprise ozone, chlorine dioxide, and chloramine. Often, combinations are used for the purposes of oxidation and disinfection in drinking water treatment. Different disinfectants, however, also have their own advantages and disadvantages. However, THM – a well-documented and regulated DBP indicator of chlorination –is not an appropriate indicator of DBP formation for all disinfectants. Notably ClO_2 – if generated and applied without any residual free chlorine – does not produce halogenated DBPs to any considerable degree, but produces chlorite and organic oxidation products similar to those produced by ozone [153, 154]. Also, chloramine does not produce any appreciable amounts of halogenated DBPs, but cyanogen chloride formation is higher than with free chlorine, and dichloroacetic acid can be formed. Furthermore the formation of nitrosamines and iodinated DBPs has to be considered with chloramine. Ozonation by-products include bromate, aldehydes, aldo- and ketoacids, acids, and hydrogen peroxide. Ozone can also result in the production of brominated DBPs by oxidizing bromide to hypobromous acid, which itself is a brominating agent.

A comparative assessment of disinfectants used for pre-disinfection and post-disinfection revealed that chlorine produces the highest concentration of DBPs, whereas pre-ozonation chlorination and pre-oxidation chlorination considerably reduces DBP formation. Pre-ozonation chloramination showed the lowest DBP concentrations of all (Table 10.12) [155].

Tab. 10.12 Formation of halogenated DBPs (in µg L^{-1}) with different disinfectant combinations in distribution systems (TOC = 3 mg L^{-1}; pH = 7.6; disinfectant residual maintained for five days; data from Ref. [155]).

DBPs	Cl_2-sand-Cl_2	O_3-sand-Cl_2	ClO_2-sand-Cl_2	O_3-sand-NH_2Cl
THMs	225	154	138	3.2
HAAs	146	82	44	9.0
HANs	2.9	2.7	<0.1	<0.1
Haloketones	2.6	2.6	4.2	<0.1
Chloropicrin	1.3	7.7	1.4	<0.1
Chloral hydrate	75	55	45	<0.1
TOX	540	339	379	27

Results of the United States DBP occurrence study revealed that the use of alternative disinfectants can decrease THM concentrations, but can even increase the concentration of some other DBPs [34]. For example, the highest concentrations of dichloroacetaldehyde were found after ozonation and chloramination, and the highest levels of MX and BMX were found after chlorine dioxide was used as primary disinfectant. Halonitromethanes were found to be increased by pre-ozonation. There is no correlation between THM, HAA, and MX or other HHF analogs.

10.6
Conclusions

The central message for the minimization of DBP formation and therefore of adverse health effects associated with organic and inorganic DBPs is to use

- best available water quality (high microbial quality, low TOC and bromide concentrations),
- disinfection as one of the last steps of water treatment,
- appropriate types of disinfectants (high disinfectant efficiency at low doses),
- low disinfectant doses.

On a worldwide perspective, drinking water is still a challenging issue, because many millions of people do not have access to drinking water of adequate quality. The United Nations and member governments have therefore proclaimed their goal to halve the number of people without access to safe drinking water by 2015 [156]. This implies the important task of exporting treatment technologies to other countries. For this purpose, knowledge of current research has to be translated into practice and adapted to the requirements and capabilities of these other countries. A primary focus will certainly be on chlorination, because of its widespread, cost-effective, and simple operation [64]. For small water supply units there is still a need to develop simple and easily operated treatment systems for NOM removal to reduce DBP formation.

Acknowledgments

I would like to thank S. D. Richardson for valuable discussion and comments on the manuscript.

References

1. J. J. Rook, *Water Treatment Exam.* **1974**, 23, 234–243.
2. S. D. Richardson, in: R. A. Meyers (Ed.), *Encyclopedia of Environmental Analysis and Remediation*, Vol. 3, pp. 1398–1421, Wiley, New York, **1998**.
3. B. Holmbom, R. H. Voss, R. D. Mortimer, A. Wong, *Tappi* **1981**, 64, 172–174.
4. J. Henning, B. Holmbom, M. Reunanen, L. Kronberg, *Chemosphere* **1986**, 15, 549–556.
5. T. Sirivedhin, K. A. Gray, *Water Res.* **2005**, 39, 1025–1036.
6. R. L. Jolley, *J. Water Pollut. Contr. Fed.* **1975**, 47, 601–617.
7. W. A. Mitch, D. L. Sedlak, *Water Sci. Technol.: Water Supply* **2002**, 2, 191–198.
8. J. G. Jacangelo, R. R. Trussell, *Water Sci. Technol.: Water Supply* **2002**, 2, 147–157.
9. P. Pavelic, B. C. Nicholson, P. J. Dillon, K. E. Barry, *J. Contam. Hydrol.* **2005**, 77, 119–141.
10. G. E. Cordy, N. L. Duran, H. Bouwer, R.C. Rice, E. T. Furlong, S. D. Zaugg, M. T. Meyer, L. B. Barber, D. W. Kolpin, *Ground Water Monit. Remed.* **2004**, 24, 58–69.
11. P. C. Singer, A. Obolenski, A. Greiner, *J. Amer. Water Works Assoc.* **1995**, 87(10), 83–92.

12 Y. Bichsel, U. von Gunten, *Environ. Sci. Technol.* **2000**, 34, 2784–2791.
13 P. C. Singer, *Water Sci. Technol.* **2004**, 4 (9), 151–154.
14 X. Zhang, S. Echigo, R. A. Minear, M. J. Plewa, in: *Natural Organic Matter and Disinfection By-products*, S. E. Barrett, S. W. Krasner, G. L. Amy (Eds.), ACS Symp. Series 761, pp. 2–14, Am. Chem. Soc.,Washington, **2000**.
15 S. D. Richardson, J. E. Simmons, G. Rice, *Environ. Sci. Technol.* **2002**, 36, 198A–205A.
16 G. Kleiser, Diss. University of Karlsruhe, Germany, **2000**.
17 L. Heller-Grossmann, J. Manka, B. Limoni-Relis, M. Rebhun, *Water Sci. Technol. – Water Supply* **2001**, 1, 259–266.
18 D. A. Reckow, P. C. Singer, R. L Malcolm, *Environ. Sci. Technol.* **1990**, 24, 1655–1664.
19 H. Pourmoghaddas, A. Stevens, *Water Res.* **1995**, 29, 2059–2062.
20 W. D. King, L. D. Marrett, *Cancer Causes and Control* **1996**, 7, 596–604.
21 K. Waller, S. H. Swan, G. DeLorenze, B. Hopkins, *Epidemiology* **1998**, 9, 134–140.
22 R. Sadiq, M. J. Rodriguez, *Sci. Total Environ.* **2004**, 321, 21–46.
23 WHO World Health Organisation, Local Authorities, Health and Environment Briefing Pamphlet Series No 3, *1994*.
24 M. Palacios, J. Pampillón, M. E. Rodríguez, *Water Res.* **2000**, 34, 1002–1016.
25 W. E. Elshorbagy, H. Abu-Qdais, M. K. Elsheamy, *Water Res.* **2000**, 34, 3431–3439.
26 EEC Council Directive 98/83/EC of 3 November 1998 on the quality of water intended for human consumption. *Off. J. Europ. Comm.* **1998**, L330/32, 5.12.98.
27 German Drinking Water Ordinance, May 21 **2001**, BGBl. I, 959.
28 United States Environmental Protection Agency, Federal Register, **1998**, 63(241) 69 390–69 476.
29 O. Conio, L. Meucci, G. Ziglio, G. Premazzi, *Water Supply* **1998**, 16, 113–114.
30 B. Wricke, *gwf Wasser Abwasser* **2004**, 145, 82–88.
31 Bundesverband der Deutschen Gas- und Wasserwirtschaft, BGW, Annual Report, Berlin, **1998**.
32 K. Haberer, *gwf Wasser Abwasser* **1994**, 135, 409–417.
33 B. Wricke, IWA – World Water Congress, Berlin, **2001**.
34 H. S. Weinberg, S. W. Krasner, S. D. Richardson, A. D. Thruston Jr., Report EPA/600/R–02/068, United States Environmental Protection Agency, National Exposure Research Laboratory, Athens, GA, **2002**.
35 C. Rav-Acha, in: O. Hutzinger (Ed.), *The Handbook of Environmental Chemistry*, Vol. 5, Part C, Water Pollution, Springer, Berlin, **1998**.
36 W. Schmidt, *Acta hydrochim. hydrobiol.* **2004**, 32, 48–60.
37 U. von Gunten, *Water Res.* **2003**, 37, 1443–1467.
38 U. von Gunten, *Water Res.* **2003**, 37, 1469–1488.
39 J. Hoigne, H. Bader, W. R. Haag, J. Staehelin, *Water Res.* **1985**, 19, 993–1004.
40 M. Elovitz, U. von Gunten, H.-P. Kaiser, in: *Natural Organic Matter and Disinfection By-products*, S. E. Barrett, S. W. Krasner, G. L. Amy (Eds.), ACS Symp. Series 761, pp. 2–14, Am. Chem. Soc.,Washington, **2000**.
41 A. Grohmann, in: K. Höll (Ed.), *Wasser*, pp. 619–634, de Gruyter, Berlin, **2002**.
42 B. Legube, B. Parinet, F. Dossier-Berne, J. P. Croué, *Ozone Sci. Eng.* **2002**, 24, 293–305.
43 W. R. Haag, J. Hoigné, *Environ. Sci. Technol.* **1983**, 17, 261–267.
44 U. von Gunten, J. Hoigné, *Environ. Sci. Technol.* **1984**, 28, 1234.
45 M. J. Plewa, E. D. Wagner, P. Jazwierska, S. D. Richardson, A. D. Thruston, Y.-T. Woo, A. B. McKague, *Environ. Sci. Technol.* **2004**, 38, 4713–4722.
46 J. N. Jenson, J. D. Johnson, J. J. S. Aubin, R. F. Christman, *Org. Geochem.* **1983**, 8, 71–76.
47 P. C. Singer, *J. Environ. Eng.* **1994**, 120, 727–744.
48 S. Barrett, S. W. Krasner, G. L. Amy, in: *Natural Organic Matter and Disinfection By-products*, S. E. Barrett, S. W. Krasner, G. L. Amy (Eds.), ACS Symp. Series 761,

pp. 2–14, Am. Chem. Soc.,Washington, **2000**.
49 A. D. Nikolaou, T. D. Lekkas, *Acta Hydrochim. Hydrobiol.* **2001**, 29, 63–77.
50 J. A. Leenheer, C. E. Rostad, P. M. Gates, E. T. Furlong, I. Ferrer, *Anal. Chem.* **2001**, 73, 1461–1471.
51 E. B. Kujawinski, P. G. Hatcher, M. A. Freitas, *Anal. Chem.* **2002**, 74, 413–419.
52 A. C. Stenson, A. G. Marshall, W. T. Cooper, *Anal. Chem.* **2003**, 75, 1275–1284.
53 T. Reemtsma, A. These, *Anal. Chem.* **2003**, 75, 1500–1507.
54 R. F. Christman, D. L. Norwood, *Environ. Sci. Technol.* **1983**, 17, 625–628.
55 M. L. Trehy, R. A. Yost, C. L. Miles, *Environ. Sci. Technol.* **1986**, 20, 1117–1122.
56 S. K. Golfinopoulos, N. K. Xylourgidis, M. N. Kostopopoulou, T. D. Lekkas, *Water Res.* **1998**, 32, 2821–2829.
57 Z. Huixian, Y. X. Sheng, X. Ouyong, *Water Res.* **1997**, 31, 1536–1541.
58 J. P. Croué, D. Violleau, L. Labouyrie, in: *Natural Organic Matter and Disinfection By-products*, S. E. Barrett, S. W. Krasner, G. L. Amy (Eds.), ACS Symp. Series 761, pp. 2–14, Am. Chem. Soc.,Washington, **2000**.
59 P. Westerhoff, J. Debroux, G. Amy, D. Gatel, V. Mary, J. Cavard, *J. Am. Water Works Assoc.* **2000**, 92, 89–102.
60 W. Sung, B. Reilley-Matthews, D. O'Day, K. Horrigan, *J. Am. Water Works Assoc.* **2000**, 92, 53–63.
61 T. D. Lekkas, A. D. Nikolaou, *Water Qual. Res. J. Canada* **2004**, 39, 149–159.
62 T. Knepper, I. Schönsee, M. O. Glocker, U. Petersen, J. Driemler, K. Haberer, *Vom Wasser* **1997**, 88, 329–350.
63 R. Chinn, S. E. Barrett, in: *Natural Organic Matter and Disinfection By-products*, S. E. Barrett, S. W. Krasner, G. L. Amy (Eds.), ACS Symp. Series 761, pp. 96–108, Am. Chem. Soc.,Washington, **2000**.
64 G. Amy, R. Bull, G. F. Craun, R. A. Pegram, M. Siddiqui, in: *Disinfectants and Disinfectant By-products*, World Health Organization, Geneva, **2000**.
65 M. S. Siddiqui, G. L. Amy, B. D. Murphy, *Water Res.* **1997**, 31, 3098–3106.
66 C. Zwiener, S. D. Richardson, *Trends Anal. Chem.* **2005**, 24, 613.
67 S. D. Richardson, *Trends Anal. Chem.* **2003**, 22, 666–684.
68 T. Nissinen, I. Miettinen, P. Martikainen, T. Vartiainen, *Chemosphere* **2002**, 48, 9–20.
69 C. M. Villanueva, M. Kogevinas, J. O. Grimalt, *Water Res.* **2003**, 37, 953–958.
70 J. W. A. Charrois, D. Graham, S. E. Hrudey, K. Froese, *J. Toxicol. Environ. Health A* **2004**, 67, 1797–1803.
71 M. Rodriguez, J.-B. Sérodes, P. Levallois, *Water Res.* **2004**, 38, 4367–4382.
72 R. J. Garcia-Villanova, C. Garcia, J. A. Gomez, M. P. Garcia, R. Ardanuy, *Water Res.* **1997**, 31, 1405–1413.
73 W. J. Chen, C. P. Weichsel, *J. Am. Water Works Assoc.* **1998**, 90, 151–163.
74 S. K. Golfinopoulos, *Chemosphere* **2000**, 41, 1761–1767.
75 H. Baribeau, S. W. Krasner, R. Chinn, P. C. Singer, *Proc. Am. Water Works Assoc. WQTC*, Salt Lake City, **2000**, 1178–1200.
76 S. L. Williams, R. L. Williams, D. F. Rinffleisch, *Proc. Am. Water Works Assoc. WQTC Conf.* **1994**, 1053–1058.
77 H. J. Zhou, Y. F. F. Xie, *J. Am. Water Works Assoc.* **2002**, 94, 194–200.
78 X. Zhang, R. A. Minear, *Water Res.* **2002**, 36, 3665–3673.
79 E. S. Hunter, J. A. Tugman, *Teratology* **1995**, 52, 317–323.
80 R. J. B. Peters, E. W. B. de Leer, L. de Galan, *Water Res.* **1990**, 24, 797–800.
81 S. W. Krasner, M. J. McGuire, J. G. Jacangelo, N. L. Patania, K. M. Reagan, M. E. Aieta, *J. Am. Water Works Assoc.* **1989**, 81, 41–53.
82 E. C. Nieminski, S. Chauduri, T. Lamoreaux, *J. Am. Water Works Assoc.* **1993**, 85, 98–105.
83 R. J. Bull, F. C. Kopfler, AWWA Research Foundation, Denver, CO, **1991**.
84 United States Environmental Protection Agency, Federal Register, **1991**, 56, 1470–1474.
85 United States Environmental Protection Agency, Federal Register, **1996**, 61, 24353–24388.
86 C. Na, T. M. Olson, *Environ. Sci. Technol.* **2004**, 38, 6037–6043.

87 E. J. Pedersen, E. T. Urbanski, B. J Marinas, D. W. Margerum, *Environ. Sci. Technol.* **1999**, 33, 4239–4249.
88 I. N. Najm, S. W. Krasner, *J. Am. Water Works Assoc.* **1995**, 87, 106–115.
89 W. Lykins, R. M. Clark, United States Environmental Protection Agency EPA 600/9–88/004, Washington, DC, **1988**.
90 R. Paode, G. L. Amy, S. Krasner, S. Summers, E. W. Rice, *J. Am. Water Works Assoc.* **1997**, 89, 79–93.
91 B. Legube, in: O. Hutzinger, *The Handbook of Environmental Chemistry*, pp. 95–116, Springer, Berlin, **2003**.
92 R. J. Miltner, H. M. Shukairy, R. S. Summers, *J. Am. Water Works Assoc.* **1992**, 84, 53–62.
93 G. Gilli, *J. Ozone Sci. Eng.* **1990**, 12, 116–125.
94 S. D. Richardson, T. V. Caughran, A. D. Thruston, T. W. Collette, K. M. Schenck, B. W. Lykins, in: *Proc. Int. Conf. Disinfection By-products: The way forward* **1998**, Vol. 245, pp 46–53, Royal Soc. Chem, Cambridge, **1999**.
95 S. D. Richardson, A. D. Thruston, T. V. Caughran, P. H. Chen, T. W: Collette, T. L. Floyd, K. M. Schenck, B. J. Lykins, G.-R. Sun, G. Majetich, *Environ. Sci. Technol.* **1999**, 33, 3368–3377.
96 S. D. Richardson, T. V. Caughran, T. Poiger, Y. Guo, F. G. Crumley, *Ozone Sci. Eng.* **2000**, 22, 653–675.
97 C. Zwiener, F. H. Frimmel, in: P. A. Wilderer, J. Zhu, N. Schwarzenbeck (Eds.), *Water in China – Water and Environmental Management Series*, pp 59–67, IWA Publishing, London, **2003**.
98 C. Zwiener, T. Glauner, F. H. Frimmel, in: I. Ferrer, E. M. Thurman (Eds.), *Liquid Chromatography/Mass Spectrometry, MS/MS and Time-of-Flight MS – Analysis of Emerging Contaminants*. ACS Symp. Series, Vol. 850, Chapter 21, pp 356–375, Am. Chem. Soc. and Oxford Univ. Press, Washington, **2003**.
99 C. Ventresce, G. Bablon, A. Jadas-Hécart, *Ozone Sci. Eng.* **1991**, 13, 91–107.
100 J. Hemming, B. Holmbom, M. Reunanen, L. Kronberg, *Chemosphere* **1986**, 15, 549–556.
101 A. Smeds, T. Vartiainen, J. Mäki-Paakkanen, L. Kronberg, *Environ. Sci. Technol.* **1997**, 31, 1033–1039.
102 D. T. Shaughnessy, T. Ohe, S. Landi, S. H. Warren, A. M. Richard, T. Munter, R. Franzen, L. Kronberg, D. M. de Marini, *Environ. Mol. Mutagen.* **2000**, 35, 106–113
103 R. T. Lalonde, B. Lin, A. Henwood, J. Fiumano, L. Zhang, *Chem. Res. Toxicol.* **1997**, 10, 1427–1436.
104 N. Guttmann-Bass, M. Bairey-Albuquerque, S. Ulitzur, A. Chartrand, C. Rav-Acha, *Environ. Sci. Technol.* **1987**, 21, 252–260.
105 J. R. Meier, R. D. Lingg, R. J. Bull, *Mutat. Res.* **1983**, 118, 25–41
106 P. Andrzejewski, J. Nawrocki, in: O. Hutzinger, *The Handbook of Environmental Chemistry*, pp. 61–94, Springer, Berlin, **2003**.
107 J. M. Wright, J. Schwartz, T. Vartiainen, J. Maki-Paakkanen, L. Altshul, J. J. Harrington, D. W. Dockery, *Environ. Health Perspect.* **2002**, 110, 157–164.
108 Umweltbundesamt UBA (Environmental Protection Agency), Germany, internet: http://www.umweltbundesamt.de/wasser/themen/trink7.htm, accessed **12/2004**.
109 M. J. Plewa, E. D. Wagner, P. Jazwierska, S. D. Richardson, P. H. Chen, A. B. McKague, *Environ. Sci. Technol.* **2004**, 38, 62–68.
110 S. W. Krasner, R. Chinn, C. J. Hwang, S. E. Barrett, Proc. 1990 AWWA Water Qual. Technol. Conf., AWWA, Denver, CO, **1991**.
111 J. Hoigne, H. Bader, *Water Res.* **1988**, 22, 313–319.
112 D. B. Jobb, R. B. Hunsinger, O. Meresz, V. Y: Taguchi, Proc. AWWA Water Qual. Technol. Conf., Denver, CO, **1995**.
113 W. A. Mitch, J. O. Sharp, R. R. Trussell, R. L. Valentine, L. Alvarez-Cohen, D. L. Sedlak, *Environ. Eng. Sci.* **2003**, 20, 389–404.
114 J. W. A. Charrois, M. W. Arend, K. L. Froese, S. E. Hrudey, *Environ. Sci. Technol.* **2004**, 38, 4835–4841.
115 J. H. Choi, S. E. Duirk, R. L. Valentine, *J. Environ. Monit.* **2002**, 4, 249–252.
116 J. H. Choi, R. L. Valentine, *Water Res.* **2002**, 36, 817–824.

117 A. C. Gerecke, D. L. Sedlak, *Environ. Sci. Technol.* **2003**, 37, 1331–1336.
118 S. D. Richardson, *personal communication*, June **2005**.
119 A. D. Nikolaou (Volume Ed.), in: O. Hutzinger (Ed.), *The Handbook of Environmental Chemistry*, Springer, Berlin, **2003**.
120 A. A. Stevens, L. A. Moore, R. Miltner, *J. Am. Water Works Assoc.* **1989**, 81, 54–60.
121 S. W. Krasner, B. M. Coffey, P. A. Hacker, C. J. Hwang, C.-Y. Kuo, A. A. Mofidi, M. J. Sclimenti, Proc. 4th Int. BOM Conf.; Int. Working Group on Biodegradable Organic Matter in Drinking Water, Waterloo, Ontario, Canada, **1996**.
122 F. C. Kopfler, H. P. Ringhand, W. E. Coleman, J. R. Meier, in: R. L. Jolley, R. J. Bull, W. P. Davis, S. Katz, S. Roberts, M. H. Jr. Roberts, V. A. Jacobs (Eds.), *Water Chlorination: Chemistry, Environmental Impact and Health Effects*, Vol. 5, Lewis Publishers, Chelsea, MI, USA, **1984**, pp. 161–173.
123 D. Khiari, S. W. Krasner, C. J. Hwang, R. Chinn, S. E. Barrett, Proc. 1996 Am. Water Works Assoc. Water Qual. Technol. Conf.; American Water Works Association: Denver, CO, USA, **1997**.
124 H. S. Weinberg, *Anal. Chem.* **1999**, 71, 801A–808A.
125 X. Zhang, R. A. Minear, Y. Guo, C. J. Hwang, S. E. Barrett, K. Ikeda, Y. Shimizu, S. Matsui, *Water Res.* **2004**, 38, 3920–3930.
126 X. Zhang, R. A. Minear, S. E. Barrett, *Environ. Sci. Technol.* **2005**, 39, 963–972.
127 A. Lopez, G. Mascolo, G. Tiravanti, R. Passino, *J. Anal. Chem.* **1998**, 53, 856–860
128 C. G. Zambonin, I. Losito, F. Palmisano, *Rapid Commun. Mass Spectrom.* **2000**, 14, 824–828.
129 C. H. Lin, R. N. Lerch, H. E. Garrett, M. F. George, *J. Agric. Food Chem.* **2003**, 51, 8011–8014
130 G. Mascolo, A. Lopez, R. Passino, G. Ricco, G. Tiravanti, *Water Res.* **1994**, 28, 2499–2506
131 S. Nelieu, L. Kerhoas, J. Einhorn, *Environ. Sci. Technol.* **2000**, 34, 430–437.
132 J.-Y. Hu, T. Aizawa, S. Ookubo, *Environ. Sci. Technol.* **2002**, 36, 1980–1987.
133 K. Moriyama, H. Matsufuji, M. Chino, M. Takeda, *Chemosphere* **2004**, 55, 839–847.
134 J. Hu, S. Cheng, T. Aizawa, Y. Terao, S. Kunikane, *Environ. Sci. Technol.* **2003**, 37, 5665–5670.
135 M. M. Huber, T. A. Ternes, U. von Gunten, *Environ. Sci. Technol.* **2004**, 38, 5177–5186.
136 M. Petrovic, A. Diaz, F. Ventura, D. Barcelo, *Environ. Sci. Technol.* **2003**, 37, 4442–4448.
137 M. C. Dodd, C.-H. Huang, *Environ. Sci. Technol.* **2004**, 38, 5607–5615.
138 T. P. J. Kull, P. H. Backlund, K. M. Karlsson, J. A. O. Meriluoto, *Environ. Sci. Technol.* **2004**, 38, 6025–6031.
139 M. Sörensen, H. J. Reichert, F. H. Frimmel, *Acta hydrochim. hydrobiol.* **2001**, 29, 301–308.
140 T. Glauner, C. Zwiener, F. H. Frimmel, 70th Meeting of the Waterchemical Society, Bad Saarow, May 17–19, **2004**.
141 E. Böhler, H.-J. Brauch, U. Müller, W. Schmidt, Technologiezentrum Wasser Karlsruhe (TZW) **2003**.
142 S. A. Andrews, M. J. Ferguson, in: R. A. Minear, G. L. Amy (Eds.), *Disinfection By-products in Water Treatment*, Chelsea, Michigan, Lewis Publishers, **1995**.
143 W. Schmidt, *Ozone Sci. Engin.* **2000**, 22, 215–226.
144 M. S. Siddiqui, G. L. Amy, *J. Am. Water Works Assoc.* **1995**, 87, 58–70.
145 C. Douville, G. L. Amy, in: *Natural Organic Matter and Disinfection By-products*, S. E. Barrett, S. W. Krasner, G. L. Amy (Eds.), ACS Symp. Series 761, pp. 282–298, Am. Chem. Soc.,Washington, **2000**.
146 C. A. Delcomyn, H. S. Weinberg, P. C. Singer, Proc. Am. Water Works Assoc. Water Qual. Technol. Conf., Am. Water Works Assoc., Denver, CO, **2000**.
147 F. H. Frimmel, in: A. Grohmann, U. Hässelbarth, W. Schwerdtfeger (Eds.), pp. 625–635, Erich Schmidt, Berlin, **2002**.
148 P. C. Singer, G. W. Harrington, Proc. Am. Water Works Assoc. Water Qual. Technol. Conf., Denver, CO, **1993**.

149 G. Kleiser, G. Schmit, F. H. Frimmel, *gwf Wasser Abwasser* **1999**, 140, 396–403.
150 F. H. Frimmel, S. Hesse, G. Kleiser, in: *Natural Organic Matter and Disinfection By-products*, S. E. Barrett, S. W. Krasner, G. L. Amy (Eds.), ACS Symp. Series 761, pp. 84–95, Am. Chem. Soc.,Washington, **2000**.
151 W. Kühn, B. Wricke, *gwf Wasser Special* **1995**, 136, S92–S98.
152 B. Wricke, Bd. 17, *Veröff. Technologiezentrum Wasser*, Karlsruhe, **2002**.
153 C. Rav-Acha, A. Serri, E. Choshen, B. Limoni, *Water Sci. Technol.* **1985**, 17 (4–5), 611–621
154 S. D. Richardson, A. D. Thruston, T. W. Collette et al., *Environ. Sci. Technol.* **1994**, 28, 592–609.
155 R. M. Clark, J. Q. Adams, B. W. Kyjins, *J. Environ. Eng.* **1994**, 120, 759–771.
156 United Nations, Water for Life 2005–2015, internet: http://www.un.org, accessed June 2005.

11
Toxicology and Risk Assessment of Pharmaceuticals

Daniel R. Dietrich, Bettina C. Hitzfeld, and Evelyn O'Brien

11.1
Introduction

For many years, the main focus of environmental pollution has been on chemical and pharmaceutical manufacturers. Spectacular accidents, such as that at the BASF facility in Ludwigshafen, Germany in 1948, the Union Carbide plant in Bhopal, India in 1984 and the Sandoz facility in Schweizerhalle, Switzerland, 1986, among others, provided the impetus for improving safety and spillage regulations, carrying out more environmental modeling, and improving the analysis of environmental samples. More recently, attention has turned to the far less dramatic but nonetheless important consideration of the potential effects of medication residues on the aquatic environment. As a complete analysis of national and international regulations governing all polar organic pollutants is beyond the scope of this chapter, only regulations and concerns with specific relevance for pharmaceuticals will be discussed here. Waste Water Treatment Plants (WWTPs) [1] have been identified as the major point source of pharmaceuticals in surface waters, but diffuse intake via run-off can also be an important route for veterinary pharmaceuticals [2]. The realization of the latter situation has led to a parallel increase in the number of surveys of levels of contamination of surface waters, accompanied by attempts to characterize the potential risks posed by these substances to the aquatic environment [3–7]. These studies have led to the recognition that while the concentrations in the WWTPs are generally related to the population equivalents of the respective regions and the highest concentrations of pharmaceuticals measured at sites where WWTP effluent is introduced into the receiving water body, the concentration of the substances found downstream of the WWTPs is a function of the tributary rivers, which can cause either an increase or a decrease in the concentration of a given substance in the main river, depending on their own pre-loads [1]. While many pharmaceutically active substances can indeed be broken down in modern sewage treatment plants, by their very nature and, perhaps more importantly, because of their continual infusion into the environment, several of them can be classified as persistent polar organic pollutants. The efficiency of the remov-

Organic Pollutants in the Water Cycle. T. Reemtsma and M. Jekel (Eds.).
Copyright © 2006 WILEY-VCH Verlag GmbH & Co. KGaA, Weinheim
ISBN: 3-527-31297-8

al of pharmaceuticals from wastewater in WWTPs can vary in accordance with a number of factors, including the population served, precipitation volumes, ambient temperature, and nutrient loads. The particular substance classes delivered in sewage to the WWTP can also influence the extent of substance removal. For example antimicrobials and antineoplastics in sewage are by their very nature toxic to the microbes present in the sludge. Therefore, some pharmaceuticals not only reduce the efficiency of their own removal, but can also impair the removal of all other substances. Metabolically conjugated derivatives of pharmaceuticals can be released from WWTPs and, as such, are not detected or accounted for in the determination of the concentrations of the parent compounds. Consequently, unless metabolically converted parent compounds have been specifically investigated and analytically accounted for, the absolute concentration of pharmaceuticals can be higher than that anticipated with present analytical schemes. Furthermore, as many of these conjugated compounds can be converted back to the parent substance, any derived environmental effect or risk assessment should consider the concentrations of both the parent compound and of reversibly-conjugated parent compounds.

In recent years, over eighty compounds have been found by researchers in rivers, lakes, and sediments in several countries [8], and it is likely that the number of compounds found will rise exponentially with improvements in analytical methods. Despite the plethora of reports of pharmaceuticals being found (a summary of which can be found in an article by Daughton and Ternes published in 1999 [9]), these reports represent only a fraction of the pharmaceuticals on the market, and most agents have not as yet been investigated. Data on maximal environmental concentrations measured in surface waters is not available for every country or in certain cases is not in the public domain. For example, data on the occurrence and concentration of several common surface water contaminants could be found for Germany but not for French surface waters and vice versa [10]. This dearth of information in the public domain makes risk assessment dependent on the availability from the manufacturer of correct estimates of market share and total amounts produced.

Environmental risk assessment is usually performed in a tiered fashion, as described in the following sections and shown in Figs. 11.3, 11.4, and 11.5, and requires data on biological effects and expected substance exposure levels. Exposure assessment looks at the release and fate of the substance into the environment and determines a predicted environmental concentration (PEC). The effect is extrapolated from the lowest NOEC of toxic effects in three trophic levels (algae, invertebrate, fish), with the inclusion of safety factors (usually a total of 1000 is assumed). In an initial assessment (first tier), a PEC/PNEC ratio is calculated, where PNEC is the predicted no effect concentration, and the decision on the safety of a substance is based on this ratio. If the PEC/PNEC ratio is <1, then it is assumed that the substance poses little environmental risk. If the PEC/PNEC ratio is >1, then a risk is assumed and a refined risk assessment is usually performed. In an effort to collate data on PEC/PNEC ratios, Webb [11] collected data on consumption of pharmaceuticals in order to derive predicted environmental concentrations. No metabolism, 100% loss to drain after use by humans, no removal in wastewater treatment plants, and no dilution in surface waters were assumed. This very crude assessment showed that out of all 60 investigated drugs, which account for about 50% of known pharmaceutical usage in tons, all but eight have PEC/PNEC ratios less than one. The eight substances were paracetamol (analgesic, antipyretic; PEC/PNEC 39.92), aspirin (analgesic, antipyretic, antiinflammatory; PEC/PNEC 1.00), dextropropoxyphene (narcotic analgesic; PEC/PNEC 2.06), fluoxetine (antidepressant; PEC/PNEC 14.19), oxytetracycline (antibiotic; PEC/PNEC 26.8), propanolol (antihypertensive, antianginal, antiarrhythmic; PEC/PNEC 1.16), amitriptyline (antidepressant; PEC/PNEC 1.29) and thioridazine (antipsychotic; PEC/PNEC 2.59). Based on the current draft EU technical guidance document (see below), a refined risk assessment would therefore be advised for these compounds. By taking into account elimination in wastewater treatment and dilution in surface water (a factor 10 is normally assumed), this in most cases would result in refined PEC/PNEC ratios below one. Paracetamol, for example, is eliminated by up to 98% in wastewater treatment plants, mostly by biodegradation, and this would lead to a refined PEC/PNEC ratio of 0.08. Furthermore, paracetamol is usually not detected in surface waters, and the calculation of a MEC/PNEC ratio (where MEC is the maximum environmental concentration) was given as <0.02. Equally, for all of the other eight substances, such a refined risk assessment would result in PEC/PNEC ratios below one.

A similar study using a slightly different approach (no metabolism in man, no elimination in sewage treatment, but 10-fold dilution factor in surface waters) came to similar conclusions [12]. In this study, assessment of paracetamol (PEC/PNEC 1.29), amoxycillin (PEC/PNEC 588.02), oxytetracycline (PEC/PNEC 3.60), diclofenac (PEC/PNEC 3.16) and mefenamic acid (anti-inflammatory; PEC/PNEC 1.03) gave rise to PEC/PNEC ratios >1 when using the most sensitive endpoint in acute toxicity testing. Of course, acute endpoints are not very sensitive, especially when we consider compounds with specific mechanisms as discussed below.

Unfortunately, aquatic toxicity data have not been published for the majority of those substances which have been confirmed to be present in the aquatic environment. However, some theoretical estimation of possible toxic interactions can be gleaned from a comparison of mammalian toxicity data (kinetics and dynamics) with human therapeutic plasma levels and those expected to occur in fish plasma [13]. Huggett and co-workers used the ratio of human therapeutic plasma concentration to steady state fish plasma concentration as an indicator of potential interaction in fish, the lower the ratio the higher being the likelihood of a possible long-term effect on fish upon chronic exposure. Although such calculations are initially helpful, only experimental validation can strengthen the predictive capability of such estimations. In view of the paucity of data available for such model estimations, current risk assessment strategies rely on much cruder risk estimation schemes for prioritizing pharmaceuticals and are primarily dependent on a drug-by-drug detailed evaluation of environmental risk as well as on rather arbitrary non-scientifically validated cut-off or decision "triggers". In the following sections, the different risk assessment procedures under discussion or already implemented in various countries and regions will be presented. The European and Canadian systems are summarized, and specific differences between these and the United States and Japanese regulations are outlined.

11.2
A Comparison of International Risk Assessment Procedures

11.2.1
The European Union Technical Guidance Document (TGD)

A council directive of the European Commission (2001/83/EC) stipulates that an environmental risk assessment (ERA) must be carried out before marketing authorization can be given for medicinal products for human use [14]. In contrast to the Canadian system (see below), this regulation deals with the use, storage, and disposal of the product, and synthesis and manufacture are treated under separate legislation, as are veterinary products. Furthermore, whereas the Canadian regulations govern new substances and medications with new indications, the guidelines of the EU require the generation of an ERA for new substances, those with new indications (type two variations), and existing active pharmaceutical ingredients (APIs) with a specific mode of action, effectively including all pharmaceutically-active substances. The two-phased stepwise procedure for the generation of this ERA is described in the "Guideline on Environmental Risk Assessment of Medicinal Products for Human Use" [15], written by the European Agency for the Evaluation of Medicinal Products, which was re-released in draft form for consultation in January 2005.

In this draft guidance document, phase I aquatic risk assessment begins with an estimation of the concentration to which aquatic organisms may be exposed. This is generally expressed as the risk quotient of either the PEC or MEC, and the PNEC. If, as briefly described above, this ratio exceeds one, then an ecological risk is assumed. Worst-case scenarios involving no human metabolism and no degradation within the WWTP of the compound under review are generated. The total population of the country in question is factored into the equation as well as the estimated daily production of wastewater per head of population and the market penetration (fpen) of the compound. The so-called predicted environmental concentration in surface water ($PEC_{surfacewater}$) is calculated based on several factors, as indicated in Fig. 11.1. An even distribution of product use throughout the year and throughout the geographic area is assumed.

$$PEC_{surface\ water} = \frac{DOSEai * Fpen}{WASTEinhab * DILUTION * 100}$$

Fig. 11.1 Formula for $PEC_{surfacewater}$ estimation. DOSEai: maximum daily dose of active ingredient per inhabitant; Fpen: percentage of market penetration (defaults to 1% based on a survey of approx 800 APIs currently marketed); WASTEinhab: amount of wastewater per inhabitant per day (defaults to 200 L inhabitant^{-1} day^{-1}); DILUTION: dilution factor (defaults to 10).

If the initial calculated PEC$_{surfacewater}$ is less than 0.01 µg L^{-1}, then the substance is considered not to represent a risk to the aquatic environment and no actual toxicity testing is required. A PEC$_{surfacewater}$ greater than 0.01 µg L^{-1} makes the compound subject to phase II analysis. Based on this formula and the associated defaults, the maximum dosage not resulting in a PEC$_{surfacewater}$ of greater than 0.01 µg L^{-1} can be calculated as 2 mg patient^{-1} day^{-1}. Substances with a PEC of greater than 0.01 µg L^{-1} proceed to phase II: tier A of the assessment procedure (see Fig. 11.3). However, these action limits are not applicable when an expert evaluation of preclinical safety and ecotoxic potential suggests potential atypical ecotoxic effects. In phase II tier A, environmental fate and effects testing are required, and the PEC must be refined using substantiated information on predicted use and market share. This assessment should include physico-chemical parameters such as water solubility, K_{ow}, vapor pressure, etc., as well as information pertaining to the biodegradability, photolysis, hydrolysis, and aerobic and anaerobic transformation potentials. These tests should be carried out in accordance with the relevant OECD guidelines as outlined in Table 11.1. A standard base set of acute aquatic toxicity data at three trophic levels (algae, OECD 201; daphnia, OECD 211; fish early life stage, OECD 210; and activated sludge respiration, OECD 209) is also required in order to generate the PNEC via the application of an assessment factor (AF) to the

$$PNEC_{surfacewater} = \frac{\text{Lowest acute LC}_{50} \text{ or EC}_{50} \text{ or NOEC}}{}$$

Fig. 11.2 Formula for PNEC$_{surfacewater}$ estimation.

Tab. 11.1 Set of physico-chemical data and fate studies required at the start of phase II, tier A, of the EU Technical Guidance Document.

Data required/Test	OECD Guideline
Water solubility	105
Dissociation constant	112
UV-visible absorption spectrum	101
Melting temperature	102
Vapor pressure	104
K_{ow}	107 or 117
Adsorption-desorption using batch equilibrium method	106
Ready biodegradability	301
Aerobic and anaerobic transformation in aquatic sediment systems	308
Photolysis (optional)	Seek regulatory guidance or use OECD monograph No. 61
Hydrolysis as a function of pH (optional)	111

lowest determined LC_{50}, EC_{50}, or NOEC value. AFs are applied in order to account for the extrapolation from acute to chronic toxicity, interspecies variations in sensitivity, intraspecies variability, and the extrapolation from laboratory to field scenarios. Generally, when just the basic set of acute toxicity data is available, an AF of 1000 should be applied; however, when more information on the effects of long-term exposure is available, the AF may be as low as 10.

Tier A also requires the estimation of the $PNEC_{microorganisms}$, as substances which are toxic to microorganisms are likely to have detrimental effects on the potential for substance removal in the WWTP. A $PEC_{surfacewater} : PNEC_{microorganism}$ ratio of greater than one results in a requirement for tier B testing. If, on the other hand, neither $PEC_{surfacewater} : PNEC_{microorganism}$ nor $PEC_{surfacewater} : PNEC_{aquatic}$ is greater than one, and the $\log K_{ow}$ is less than three, it is assumed that the substance is unlikely to have a negative impact on the aquatic environment, and no further testing is required. A K_{oc} of greater than 10 000, signifying a high affinity of the substance for sewage sludge, indicates a need for an assessment of the substance in the terrestrial environment. Such testing schemes are also outlined in the TGD, but will not be dealt with further in this chapter. Details are available under http://www.emea.eu.int. Tier B of phase II of the risk assessment process aims to assess which compartments of the environment are particularly at risk and employs specific test systems to assess the individual compartments (biota, microorganisms, aquatic organisms, terrestrial organisms, and sediment dwellers). Tier B furthermore considers human and environmental metabolites if they constitute more than 10% of the API.

Perhaps two of the most critical aspects of the ERA procedure currently proposed by the EU are the use of a cut-off approach based on the predicted environmental concentration and the use of acute toxicity testing to generate the predicted no-effect concentration of a substance. Although these premises at first glance appear logical, closer examination reveals several problems. The former dictates that if the PEC exceeds 0.01 µg L^{-1}, a complete assessment of the potential of a substance to be problematic must be carried out. The majority of PECs overestimate the concentrations of pharmaceutically active substances reaching the aquatic environment by approximately one order of magnitude, although the PECs of substances may also indeed be greatly underestimated because of the availability of the active ingredient in other formulations and in over-the-counter preparations. One notably exceptional result is the finding that estrogen has been detected at concentrations of 130 µg L^{-1} in an area where a PEC, based on prescription contraceptives, of just 1ng L^{-1} had been predicted [16]. The higher concentration was probably due to a combination of the excretion of natural estrogen by humans combined with estrogens of animal origin. These examples illustrate the manifold uncertainties in this approach. In addition, the relationship between the PEC and the maximal environmental concentrations (MEC = measured ec) in either treatment plant effluent (MEC_{effl}) or surface water (MEC_{sw}) for any given substance is dependent on the particular sewage treatment processes employed, with longer retention times and more complicated processes generally resulting in a lower delivery of pharmaceutically active compounds to the aquatic environment. While this leads to a general

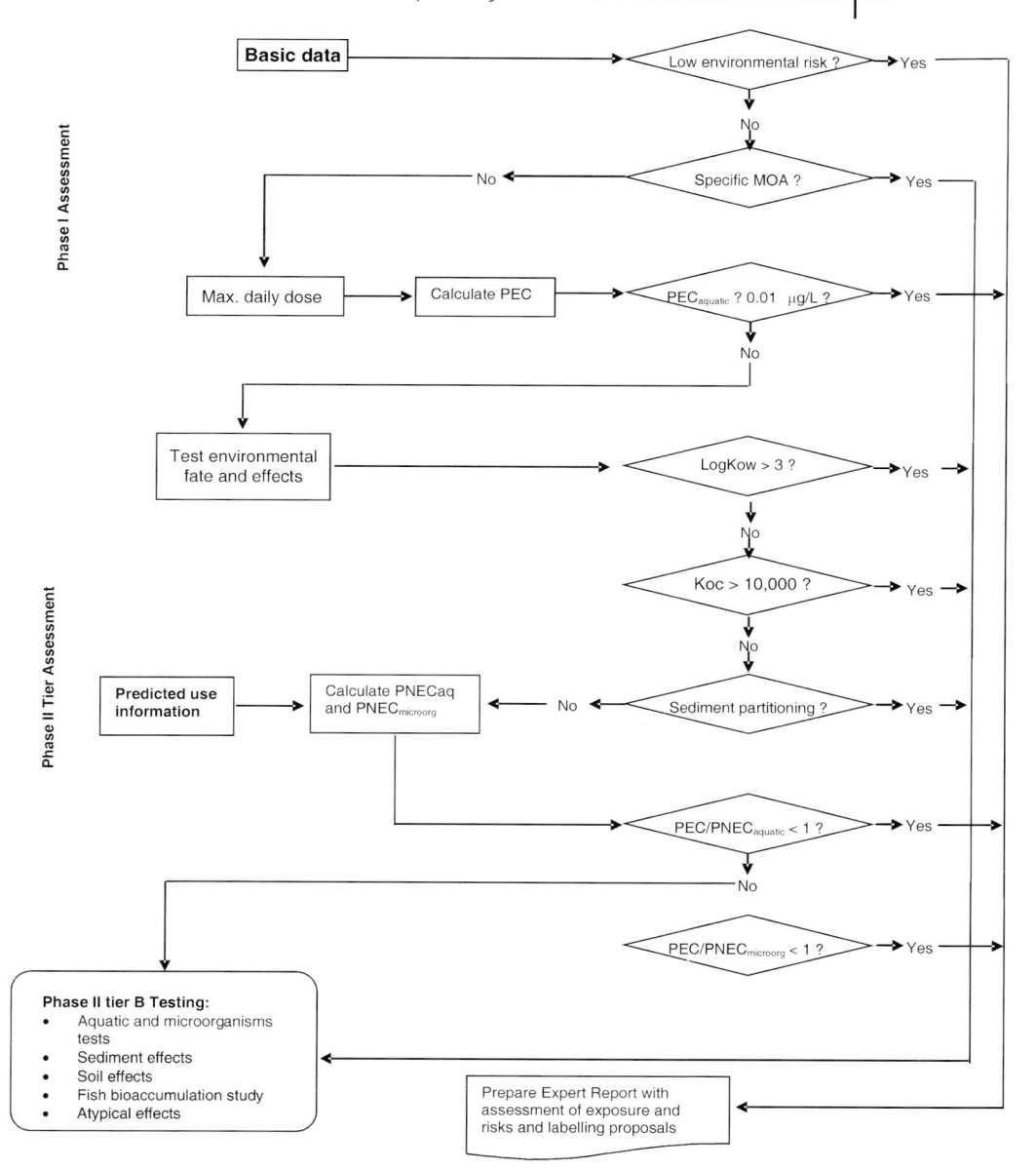

Fig. 11.3 Schematic diagram of the environmental risk assessment proposed in the EU-TGD.

increase in the level of safety built into the risk assessment, it may result in unnecessary testing of compounds which are in fact harmless.

Stuer-Lauridsen and co-workers attempted to carry out a risk assessment for the 25 most used human pharmaceuticals in Denmark [16]. In this study, the authors applied the criteria proposed by the EU draft guideline document for new pharmaceuticals. Briefly, a risk quotient was generated using the PEC and PNEC and a safety factor of 1000 (AF = 1000) was employed in the generation of the PNECs. Using this approach, all of the evolving PECs exceeded $0.01\,\mu g\,L^{-1}$, making all of these substances candidates for phase II testing under the proposed EU regulations. Interestingly, several compounds for which estimates indicated an extremely high likelihood of relevant concentrations in WWTP effluent have to date not been detected in the environment [17–19]. Two possible explanations exist for this: either (a) the substance in question cannot as yet be detected in water bodies at environmentally relevant concentrations, or (b) the substance is not to be found in the aquatic environment, regardless of the calculated predictions. The latter would call into question the validity of using estimates to determine the "safe" concentrations.

Lack of relevant data on the toxicity of most pharmaceuticals to aquatic organisms as well as of data pertaining to environmental concentrations of these substances logically prevents the calculation of risk quotients based on actual measurements. At best, only predictions can be made. This is exacerbated by the lack of standardized analytical methods for the detection of these substances at environmentally relevant concentrations, as already mentioned.

A further problem arises when one attempts to assess potential chronic effects of pharmaceuticals in the aquatic environment. The use of the maximum suggested AF of 1000 to account for acute to chronic extrapolation is, in many cases, not sufficient, as acute to chronic ratios (ACRs) ranging between 0.79 and 5000 have been demonstrated within a single species [20, 21]. This becomes even more critical when one considers certain hormonally active substances where ACRs may be far greater. Indeed an ACR of 800 000 for ethinylestradiol in rainbow trout was described by Webb [22], clearly illustrating that little if any useful information about the extrapolation of potential chronic toxicity can be gleaned from acute toxicity data. This further highlights the problems associated with using cut-off triggers and ratio approaches rather than improved scientific understanding of potential toxicological effects. Fortunately, the new release of the EU Draft guideline strongly embraces expert evaluation of mammalian toxicokinetic and toxicodynamic data as a means of deriving potential ecotoxic effects. Indeed, as many enzyme/receptor systems are highly conserved across the different phyla, specific target interactions of given pharmaceuticals in species other than the humans are to be expected, although the dynamically required doses and the array of effects may be different. Indeed, the reported nephrotoxic effects of diclofenac in rodents and other mammals were also demonstrated to occur in fish [23–25], albeit at much lower (i.e. close to environmentally relevant) concentrations. Indeed, despite the fact that the PEC calculations using the EU guideline provisions predicted an environmental concentration of diclofenac of $0.54\,\mu g\,L^{-1}$, reported values for diclofenac, with a $\log K_{ow}$ of 4.51, reach concentrations of up to $1.2\,\mu g\,L^{-1}$ and higher [26]. The no-observed-effect concentration (NOEC) in the study by Hoeger et al. [23] was determined to be $0.5\,\mu g\,L^{-1}$ for monocyte infiltration/accumulation in livers of brown trout exposed for 21 days, although mild effects were seen in two out of six animals in the $0.5\,\mu g\,L^{-1}$ group.

The calculation of a PNEC requires the application of an assessment factor of 10 to account for inter-/intraspecies variations and extrapolation from laboratory data to field impact. The "Draft Guideline on the Environmental Risk Assessment of Medicinal Products" [15] states, however, that the application of an AF of 1000 to acute data will not be protective for pharmaceuticals, especially as the acute-to-chronic calculations have been demonstrated to show extreme divergence in predictive capability within one species and even poorer prediction for multiple species. Based on this it is considered justified to base the PNEC not on acute but rather on chronic data. As the data from Hoeger et al. and Schwaiger and co-workers illustrate, [23–25] neither presents an acute or a chronic exposure scenario but rather a subchronic situation; an AF of 100 would most likely be acceptable as a basis for calculation. A PNEC of 0.005 µg L^{-1} (NOEC/100) would therefore be derived as a conservative scenario. In conjunction with a PEC of 0.54 µg L^{-1}, this results in a PEC/PNEC ratio of >100, calling for further investigations and a refined risk assessment for this substance. As diclofenac is a cycloxygenase inhibitor, pharmaceuticals in the same effect class should be investigated conjointly as single entities as well as mixtures rather than individually for risk assessment purposes. A similar premise should be applied to other pharmaceutical classes.

11.2.2
US-EPA

The environmental risk assessment process in the United States, which has been in existence and under constant review since 1977 under the auspices of the National Environmental Policy act of 1969, begins with a new drug application, which is submitted to the FDA [27]. Many of the steps involved are similar to the proposed European scheme and also display similarities to the Canadian registration system. Therefore only major differences will be outlined in this section. Within the application, the manufacturer is required to provide an estimate of the amount of the drug entering the environment. This is termed the "expected introductory concentration" (EIC) and is based on total five-year production estimates. Virtually all of the environmental assessments of pharmaceuticals in the United States have resulted in a "finding of no significant impact" (FONSI), which had been set with a cut-off threshold of 1 µg L^{-1}. This threshold was derived from the fact that no effects were observed on standard environmental test organisms at drug concentrations less than 1 µg L^{-1} in acute and chronic tests of over 60 compounds. If the EIC is lower than 1 µg L^{-1}, the drug is classified as acceptable and receives a so-called categorical exclusion, and no environmental testing is required. Remarkably, once a drug has received this status, no subsequent monitoring to confirm that the expected environmental concentrations hold true is carried out. Considerations such as the specific mode of action of the agent in question are not considered.

An EIC of greater than 1µg L^{-1} results in a requirement for a formal environmental risk assessment. Similarly to the EU system, this consists of a tiered system of ecotoxicity tests. However, in contrast to the EU TGD, the base set of data normally includes an assessment of the potential effects of the pharmaceutical on microbial respiration as well as standard test systems involving at least one algal, one invertebrate, and one fish species. This approach places more emphasis on the ability of a substance to inhibit the activity of microbes present in sewage treatment works and hence to reduce the removal capability of the facility for this and other substances.

In contrast to the EU-TGD, chronic testing is not indicated unless the drug has the potential to bioaccumulate. This is described by the log octanol-water partition co-efficient ($\log K_{ow}$). In contrast to the European system, where a $\log K_{ow}$ of 3 indicates a requirement for chronic toxicity testing, the $\log K_{ow}$ which automatically leads to chronic toxicity testing in the United States is 3.5. Details on the course of the chronic testing scheme, are available on the EPA homepage (http://www.epa.gov/epahome/resource.htm).

One of the largest differences from the European system is that the US-EPA suggests the use of a watershed-based ERA approach, i.e. the use of a geographically defined area and local information such as the definition of activities within the watershed area and assessment of stress or transfer across watershed boundaries. This approach could have high relevance for the European mainland where countries may possess several land borders yet may be joined by a water body or river, e.g., the river Rhine, which forms both borders and links between several European countries. Similar approaches within individual countries could also prove useful.

11.2.3
Japan

Based on recommendations of the OECD and in consultation with the national Industrial Structure, Health Sciences and Central Environment councils, the Japanese government concluded that the stipulations of the Japanese Chemical Substances Control Law of 1973 should be extended to include an evaluation and regulation of the adverse effects of chemical substances on living organisms in the environment. Further, the efficiency and effectiveness of risk management strategies should be improved [28]. The new regulations were promulgated in May 2003. As in the case of the US-EPA, only the major differences with the European and Japanese processes will be outlined in the following section.

New chemical substances with a total yearly domestic manufacture or import volume of less than one ton are exempt from prior evaluation processes, as are substances with a total yearly domestic manufacture or import volume of up to ten tons if these substances are judged to be persistent but without the potential to bioaccumulate.

In recognition of the importance of endocrine-active compounds, the Ministry of Economy, Trade, and Industry (METI) has started a program of three-dimensional QSAR analyses in order to generate a screening system for compounds with the potential for endocrine disruption (EDC). *In-vitro* screening techniques are currently being evaluated in parallel [29] and are subject to a harmonized pre-validation and validation exercise under the auspices of the OECD [30]. These pre-screen test scenarios focus on the mode of action of the putative EDC, for example, sex hormone receptor recognition or binding, arylhydrocarbon receptor (AhR) recognition or binding, or effects on aromatase activity. Although primarily directed on human risk assessment, these *in-vitro* methodologies as well as QSARs are considered to be transferable for use in environmental risk assessment. Despite these recent and promising developments it has been difficult to assess whether or not METI has regulated any pharmaceutical with regard to its inherent and specific pharmacological properties and consequently putative toxicological properties in the environment, e.g., endocrine modulating capabilities.

11.2.4
Canada

Several considerable differences exist between the Canadian regulations and those described for the European Union, the United States, and Japan. These differences and their benefits/disadvantages are discussed in this section. In Canada, the regulation of pharmaceuticals falls under the control of the Canadian Environmental Protection Act, (CEPA) in combination with the Food and Drug Act (F&DA). These laws aim to ensure that all new substances are assessed with respect to their potential to harm the environment before introduction onto the Canadian market, and systematic assessment of substances began in September 2001 [31]. The CEPA classifies a substance as being toxic if "it enters or may enter the environment in a quantity or concentration or under conditions that (a) have or may have an immediate or long-term harmful effect on the environment or its biological diversity, (b) constitute or may constitute a danger to the environment on which life depends, or (c) constitute a danger in Canada to human life or health". In contrast to other sovereign regions, where environmental and human toxicity concerns are dealt with by two separate government agencies, these assessments are carried out by Health Canada in co-operation with Environment Canada. This is likely to improve both the availability and transfer of information as well as to expedite the evaluation process.

The principal component of the CEPA consists of the New Substance Notification Regulations (NSNRs). These regulations aim to prevent the entry of new substances into the Canadian marketplace before an assessment of their toxic potential has been carried out. In contrast to the European Union guidelines, where industrial chemicals and even veterinary and human pharmaceuticals are regulated by different documents, the CEPA encompasses all new chemical substances including those destined for research and development, for export only, and for the Canadian market, as well as site-only intermediates during the manufacture of other compounds. A new substance is defined as any chemical, polymer, or living organism which is not on the Domestic Substance List (DSL). The DSL is a list of approximately 24 000 substances known to be or have been on the Canadian market. Substances contained in this list are not subject to notification. A list of substances in products regulated under the Food and Drugs Act (F&DA) that were in commerce between January 1, 1987 and September 13, 2001 is available at www.hc-sc.gc.ca/ear-ree/1987–2001_webpost_e.html. These substances are eligible for addition to the DSL and are not subject to notification under NSNR.

Substances not on the DSL but listed on the Non-Domestic Substance List (NDSL: substances not on the DSL but believed to be in international commerce) are subject to notification; however, the comprehensiveness of the information required is reduced in consideration of information and experience in the United States. The Unites States Toxic Substances Control Act acts as a basis for this list. In co-operation with the US-EPA, the NDSL has undergone annual revision to add or remove substances. However, substances restricted, either in their manufacture or import, may not be added to the NDSL.

Notification is required for any substance not on the DSL and also for substances where a Significant New Activity (SNAc) is proposed. As in the system proposed by the European Union, the assessment consists of a stepwise process, which can be represented by a flow chart (see Fig. 11.4). However in contrast to the EU procedures, the dossier of information to be submitted for assessment does not require an estimation of the PEC, although certain information on the potential for environmental release or delivery to municipal WWTPs and identification of the facilities and water bodies is requested. Decisions are instead based on the yearly or accumulated amount of production or import. New substances are allocated into one of three classes (polymers and biopolymers, living organisms, or chemicals and biochemicals) each with specific requirements for their registration. Pharmaceuticals fall under the category of chemicals and biochemicals, which is then further divided based on the volume of manufacture or import and proposed use (see Fig. 11.5). The regulations governing transitional substances, substances produced for research and development, export-only chemicals, and site-limited intermediates are dealt with under separate sections (Schemes A, B, C and D, respectively) of the guideline document. Pharmaceuticals fall by default into the class of "all others" (Scheme E), and, as the analysis of the steps involved in the other categories mentioned is not the intention of this chapter, only the testing requirements for Scheme E will be outlined here. The reader is directed to the guideline document [31] and the CEPA text (available at http://laws.justice.gc.ca/en/C-15.31/SOR-94-260/69450.html#rid-69496) for more detailed information.

Part A of the notification submission consists of administrative and substance identity information. Part B contains technical information, which must include (but is not limited to) physical-chemical properties, mammalian toxicity, ecotoxicity, exposure information, and a list of other agencies notified. Within Scheme E, the actions taken and level of investigation required for the NSR depends solely on the amount of the substance to be produced either within one calendar year or the total accumulated production (see Fig. 11.5). Substances are allocated to one of a total of nine Schedules (I–IX), depending on the type of chemical in question and the estimated production or import. Of these, Schedules I–III are those with relevance for pharmaceuticals, and the stringency of the criteria and information to be provided increases with increasing schedule number, based on total substance amounts. Schedules I–II require standard information such as physico-chemical data, octanol-water partition coefficient, vapor pressure, water solubility, etc., as

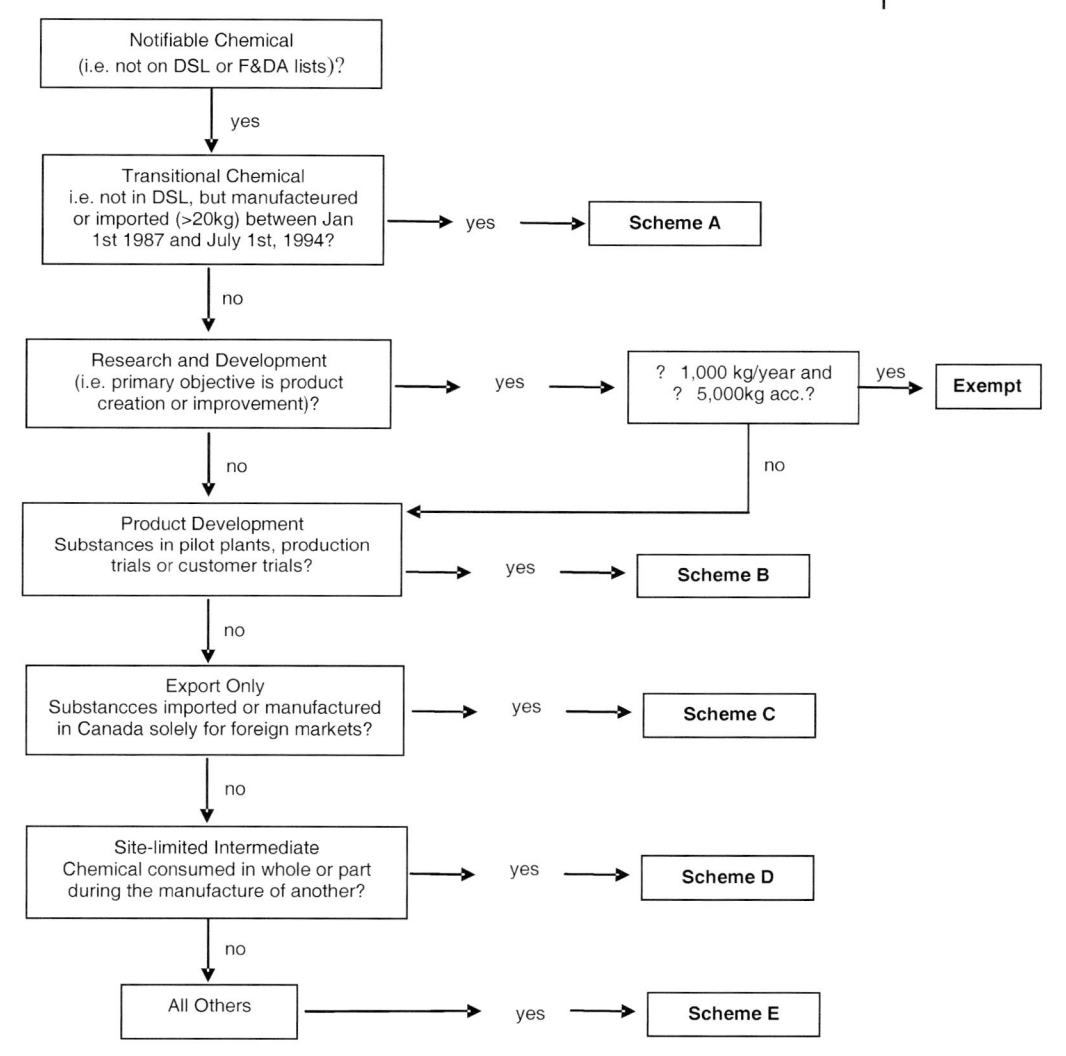

11.2 A Comparison of International Risk Assessment Procedures

Fig. 11.4 Notification requirements for new chemicals. Adapted from the Guidelines for the Notification of New Substances: Chemicals and polymers [31].

well as acute and repeated-dose mammalian toxicity, skin irritation and sensitivity data, and mutagenicity testing (details can be taken from the original source [31], p. 58 ff.). Schedule III (production/import >10 000 kg year^{-1} or >50 000 kg accumulated) is the only schedule requiring aquatic toxicity testing, although the number of tests required is minimal, comprising just fish and *Daphnia* acute toxicity tests with LC$_{50}$ determination and biodegradation tests. All testing should be carried out in accordance with the OECD guidelines valid at the time of testing (the specific OECD standard protocols for each test are listed in the guideline document), and

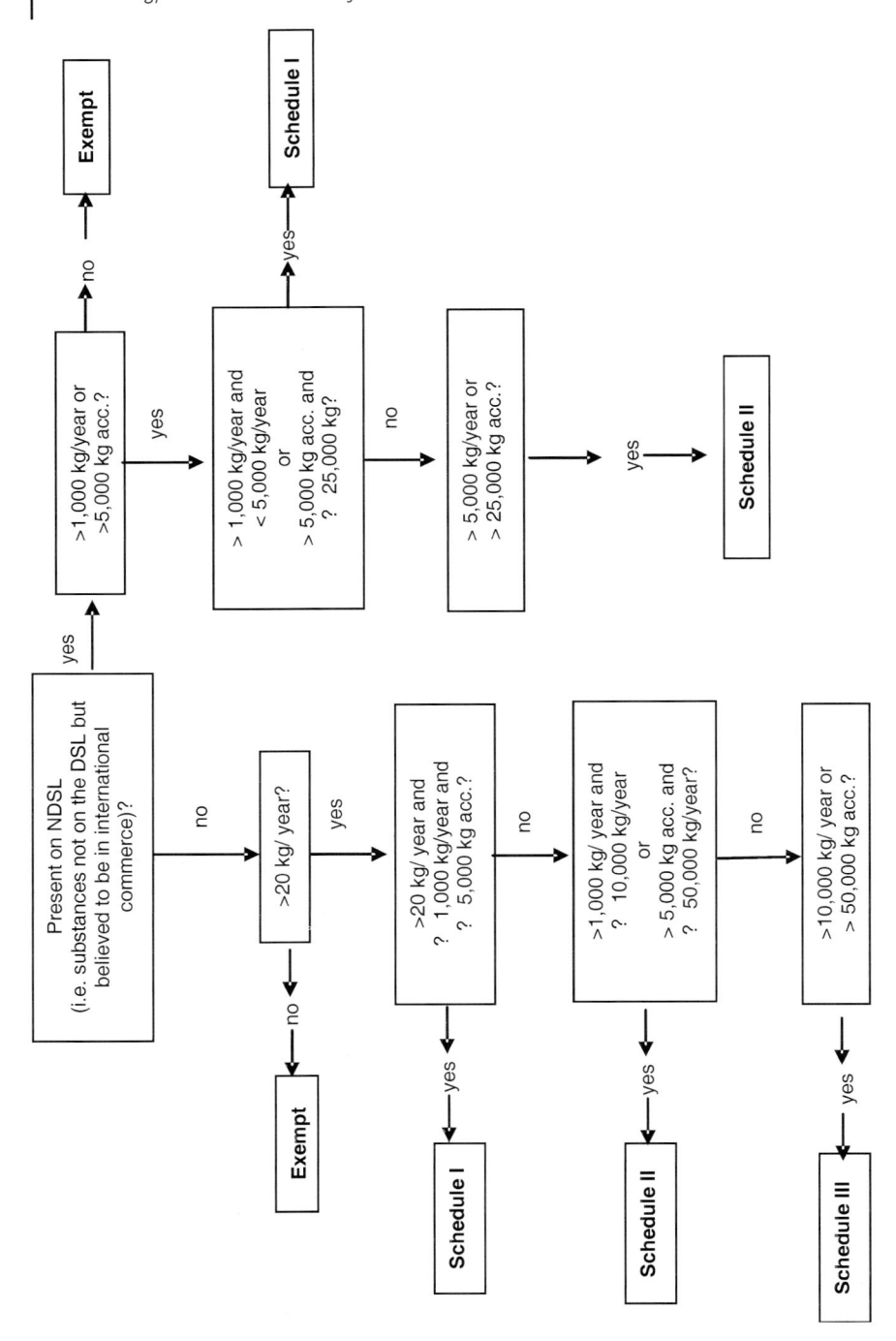

Fig. 11.5 Testing schedule for new substances as stipulated by Canadian law.

the conditions of the study must be applicable to Canada (e.g. species tested, soil or sediment used). All information relevant to the environmental toxicity of the substance should be detailed including literature reviews and database searches, and data generated for notifications in other countries is acceptable. Structure-activity relationship analyses and environmental fate modeling should also be provided. The NSN must also contain a detailed description of an analytical protocol for the detection of the substance at concentrations at or below the lowest reported LC_{50} value in fish or *Daphnia*. This illustrates a more focused approach on the part of Environment Canada to the development of appropriate analytical techniques than do any of the other regulatory documents.

In contrast to other regulations, where the trigger for testing is the PEC, under Canadian law it is the quantity of substance imported or manufactured which, if exceeded, requires the notifier to provide an NSN. Based on the information provided by the notifier, the government authorities make their assessment of the substance. If the substance is assessed as being innocuous, then production and/or import can proceed, and, if all criteria are met, the substance may then be added to the DSL as described above and thus become exempt from future notification requirements. If on the other hand the substance is suspected to have toxic potential, the government may request additional information from the notifier, impose specific conditions on the manufacture and import of the substance, or indeed prohibit its use.

11.3
The Toxicological Data Set for Environmental Risk Assessment

Although many countries now have or are in the process of preparing regulations pertaining to aquatic toxicity testing for pharmaceuticals, several problems regarding the practicality and applicability of these regulations still exist. Some of these problems and some approaches for improvement are outlined in the following section.

11.3.1
Extrapolation from Acute to Chronic Toxicity

Extrapolation from acute to chronic for risk estimation is a particularly critical aspect of current guidelines on aquatic toxicity testing. This is strikingly illustrated by the findings of Ferrari and co-workers, who assessed the chronic and acute toxicity of three pharmaceuticals (carbamazepine, clofibric acid, and diclofenac) in several different standard test systems [21]. Enormous differences in the EC_{50}, NOEC, and LOEC values for carbamazepine were apparent even in the acute assays (microtox, 30 min, *Daphnia magna*, 48 h, *C. dubia*, 48 h) depending on the type of test employed. Furthermore, test species were found to display stark variation in their relative sensitivity to the three substances tested, further illustrating that the choice of test system is critical for the generation of data which can be used to perform a reliable risk assessment for the aquatic environment. These differences were even more pronounced in the chronic test systems, with NOECs for carbamazepine ranging from >100 000 µg L^{-1} in the *Pseudokirchneriella subcapitata* 96-h test to 25 µg L^{-1} for the *Ceriodaphnia dubia* seven-day test. In contrast, Pfluger and Dietrich have reported moderate toxicity for carbamazepine (74–138 mg L^{-1}) in *Daphnia magna* (48-h exposure), *Danio rerio* (96 h), and *Xenopus laevis* (96 h) [32], while Ferrari and co-workers [21] reported an NOEC of 25 mg L^{-1} in a *D. rerio* early life stage test (ten days). Indeed, a recent review of available acute and chronic aquatic ecotoxicity data by Webb [11] has shown that >90% of the observations of acute ecotoxicity for more than 100 pharmaceuticals were at concentrations above 1 mg L^{-1} and that all environmental values were at concentrations of <1 µg L^{-1}. Ten compounds showed acute toxicity, with specific test systems at concentrations lower than 1 mg L^{-1} (Table 11.2).

Interestingly, four out of the ten substances belong to the antidepressant or antipsychotic classes. The range of reported ecotoxicity effect concentrations from >15 000 mg L^{-1} for atropine sulfate (anticholinergic/mydriatic; LC$_{50}$ *Artemia salina*) to 0.003 mg L^{-1} for fluvoxamine (antidepressant; LOEC parturition *Sphaerium striatinum*) is extremely wide and illustrates that pharmaceutical compounds cannot be treated as a general class but must be looked at from the perspective of their mode of action and chemical properties. Indeed, the mode of action as suggested by the Specific Serotonin Reuptake Inhibitors (SSRI), fluoxetine and fluvoxamine, is maintenance of an unusually high level of serotonin over prolonged periods, thus largely suppressing daily and/or seasonal fluctuations [33].

As previously suggested by many authors [32, 34–36], it is of the essence not only to understand the mode of action (MOA) but also the ramifications of the presence of similar or identical enzymes and receptors across different phyla. The usefulness of acute ecotoxicity data in the risk assessment of pharmaceutical com-

Tab. 11.2 Ten compounds showing acute toxicity at concentrations lower than 1 mg L^{-1}.

Pharmaceutical	Class/indication	Test System and Result
Alendronate	Metabolic bone disease	MIC green algae: >0.5 mg L^{-1}
Amitriptyline	Antidepressant	LC$_{50}$ 24 h *B. calyciflorus*: >0.5 mg L^{-1}
Carvediol	Antihypertensive/antianginal	LC$_{50}$ Fish: 1 mg L^{-1}
Ethinylestradiol	Estrogen	EC$_{50}$ Algae: 0.84 mg L^{-1}
Fluticasone	Antiasthmatic	EC$_{50}$ *Daphnia spp.*: 0.55 mg L^{-1}
Fluoxetine	Antidepressant	EC$_{50}$ Algae: 0.031 mg L^{-1} EC$_{50}$ *Daphnia spp.*: 0.94 mg L^{-1}
Fluvoxamine	Antidepressant	LOEC 4 h *S. striatinum*: 0.003 mg L^{-1}
Midazolam	Anesthetic	EC$_{50}$ *D. magna*: 0.2 mg L^{-1}
Paclitaxel	Antineoplastic	LC$_{50}$ *D. spp.*: >0.74 mg L^{-1}
Thioridazine	Antipsychotic	EC$_{50}$ 24 h *D. magna*: 0.69 mg L^{-1}

pounds is therefore questionable, and effects that might occur because of specific mechanisms will remain undetected. Chronic bioassays performed over the full life cycle of the organisms or covering several trophic levels are considered to be more appropriate, though probably prohibitive in terms of financial and temporal costs. A survey of the currently available data showed that chronic ecotoxicity data is available for approximately twenty substances, of which ethinylestradiol is the most intensively studied. Ethinylestradiol is a special case among the pharmaceuticals, as it is intended to be active in the low ng L^{-1} concentration range in mammals, so that it is not surprising to find similar endocrine-mediated effects in non-mammalian species such as fish and other aquatic species [37]. This is a typical compound for which the trigger values of the EU TGD would not function. Nevertheless, the EU TGD does specify MOA as a separate criterion. However, the risk assessment process does not envision the fact that other naturally occurring estrogens in the environment may have an additive or more than additive effect [36]. It should, however, also be remembered that many substances have been demonstrated to have a variety of completely unforeseen effects in non-target organisms, for example, the inhibition of estrogen-induced vitellogenin production in isolated trout hepatocytes by micromolar concentrations of paracetamol [38].

Despite this, MOA considerations rather than tiered testing would arguably provide more credible and useful results. For example the use of antibiotics in a *Vibrio fischeri* (Microtox) luminescence assay not surprisingly will provide toxic effects. Similarly, blue green algae (cyanobacteria) employed for first-tier testing of antibiotics, e.g., ciprofloxacin, will yield effects at lower concentrations than the corresponding test systems with green algae. As effects at low concentrations are expected in these specific test systems and essentially provide only confirmatory data that the test systems are working, little information can be gained for environmental risk assessment purposes. These findings support consideration of the MOA of the pharmaceutical in question in the definition of the tests to be carried out. Indeed, Henschel and colleagues reported non-standard tests, which take the MOA into account, to be more sensitive than the standard test constellation of algae, *Daphnia*, and fish, for three out of four tested substances [39].

Ferrari et al. [10] have considered chronic and acute endpoints and have performed an environmental risk assessment for carbamazepine, clofibric acid, diclofenac, ofloxacin, propanolol and sulfamethoxazole, taking into consideration conditions in both Germany (D) and France (F). While ofloxacin (PEC/PNEC 8.75), propanolol (PEC/PNEC 4.25) and sulfamethoxazole (PEC/PNEC 11.4) may pose an acute risk under conditions in France, only sulfamethoxazole (PEC/PNEC 59.3) would show a PEC/PNEC ratio >1 in Germany. On the other hand, carbamazepine (PEC/PNEC 2.4 (F), 3.82 (D)), propanolol (PEC/PNEC 104 (F), 11.9 (D)) and sulfamethoxazole (PEC/PNEC 2.72 (D)) would also display increased risk when considering chronic endpoints.

A further study looked in greater depth at the specific situation in the German federal state of Brandenburg, taking into account on the exposure side yearly consumption, number of inhabitants, amount of wastewater per inhabitant, human metabolism, and elimination in the sewage treatment plant. The effect side considered the lowest known effect concentration and/or $\log K_{ow}$, half-life in surface waters, biological degradation, elimination in the WWTP, and possible carcinogenic/mutagenic/reprotoxic and endocrine effects in mammals [40]. For some pharmaceuticals, QSAR calculations were used. Using this type of assessment, eight substances were expected, based on their respective PECs, to be found at concentrations greater than 1 µg L^{-1} in surface waters: metaformin-HCl (antidiabetic), phenoxypropanol isomers, cocospropylenediamine guaniacetate, glucoprotamine, polyvidon-iodine (all antiseptics), iodixanol (contrast media), metoprololtartrate (β-blocker, anti-hypertensive), and furosemid (diuretic). PNEC values, on the other hand, were shown to be low (<1 µg L^{-1}) for ethinylestradiol, ciprofloxacin-HCl (antibiotic), carbamazepin (anti-epileptic), clofibric acid (lipid lowering agent metabolite), benzalkonium chloride, and glucoprotamine (all antiseptics). A comparison of PEC and PNEC values to assess the risk to organisms in surface waters shows that 11 substances show PEC/PNEC ratios >1. These substances are listed in Table 11.3.

Unfortunately, all of the PEC/PNEC calculations depend on the determination of the PNEC as described earlier. Consequently all of the PNEC calculations (including the application of the assessment factors) completely rely on the initial quality of the original data and the relevance of the endpoint determined.

At this point one may ask the question what value endpoint determinants, e.g., NOEC and LOEC, really have for environmental risk assessment. Indeed both NOEC and LOEC are entirely determined by the experimental design of the study at hand and thus largely arbitrary. A much more conclusive and reliable approach is to use the EC$_5$ or EC$_{10}$ determinations while applying the strictest statistical quality criteria to the curve determinants [41]. Use of EC$_5$ or EC$_{10}$ values for the LOEC calculation would certainly merit more trust, as the subsequent assessment factors would ensure limitation of uncertainties to the uncertainty factors for which they were envisioned, e.g., intra- and interspecies differences etc., and not to experimental design flaws of the original study.

Tab. 11.3 Selected substances with PEC/PNEC ratios greater than one.

Substance class	Specific substance
Antibiotic	Ciprofloxacin-HCl, clarithromycin
Antiepileptic	Carbamazepine
Antiseptic	Benzalkonium chloride
	Cocospropylenediamine guaniacetate
	Glucoprotamine
	Laurylpropylenediamine
	Polyvidon-iodide
Hormone	Ethinylestradiol
Antidiabetic	Metaformin-HCl
Lipid lowering metabolite	Clofibric acid

11.3.2
QSARs

As mentioned above, ecotoxicity data are often non-existent or at least not in the public domain, and as it not feasible to perform the necessary tests in an adequate time frame, other approaches have been used or should be developed to assess the environmental risk of pharmaceuticals. One of these approaches is using (Q)SARs [(quantitative) structure activity relationships], at least for prescreening or prioritizing substances. Further approaches, such as (Q)PPRs [(quantitative) property property relationships], (Q)AARs [(quantitative) activity activity relationships], (Q)SPRs [(quantitative) structure property relationships] or even (Q)SBRs [(quantitative) structure bioaccumulation relationships] consider other properties of the chemical and are also employed. QSARs are currently being used, among others, by industry in internal screening programs and by the EU in risk assessments of biocides and new and existing chemicals (http://ecb.jrc.it/Documents/TECHNICAL_GUIDANCE_DOCUMENT/EDITION_2/tgdpart3_2ed.pdf) and by US EPA in risk screenings, for example, for premanufacture notices. In ecotoxicology, one of the most widely used QSARs is ECOSAR, which was developed by US EPA and has been validated by US EPA, OECD, and the EU. ECOSAR uses both baseline toxicity as well as expert-system-based ecotoxicity assessment for certain chemical groups. Furthermore, by combining ECOSAR with other models in EPISuite such as BIOWIN (biological degradation), BCF (logK_{ow}-based extrapolation of the bioconcentration factor) and the WWTP fugacity models (modeling of fate in a three-step sewage treatment plant), one may characterize the ecotoxic properties of substances and their environmental fate. (EPISuite can be downloaded free of charge at http://www.epa.gov/oppt/exposure/docs/episuitedl.htm.) The EU proposes to use QSARs for modeling biocides and new and existing chemicals based on nonpolar narcosis in acute and (sub)chronic algae, *Daphnia*, and fish assays, and on polar narcosis in acute *Daphnia* and fish assays and BCF for substances with log-K_{ow} <6. Pharmaceuticals are complex molecules, and relatively little experience has been gained to date concerning the predictability of ecotoxicological risks of pharmaceuticals or pharmaceutical classes [42]. ECOSAR classifies pharmaceuticals as neutral organics and uses mainly lipophilicity to develop models for toxicity prediction. Comparisons of effect concentrations of pharmaceuticals with modeled values are therefore necessary. Performing such a comparison with 20 pharmaceuticals present in surface waters and for which experimental data for at least algae, *Daphnia*, or fish are available, Sanderson et al. [43] were able to show that in 80% of the cases the modeled ECOSAR EC_{50} values (the lowest predicted ecotoxicology value was chosen) were lower than the measured effect concentration and therefore over-protective. This, however, also implies that the predictive value of the QSAR is not high. By applying the same methodology on a larger set of pharmaceuticals with the expectation that ECOSAR is applicable to pharmaceuticals for ranking purposes, almost 3000 substances in 51 classes were screened [44]. Cardiovascular, gastrointestinal, antiviral, anxiolytic sedative hypnotics and antipsychotics, corticosteroids, and thyroid pharmaceuticals were predicted to be the most hazardous classes. This of course does not imply that the other pharmaceutical classes will be acquitted based solely on the QSAR predictions [44]. Pharmaceuticals generally have very specific modes of action (MOA), and a QSAR taking into account the MOA might give a better predictivity in terms of hazard identification. This implies, however, that the QSAR is well characterized, as it was shown that more complex QSARs that take MOA into account are outperformed by more simple QSARs in terms of prediction accuracy [45]. In conclusion, QSARs may be a useful tool for preliminary screening and prioritization for environmental risk assessment, but should always be coupled with experimental data and expert knowledge. This very conclusion is also recognized in the recent revision of the EU TGD, where expert evaluation of available data has gained a central role in the risk assessment process. Furthermore, the term "expert" is substantially defined, in that the background and experience of experts employed in the risk assessment process are required entities for the provision of high stringency and consistency in expert knowledge.

11.3.3
"Omics"

"Omic" technologies have not halted before environmental risk assessment, and their use and applicability are currently being discussed in regulatory bodies and intergovernmental organizations such as the OECD. "Omics" technologies, including genomics, proteomics, transcriptomics, and metabolomics/metabonomics, are tools being evaluated for chemical hazard and risk assessment in order to better understand species and subgroup sensitivities, assess mixtures and combinations, and offer the long-term possibility to reduce animal-intensive methods for screening and testing. It should, however, be understood, that current approaches in the research community are very varied and at different stages of methodological development. A further critical issue is the development of bioinformatics tools that can handle, analyze, and distil the huge amount of data that can be generated using omics techniques. A valuable effort has been made by the MGED (Microarray Gene Expression and Data) Society (http://www.mged.org) in initiating a standardized format for reporting ecotoxicogenomics data, namely MIAME (Minimum Information About a Microarray Experiment), which is already a prerequisite for publication of (eco)toxicogenomics data in many scientific journals. Although (toxico)genomics techniques are currently being used in the development of pharmaceutical compounds, there are currently no publications relating to ecotoxic effects of pharmaceuticals using genomic techniques. In all this one must never forget that all the "omics" techniques have one essential limitation, in that they all describe very short-time windows in the exposure experiment of a given species. In order to provide more insight into the processes involved, numerous time points would be necessary. However, this is currently prohibitive because of the sheer amount of data and huge financial costs involved.

11.3.4
Toxicity of Mixtures

The problems associated with assessing the aquatic toxicity of a single compound burgeon when one considers the possibility of interactive/synergistic effects of the myriad of different pharmaceuticals currently on the market. In reality, aquatic organisms are exposed to a veritable cocktail of contaminants, and synergistic effects have indeed been reported for compounds with the same mode of action, for example the non-steroidal anti-inflammatory COX inhibitors. A study carried out by Cleuvers and colleagues [46] using standard *Daphnia magna* toxicity tests showed that the toxicity displayed by a mixture of diclofenac, ibuprofen, naproxen, and acetylsalicylic acid was far higher than that predicted by simple addition of the effects of the individual substances. In contrast, concentration addition was shown to accurately predict the toxicity of NSAID mixtures in acute algal (*Desmodesmus subspicatus*) toxicity testing. The authors suggested that non-polar narcosis rather than a specific mode of action was responsible for the observations and supported this with quantitative structure activity relationship (QSAR) analysis, and further suggested that significant combination effects could also be expected for similar compounds at concentrations below their NOECs. Indeed a study by Renner demonstrated a mixture of ibuprofen, prozac, and ciprofloxacin to be toxic to plankton, aquatic animals, and fish at concentrations up to 200 times lower than the standard human dose [47]. Such combination effects are probably even more pronounced in a chronic exposure situation. Thus, the use of individual NOECs as a basis for risk assessment is questionable.

A further class of compounds could enhance the toxic potential of substances not normally toxic in single exposure regimens. Verapamil, reserpine, and cyclosporine exert some of their therapeutic effects by functioning as so-called efflux pump inhibitors. Similarly, non-pharmaceutical compounds, e.g., nitromusk and polyclic musk compounds and personal care products, can inhibit multidrug/multixenobiotic resistance (MDR/MXR) efflux transporters [48, 49]. This could potentially also have serious consequences with respect to mixture toxicity. Indeed the teratogenic effects of a range of compounds (vinblastine, mitomycin C, cadmium chloride, methylmethanesulfonate, chloroquine, and colchicines) in mussel larvae have been shown to be exacerbated with verapamil (20 µM) co-exposure [50]. The question, however, remains as to the doses that are required for transport inhibition and whether these doses realistically are found in the environment at the same time as other pharmaceuticals or polar organic pollutants.

11.4
Conclusions

In conclusion, estimation of the potential of an individual pharmaceutical to cause detrimental effects in the aquatic ecosystem does not include an assessment of the potential of the compound to interact and possibly cause additive, more than additive, or even synergistic effects with other formulations of the same substance or with other substances with similar modes of action. By the same token, counteractive, i.e. inhibitive, effects by competition for the same enzyme/receptor must also be taken into consideration. The analysis of structure activity relationships (SARs) could help to abrogate part of this problem, as discussed above. However, only high quality data sets using realistic compound concentrations, coupled with simultaneous analysis of the bioavailable compound concentrations and accounting for the possible MOA as well as cross-reactions and interacting enzyme/receptor systems can shed light on this challenging field. As of now, despite the large and exponentially increasing number of publications, there is a great paucity of qualitative acceptable data sets that can be readily employed for environmental risk assessment purposes.

References

1 S. Wiegel, A. Aulinger, R. Brockmeyer, H. Harms, J. Löffler, H. Reincke, R. Schmidt, B. Stache, W. von Tümpling, A. Wanke. *Chemosphere* **2004**, 57, 107–126.

2 A. Boxall, C. Long. *Environ. Toxicol. Chem.* **2005**, 24, 759–760.

3 M. L. Richardson, J. M. Bowron. *J. Pharm. Pharmacol.* **1985**, 37, 1–12.

4 T. Ternes, M. Stumpf, J. Mueller, K. Haberer, R. Wilken, M. R. Servos. *Sci. Total Environ.* **1999**, 225, 81–90.

5 T. A. Ternes. *Water Res.* **1999**, 32, 3245–3260.

6 M. Stumpf, T. A. Ternes, R. D. Wilken, S. V. Rodrigues. *Sci. Total Environ.* **1999**, 225, 135–141.

7 T. Heberer. *Toxicol. Lett.* **2002**, 131, 5–17.

8 J. P. Bound, N.Voulvoulis. *Chemosphere* **2004**, 56, 1143–1155.

9 C.D. Daughton, T.A.Ternes. *Environ. Health Perspect.* **1999**, 107, 907–938.

10 B. Ferrari, R. Mons, B. Vollat, B. Fraysse, N. Paxéus, R. Lo Giudice, A. Pollio, J. Garric. *Exp. Toxicol. Chem.* **2004**, 23, 1344–1354.

11 S. Webb. In *Pharmaceuticals in the Environment, Sources, Fate, Effects and Risks*, 2nd. Edn. (Ed. K. Kümmerer) Springer Verlag, Berlin, Heidelberg **2004**, pp. 317–343.

12 O. A. H. Jones, N. Voulvoulis, J. N. Lester. *Water Res.* **2002**, 36, 5013–5022.

13 D. B. Huggett, J. C. Cook, J. F. Ericson, R. T. Williams. *Hum. Ecol. Risk Assess.* **2003**, 9, 1789–1799.

14 European Commission **2001**. Directive 2001/83/EC.

15 EMEA CPMP/SWP/4447/00draft.

16 F. Stuer-Lauridsen, M. Birkved, L. P. Hansen, H.-C. Holten Lützhoft, B. Halling-Sorenson. *Chemosphere* **2002**, 40, 783–793.

17 N. J. Ayscough, J. Fawell, G. Franklin, W. Young. *Bristol: Environmental Agency* **2000**, 390.

18 S. F. Webb. In *Pharmaceuticals in the environment: sources, fate, effects and risks*. (Ed. K. Kümmerer) Springer-Verlag **2001**, 204–219.

19 O. A. H. Jones, N. Voulvouli, J. N. Lester. *Water Res.* **2002**, 36, 5013–5022.

20 V. Forbes, P. Calow. *BioScience* **2002**, 52, 249–257.

21 B. Ferrari, N. Paxéus, R. Lo Giudice, A. Pollio, J. Garric. *Ecotoxicol. Environ. Safety* **2003**, 55, 359–370.

22 S. Webb. In *Pharmaceuticals in the environment*. (Ed. K. Kümmerer) Springer Verlag **2001**, 203–219.

23 B. Hoeger, B. Koellner, D. R. Dietrich, B. C. Hitzfeld. *Aquat. Toxicol.* **2005**, 75, 53–64.

24 J. Schwaiger, H. Ferling, U. Mallow, H. Wintermayr, R. D. Negele. *Aquat. Toxicol.* **2004**, 68, 141.

25 R. Triebskorn, H. Casper, A. Heyd, R. Eikemper, H. R. Kohler, J. Schwaiger. *Aquat. Toxicol.* **2004**, 68, 151.

26 T. A.Ternes. *Wasser & Boden* **2001**, 53, 9–14.

27 FDA Food and Drug Administration; **1995**.
28 Outline of the partial amendment to the chemical substances control law. World Wide Web URL: www.meti.go.jp/english/policy/index_environment.html
29 Risk assessment of endocrine disruptors. World Wide Web URL: www.meti.go.jp/english/policy/index_environment.html
30 Prevalidation and validation of hormone receptor assays. World Wide Web URL: www.oecd.org/env/ehs/
31 Canada. **2001**.
32 P. Pfluger, D. R. Dietrich. In *Pharmaceuticals in the environment – sources, fate, effects and risks* (Ed. K. Kümmerer) Springer; **2001**, 11–17.
33 B. W. Brooks, C. M. Foran, S. M. Richards, J. Weston, P. K. Turner, J. K. Stanley, K. R. Solomon, M. Slattery, T. W. La Point. *Toxicol. Lett.* **2003**, 142, 169–183.
34 D. R. Dietrich, S. Webb, T. Petry. *Toxicol. Lett.* **2002**, 131, 1–4.
35 R. Länge, D. R. Dietrich. *Toxicol. Lett.* **2002**, 131, 97–104.
36 E. O'Brien, D. R. Dietrich. *Trends Biotechnol.* **2004**, 22, 326–330.
37 J. P. Nash, D. E. Kime, L. T. M. Van der Ven, P. W. Wester, G. Maack, P. Stahlschmidt-Allne, C. R. Tyler. *Environ. Health Perspect.* **2004**, 112, 1725–1733.
38 M. R. Miller, E. Wentz, S. Ong. *Tox. Sci.* **1999**, 48, 30–37.
39 K. P. Henschel, A. Wenzel, M. Didrich, A. Fliender. *Reg. Toxicol. Pharmacol.* **1997**, 18, 220–225.
40 B. Hanisch, B. Abbas, W. KratzG. Schüürmann. *UWSF – Z. Umweltchem. Ökotoxikologie* **2004**, 16, 223–238.
41 R. Ritz, N. Cedergreen, J. C. Streibig. In *Annual Conference of SETAC Europe*; Lille, Edited by SETAC-EU: SETAC EU: **2005**.
42 K. Kümmerer. In *Pharmaceuticals in the environment. Sources, Fate, Effects and Risks.* edn 2. (Ed. K. Kümmerer) Springer; **2004**, 387–389.
43 H. Sanderson, D. J. Johnson, C. J,Wilson, R. A. Brain, K. R. Solomon. *Toxicol. Lett.* **2003**, 144, 383.
44 H. Sanderson, D. J. Johnson, T. Reitsma, R. A. Brain, C. J. Wilson, K. R. Solomon. *Reg. Toxicol. Pharmacol.* **2004**, 39, 158.
45 S. Ren. *Chemosphere* **2003**, 53, 1053.
46 M. Cleuvers. *Ecotoxicol. Environ. Safety* **2004**, 59, 309–315.
47 Renner R: Drug mixtures prove harmful. *Environ. Sci. Technol.* **2002**, 36, 268A–269A.
48 T. Luckenbach, D. Epel. *Environ. Health Perspect.* **2005**, 113, 17–24.
49 T. Smital, T. Luckenbach, R. Sauerborn, A. M. Hamdoun, R. L. Verga, D. Epel. *Mutat. Res.* **2004**, 552, 101–117.
50 I. McFadzen, N. Eufemia, C. Heath, D. Epel, M. Moore, D. Lowe. *Mar. Environ. Res.* **2000**, 50, 319–323.

12
Assessment and Management of Chemicals –
How Should Persistent Polar Pollutants be Regulated?

Klaus Günter Steinhäuser and Steffi Richter

12.1
Chemicals Assessment and Management Today

12.1.1
Basic Legislation and Current Guidelines for Risk Assessment and Risk Management of Chemicals in Europe and Germany

Risk assessment and risk management of chemicals, pesticides, biocides, and human and veterinary drugs in Europe are based almost completely on EU regulations and directives. Since a main objective of this legislation is to avoid barriers in the internal market, it is based on Article 133 in conjunction with Article 95 of the EC Treaty [1], thus ensuring an equal level of compulsion in all Member States and for all industrial players in the Community, and not simply setting minimum standards, which can be exceeded by national legislation. A high level of protection of health and the environment must be achieved at the same time.

National legislation, e.g., the German "Chemicals Act" (Chemikaliengesetz) [2] and its accompanying ordinances, deals mainly with the implementation and enforcement of EU directives in the Member State, but also contains a small number of provisions which are unique to it, being deemed necessary to regulate some additional aspects of chemicals management, in particular, aspects relating to occupational health.

12.1.1.1 Notification of New Substances
Directive 92/32/EEC of 30 April 1992 [3] is the 7th amendment of Directive 67/548/EEC [4] on the approximation of the laws, regulations, and administrative provisions relating to the classification, packaging, and labeling of dangerous substances. It obliges the Member States (among others) to give notification of chemicals placed on the market for the first time. New substances are all substances which are not listed in the European Inventory of Existing Commercial Chemical Sub-

stances (EINECS) [5] and have been placed on the market after 18 September 1981. Member States receive and examine the notifications on behalf of the entire Community, following harmonized guidelines and rules.

The amount of evidence to be provided increases with the quantity put on the European market by a manufacturer or importer. Once the next tonnage threshold is reached, further documents must be submitted by the notifier (Table 12.1).

Article 6, paragraph 1 of the Chemicals Act, "Contents of Notification", requires the notifier to present the following information concerning the substance, methods of identification, use, and risks, etc.:

1. name and address of the notifying party, and, in the case of import, also the name and address of the manufacturer and location of the production site;
2. identifying characteristics, including nature and percentage by weight of the auxiliary agents, the main impurities, and the other impurities and decomposition products of which he is aware;
3. methods of detection and determination;
4. methods of analysis known to him to determine human exposure and levels found in the environment;
5. information on manufacture, application, exposure, and retention;
6. hazardous effects in application;
7. information on toxicokinetics;
8. the planned classification, packaging, and labeling;
9. recommendations on the precautions to be taken during application and emergency measures in the case of accident;
10. the amount of the substance which he wishes to place on the market or import per year;
11. procedures for the proper disposal or possible reuse of the substance or other means of rendering it harmless;
12. evidence of testing as required under Article 7 (basic testing).

According to Article 7, the following basic testing is required:

1. physical, chemical, and physico-chemical properties;
2. acute (mammalian) toxicity;
3. any reasons to suspect carcinogenic or mutagenic properties;
4. evidence of properties toxic for reproduction;
5. irritant and corrosive properties;
6. sensitizing properties;
7. sub-acute toxicity;
8. abiotic and ready biological degradability;
9. toxicity to water organisms after short-term exposure;
10. inhibition of algal growth;
11. bacterial inhibition;
12. adsorption and desorption.

If an annual quantity of 100 tons, or 500 tons in total, is exceeded since the beginning of manufacture or import into the European Community, additional evidence

Tab. 12.1 Requirements for notification of new substances [6].

Quantity placed on the market per year	Total quantity placed on the market	Type of notification
10 kg – < 100 kg	–	Reduced notification
100 kg – < 1000 kg	≥ 500 kg	Reduced notification
≥ 1000 kg	≥ 5000 kg	Base set
≥ 10 000 kg	≥ 50 000 kg	Early level I
≥ 100 000 kg	≥ 500 000 kg	Level I
≥ 1 000 000 kg	≥ 5 000 000 kg	Level II

has to be provided (e.g., additional tests on reprotoxic, carcinogenic, or mutagenic properties and basic toxicokinetic properties, data on long-term toxicity to water organisms, toxicity to soil organisms and plants, and data on bioaccumulation potential).

If an annual quantity of 1000 tons, or 5000 tons in total, is exceeded since the beginning of manufacture or import into the European Community, even more additional data have to be provided (e.g., data on perinatal and postnatal effects and on toxicokinetic properties including biotransformation). The data necessary for these volumes depend upon the results of the tests pursuant to Articles 6 and 7 and are a matter of advance dialogue between the authorities and the notifier.

12.1.1.2 Existing Substances Legislation and Management

The EU Existing Chemicals Regulation applies to substances which were put on the market before 1981 and aims to evaluate and, where necessary, to reduce the risks of these substances to man and environment. Council Regulation No. 793/93 [7] on the Evaluation and Control of the Risks of Existing Substances deals with priority setting and the systematic collection of data on priority substances whose risks are to be evaluated. The principles for the assessment of risks that existing substances pose to man and environment are laid down in Commission Regulation No. 1488/94 [8]. So-called high-production-volume chemicals (HPVC) (which are marketed in volumes greater 1000 tons per year) were regarded as having potentially higher risks than the other 98 000 substances listed in EINECS. Using a rather cumbersome priority-setting procedure, 141 chemicals were selected and placed in four priority lists specifying the first chemicals to be assessed comprehensively.

For each priority substance, a Member State is nominated as rapporteur. The result is a comprehensive risk assessment, which is agreed Europe-wide and finally published in the Official Journal. The producers and importers of the substances to be assessed have to provide all available data on the substances' intrinsic properties and data to evaluate exposure to them (uses, amounts, technical conditions etc.). The procedure to produce a risk assessment report takes several years, because data gaps may be identified, additional information from industry may be needed to be generated and submitted (which is not legally enforceable), and comments and data from other EU Member States must be taken into account before all technical and political bodies reach agreement. If risks to human health or the environment are identified in the risk assessment, a risk reduction strategy has to be elaborated by the authorities of the Rapporteur Member State. In the risk reduction strategy, adequate risk reduction measures are assessed with regard to their effectiveness, practicability, economic impact, and monitorability. Finally, the most appropriate combination of measures is recommended (e.g., marketing and use restrictions, technical measures, classification and labeling, voluntary schemes, etc.).

Directive 76/769/EEC aims to harmonize rules relating to restrictions or prohibitions applied to dangerous substances, preparations, and associated finished products. Restrictions imposed by individual Member States must be in accordance with Community law and are a rare exception. For example, in Germany a more extensive prohibition than that in the other Member States of Europe is in force for pentachlorophenol (PCP), and in Scandinavian countries some prohibitions from the time before accession were agreed to be prolonged. As a rule, restrictions relate to particular uses or exempt particular uses. Complete prohibitions on the marketing of substances and their preparations are the exception (e.g., PCBs and asbestos).

The European prohibitions and restrictions on chemicals have been incorporated into German law by the Ordinance on the Prohibition of Certain Chemicals [9] and the Ordinance on Hazardous Substances [10].

12.1.1.3 Technical Guidelines for Risk Assessment of Chemicals

The basis for risk management measures, in particular for restrictions on the marketing of existing substances, should be a science-based and harmonized risk assessment procedure. The "Technical Guidance Document in support of Commission Directive 93/67/EEC on Risk Assessment for new notified substances, Commission Regulation (EC) No 1488/94 on Risk Assessment for existing substances, and Directive 98/8/EC of the European Parliament and of the Council concerning the placing of biocidal products on the market" [11] provides a scientifically sound basis reflecting a broad range of findings from toxicology, ecotoxicology, and environmental chemistry as well as technical experience in production and use. A stepwise procedure is used to assess risks to the environment:

- Using basic data on the volume produced or processed, estimated releases to the environment, and properties of the chemical (e.g., volatility, water solubility, (bio)degradation, water/air partitioning behavior), the environmental distribution of a chemical can be modeled or calculated. This results in a series of **Predicted Environmental Concentrations (PEC)** at each industrial site, and also over a defined region, for each environmental compartment (air, water, soil). Where available, representative measured data are used to refine or replace the modeling results. Realistic worst case scenarios are used to fill information gaps on the exposure side.

- For the environmental effect assessment a **Predicted No Effect Concentration (PNEC)** is calculated for species representative of the environmental compartment under investigation, using acute or chronic toxicity data and an assessment factor. The PNEC represents a concentration which is predicted to have no effects within the relevant compartment.

- The PNEC is determined on the basis of available **No Observed Effect Concentrations (NOEC)** and dose-response assessment of ecotoxicological tests including assessment factors. For each compartment, the data for the most sensitive species tested have to be applied.

- Quantitative Risk Assessment: A risk is identified when the PEC for a compartment is higher than the PNEC. However, since sometimes the **PEC/PNEC ratio** of 1 is exceeded because of high default values or high assessment factors due to missing data on chronic (eco)toxicity, the risk assessment is an iterative process which may be very time-consuming. More detailed data on exposure or hazardous effects may cause a preliminarily assumed risk to be revised to "no concern". With this iterative mechanism, manufacturers may win time before risk reduction measures are decided.

This type of risk assessment is characterized by the following main principles [12, 13]:

- Ecological and toxicological effects are considered separately.
- Each single substance is assessed separately for each step of its lifecycle and for each use.
- Each environmental compartment is considered separately.
- The assessment refers to a standardized "generic" European environment, not taking into account specific effects depending on climate, etc.
- The risk assessment should cover all uses, emissions and compartments, and in principle the whole lifecycle.

A similar procedure is used to assess risks to human health, i.e. comparison of no effect levels with exposure levels and application of margins of safety.

12.1.2
Problems Impeding Effective Chemicals Management

When looking at industrial chemicals, the present legislation distinguishes between "existing chemicals" and "new substances". While assessment of "new substances" is based on a mandatory data set whose size depends on the volume placed on the market, gathering information on "existing chemicals" is time consuming and ineffective. There is only limited information available on existing chemicals. The official procedure according to the Existing Chemicals Regulation (793/93/EC) focuses on high-production-volume chemicals. The rules require a comprehensive risk assessment, which means that every aspect of potential risk to health and environment has to be evaluated. For the authorities required to prove that a chemical poses a risk, it takes years to get a comprehensive view and to conclude the assessment procedure.

Many discussions at national and European level in different groups are time consuming and tie up capacities. Furthermore the rules are only directed at manufacturers and importers, with the result that often information on where and how the chemical is used in the supply chain is missing. Downstream users of a substance are not legally required to provide data on the uses of the substance and the associated exposures. For many substances it is not known in which preparations they are used and to what extent they are disposed of as waste. This lack of information makes risk assessments difficult and sometimes impossible. 141 priority chemicals out of 100 000 existing chemicals compared with only 5000 notified new chemicals in Europe demonstrate that the current system gives no incentives for the development of new substances. On the other hand, the system rewards ignorance, because only those existing substances are regulated for which a hazard or risk has been identified [13].

Some experts have criticized the very resource-demanding risk assessment procedure in the EU by making the following comparison [14]:

"Rio Declaration: Where there are threats of serious or irreversible damage, lack of full scientific certainty shall not be used as a reason for postponing cost-effective measures to prevent environmental degradation.

EU practice: Where there is lack of scientific certainty of serious or irreversible damage, even cost-effective measures will be postponed."

It has to be emphasized that several chemicals producers have assumed responsibility and provide core data sets. For example, the German Association of the Chemical Industry (VCI) pledged in 1997 to make available core data on all substances above 1 ton per year by 2002 [15].

A further drawback of the Existing Substance Regulation is that the risk assessment considers only the actual situation and disregards potential changes, e.g., in production volume, new uses of the chemical, and differing circumstances in the new Member States.

The scientific assessment procedure for deriving a PEC/PNEC relationship (cf. Section 12.1.1.3) is adequate for most chemicals. However, it fails to sufficiently address cases where the effects of the exposure are irreversible. Substances which are persistent and bioaccumulative (persistent, bioaccumulative, and toxic substances = PBT substances, and very persistent and very bioaccumulative substances = vPvB substances) cannot be assessed adequately through determination of a PEC/PNEC equation, as persistence causes irreversible environmental exposure and – when emissions continue – increasing levels in the environment, in particular in biota. Bioaccumulation enhances the probability of toxic effects. Therefore, unpredictable adverse effects may appear in the long term. The criteria for PBT/vPvB substances are listed in Table 12.2. Some of the PBT/vPvB substances – the so-called persistent organic pollutants (POPs) – are transported over large distances, resulting in high concentrations far away from the source of release (cf. Section 12.3.1). PBT and vPvB substances pose a risk because of their intrinsic properties alone. This risk is not, however, independent of the extent of long-term exposure.

Tab. 12.2 Criteria of PBT and vPvB substances (according to Ref. [16], Annex XII).

Criterion	PBT criteria	vPvB criteria
P	$t_{1/2} > 60$ d marine water or $t_{1/2} > 40$ d freshwater or $t_{1/2} > 180$ d marine sediment or $t_{1/2} > 120$ d freshwater sediment or $t_{1/2} > 120$ d soil	$t_{1/2} > 60$ d marine or freshwater, or $t_{1/2} > 180$ d marine or freshwater sediment or $t_{1/2} > 180$ d soil
B	BCF > 2000	BCF > 5000
T	Chronic NOEC < ,01 mg L^{-1} for marine or freshwater organism or classification T – R45, R46, R48, R60, R61 or Xn – R48, R62, R 63 or endocrine disrupting effects	Not applicable

Therefore, the amounts released to the different environmental compartments during manufacture and use as well as the distribution between environmental compartments have to be considered and estimated. A precautionary approach should be applied to take into account the high level of uncertainty in predicting reliable exposure concentrations, i.e. only negligible releases are acceptable.

It has often been discussed whether an assessment of single substances is an adequate approach, given that multi-exposure is usual in the environment. Several attempts have been made to describe the risks caused by multiple stressors (simultaneously or consecutively), which can be subsumed under the term cumulative risk assessment [17]. In recent years intensive efforts have been made, in particular in the field of aquatic toxicology, to improve the scientific basis. In essence, there are two fundamental concepts to describe combined effects: concentration addition and independent action [18]. The general assumption in the concentration addition model is a similar mode of action of the components of the mixture, whereas the independent action approach assumes different mechanisms. Real synergisms, i.e. more than additive combined effects, are apparently a rare occurrence. In most cases, the concentration addition model predicts higher toxicity and is therefore more precautionary [19]. It may be adequate, whenever the necessary data are available, to apply this model when conducting risk assessment. However, if reliable exposure and effects data for the components of the mixture are lacking, adequate safety factors are the only possible way to overcome this deficiency. The possibility that a mixture of several substances in concentrations below their individual NOECs may be significantly toxic [20] demonstrates the necessity to apply safety factors to the results of higher tier studies also.

Another deficiency of the current European assessment scheme for industrial chemicals concerns persistent polar compounds. The risk to groundwater posed by these substances is not assessed adequately. In the past, the origin of groundwater contamination was mostly either pesticide application or contaminated sites. In pesticides assessment, the risk to groundwater is assessed thoroughly, also covering the most significant and relevant metabolites (cf. Section 12.1.4.1). Contaminated sites are mostly the result of unintended releases or improper and irresponsible handling of chemicals and wastes. These aspects are not considered in the risk assessment of industrial chemicals. However, there is evidence that regular intended use of industrial chemicals may also result in groundwater contamination. Some well-known examples, some of which are mentioned in this book, are the gasoline additive methyl tertiary butyl ether (MTBE), the corrosion inhibitor benzotriazole, and the complexing agent EDTA [21]. An interesting example is sarcosine-N-phenylsulfonyl (SPS). First suspected of occurring as a residue of an antibiotic drug, it was identified to be a stable metabolite of the corrosion inhibitor 6-[Methyl-(phenylsulfonyl)amino] hexanoic acid, occurring in surface waters, wastewater effluent, and bank filtered water in concentrations above 1.0 µg L^{-1} [22, 23]. Protection of groundwater is of outstanding importance because contamination may be irreversible and difficult or impossible to remedy.

Groundwater protection is an issue that demonstrates that separation of environmental and human health assessment as commonly suggested in current guidance documents is not always satisfactory. Environmental exposure via air, water, or biota may result in a threat to human health. In addition, several toxicological effects like endocrine disruption are of ecological significance, too, as well as ecotoxicological effects may be of importance for human health. This is why better communication between risk assessors in both fields is necessary in order to identify common problems of concern and to develop approaches to integrated risk assessment. Integrated risk assessment (IRA) is a concept under construction. The aim is to combine the different lines of evidence. Commonalities in pollutant sources, exposure routes, and modes of action provide a starting point for describing human health risks and ecological threats in an integrated manner [24].

The last problem to be mentioned is that risk reduction is often inadequate. Although it should be implemented in all cases where a risk by an industrial chemical is identified, the necessary measures cannot be enforced. The EU chemicals legislation provides for restrictions on marketing and use only. Other risk reduction measures (e.g., emission limits at point sources or emission abatement techniques) are regulated by EU legislation in other areas (e.g., IPPC Directive or Water Framework Directive). An instrumental gap exists which prevents the results of chemical risk assessment from being acted upon under other legislation. Provisions interlinking the two are absent in both chemicals legislation and environmental legislation [25].

12.1.3
Chemicals Management in the United States of America

The Toxic Substances Control Act (TSCA) of 1976 was enacted by Congress and is the centerpiece of United States Chemicals regulation. By this legal act the ability to regulate existing industrial chemicals currently produced or imported was given to the Environmental Protection Agency (EPA). The TSCA inventory includes

75 000 chemicals [26] and is thus smaller than the EINECS inventory of the EU (100 000 chemicals). 60 000 substances are reported to be on the United States market today [26]. The EPA can screen these chemicals, e.g., by using Quantitative Structure Activity Relationships (QSARs), and can require reporting or testing of those that may pose an environmental or human-health risk. The EPA can ban the manufacture and import of those chemicals that pose an unreasonable risk. For substances which are "existing", the inventory can be used to determine if there are restrictions on manufacture or use under the TSCA.

Chemicals placed on the market after December 1979 are regarded as new chemicals and not listed in the inventory. They need a premarketing notification (PMN), which contains information on chemical identity, physical characteristics, processing and use, and *available* toxicity data. The information the EPA receives through the PMN is significantly less than the information required in Europe. The EPA can subsequently control these chemicals as necessary for the protection of human health and the environment.

Like the European Union, the EPA treats chemical substances as either "existing" or "new", with more requirements for new chemicals. As in Europe, this differentiation has the effect that the development of new substances is not encouraged by the regulatory framework. However, the difference is smaller.

If a substance is "new", it may be subject to an exemption from reporting (for example, a low volume exemption, or exclusion as a naturally occurring material) and can still be marketed for commercial purposes.

After years of experience with the TSCA and great hopes that it would help eliminate substantial gaps in the regulation of toxic substances, its implementation seems to have run into problems similar to those experienced by the chemicals management in Europe, particularly for existing chemicals [27]. The TSCA places an extremely large burden on the EPA for the implementation of restrictions on the manufacture or use of toxic chemicals, with the result that restrictions on chemicals are few.

To restrict such chemicals, the EPA must prove that the chemical "will present an unreasonable risk", that it is choosing the least burdensome regulation to reduce risks to a reasonable level, and that the benefits of regulation outweigh the costs to industry. The EPA must do this on a chemical-by-chemical basis. A 1994 report by the United States Government Accounting Office (GAO) found that throughout its existence the EPA has restricted only five chemicals (PCBs, chlorofluorocarbons, dioxin, asbestos, and hexavalent chromium) [28]. PCBs were banned under TSCA in 1976 and the asbestos ban was overturned by the Fifth Circuit Court of Appeals. That number has only slightly increased in the past decade. The EPA has in recent years concluded consent agreements with individual companies to stop production of problematical chemicals on a voluntary basis, such as the penta- and octabromo diphenyl ethers [29] and perfluoroalkyl sulfonate. Overall, while the EPA is important, its lack of power to regulate existing chemicals actually provides a disincentive to bringing safer chemicals to market.

With regard to persistent and bioaccumulative substances, the EPA's Persistent and Bioaccumulative Toxics (PBT) programme should be mentioned. It focuses on multi-media, cross-programme management of PBT substances. Through the programme, precautionary guidelines can be established for chemical manufacturers to avoid introduction of new PBTs to the market and to develop Internet-based tools for assessing chemicals for their potential persistence and capacity to bioaccumulate. The EPA is currently working on the reduction of priority PBTs – in conjunction with the Binational Toxics Strategy on the Great Lakes – such as mercury and dioxins [30].

In addition to that programme, a Chemical Right-to-Know Initiative exists in order to enhance understanding of the hazards posed by high-production-volume chemicals (HPVC), to improve reporting of persistent, bioaccumulative, or toxic substances, and to facilitate public awareness of the dangers which these chemicals may pose to children [31]. Three programmes were designed to achieve the goals of this initiative: (a) the HPV Challenge Programme, (b) the Voluntary Children's Assistance Programme, and (c) the PBT Chemical Reporting Programme.

The HPV Challenge Programme of 1998 is a voluntary agreement with the chemical industry to collect data on 2800 chemicals with an annual production volume above 500 tons. It was fairly successful, but about 500 chemicals were not adopted by industry consortia, and the programme does not cover substances which reached HPV status afterwards.

Chemicals management at federal level is supported by management at regional or state level. Often, regions are innovators and drivers in environmental policy in the United States. One of the best-known examples is the Great Lakes region, which very early on gained much world attention as a highly contaminated region in Northern America subject to the impact of persistent and bioaccumulative toxics. To address contamination of the Great Lakes region, the United States and Canada signed the Great Lakes Water Quality Agreement in 1977 to express a joint commitment to restore and maintain the chemical, physical, and biological integrity of the Great Lakes Basin Ecosystem [32]. The Agreement calls for the virtual elimination of persistent and bioaccumulative pollution in the region, and the subsequent Great Lakes Binational Toxics Strategy provides a mechanism for voluntary action to reduce priority PBT chemicals [33].

Further differences between the United States chemicals management system and that of the EU are that the United States has more postmanufacture monitoring activities (in which the EPA collects data to identify potentially harmful substances) and a better linkage to pollution prevention and reduction initiatives, e.g., by the establishment of and connection with the Toxic Release Inventory (TRI).

However, the problems and limitations of chemicals management in the United States are largely the same as those in Europe:

- a lack of information on most chemicals in commerce;
- no connection between the regulation of existing and new chemicals;
- a slow and resource-intensive risk assessment process for substances suspected to be harmful;
- a lack of incentives to substitute and innovate in order to move from problem chemicals to safer alternatives.

12.1.4
Management of Specific Chemicals

Some chemicals with specific uses and effects on health and the environment need to be assessed and managed according to specific rules. They are covered by specific legislation, mostly require authorization before being marketed, and are examined in more detail than industrial chemicals. Four groups of specific chemicals will be discussed in this section:

- pesticides,
- biocides,
- pharmaceuticals,
- detergents and cleansing agents.

12.1.4.1 Pesticides

Pesticides are used in order to protect plants or plant products against harmful organisms or prevent the action of such organisms (see Chapter 6). Many of them are not only fatal for the target organisms but may also harm organisms in the environment and in this way affect the natural balance. In general, they are released directly and deliberately into the environment. Moreover, agricultural products are used as food. Therefore, pesticides are examined more thoroughly for their risks than other chemicals. Directive 91/414/EEC [34] concerning the placing of plant protection products on the market provides common rules for the authorization of plant protection products in the Member States. Companies applying for an authorization have to provide extensive and detailed documents which demonstrate that the plant protection product concerned, when used properly and for the intended purpose,

- is sufficiently effective,
- has no unacceptable effect on plants or plant products,
- does not cause unnecessary suffering and pain to vertebrates to be controlled,
- has no harmful effect on human or animal health, directly or indirectly (e.g., through drinking water, food or feed) or on groundwater,
- has no unacceptable influence on the environment, having particular regard to the following considerations:
  – its fate and distribution in the environment, particularly contamination of water including drinking water and groundwater;
  – its impact on non-target species.

Authorization can be granted only if all these conditions are met.

Directive 91/414/EEC has been incorporated into German law by the Plant Protection Act [35], which provides detailed rules for the authorization procedure, enforcement, and monitoring in Germany. The Federal Office of Consumer Protection and Food Safety (BVL) is responsible for granting authorizations with the consent of the Federal Environment Agency (UBA), which is responsible for all environmental issues regarding risk assessment and management. The Federal Institute for Risk Assessment (BfR) and the Federal Biological Research Centre for Agriculture and Forestry (BBA) have to be consulted.

Authorization of plant protection products (i.e. the preparations applied to the field) is within the responsibility of the Member States, based on harmonized rules. In contrast, the European active substances programme for pesticides has been established to include active substances in Annex I to Directive 91/414/EEC. Under the leadership of the European Food Safety Authority (EFSA), active substances in plant protection products are evaluated by Member States through a Community-wide procedure at EU level. Decisions on the acceptance of an active substance and its listing in Annex I to the Directive are taken by the Member States after proposal by the European Commission. Inclusion in this positive list is a prerequisite for the authorization of plant protection products containing the respective active substance. In most cases, such listing is connected with conditions concerning risk mitigation measures or restrictions for national authorizations, e.g., application to certain cultures only. Plant protection products containing substances which are not accepted must no longer be authorized in the Member States. In 2005, 46 existing active substances were listed, while 31 substances were rejected. 426 out of 984 existing active substances were not defended by a notifier and thus not listed, while the remaining 481 are still in the programme, which is due to be completed by the end of 2008. In addition, about 100 new active substances are being examined, of which 53 have been included in Annex I and 7 have been rejected.

In order to guarantee the necessary level of safety, users of plant protection products are obliged to observe certain requirements and operating conditions as stipulated in the authorization process and included in mandatory labeling on the packages or as part of the instructions for use. In addition, they must comply with common rules like Good Agricultural Practice, including instructions on integrated pest management [36]. The EU furthermore tries to reduce the risks and amounts of pesticide application through certain programmes. The Communication from the European Commission "Towards a Thematic Strategy on the Sustainable Use of Pesticides" [37] is partly being implemented in Germany through the "Pesticide Reduction Programme" of the Federal Ministry for Consumer Protection, Nutrition, and Agriculture [38].

The thorough assessment of pesticides includes an evaluation of the risks to groundwater. Pesticides must not be harmful to groundwater, which means that neither the active substance nor its relevant metabolites shall occur in groundwater in concentrations of more than 0.1 µg L^{-1} after application. The first step of this part of the evaluation is a computer-aided model calculation (PELMO: **PE**sticide **L**eaching **MO**del) which uses physico-chemical, adsorption/desorption and degradation data on the chemical and takes into account realistic worst case soil and climatic conditions in Germany. It is followed, where necessary, by lysimeter studies as a second step. This approach takes into consideration not only the active substances, but also the relevant metabolites.

An intensive discussion is ongoing as to which metabolites are relevant and which are not. Relevance within the authorization procedure means those degradation products which are toxicologically or ecotoxicologically effective and occur in testing in quantities over 10% of the residues [39, 40, 41]. However, the question remains how persistent pesticide metabolites with no specific efficacy or which are not as effective as their parent substances should be assessed. One prominent example is 2,6-dichlorobenzamide, a metabolite of dichlobenil.

Currently an amendment of Directive 91/414/EEC is under discussion. DG SANCO of the European Commission [42] presented a draft of a new regulation which provides for the following, among other things:

- mutual recognition of authorizations in three geographical zones,
- stricter criteria for inclusion of active substances in different annexes,
- introduction of a comparative assessment for substances where a risk cannot be excluded but which are necessary for agriculture,
- inclusion of safeners, synergists, and adjuvants,
- need to register low risk products only,
- stronger role for Community authorities and fewer competences for member states.

It remains to be seen how fast negotiations will proceed and when the new legislation will come into force.

12.1.4.2 Biocides

Biocidal products are active substances and preparations or microorganisms which serve to repel, render harmless, or destroy harmful organisms (pests like moths, woodworm, mice, pathogens, molds, etc.) and are not applied in agriculture. The aim of biocide use is to prevent damage by harmful organisms to food, commodities, building materials (wood), and other products and to guarantee hygiene in buildings. Biocidal products may contain one or more biocidally active substances. The biocidal agents are grouped into 23 different product types according to the different fields of application, e.g., disinfectants, wood preservatives, rodenticides, etc.

The European Parliament and the Council adopted Directive 98/8/EC concerning the placing of biocidal products on the EU market [43]. The background for this directive was the need to harmonize the legislation of the Member States regarding this type of substances. Some Member States already had some legislation for distinct groups of biocides in force; others did not have any administrative procedure for such products.

The Directive requires an authorization process for biocidal products similar to that for plant protection products. Authorization is the responsibility of the Member States. As with pesticides, a European ten-year programme was launched to assess existing active substances for inclusion in positive lists. Member States may only grant authorization to biocidal products which contain active substances listed in the positive lists of Annex I.

Existing biocidally active substances are substances which were on the EU market for biocidal purposes before 14 May 2000. Any substance which was not on the market before this date is regarded as a new active substances and has to be approved by the Member States before it will be listed in the Annex to the Directive. Notification of existing biocidally active substances started in 2000 and was finalized on 31 January 2003 with ca. 940 substances [44]. The evaluation procedure for existing active substances started with wood preservatives and rodenticides, with Member States serving as rapporteurs. Whenever a risk assessment results in the conclusion that a substance may pose an unacceptable risk, a comparative assessment has to be carried out with the aim of identifying the least problematical alternative.

According to Germany's Biocides Act [2], which is part of the Chemicals Act, national authorization of biocidal products containing existing substances will not start until a positive decision has been taken on the inclusion of the active substance in Annex I. Therefore, regardless of products with new active ingredients, authorization will not begin before 2007 in Germany. All products containing notified active ingredients can be produced and sold freely as biocidal products without authorization until at least 1 September 2006. They can stay on the market until 14 May 2010 unless the decision on the inclusion of the active substances is reached before that date.

A risk assessment has to be carried out for biocidal products and their active ingredients. This risk assessment covers environment, consumer health, and occupational safety and comprises the following steps:

- hazard identification,
- dose (concentration) – response (effect) assessment,
- exposure assessment,
- risk characterization.

The environmental risk assessment has to be carried out as described in the Technical Guidance Document, which covers not only new and existing industrial chemicals but also biocides [11]. The principles are similar to those for pesticides, but experience is still lacking. In particular, exposure scenarios for the various groups of biocidal agents need to be developed and validated.

12.1.4.3 Pharmaceuticals

Pharmaceuticals and their metabolites are an emerging issue in modern ecotoxicology and water chemistry. Several contributions in this book discuss pollution of waters (see Chapters 2 and 3) with human and veterinary drugs (see Chapter 5). Management of environmental risks related to the human and veterinary use of pharmaceuticals is particularly difficult for the following reasons:

- Pharmaceuticals are biologically highly active substances with specific effect mechanisms. At present little is known about the environmental impact of these various modes of action related to pharmaceuticals.
- Compared to pesticides, drugs are often more polar because they are designed to be excreted as parent substances or metabolites.

- In most cases, exposure of the environment to pharmaceuticals occurs after body passage. In many cases, not the parent substances is of concern but its (polar) metabolites or conjugates.
- The use of human medicinal products results in regional long-term exposure of the environment via the effluent of sewage treatment plants.
- Where environmental risks are identified, risk management is hardly possible. This holds especially for human pharmaceuticals, whose therapeutic benefit almost always outweighs the environmental risk arising from their use.

Environmental risk assessment of drugs is a relatively new issue. In the past, it was common to think that what is good for human health cannot be dangerous for the environment when it is at low concentrations. Based on EU Directives 2004/28/EC (veterinary drugs) and 2004/27/EC (human drugs), the German Medicines Act [45] regulates the authorization of pharmaceuticals and the registration of homeopathic remedies. Finished pharmaceutical products may only be placed on the market if they have been licensed or registered by the competent authority.

In the EU, there are four established authorization procedurese:

- centralized authorization,
- decentralized authorization,
- mutual recognition procedure,
- national authorization.

In the case of the centralized authorization procedure for pharmaceuticals which are either produced by the application of biotechnology or are highly innovative in character, the marketing authorization is granted by the European Medicines Agency (EMEA), which is supported by scientific committees containing experts from all Member States. Decentralized authorization is the extension of national authorizations to all member states. Mutual recognition is a procedure to extend existing national authorizations to individual other EU member states. National authorization is granted in one Member State only.

In Germany, the authorization agency for human pharmaceuticals is the Federal Institute for Pharmaceuticals and Medical Products (BfArM). The Federal Agency for Consumer Protection and Food Safety (BVL) is the corresponding institution for veterinary drugs, and the Federal Office for Sera and Vaccines (PEI) is responsible for sera, vaccines, test allergens, test sera, test antigens, and blood coagulation preparations. Medicinal products containing or consisting of genetically modified organisms are authorized by the Robert Koch Institute (RKI).

In Germany, the legal basis for an environmental risk assessment of veterinary medicinal products was established in 1993; human pharmaceuticals followed in 1998. Since 1998, the German Environment Agency has been the competent authority for environmental risk assessment as part of the authorization process. In the same year, the EMEA guideline for the environmental risk assessment of veterinary pharmaceuticals came into force [46]. In November 2005, a new guideline, established by the International Cooperation on Harmonization of Technical Requirements for Registration of Veterinary Medicinal Products (VICH), has superseded the EMEA document and will be applied not only in Europe but also in North America and Japan [47]. In this two-phased concept, the first phase – identification and quantification of environmentally relevant exposure – is separated from the second phase – generation of experimental ecotoxicity data – by a so-called action limit. If the predicted environmental concentration generated for one of three scenarios – pasture, grassland, and aquaculture – exceeds the action limit, then a Phase II assessment must follow. The action limits given in the VICH guidance are 100 µg kg^{-1} in soil (which is 10 times more than the value in the previous EMEA guideline) and 1 µg L^{-1} in aquaculture effluent. If the PEC does not exceed the action limit, the risk is considered to be acceptable, and the assessment is not continued. Phase II starts with a base data set on fate and effects of the active substance, through which hazards can be identified and exposure levels can be compared with effect thresholds for risk characterization. Higher tier assessment could not be completely harmonized within the VICH process. Guidance for higher tier assessment, modeling of soil concentrations, exposure of surface waters via runoff, drainage, and leaching to groundwater will become available with the VICH technical guidance documents currently under development, which will take into account different approaches and conditions in Europe, North America, and Japan [48]. Higher-tier assessment may include data on biodegradability and metabolization in different environmental media and manure.

In 2004 the EU established the legal basis for the environmental risk assessment of medicinal products for human use (Directive 2004/27/EC). It should be noted that, in contrast to the situation for veterinary medicinal products, risk to the environment is excluded from the overall risk-benefit analysis: authorizations cannot be refused on environmental grounds. However, in a Janus-faced attempt, the need for both the environmental risk assessment itself and risk mitigation measures where an unacceptable risk is identified was stressed. Not only for ecological risk assessors but also for risk managers and risk communicators, this is a new field in which experience must be gained over the coming years.

After more than 12 years of development, the EMEA guideline for environmental risk assessment of human medical products is nearing completion [49]. This guideline also uses the two-phased assessment methodology in combination with an action limit. The action limit for a Phase II risk assessment is set at 0.01 µg L^{-1} of surface water. Following the dominant emission route of pharmaceuticals, the base data set focuses on the fate and effects of the active substance in sewage treatment and in surface water. For lipophilic substances, the methodology provides for additional assessment of risks to the terrestrial compartment, since spreading of sewage sludge on agricultural lands is still common practice in the EU.

Some drugs require specific considerations for an appropriate assessment. On the one hand, amino acids or vitamins can be exempted from standardized assessment, whereas receptor-mediated pharmacological action needs to be given particular attention. For β-ethinylestradiol the lowest observed effect concentration (LOEC) and the no observed effect concentration (NOEC) for gonad histology are 4 and 1 ng L^{-1} respectively. This is significantly below the Phase I action limit. High acute-to-chronic effects ratios (ACRs) have been demonstrated for a range of pharmaceuticals, e.g., antiepileptics, lipid regulators, β-blockers, and benzimidazoles, underlining the need for long-term effects studies as envisaged in the draft guideline. The demanding future task is to devise a methodology for extrapolation from the intended pharmacological action of the pharmaceutical product in mammals, i.e. patients, food-producing, or companion animals, to unintended receptor-mediated action in non-mammalian biota, e.g., fish and aquatic invertebrates. Antibiotics exhibit effects which are not covered by available standardized tests. They inhibit bacterial growth, can alter microbial communities, and may possibly induce resistance at concentrations below their inhibition thresholds. These efforts should go hand in hand with the establishment of appropriate endpoints for the effects assessment to provide a targeted risk assessment approach.

One last point should be mentioned: environmental risk assessment of pharmaceuticals as part of authorization processes has up to now been restricted to new medicinal products. This will change for veterinary drugs in 2006 by the new legislation. In the medium term it is necessary to also set up a prioritized programme for existing human pharmaceuticals, as established for pesticides and biocides, in order to avoid a distortion like that experienced for new and existing industrial chemicals.

12.1.4.4 Detergents and Cleansing Agents

Laundry products and cleansing agents are mixtures of substances used in cleaning or in aiding the cleaning process, and they may subsequently enter the aquatic environment. This category includes products containing surface-active substances (surfactants) or organic solvents if these are used for cleaning purposes and may enter waters after use. In the past, persistent detergents like tetrapropylenebenzenesulfonate (TPBS) posed a severe problem because they were toxic to water organisms and formed foams on the water surface (see Chapter 9). This was the reason why specific legislation was developed for such products, which are produced and used in quantities of more than 1 million tons annually. The German Washing and Cleansing Agents Act [50] represents a piece of precautionary legislation which lays down additional requirements for these products. The Act provides that washing and cleansing products may only be placed on the market in such a form that, "their use will not have any avoidable detrimental effects on the quality of waters, particularly with regard to the ecosystem and to the supply of drinking water". The aim of this act is not, however, limited to environmental protection, but also covers consumer protection.

Prior to placing such products on the market for the first time, the producer, importer, or person marketing them must inform the Federal Environment Agency (UBA), in writing, of the frame formulation of the product and provide data on the product's environmental compatibility. The Federal Environment Agency then examines the data submitted relating to environmental compatibility, ensuring that the constituents of the product meet the requirements of the law, and then informs the regional control authorities about its findings [6].

The Act is accompanied by ordinances (Surfactants Ordinance, Ordinance on Maximum-permissible Quantities of Phosphate) giving more detailed requirements. The Surfactants Ordinance calls for ready primary aerobic degradability of surfactants in washing and cleansing agents. The legislation is supplemented by voluntary commitments in which the industry unilaterally agrees, for example, to dispense with the use of alkyl phenol ethoxylates and to minimize the use of EDTA and NTA. The obligation on the producers of laundry products to indicate, on the packaging, recommended dosages as well as the most important constituents enables consumer and environmental organizations to provide recommendations about the purchase of more environmentally compatible laundry products.

The EU Detergents Regulation EC 648/2004 [51] came into force in October 2005 and replaces most parts of the German Washing and Cleansing Agents Act. Most obligations are very similar, but the current German register listing all washing and cleansing agents will no longer be maintained. There will be less room and incentive for voluntary agreements. The requirements with respect to the biodegradability of surfactants will be stricter, however. These must pass tests for ultimate degradation (mineralization), not for primary degradation only. If a surfactant does not meet these criteria and recalcitrant degradation products occur, use can nevertheless be permitted if an environmental risk assessment is carried out based on the principles of the EU Risk Assessment of chemicals according to Directive 93/67/EEC or Regulation (EEC) No. 793/93, i.e. assessment of the exposure and effects of chemicals in the (aquatic) environment.

12.1.5
Reflections on Current Chemicals Management of Persistent Polar Pollutants

As discussed in previous sections, PBT and vPvB substances (cf. Section 12.1.2) are currently the only substances to which the common quantitative comparison of exposure and effect thresholds is not applied. Most PBT and vPvB substances are rather lipophilic, hardly soluble in water, and therefore transported by water only over short distances. They pose a problem to waters either locally, by contamination of sediment, or by deposition after long-range atmospheric transport, like most POPs which are distributed in the environment in all three main compartments: soil, air and water.

Most persistent substances that are polar do not fulfil the bioaccumulation criterion and are therefore assessed conventionally. According to current rules they only pose a risk when environmental concentrations exceed (eco)toxicity thresholds. However, they share one undesired property with PBTs: once released into the environment they cannot be retrieved. Concentrations may increase and cause an unknown long-term problem. Their mobility in the water phase makes them likely to be distributed and, by transport via groundwater, to appear in drinking water as a hygiene problem.

When comparing the various assessment schemes for industrial chemicals and others, it can be seen that, with the exception of pesticide assessment, all have significant deficiencies as regards the assessment of risks to groundwater. One reason may be the separation of environmental and human risk assessment. Water suppliers are arguably the most interested in groundwater protection, since they want to obtain raw water which poses no hygienic or toxicological risk. As an ecosystem, groundwater is only of limited scientific interest to ecologists, because the biocoenosis is difficult to explore and much less diverse than that in surface water.

As already mentioned, pesticides assessment (cf. Section 12.1.4.1) is the only assessment scheme which takes a close look at the properties of substances which are persistent and polar. This also includes consideration of the risks of metabolites, which often are more polar than their parent substances and therefore, if persistent, present a greater hazard for the water cycle. However, for practical reasons and because of lack of knowledge of the hazardous properties of the metabolites, this is mostly limited to metabolites assumed to have a similar mode of action. It should be remembered that analytical identification of polar metabolites is resource-intensive and cannot be carried out for low-volume industrial chemicals, not to mention the difficulty of isolating those metabolites in order to examine their properties. One possible approach to solving this problem is first to apply quantitative structure-activity relationships (QSAR) to identified metabolites to select those warranting a more detailed examination. However, the limitations of the QSAR method in predicting toxicological and ecotoxicological properties should not be forgotten. This approach may improve risk assessment of substances that have toxic persistent metabolites, but leaves some persistent polar metabolites that are not identified as harmful but could accumulate in waters.

A few persistent polar pollutants may also meet the PBT criteria. The most prominent examples are the perfluorinated compounds with amphiphilic properties. Perfluorooctylsulfonate (PFOS), a persistent compound used in fire-fighting foams, as an anticorrosive, and in the textile industry, together with other similar substances, is present in environmental media, blood serum, and biota in relevant concentrations, also in remote areas [52, 53]. The main distribution pathway seems to be the water phase, and not atmospheric transport. PFOS is being discussed as a POP candidate under the POP Protocol and the Stockholm Convention.

12.2
Future Chemicals Management in Europa with REACH

On 27 February 2001, the European Commission presented a White Paper on a "Strategy for a Future Chemicals Policy" [54]. The centerpiece of this future chemicals policy will be a European regulation on the registration, evaluation, and authorization of chemicals (REACH). This concept was created to achieve the following goals, among others:

- to protect people and the environment from the damaging effects of the production, processing, application, and disposal of chemical substances;
- to increase the competitiveness of the chemicals industry;
- to prevent the Common Market from disintegrating, for instance as a result of different standards for risk assessment and communication in legislation on chemicals;

- to promote greater transparency, for instance, with respect to the dangerous properties of chemicals and the consequences of possible exposure to them;
- to integrate European policy in international programmes such as GHS (Globally Harmonized System of Classification and Labeling of Chemicals) and the objectives of the Johannesburg Summit;
- to promote research methods not based on experiments with vertebrates;
- to conform with the EU's international obligations (such as towards the World Trade Organization – WTO).

On 29 October 2003, after an extensive Internet consultation on a pre-draft, with over 6000 contributions, the European Commission released a draft regulation for a new EU chemicals policy [16]. This draft was being discussed in depth in the Council and the European Parliament. The Council reached political agreement in December 2005 and the Parliament voted in November 2005. Both opinions alleviate the duties for the registration of low volume substances. Because there are minor differences, it is expected that REACH will come into force in 2007.

The new REACH system represents a paradigm shift for the production and marketing of chemicals. While in the current system the authorities have to identify and evaluate critical substances and prove whether they are safe or not, the responsibility to prove that a substance is safe will shift to the manufacturers and importers (reversal of the burden of proof). Manufacturers and importers are obliged to register all their chemicals above a volume of 1 ton per year. All chemicals have to be registered according to a binding schedule depending on the volume (from 3 to 11 years after the REACH Regulation takes effect) at a central EU agency which will be established in Helsinki. A defined package of information must be submitted, whose extent also depends on volume. An important target is first to clarify the risks associated with high-production-volume chemicals to which the environment and humans are most exposed.

REACH no longer distinguishes between existing and new substances. The information to be provided will be the same for both. The 30 000 substances, not including polymers and internal intermediates, will be phased in continuously.

Substances with an annual production volume exceeding 10 t per manufacturer (some 11 000 substances, not including polymers and internal intermediates) require a Chemical Safety Assessment (CSA) which includes a detailed evaluation of the hazardous properties of the substance. If the substance is judged to be hazardous according to the rules for classification and labeling, an exposure assessment and a risk characterization are additionally required for all stages of the substance's life cycle. The CSA must be documented in a Chemical Safety Report (CSR), which has to be submitted to the central agency. The substance manufacturer has to derive from this assessment practical, plausible information and specifications for the safe use of the substance by his customers (formulators of preparations and other commercial users of the chemical) and has to communicate this information and these specifications down the supply chain through safety data sheets.

The main responsibility will be with the manufacturers and importers, but the so-called downstream users will be included in the legislation. They have to prove that their uses are included in the CSAs of their suppliers. If not, they are obliged to inform their manufacturer of that use or to adapt the CSA to their own needs.

Member State authorities will be involved in the evaluation of the registration dossiers. In principle, the central agency only performs a completeness check and furthermore will be responsible for the evaluation of the dossiers in several cases. However, based on common rules, Member State authorities may evaluate the quality, the contents, and the conclusions of a small number of registration dossiers. If necessary, regulatory action can be taken by the central agency.

Substances of very high concern will be subject to authorization. It has to be demonstrated that they are controlled adequately or, if not, that no better alternatives are available and a socioeconomic analysis demonstrates the need for that use. Carcinogenic, mutagenic, and reprotoxic (CMR) substances, persistent, bioaccumulative and toxic (PBT) substances and very persistent and very bioaccumulative (vPvB) substances are regarded as substances of very high concern. Substances with an equal level of concern, e.g., endocrine disrupters, may be included in this procedure. The EU has sent out a clear signal that some very hazardous substances should be phased out wherever possible.

The new chemicals policy is, in the first place, an instrument to gain knowledge about the properties and effects of chemicals. This is a prerequisite for the safe handling of chemicals and ensures that "toxic ignorance" is no longer an advantage in the manufacture and use of chemicals but will be a disadvantage.

REACH will overcome some main drawbacks of current legislation:

- Existing and new chemicals will be treated equally.
- Downstream users are included.
- Substances with irreversible effects or exposure (incl. PBTs and vPvBs) are subject to a cumbersome authorization procedure.
- It clearly provides that the main responsibility for chemical safety lies with the industry and not with the authorities.

However, there are also some problems which have not been solved sufficiently.

One drawback of the draft regulation on REACH is that data requirements for substances manufactured in quantities of 1–10 t year^{-1} will depend how these substances are used and therefore in most cases are not sufficient to evaluate potential risks to environment and health. With this poor data base, classification and labeling for human health and environmental hazards in many cases will not be possible, nor can substances be identified as PBT/vPvB candidates. It should be taken into account that the data requirements listed in Annex V of REACH are significantly reduced compared with the current requirements for new chemicals (cf. Section 12.1.1.1).

Furthermore, the instrumental gap between chemicals legislation and sectoral environmental legislation (e.g. IPPC Directive) will remain. Technical risk reduction measures derived under REACH for certain substances will be difficult to put into effect.

With regard to persistent polar pollutants, REACH will not consider them as substances of very high concern. Therefore, these substances will not be subject to an authorization procedure. Potential risks that they pose to human health and the environment must be evaluated in a classical way. This may be the subject of further expert discussions leading to the development of technical guidance, clarifying what is meant by "equal level of concern", in order to raise awareness of these risks of persistent polar pollutants.

12.3
Persistent Organic Pollutants (POPs) and Persistent Polar Pollutants (PPPs) – A Comparison

12.3.1
Persistent Organic Pollutants (POPs)

A clear message on chemicals management was sent out by the World Summit on Sustainable Development in Johannesburg (WSSD) in 2002, which, in the chapter "Sustainable Consumption and Production" of its declaration, called for sound management of chemicals over their whole life cycle and minimization of significant negative impacts from chemicals to be achieved by 2020 [55].

One of the main pillars of international chemicals management is the Stockholm Convention on Persistent Organic Pollutants (POPs) [56]. Under the auspices of the United Nations Environment Programme (UNEP), negotiations for a Convention on Persistent Organic Pollutants were concluded in December 2000. The Convention was adopted and opened for signature in May 2001 in Stockholm, and came into force in May 2004. The Convention provides a framework, based on the precautionary principle, for the elimination of production, use, import, and export of the initial twelve priority Persistent Organic Pollutants (POPs), their safe handling and disposal, and elimination or reduction of releases of certain unintentional Persistent Organic Pollutants. The Stockholm Convention is derived from the POP Protocol to the Convention on Long Range Transboundary Air Pollution of the UN ECE (United Nations Economic Commission for Europe), a regional predecessor which came into force in November 2003 [57].

The criteria which characterize a POP are outlined in Annex D to the Stockholm Convention. A POP exhibits the following features:

(a) **Persistence**
   (i) Evidence that the half-life of the chemical in water is greater than two months, or that its half-life in soil is greater than six months, or that its half-life in sediment is greater than six months, or
   (ii) Evidence that the chemical is otherwise sufficiently persistent to justify its consideration within the scope of this Convention

(b) **Bioaccumulation**
   (i) Evidence that the bioconcentration factor or bioaccumulation factor in aquatic species for the chemical is greater than 5000 or, in the absence of such data, that the log K_{ow} is greater than 5, or

(ii) Evidence that a chemical presents other reasons for concern, such as high bioaccumulation in other species, high toxicity or ecotoxicity, or

(iii) Monitoring data in biota indicating that the bioaccumulation potential of the chemical is sufficient to justify its consideration within the scope of this Convention

(c) **Potential for long-range environmental transport**
   (i) Measured levels of the chemical in locations distant from the sources of its release that are of potential concern, or
   (ii) Monitoring data showing that long-range environmental transport of the chemical, with the potential for transfer to a receiving environment, may have occurred via air, water or migratory species, or
   (iii) Environmental fate properties and/or model results that demonstrate that the chemical has a potential for long-range environmental transport through air, water, or migratory species, with the potential for transfer to a receiving environment in locations distant from the sources of its release. For a chemical that migrates significantly through the air, its half-life in air should be greater than two days.

(d) **Adverse effects**
   (i) Evidence of adverse effects to human health or to the environment, or
   (ii) Toxicity or ecotoxicity data that indicate the potential for damage to human health or to the environment.

According to this profile, POPs are characterized *inter alia* by their capacity to bioaccumulate, which is usually higher with hydrophobic than with hydrophilic substances. High hydrophobicity of non-polar substances is linked with high adsorption to soil and sediments. This means that most POPs are not mobile in water.

An important application of Annex D is found in Article 3, paragraphs 3 and 4 of the Stockholm Convention. These paragraphs call on Parties to apply the criteria for POPs in Annex D to new chemicals and pesticides in order to avoid the development of new POP-like substances. This gives a strong incentive for early preventive action.

Currently, only twelve chemicals are covered by the Stockholm Convention: aldrin, chlordane, dieldrin, endrin, heptachlor, hexachlorobenzene, mirex, toxaphene, DDT, polychlorinated biphenyls (PCBs), and dioxins/furans. Most of these substances have already been phased out in industrial countries but are still used in some developing countries. An important representative of the so called "dirty dozen" is DDT, which is still needed in several countries for malaria control. Initiatives to promote reasearch have been launched in cooperation with the World Health Organization (WHO) with the aim of phasing out DDT as soon as other effective and feasible alternatives for malaria vector control are available.

The Stockholm Convention is the most important instrument for international action to protect man and the environment against chemical-induced damage. It shows the Parties' real commitment to find compromises and that global action is needed to tackle the most urgent problems caused by chemicals. The Convention also gives a signal for the inclusion of further substances with POP characteristics. The criteria listed in Annex D have significantly influenced the discussion on hazardous properties of chemicals and were an important incentive for the PBT/vPvB strategy in the EU and other countries (cf. Section 12.1.2). Reaching consensus on further POPs to be covered by the Convention will be a difficult and lengthy process, but awareness of the fact that persistence and bioaccumulation are extremely problematic characteristics has increased through this negotiation process. Persistence, a property previously rather neglected by risk assessors, has become a central criterion for the identification of hazardous, non-sustainable substances [58].

The Rotterdam Convention on Prior Informed Consent (PIC) is another global treaty to promote safe management of chemicals globally. This Convention has no direct relationship to POPs. Its aim is to facilitate information exchange on the hazards and risks of dangerous substances when exported and to control the amounts of these substances in international trade [59].

12.3.2
Persistent Polar Pollutants (PPPs) in the Water Cycle

As already mentioned in Section 12.1.5, in most cases persistent polar pollutants (PPPs) in the water cycle do not meet the POP criteria. Their polarity makes them mobile in aqueous systems but prevents them accumulating in biota. For PPPs, this difference means that in general the risk potential is lower than that for POPs, because bioaccumulation enhances the probability of toxicologically relevant internal concentrations in organisms. Moreover, only in exceptional cases are PPPs detected in remote regions far from the emission sources. This means that global action to reduce the risks of substances with these characteristics seems not to be adequate.

Nonetheless, currently the risks of persistent polar pollutants tend to be underestimated. Awareness should be raised of the specificity of the risks which these substances pose to man and environment. As for PBTs, the risks of persistent polar pollutants are not adequately described by a quantitative comparison of exposure and effect thresholds. Long-term risks tend to be underestimated and knowledge about effects is never complete. In 1998 the German Advisory Council on Global Change published a classification of risks [60]. Risk should not only be described by the extent of damage and the probability of its occurrence, but also by other characteristics [61]:

- *Ubiquity:* Is the risk global in scale or limited to local level?
- *Persistence:* Does the risk persist for a long time or does it subside rapidly?
- *Irreversibility:* Is the damage that has occurred reversible or can it not be remedied?
- *Delay effect:* Do the adverse consequences occur spontaneously or after a considerable delay?
- *Mobilization potential:* Is the risk of high political relevance or does it tend to be ignored?

For PPPs, persistence, irreversibility, and potential delay effects are characteristic. Because of their mobility in the water phase their exposure is often not limited to a certain area. PPPs are long-range chemicals [62] which pose long-lasting risks and may affect large areas. A criterion which is often not adequately assessed is the mobility of a chemical. Mobility enhances the probability of exposure over large areas. Polar chemicals may be distributed by hydrospheric transport over long distances without being eliminated. The mobility and persistence of these chemicals also causes the problem that they cannot be effectively eliminated from the water phase by biological treatment or by adsorption, e.g., on charcoal.

That high persistence and high mobility are a combination of properties which deserve more attention can be demonstrated for gases and very volatile compounds which are persistent in the atmosphere. Ozone-depleting CFCs and fluorinated greenhouse gases like HFCs do not bioaccumulate but are distributed globally and negatively affect atmospheric processes. Consequently they are regulated by global protocols.

Since risks of PPPs are not of a global scale they should be managed regionally. The first step should be the development of criteria which describe a common understanding of what PPPs are. Persistence and toxicity may be defined in the same manner as they are in the case of PBTs. However, bioaccumulation has to be replaced by (physico-chemical) parameters which describe the fact that the substance is not likely to be transferred to other environmental media and is mobile in the water phase. The Henry coefficient and sorption coefficients to soil and/or sediment could be first proposals to fill the gap.

A common set of criteria may also be helpful for further development of the assessment of chemical risks to the groundwater compartment. The experience gained with the assessment schemes for pesticides suggests that it may be possible to calculate these risks without additional testing, at least in a first tier. Groundwater risk assessment will be a touchstone for the integration of human and ecological risk assessment [24], because risk deriving from exposure of chemicals to groundwater comprises health and ecology as well (cf. Section 12.1.2). In principle, REACH will offer the possibility to consider persistent polar pollutants as substances of high concern which are subject to authorization. Not only CMR and PBT/vPvB substances may fall under these provisions, but also chemicals with an equal level of concern. Once awareness of the risks of PPPs has been raised and a common set of criteria developed, the most problematic persistent polar pollutants could be included in the REACH authorization scheme, thus providing strong incentives to replace these chemicals by less problematic alternatives.

Acknowledgments

This contribution was made possible by helpful discussions with Albrecht-Wilhelm Klein, Jan Koschorreck, and Ingrid Nöh of the Federal Environment Agency. We wish to thank them for their assistance in revising the text.

References

1. European Community: Consolidated Version of the Treaty Establishing the European Community, Official Journal of the European Communities C 325/33, 24.12.2002.
2. Chemikaliengesetz; Gesetz zum Schutz vor gefährlichen Stoffen; BGBl I 2002, 2090, last amended on 13.05.2004 (BGBl I 2004, 934).
3. Council Directive 92/32/EEC of 30 April 1992 amending for the seventh time Directive 67/548/EEC on the approximation of the laws, regulations and administrative provisions relating to the classification, packaging and labelling of dangerous substances; Official Journal of the European Community L 154, 05/06/1992 P. 0001–0029.
4. Council Directive (EC) 67/548/EEC of 27 June 1967 on the approximation of laws, regulations, and administrative provisions relating to the classification, packaging and labelling of dangerous substances, Official Journal of the European Community P 196.
5. EINECS (European inventory of existing commercial chemical substances); Official Journal of the European Communities C 146 A of 15 June 1990.
6. BAuA: National Profile – Chemicals Management in Germany, Schriftenreihe der Bundesanstalt für Arbeitsschutz und Arbeitsmedizin, 2004.
7. Council Regulation (EEC) No 793/93 of 23 March 1993 on the evaluation and control of the risks of existing substances, Official Journal of the European Community L 84/ 1 05/04/93.
8. Commission Regulation (EC) No 1488/94 of 28 June 1994 laying down the principles for the assessment of risks to man and the environment of existing substances in accordance with Council Regulation (EEC) No 793/93, Official Journal of the European Community L 161, 29/06/1994, p. 3–11.
9. Chemikalien-Verbotsverordnung; Verordnung über Verbote und Beschränkungen des Inverkehrbringens gefährlicher Stoffe, Zubereitungen und Erzeugnisse nach dem Chemikaliengesetz vom 14.10.1993; BGBl I 1993, 1720, Status: Neugefasst durch Bek. v. 13.6.2003 I 867, last amended by Art. 1 V v. 23.12.2004 BGBL I 3855.
10. Gefahrstoffverordnung; Verordnung zum Schutz vor gefährlichen Stoffen vom 26.10.1993; BGBl I 1993, 1783, Status: Neugefasst durch Bek. v. 15.11.1999 I 2233; 2000 I 739; last amended by Art. 2 V v. 25. 2.2004 BGBl I 328.
11. European Chemicals Bureau: Technical Guidance Document in support of Commission Directive 93/67/EEC on Risk Assessment for new notified substances, Commission Regulation (EC) No 1488/94 on Risk Assessment for existing substances and Directive 98/8/EC of the European Parliament and of the Council concerning the placing of biocidal products on the market. (http://ecb.jrc.it/biocides/).
12. Umweltbundesamt: "Action Areas and Criteria for a precautionary, sustainable Substance Policy using the example of PVC" Erich Schmidt Verlag, ISBN 3-503-06006-5, Berlin 2001.
13. Umweltbundesamt: Texte 30/01 "Nachhaltigkeit und Vorsorge bei der Risikobewertung und beim Risikomanagement von Chemikalien" ("Sustainable and Precautionary Risk Assessment and Risk Management of Chemicals"), Umweltbundesamt Berlin 2001.
14. Van der Kolk, EU workshop on "Industrial Chemicals: Burden of the Past, Challenge for the Future" on 24th/25th February 1999, Brussels.
15. VCI: Freiwillige Selbstverpflichtung der deutschen chemischen Industrie zur Erfassung und Bewertung von Stoffen (insbesondere Zwischenprodukten) für die Verbesserung der Aussagefähigkeit, Dr. Jürgen Strube, Frankfurt, den 23. September 1997.
16. European Commission: Proposal for a Regulation of the European Parliament and of the Council Concerning the Registration, Evaluation, Authorization and Restriction of Chemicals (REACH), COM 2003 0644 (03) of 29.10.2003.
17. USEPA – Environmental Protection Agency. Framework for cumulative risk

18. Bödeker W., Altenburger R., Faust M., Grimme L. H., Synopsis of concepts and models for the quantitative analysis of combination effects: from biometrics to ecotoxicology, *Archives of Complex Environmental Studies* 4 (**1992**), 45–53.
19. Faust M., Altenburger R., Backhaus T., Bödeker W., Scholze M., Grimme L. H.: Predictive assessment of the aquatic toxicity of multiple chemical mixtures, *J. Environ. Qual.* 29, **2000**, 1063–1068.
20. Kortenkamp A.: Kombinationswirkungen von endokrin wirksamen Substanzen im Niedrigdosisbereich, Vortrag beim 3. Statusseminar des Umweltbundesamtes "Chemikalien in der Umwelt mit Wirkung auf das endokrine System", 02.06.2005, Berlin.
21. Giger W., Alder A. A., Göbel A., Golet E., McArdell C., Molnar E., Schaffner C.: Polar POPs ante Portas – Antibiotics and Anticorrosives as Contaminants in Wastewaters, Ambient Water and Drinking Water, Lecture at CEEC Workshop "Classical and Environmental Contaminants: from Lakes to Ocean, Dübendorf CH, January 21, 2005.
22. Werres F., Balssaa P., Overath H.: Nachweis von Industriechemikalien (HPS, BPS und SPS) im Oberflächenwasser, *Acta Hydrochim. Hydrobiol.* 29, **2001**, 16–21.
23. Knepper Th., Haberer K.: Auftreten von Phenylsulfonamiden in Kläranlagen-, Oberflächen- und Trinkwässern. *Vom Wasser* 86, **1996**, 263–276.
24. World Health Organization: Integrated Risk Assessment, Report Prepared for the WHO/UNEO/ILO, International Programme of Chemical Safety, http://www.who.int/pcs/emerg_site/integr_ra.
25. Führ M., Merenyi S.: Interface Problems between EC Chemicals Law and sector specific Environmental Law, Report to the German Environment Agency, Berlin 30th September 2004.
26. USEPA – Environmental Protection Agency 2005. Information on "New Substances: What is the TSCA Chemical Substance Inventory?" http://www.epa.gov/opptintr/newchems/inventory.htm.
27. Tickner J. A., Geiser K., Coffin M., The U.S. Experience in Promoting Sustainable Chemistry; ESPR – *Environ. Sci. Pollut. Res.* 12 (2), **2005**, 115–123.
28. Percival C., Schroeder C., Leape J. Environmental Regulation: Law, Science, and Policy. New York: Little Brown and Company 1992.
29. USEPA – Environmental Protection Agency. 2004. Polybrominated Diphenylether (PBDE) Significant New Use Rule (SNUR) Questions and Answers. Available at www.epa.gov/opptintr/pbde/qanda.htm.
30. USEPA – Environmental Protection Agency. 2004. Green Chemistry Institute: Environmental Protection Agency. Available at www.epa.gov/opptintr/greenchemistry/gcinstitute.html.
31. USEPA – Environmental Protection Agency. 2003. Chemical Right to Know Initiative. Available at www.epa.gov/chemrtk/index.htm..
32. USEPA – Environmental Protection Agency. 2004. Great Lakes Water Quality Agreement: Environmental Protection Agency. Available at www.epa.gov/glnpo/glwqa/usreport/about.html.
33. USEPA – Environmental Protection Agency. 2003. Binational Toxics Strategy: Environmental Protection Agency. Available at www.epa.gov/glnpo/bns/index.html.
34. Council Directive 91/414/EEC concerning the placing of plant protection products on the market, Official Journal of the European Community L 230; Status: last amendment: Commission Directive 2003/39/EC of 15 May 2003 amending Council Directive 91/414/EEC to include propineb and propyzamide as active substances; Official Journal of the European Union L 124/30 of 20.05.2003.
35. Pflanzenschutzgesetz; Gesetz zum Schutz der Kulturpflanzen vom 14.05.1998, BGBl I 1998, 971, last amended on 19.08.2004, BGBL I 2004, 1154.

36 International Union of Pure and Applied Chemistry: Glossary of Terms relating to Pesticides, IUPAC Report on Pesticides 36, Pure Appl. Chem, 68, 1996, 1167–1193.
37 European Commission. Communication from the Commission to the Council, the European parliament and the Economic and Social Committee. Towards a Thematic Strategy on the Sustainable Use of Pesticides, No 349 from 01.07.2002.
38 Bundesministerium für Verbraucherschutz, Ernährung und Landwirtschaft: Reduktionsprogramm Chemischer Pflanzenschutz, Berlin 2004.
39 DG SANCO: Guidance Document on the Assessment of the Relevance of Metabolites in Groundwater of Substances Regulated under Council Directive 91/414/EEC (SANCO 221/2000 of 25.02.2003).
40 Council of Europe: Pesticides and ground water, Council of Europe Press 1993.
41 Winkler R., Stein B., Gottschild D., Streloke M.: Prüfung und Bewertung des Eintrags von Pflanzenschutzmitteln in das Grundwasser sowie deren Bedeutung für die Entscheidung über die Zulassung, *Nachrichtenbl. Deut. Pflanzenschutz*. 51, **1999**, 38–43 .
42 DG SANCO: Draft Working Proposal for Council and Parliament Regulation concerning the placing of plant protection products on the market, EC Commission, DG SANCO /10159/2005, 06 April 2005.
43 Directive 98/8/EC of the European Parliament and of the Council of 16 February 1998 concerning the placing of biocidal products on the market, Official Journal of the European Communities L 123/1 of 24.04.98.
44 Regulation (EC) No. 2032/2003 from the 4th November 2003 on the second phase of the 10-year work programme referred to in Article 16(2) of Directive 98/8/EC of the European Parliament and of the Council concerning the placing of biocidal products on the market, and amending Regulation (EC) No 1896/2000 Official Journal of the European Community L 107/1 24/11/2003.
45 Arzneimittelgesetz; Gesetz über den Verkehr mit Arzneimitteln, BGBl I 1976, 2445, 2448; neu gefasst durch Bek. v. 11.12.1998 BGBl I 3586; last amended by Art. 2 G v. 10. 2.2005 BGBl I 234.
46 EMEA: Note for guidance: environmental risk assessment for veterinary medicinal products other than GMO-containing and immunological products. European Agency for the Evaluation of Medicinal Products, EMEA/CVMP/055/96, London UK, 1998.
47 EMEA: CVMP/VICH Topic GL6 (Ecotoxicity Phase I) Step 7: Guideline on environmental impact assessment (EIAs) for veterinary medicinal products – Phase I. European Agency for the Evaluation of Medicinal Products, Report No. CVMP/VICH/592/98, London UK, 2000.
48 Koschorreck J., de Knecht J.: Environmental Risk Assessment of Pharmaceuticals in the EU – a Regulatory Perspective, in: Pharmaceuticals in the Environment – Sources, Fate, Effects and Risks, Klaus Kümmerer (Ed.), Springer-Verlag Berlin- Heidelberg 2004.
49 EMEA: Note for guidance: environmental risk assessment for medicinal products for human use. European Agency for the Evaluation of Medicinal Products, Draft http://www.emea.eu.int/pdfs/human/swp/444700en.pdf).
50 Gesetz über die Umweltverträglichkeit von Wasch- und Reinigungsmitteln; BGBl I 1987, 875, last amended on 25.11.2003 (BGBl I 2003, 2304).
51 Regulation (EC) No. 648/2004 of the European Parliament and of the Council of 31 March 2004 on detergents; Official Journal of the European Union L 104 of 08.04.2004, 1.
52 de Voogt, P.: Perfluorinated Organic Compounds, Lecture at CEEC Workshop "Classical and Environmental Contaminants: from Lakes to Ocean", Dübendorf C. H., January 21, 2005.
53 Fricke, M., Lahl U.: Risikobewertung von Perfluortensiden als Beitrag zur aktuellen Diskussion zum REACH-Dossier der EU-Kommission. *UWSF-Z Umweltchem Ökotox* 17(1), **2005**, 36–49.
54 European Commission: White Paper – Strategy for a Future Chemicals Policy, European Commission, Brussels, 27.02.2001.

55 UN – United Nations (2002): Report on the World Summit on Sustainable Development in Johannesburg.
56 Stockholm Convention on Persistent Organic Pollutants: http://www.pops.int/documents/convtext/convtext_en.pdf.
57 Convention on Long-Range Transboundary Air Pollution – Protocol on Persistent Organic Pollutants (http://www.unece.org/env/lrtap/pops_h1.htm) .
58 Steinhäuser K. G., Richter S., Greiner P., Penning J., Angrick M.: Sustainable Chemistry – Principles and Perspectives, *ESPR – Environ. Sci. Pollut. Res.* 11 (5), **2004**, 284–290.
59 Rotterdam Convention on the Prior Informed Consent Procedure for Certain Hazardous Chemicals and Pesticides in International Trade: http://www.pic.int/en/ViewPage.asp?id=104.
60 Wissenschaftlicher Beirat der Bundesregierung für globale Umweltveränderungen: Welt im Wandel – Strategien zur Bewältigung globaler Umweltrisiken, Jahresgutachten 1998, Springer-Verlag, Berlin-Heidelberg 1999.
61 Steinhäuser K. G.: Environmental Risks of Chemicals and Genetically Modified Organisms – a Comparison, Part I, *ESPR – Environ. Sci. Pollut. Res.* 8 (2), **2001**, 120– 126.
62 Scheringer M.: Persistenz und Reichweite, Wiley Verlag 1999.

Subject Index

a
acetylation 16
activated carbon filtration 59, 145f., 174
– aminopolycarboxylate 174
– pharmaceutical 56
acute-to-chronic effect ratio (ACR) 294, 327
acylation 16
adsorbable organic halogen (AOX, AOBr, AOCl, AOI) *see also* total organic halogen 88, 254, 266ff.
advanced oxidation process (AOP) 261
alcohol ethoxylate (AEO) 24, 226
– biodegradation 226
– liquid chromatography-mass spectrometry 24
algal toxicity 82
alkane sulfonate
– secondary 220
alkyl ether sulfate (AES) 215ff.
alkyl glucamide 223
– biodegradation 223
alkyl phosphate 4
alkyl polyglucoside (APG) 223
– biodegradation 223
alkyl sulfate (AS) 215ff.
– biodegradation 220
alkylamine 185ff.
– biodegradation 198f.
– groundwater 201
– sorption 203
– surface water 193
– use 185
– wastewater 190
alkylation 15
– *in-port* 15
– *on-column* 15
alkylbenzene sulfonate (ABS) 211
alkylphenol ethoxycarboxylate (APEC) 21, 238
– groundwater 238

alkylphenol ethoxylate (APEO) 22f., 237
alkylphenol polyethoxylate (APEO) 226
alkylpolyglycoside 24
– liquid chromatography-mass spectrometry 24
amine 16, 181ff.
– abiotic process 195
– aromatic 188
– bank filtration 203
– biodegradation 196
– biological treatment 189
– carcinogenic 188
– chlorination 205
– derivatization 16
– drinking water 203
– emission 186ff.
– groundwater 200
– octanol-water partition coefficient 184
– ozonation 191, 204
– photolysis 195
– property 182
– sorption 201
– surface water 192
– toxicity 183
– use 185
– wastewater 189
aminoglycoside 66, 80, 101
– biodegradation 80
– octanol-water partition coefficient 101
– use 66, 101
aminomethylphosphonic acid (AMPA) 25, 138ff.
– bank filtration 148
– liquid chromatography-mass spectrometry 25
– surface water 138
aminopolycarboxylate 158ff.
– activated carbon filtration 174
– bank filtration 173
– complexing agent 155ff.

342 | Subject Index

- drinking water   175
- groundwater   172
- ozonation   174
- toxicity   172
- use   158f.
- wastewater   167

aminopolycarboxylic acid   21
- liquid chromatography-mass spectrometry   21

ammonium compound
- quaternary   133

aniline   185ff.
- biodegradation   197
- groundwater   201
- ozonation   191
- photolysis   195
- sorption   202f.
- surface water   193
- use   185
- wastewater   190

antibiotic   5ff., 26, 65ff., 80
- biodegradation   79
- excretion   69
- hospital wastewater   69
- groundwater   75
- ozonation   49
- pathway to environment   110
- photolysis   78
- resistance   81
- risk assessment   304
- sorption   76
- surface runoff   115
- surface water   74
- toxicity   80
- use   65ff.
- wastewater   71

antimicrobial resistance   81, 115
AOP, see advanced oxidation process
AOX, see adsorbable organic halogen
APCI (atmospheric pressure chemical ionization)   27
API (atmospheric pressure ionization)   19
aquaculture   113
atrazine   127, 137ff.
- activated carbon filtration   146
- groundwater   141
- membrane filtration   148
- ozonation   147
- regulation   127
- surface water   137
- use   126

azo dye   188f.

b

bank filtration   53ff., 93, 147f., 173, 203
- aminopolycarboxylate   173
- pharmaceutical   53

benzidine   188
benzonitrile   128
beta-blocker   23, 45ff.
- liquid chromatography-mass spectrometry   23
- ozonation   49

beta-lactam   25, 79f., 102
- biodegradation   80
- liquid chromatography-mass spectrometry   25
- octanol-water partition coefficient   102
- wastewater treatment   45
- use   102

bezafibrate   55f.
- activated carbon filtration   56
- bank filtration   55
- ozonation   56

bioaccumulation   332
biocide   323
- regulation   323

biodegradable dissolved organic carbon (BDOC)   271
biodegradability   215, 235
biodegradation   52, 79, 162ff., 216
- amine   196f.
- aminopolycarboxylate   162f.
- antibiotic   79
- linear alkylbenzene sulfonate (LAS)   216
- nonylphenolethoxylate   221
- surfactant   213ff.

biofilm   81
biotransformation   26, 79
bismuth iodide-active substance (BiAS)   215
blood lipid regulator   43ff.
bromate   256, 277
- drinking water   277
- ozonation   277
- regulation   256

c

carbamazepine (CBZ)   23f., 52ff.
- activated carbon filtration   56
- bank filtration   53ff.
- liquid chromatography-mass spectrometry   23f.
- ozonation   56f.
- photolysis   52
- risk assessment   303
- toxicity   301

chemical
- existing 311ff.
- management 312ff.
- risk assessment 311ff.
- technical guideline 314
chiral selector 26
chiral separation 26
chloramine 261
- use 261
chlorate 277
chlorination 205, 275
- herbicides 147
chlorine
- regulation 258
- use 259
chlorine dioxide 258
- regulation 258
- use 260
chlorite 277
chromatography
- gas (GC), *see* gas chromatography
- ion chromatography (IC) 20
- liquid (LC), *see* liquid
ciprofloxacin 25, 72ff.
- liquid chromatography-mass spectrometry 25
- octanol-water partition coefficient 77
- sorption 77
- surface water 72
- toxicity 307
- wastewater 72
- wastewater treatment 78
clofibric acid 43ff.
- activated carbon filtration 56
- bank filtration 53ff.
- ozonation 49
- risk assessment 303f.
- toxicity 301
- wastewater 43
- wastewater treatment 47
CMR (carcinogenic, mutagenic, and reprotoxic) substance 331
cocamidopropyl betaine (CAPB) 226
- liquid chromatography-mass spectrometry 25
complexing agent, *see also* aminopolycarboxylate 155ff.
contrast agent, *see* X-ray contrast media

d

dehalogenation 95
- reductive 95
derivatization 11ff.
desethylatrazine (DEA) 141

- surface water 138f.
- groundwater 141
detector
- ECD (electron capture detector) 16
- ITD (ion trap detector) 17
- NPD (nitrogen phosphorus detector) 17
detergent, *see also* surfactant 157, 327
diatrizoate 89ff.
- bank filtration 93
- groundwater 92
- hospital wastewater 89
- ozonation 94
- surface water 91
- wastewater treatment 89
dicamba (3,6-dichloro-2-methoxybenzoic acid) 127
- use 128
2,4-dichlorophenoxyacetic acid (2,4-D) 5, 137ff.
- activated carbon filtration 146
- bank filtration 146
- groundwater 142
- ozonation 147
- surface water 137
- use 125
dichlorprop (2,4-DP) 125, 142ff.
- activated carbon filtration 146
- groundwater 142
- use 125
diclofenac 44ff.
- activated carbon filtration 56
- bank filtration 53ff.
- ozonation 49ff.
- photolysis 52
- risk assessment 303
- toxicity 294ff., 307
- wastewater 44
- wastewater treatment 47
diethylenetriamine pentaacetate (DTPA) 155ff.
- advanced oxidation 166
- bank filtration 173
- drinking water 175
- use 161
- wastewater 168
dimethylamine 193
- chlorination 205
- surface water 193
disinfectant 278
- regulation 258
- use 257ff.
disinfection 146, 251ff.
- by-product (DBP) 9ff., 251ff.
- control 278f.

- DBPFP (disinfection by-product formation potential) 252
  - drinking water 253
  - groundwater 253
  - inorganic 277
  - organic 262ff.
  - regulation 255
  - sand filtration 280
  - solid phase extraction 9
  - wastewater 252
dissolved organic carbon (DOC) 252ff.
  - biodegradable dissolved organic carbon (BDOC) 271
dissolved organic matter (DOM) 196
diuron 131ff., 146
  - activated carbon filtration 146
  - membrane filtration 148
  - ozonation 147
  - surface water 137
  - use 131
drinking water 56, 145, 175, 241, 251ff.
  - aminopolycarboxylate 175
  - disinfection by-product 253
  - herbicide 145
  - surfactant 241
drug, see also pharmaceutical 9
  - acidic 9ff.
  - amide type 16
  - analgesic 44
  - antiepileptic 43ff., 304
  - antiphlogistic 43ff.
  - antirheumatic 44
  - cytostatic 45
  - neutral 14
  - non-steroidal 22ff.
DTDMAC (ditallowdimethylammomium chloride) 236

e
ECD (electron capture detector) 16
EI-MS (electron ionization-MS) 16
electrospray ionization (ESI) 28ff.
enantio-separation 26
endocrine disruptor 15ff.
environmental risk assessment (ERA) 41
esterification 15
17β-estradiol 275
estrogen 275
ethylenediamino tetraacetate (EDTA) 5, 21, 155ff.
  - activated carbon filtration 174
  - advanced oxidation 166
  - bank filtration 173
  - biodegradation 164f.
  - drinking water 175
  - groundwater 173
  - liquid chromatography-mass spectrometry 21
  - ozonation 174
  - photolysis 170
  - use 161
  - wastewater 168
excretion 42, 69
  - antibiotic 69
  - pharmaceutical 42
existing chemical
  - regulation 311
extraction 3f.
  - ion-pair-solid phase extraction (IP-SPE) 8
  - liquid liquid extraction (LLE) 6
  - liquid phase microextraction (LPME) 13
  - microwave assisted extraction (MAE) 4
  - pressurized liquid extraction (PLE) 4
  - solid phase extraction (SPE) 6ff.
  - solid phase microextraction (SPME) 10f.
  - Soxhlet 3
  - supercritical fluid extraction (SFE) 4
  - ultrasound assisted extraction (USE) 4

f
filtration
  - activated carbon 59
  - anthracite 271
  - bank 53ff., 93, 147f., 173, 203
  - granular activated carbon (GAC) 59, 242, 271ff.
  - membrane 58, 98
  - nanofiltration (NF) 58
  - sand 271ff.
fixed-bed bioreactor (FBBR) 215
flocculation 56
fluorescence *in situ* hybridization (FISH) 80
fluoroquinolone 4, 25f., 72ff., 104
  - biodegradation 80
  - extraction 4
  - liquid chromatography-mass spectrometry 25
  - octanol-water partition coefficient 104
  - photolysis 78
  - surface water 72
  - use 104
  - wastewater 72

g
GAC, see granular activated carbon filtration
gas chromatography (GC) 11ff.
  - derivatization 14ff.
  - electron capture detection (ECD) 17

- GC-EI-MS (gas chromatography-electron ionization-mass spectrometry) 16
- GC-MS 12ff.
- nitrogen-phosphorus detector (NPD) 17
- two-dimensional (GCxGC) 17
- ultra-fast 17
glufosinate 131
- use 131
glyphosate 25, 130, 147
- bank filtration 148
- liquid chromatography-mass spectrometry 25
- ozonation 147
- use 130
granular activated carbon (GAC) filtration 59, 242, 271ff.
graphitized carbon black (GCB) 9
groundwater 53, 75ff., 92, 116, 140, 172, 200, 233
- amine 200
- aminopolycarboxylate 172
- antibiotic 75
- pharmaceutical 53
- surfactant 233

h

haloacetic acid (HAA) 6, 21, 251ff.
- drinking water 265
- liquid chromatography-mass spectrometry 21
- regulation 258
haloacetonitrile (HAN) 269
- drinking water 269
halonitromethane 272
herbicide 5ff., 26ff.
- acidic 9ff.
- activated carbon filtration 145
- bank filtration 147
- chloracetanilide 22
- chlorination 147, 275
- classification 122
- derivatization 15
- dinitroalkylphenol 129
- drinking water 145
- groundwater 140
- liquid chromatography-mass spectrometry 26
- membrane filtration 148
- metabolite 144
- ozonation 146
- pathway to environment 134
- phosphorus compound 130
- polar 121ff.
- regulation 321
- risk assessment 311
- sand filtration 147
- solid phase extraction 7ff.
- sorption 140
- surface water 136
- use 123
- wastewater 135
- water treatment 145
high production volume (HPV) chemical (HPVC) 181ff.
high-performance liquid chromatography (HPLC), see liquid chromatography
hospital wastewater, see wastewater
HPV, see high production volume
hydrophilic interaction chromatography (HILIC) 20, 23
hydroxyfuranone
- halogenated (HHF) 271

i

ibuprofen 44ff.
- biodegradation 52
- ozonation 49ff.
- toxicity 307
- wastewater 44
ICM, see X-ray contrast media
ion chromatography (IC) 20
ion exchange (IE) 7
ion pair (IP)
- reagent 8
- reversed-phase liquid chromatography (IP-RPLC) 20
- solid phase extraction (IP-SPE) 8
ion trap detector (ITD) 17
ionization 27
iopamidol 90f.
- bank filtration 93
- groundwater 92
- surface water 91
- wastewater treatment 90
iopromide 56, 88ff.
- bank filtration 93
- groundwater 92
- hospital wastewater 89
- ozonation 56, 94
- reductive dehalogenation 95
- risk assessment 88
- surface water 91
- wastewater treatment 90
isocyanate 188
isoproturon 137ff.
- activated carbon filtration 146

- groundwater   142
- ozonation   147
- surface water   137

k

K_{OW} (octanol-water partition coefficient)   77, 88, 184, 296

l

large volume injector (LVI)   19
LC, see liquid chromatography
linear alkylbenzene sulfonate (LAS)   20, 215f., 232ff.
- biodegradation   216ff.
- groundwater   233
- liquid chromatography-mass spectrometry   20
- risk assessment   243
- surface water   232
- wastewater treatment   229
linuron   147
- ozonation   147
lipid regulator   43ff.
liquid chromatography (LC)   20ff.
- chiral separation   26f.
- hydrophilic interaction LC (HILIC)   20, 23
- ion-pair LC (IPLC)   20f.
- normal phase LC (NPLC)   27
- reversed-phase LC (RPLC)   21ff.
- ultra performance LC (UPLC)   32
liquid chromatography-mass spectrometry (LC-MS)   19ff.
- amphoteric compounds   25f.
- atmospheric pressure chemical ionization (LC-APCI-MS)   27f.
- atmospheric pressure ionization (LC-API-MS)   19
- basic compounds   23
- calibration   33
- electrospray ionization (LC-ESI-MS)   27ff.
- ionic analytes   20ff.
- ionization   27
- matrix effect   28, 32f.
- non-ionic analytes   23f.
- quantitation   32
liquid-phase microextraction (LPME)   13
- three-phase system   13
- two-phase system   13
lowest observed effect concentration (LOEC)   326
LVI, see large volume injector

m

macrolide   24ff., 66ff., 80, 103
- liquid chromatography-mass spectrometry   24
- manure   111
- octanol-water partition coefficient   103
- soil   112
- surface water   72
- use   66, 103
- wastewater   73
- wastewater treatment   71
MAE, see extraction
manure   111
mass spectrometry (MS), see also liquid chromatography-mass spectrometry   12, 27ff.
- gas chromatography (GC-MS)   12ff.
- ion-trap-MS (IT-MS)   29
- multiple reaction monitoring (MRM)   22, 30
- negative chemical ionization-MS (NCI-MS)   19
- quadrupole-time-of-flight-MS (Q-TOF-MS)   30f.
- triple-stage quadrupole (QqQ-MS)   26ff., 30
- time-of-flight-MS (TOF-MS)   30f.
matrix effect   6, 32f.
mecoprop (MCPP)   125, 135ff.
- activated carbon filtration   146
- bank filtration   147
- groundwater   142
- ozonation   147
- surface water   137
- use   125
- wastewater treatment   135
membrane
- bioreactor   48
- filtration   58, 95, 148
- pharmaceutical   58
- reverse osmosis (RO)   58
methylene blue-active substance (MBAS)   215
microextraction   10
molecularly imprinted polymer (MIP)   10
monuron   131, 146
- activated carbon filtration   146
- ozonation   147
- use   131
MS, see mass spectrometry, see also liquid chromatography, and gas chromatography
MTBE   6
MTBSTFA (N-tert-butyldimethylsilyl-trifluoroacetamide)   16

Subject Index | 347

multiple reaction monitoring (MRM)   26
MX (3-chloro-4-(dichloromethyl)-5-hydroxy-2(5*H*)-furanone)   252, 271
– drinking water   271

n

naphthalene carboxylate   22
naphthalene di-sulfonate (NDSA)   20
naphthalene mono-sulfonate (NSA)   20
naproxen   307
– toxicity   307
natural organic matter (NOM)   251ff., 278
negative chemical ionization (NCI)   19
– NCI-mass spectrometry (NCI-MS)   16
NF, *see* filtration
nitrilotriacetate (NTA)   155ff.
– advanced oxidation   166
– bank filtration   173
– biodegradation   162
– drinking water   175
– use   161
– wastewater   168
nitroaromates   201
– groundwater   201
nitrogen phosphorus detector (NPD)   17
N-nitrosamine   252, 273
– drinking water   273
N-nitrosodimethylamine (NDMA)   252, 273
– wastewater   252
no observed effect concentration (NOEC)   315, 326
NOM, *see* natural organic matter
nonylphenol   231
– drinking water   241
– halogenated (XNP)   276
– surface water   233
– toxicity   244
– wastewater   231
nonylphenol ethoxylate (NPEO)   23, 221ff., 276
– biodegradation   221
– drinking water   241
– groundwater   237
– halogenated (XNPEO)   276
– liquid chromatography-mass spectrometry (LC-MS)   23
– risk assessment   244
– surface water   233
– wastewater treatment   230
nonylphenolethoxy carboxylate (NPEC)   22, 230ff.
– groundwater   238
– halogenated (XNPEC)   276

– liquid chromatography-mass spectrometry (LC-MS)   22
– wastewater treatment   230
normal phase LC (NPLC)   27

o

octanol-water partition coefficient (K_{ow})   77, 88, 184, 296
octylphenol ethoxylate (OPEO)   276
– halogenated (XOPEO)   276
OECD (Organisation for Economic Co-operation and Development)   80, 215, 299
'omics' technology (including genomics, proteomics, transcriptomics, metabolomics/metabonomics)   306
ozonation   49ff., 94, 147, 174
– amine   191
– aminopolycarboxylate   174
– pharmaceutical   49
ozone
– regulation   258
– use   261

p

PAH, *see* polycyclic aromatic hydrocarbons
PBT (persistent, bioaccumulative, and toxic) substance   316f., 328
PBT/vPvB   316
PCB, *see* polychlorinated biphenyl
PEC (predicted environmental concentration)   88, 288ff., 314
PELMO (pesticide leaching model)   322
penicilline   25, 66ff., 79
– liquid chromatography-mass spectrometry (LC-MS)   25
– surface water   72
– use   66
– wastewater   72
perfluoroalkanoic acid (PFAA)   240
perfluorooctylsulfonate (PFOS)   329
pesticide, *see also* herbicide   7ff., 21ff.
– acidic   17
pharmaceutical   7ff., 21, 41ff., 99
– acidic   15ff.
– activated carbon filtration   56
– bank filtration   53
– excretion   42
– groundwater   53
– membrane bioreactor   48
– membrane filtration   58
– ozonation   49ff.
– photolysis   52
– quantitative structure activity relationship   305f.

– regulation   290, 324
– risk assessment   287ff., 311ff., 325
– solid phase extraction   7ff.
– sorption   53
– surface water   50
– toxicity   301
– veterinary, *see also* veterinary pharmaceutical   99ff.
– wastewater   43
phenol   16
phenoxyacetic acid   138
– surface water   138
phenoxycarboxylic acid   125, 142ff.
– bank filtration   148
– groundwater   142
– use   125
phenylurea   138ff.
– bank filtration   148
– surface water   138
photolysis   52, 78
– amine   195
– pharmaceutical   52
PLE, *see* extraction
PNEC (predicted non-effect concentration)   88, 288ff., 315
polar herbicide, *see* herbicide
polar pollutant
– analytical method   1ff.
– persistent polar pollutant (PPP)   332ff.
– trace analysis   19f.
pollutant, *see also* polar pollutant
– management   328
– micropollutant   263, 275
– persistent organic pollutant (POP)   316, 328ff.
polychlorinated biphenyl (PCB)   4, 17, 333
polycyclic aromatic hydrocarbons (PAH)   4
polyethylene glycol (PEG)   226
polystyrene-divinylbenzene (PS/DVB)   6ff.
polyurethane   188
POP, *see* pollutant
PPP, *see* polar pollutant
programmed temperature vaporizer (PTV)   19
purge and trap (PT)   6

q
quantitation   32
quantitative structure activity relationship (QSAR)   297ff.
quaternary ammonium compound   8, 23
– liquid chromatography-mass spectrometry (LC-MS)   23
– solid phase extraction   8

quinolone   66ff., 78ff., 104
– biodegradation   80
– octanol-water partition coefficient   104
– photolysis   78
– surface water   72
– use   66, 104
– wastewater   73
– wastewater treatment   71

r
REACH (registration, evaluation, and authorization of chemicals)   329ff.
reductive dehalogenation   95
regulation   127, 255ff., 290, 311, 321ff.
– biocide   323
– herbicide   321
– pharmaceutical   290, 324
resistance   80
– antimicrobial   81, 115
reverse osmosis (RO)   58
reversed-phase liquid chromatography (RPLC)   20
risk assessment   88, 287ff.
– biocide   324
– pharmaceutical   290, 311, 325
– regulation   290, 324
– surfactant   243 f.
– veterinary pharmaceutical   311

s
salicyclic acid   44
– wastewater   44
sand filtration   147, 242, 271ff., 280
seawater   79
secondary alkane sulfonate (SAS)   220
– biodegradation   220
sediment   1, 52
– pharmaceutical   52
sewage sludge   81
sewage treatment plant (STP), *see* wastewater treatment plant (WWTP)
SFE, *see* extraction
silylation   16
simazine   137ff.
– activated carbon filtration   146
– groundwater   141
– membrane filtration   148
– ozonation   147
– surface water   137
sludge   2f., 17, 46, 81
soil   1ff.
– fertilization   112
– soil aquifer treatment (SAT)   253
solid phase extraction   (SPE)   6ff.

solid phase microextraction (SPME) 6ff.
– derivatization 11
– *in-tube* SPME 11
sorption 53ff., 76, 135ff., 201
– amine 201
– antibiotic 76
– herbicide 140
– pharmaceutical 53
source control 279
Stockholm Convention 333
STP, *see* sewage treatment plant
sulfamethoxazole 49, 71ff.
– groundwater 73f.
– ozonation 49
– risk assessment 303
– surface water 73
– wastewater 73
– wastewater treatment 71
sulfonamide 23ff., 66ff., 78ff., 105ff.
– biodegradation 80
– groundwater 73
– liquid chromatography-mass spectrometry (LC-MS) 23ff.
– manure 111
– octanol-water partition coefficient 106
– photolysis 78
– soil 112
– surface water 73, 116
– use 66, 105
– wastewater 73
– wastewater treatment 71
sulfonated naphthalene formaldehyde condensate (SNFC) 20
sulfonylurea 23, 132, 148
– bank filtration 148
– liquid chromatography-mass spectrometry (LC-MS) 23
– use 132
sulfophenyl carboxylate (SPC) 20f., 216ff., 234ff.
– activated carbon filtration 242
– biodegradation 218
– drinking water 242
– groundwater 234
– liquid chromatography-mass spectrometry (LC-MS) 20f.
– ozonation 242
– sand filtration 242
– surface water 232
– toxicity 243
– wastewater treatment 229
surface runoff 115
surface water 50, 74ff., 91, 115, 135ff., 169, 232

– amine 192
– aminopolycarboxylate 169
– antibiotic 74
– pharmaceutical 50
– surfactant 232
surfactant 211ff., 327
– amphoteric 226
– biodegradation 213
– drinking water 241
– groundwater 233ff.
– metabolite 211
– non-ionic 23, 215ff.
– perfluorinated 240
– regulation 327
– risk assessment 243
– surface water 232
– use 211
– wastewater treatment 228

t
terbutylazine 143f.
– surface water 137
tert-butyldimethylsilyl (TBDMS) derivative 16
tetracycline 4ff., 26, 66ff., 78ff., 107ff.
– biodegradation 80
– liquid chromatography-mass spectrometry (LC-MS) 25
– manure 111
– octanol-water partition coefficient 107
– photolysis 78
– surface water 73, 116
– use 66, 107
– wastewater 73
– wastewater treatment plant 114
tetramethylammonium hydroxide (TMAH) 15
time-of flight mass spectrometry (TOF-MS), *see* mass spectrometry
total organic halogen (TOX), *see also* adsorbable organic halogen (AOX)
– drinking water 254, 266ff.
toxicity 80ff.
– algae 82
– aminopolycarboxylate 172
– antibiotic 80
– pharmaceutical 301
toxicology 287
trialkylphosphate 4ff.
– extraction 4
– solid phase extraction 9
triazine 126, 138ff.
– groundwater 142
– surface water 138

– use   126
tributylamine (TrBA)   20
trihalomethane (THM)   251ff., 265
– drinking water   265
– iodinated   273
– regulation   255
trimethylsilyl (TMS) derivative   16
trimethylsulfonium hydroxide (TMSH)   15

u
ureas   131

v
veterinary pharmaceutical   99ff., 311
– classification   100
– groundwater   116
– risk assessment   311
– soil   112
– surface water   115
vPvB (very persistent and very bioaccumulative substance)   316f., 328

w
wastewater   43, 70ff.
– amine   189
– aminopolycarboxylate   167
– antibiotic   70f.
– herbicide   135
– hospital   70
– linear alkylbenzene sulfonate (LAS)   229
– pharmaceutical   43ff.
– surfactant   228
– total organic halogen   88
– treatment   46, 70ff., 89f.
– treatment plant (WWTP)   89, 114, 134, 287f.
water
– antibiotic   74f.
– drinking water   56, 145, 175, 241, 251ff.
– groundwater   53, 75ff., 92, 116, 140, 172, 200, 233
– seawater   79
– surface water   50, 74ff., 91, 115, 135ff., 169, 232
– treatment   117, 158
– wastewater   43ff., 70ff., 88ff., 114, 134f., 167, 189, 228f., 287f.

x
X-ray contrast media   22, 87ff.
– bank filtration   93
– excretion   88
– groundwater   92
– iodinated (ICM)   87ff.
– membrane filtration   95
– octanol-water partition coefficient   88
– ozonation   94
– reductive dehalogenation   95
– surface water   91
– use   88
– wastewater treatment   89